Holographic Data Storage

Holographic Data Storage

From Theory to Practical Systems

Kevin Curtis, Lisa Dhar, Adrian Hill,
William Wilson and Mark Ayres

InPhase Technologies, Longmont, CO, USA

WILEY

A John Wiley and Sons, Ltd, Publication

This edition first published 2010

Registered office
John Wiley & Sons Ltd, The Atrium, Southern Gate, Chichester, West Sussex, PO19 8SQ, United Kingdom

For details of our global editorial offices, for customer services and for information about how to apply for permission to reuse the copyright material in this book please see our website at www.wiley.com.

Library of Congress Cataloging-in-Publication Data

Holographic data storage : from theory to practical systems / Kevin Curtis . . .
[et al.].
p. cm.
Includes bibliographical references and index.
ISBN 978-0-470-74962-3 (cloth : alk. paper)
1. Holographic storage devices (Computer science) I. Curtis, Kevin R.
TA1632.H6635 2010
004.5'65–dc22

2010008437

A catalogue record for this book is available from the British Library.

ISBN: HB: 9780470749623

Set in 10/12pt, Times Roman by Thomson Digital, Noida, India
Printed and bound in Great Britain by CPI Antony Rowe, Chippenham, Wiltshire

Contents

Foreword

The modern field of nonvolatile digital information storage is a bit more than a half-century old. During its history, the field has seen a small handful of technologies (magnetic tape, magnetic hard disk, magnetic flexible disk, consumer-derived optical disk) take root as 'mainstream'. These technologies have persisted in the market place from their earliest introduction until today, experiencing commercial success for several decades or longer. A few other digital storage technologies have emerged and been successfully commercialized in less broad based applications (e.g. magnetic drums, optical cards, magneto-optical disk) and, typically, their success has been of shorter duration. A third, quite numerous, category of storage technologies involves new approaches that have not yet achieved commercial success. The subject of this book, holographic data storage, is in this third category.

Among the many examples of promising storage technologies that have been worked on over the past 50 years, holographic storage offers a unique combination of desirable attributes: extremely high density of stored information, a disk format capable of rapid random access, and a removable nonvolatile storage medium that may be inexpensively manufactured. While these attributes are simple to discuss in general, putting them all together in a practical and reliable storage system represents a herculean engineering task. This is exactly what InPhase Technologies has accomplished, and what the company's contributors describe in detail in this book.

As the authors point out in the Introduction, with other optical storage technologies facing obstacles to significant performance improvements, interest in holographic data storage has dramatically increased in recent years. Although the increased interest is recent, the history of research and development in holographic storage extends back nearly 50 years, with contributions from many universities, government- and industry-sponsored consortia, and industrial companies. The uniqueness of the InPhase contribution is that they have progressed far beyond the basic research aspects of holographic storage and have persevered to address the multitude of materials, optical, mechanical and electrical engineering challenges necessary to develop a fully integrated drive-media storage system.

This book takes the reader through many details of the technical challenges encountered along this development path and of the often creative design solutions implemented to meet those challenges. Undoubtedly, the complexities of implementing a fully practical commercial system proved greater than anticipated by the InPhase team at the outset, but that is a natural occurrence when attempting to break new ground and introduce a new technology that represents so significant a departure from the incumbent approaches. To the team's credit, their approach over the years has been very open and forthcoming in describing the difficult challenges in their technical conference and journal papers. The team continues that practice in the detailed chapters of this book. Because of this openness, the book can serve as an excellent reference to developers of future holographic data storage systems and enable them to build upon and improve the technology.

From an applications perspective, the book focuses on professional archival storage, with some treatment of devices for consumer applications. Professional archival storage is a domain currently dominated by magnetic tape, and the tape technology community continues to aggressively advance its technology, following a roadmap that doubles storage capacity every 2 years. This competitive environment poses an additional challenge for InPhase, beyond the strictly technical challenges noted above.

At the same time, the demand for archival storage capacity is growing at a rapid rate, as the world continues to generate an exploding quantity of digital information. Much of this information is 'fixed content' and needs to be reliably retained for a decade or longer. As one measure of the enormous amount of digital information generated, a recently published study[1] reports that in 2008 Americans consumed more than 3.6 zettabytes of information (1 zettabyte $= 10^{21}$ bytes). Although the study emphasizes information 'flow' rather than information 'storage', it is clear that a significant fraction of the information described is stored in professional archival repositories.

This growing requirement for archival data retention presents an attractive opportunity for holographic data storage. The InPhase professional storage system with removable disk media is the first ever commercial holographic product to be introduced for such applications. It is indeed a pioneering development. While time will tell if the product becomes a commercial success, this book renders a wonderfully detailed and descriptive technical account of the path taken to reach this milestone.

Barry H. Schechtman
Executive Director Emeritus
Information Storage Industry Consortium (INSIC)

[1] R.E. Bohn and J.E. Short, How much information?, 2009 Report on American Consumers, Global Information Industry Center, University of California, San Diego, 2009; http://hmi.ucsd.edu/pdf/HMI_2009_ConsumerReport_Dec9_2009.pdf

Preface

This book is a result of over 15 years of research and development in holographic data storage, first at AT&T (then Lucent) Bell Laboratories and then at InPhase Technologies. The book's release is timed to roughly coincide with the release of the first ever commercial product using this technology: a professional archive storage drive using removable disk media. While major developments in holographic data storage outside of this effort are described, the focus is on explaining the design, components, and function of the technology used in InPhase's professional drive and two related consumer data storage products.

This book will enable end users of the technology to understand how the drive and media works, and how they are tested. Our hope is that other developers of holographic storage products can use this book as a basic blueprint for developing their own products using this technology.

A wide range of topics from polymer chemistry to error correction codes are covered in this book. The chapters are in large part independent, with a separate list of references at the end of each one. Although each chapter may refer to other chapters for additional detail, there is no assumption that later chapters require a detailed knowledge of earlier ones.

The first five chapters discuss the commercial market for holographic storage, and provide a broad overview of the drive and media technology. Chapters 6–8 discuss the media in greater depth. The technology underpinning the professional drive is considered in detail in Chapters 9–14. Chapter 15 covers read only memories and high speed replication of holographic media; topics that are central to the development of a consumer market for holographic storage. Finally, Chapter 16 concludes with a discussion of the future evolution of the technology and market applications.

A storage product is an amazingly complex device. As a simple example, the firmware controlling the InPhase drive is approximately 1.5 million lines of custom C++ code, which does not include almost another 1.5 million lines of other C and C++ code comprising the drive's operating systems.

The sum total of significant breakthroughs in media, material, control, optics, mechanics, data channel, and testing in the last 15 years is immense. As such, this book represents the work of over 200 people from different companies at various times.

InPhase Technologies was spun out of Bell Laboratories after $6\frac{1}{2}$ years of fundamental research and development. The support of management and wonderful people at Bell Laboratories enabled the start of this long and improbable journey. We sincerely thank these companies and our collaborators, and acknowledge their many contributions to this work.

We have also had significant interaction with, and help from, Hitachi Maxell, Nichia, Alps Electric, Bayer Material Science, Sanyo, Lite-on, IBM, Datarius, and Sony.

This book is dedicated to the employees, investors, and supporters of InPhase Technologies for their amazing contributions and hard work. This book truly is a result of their labor of love. Above all, we acknowledge and thank our families for their patience, understanding, and support over all these years.

Kevin Curtis
Lisa Dhar
William Wilson
Adrian Hill
Mark Ayres

List of Contributors

Ken Anderson, InPhase Technologies, Longmont, CO, USA

Fred Askham, InPhase Technologies, Longmont, CO, USA

Mark R. Ayres, InPhase Technologies, Longmont, CO, USA

Ernest Chuang, InPhase Technologies, Longmont, CO, USA

Kevin Curtis, InPhase Technologies, Longmont, CO, USA

Lisa Dhar, InPhase Technologies, Longmont, CO, USA

Tod Earhart, InPhase Technologies, Longmont, CO, USA

Edeline Fotheringham, InPhase Technologies, Longmont, CO, USA

Adrian J. Hill, InPhase Technologies, Longmont, CO, USA

Alan Hoskins, InPhase Technologies, Longmont, CO, USA

David Michaels, InPhase Technologies, Longmont, CO, USA

Liz Murphy, InPhase Technologies, Longmont, CO, USA

Lakshmi Ramamoorthy, Hitachi Global Storage Technologies Inc., San Francisco, CA, USA

Brad Sissom, InPhase Technologies, Longmont, CO, USA

Paul Smith, InPhase Technologies, Longmont, CO, USA

V. K. Vijaya Kumar, Electrical and Computer Engineering, Carnegie Mellon University, Pittsburgh, PA, USA

Aaron Wegner, InPhase Technologies, Longmont, CO, USA

William L. Wilson, InPhase Technologies, Longmont, CO, USA

1

Introduction

Kevin Curtis, Lisa Dhar and Liz Murphy

1.1 The Road to Holographic Data Storage

Digital data are ubiquitous in modern life. The capabilities of current storage technologies are continually being challenged by applications as far ranging as the distribution of content, digital video, interactive multimedia, small personal data storage devices, archiving of valuable digital assets, and downloading over high-speed networks. Current optical data storage technologies, such as the compact disk (CD), digital versatile disk (DVD), and Blu-ray disk (BD), have been widely adopted because of the ability to provide random access to data, the availability of inexpensive removable media, and the ability to rapidly replicate content (video, for example).

Traditional optical storage technologies, including CD, DVD and BD, stream data one bit at a time, and record the data on the surface of the disk-shaped media. In these technologies, the data are read back by detecting changes in the reflectivity of the small marks made on the surface of the media during recording. The traditional path for increasing optical recording density is to record smaller marks, closer together. These improvements in characteristic mark sizes and track spacing have yielded storage densities for CD, DVD, and BD of approximately 0.66, 3.2, and 17 $\mathrm{Gb\,in^{-2}}$, respectively. BD has decreased the size of the marks to the practical limits of far field recording.

To further increase storage capacities, multi-layer disk recording is possible [1], but signal to noise losses, and reduced media manufacturing yields, make using significantly more than two layers impractical. Considerable drive technology changes, such as homodyne detection and dynamic spherical aberration compensation servo techniques [2–4], have been proposed to deal with the signal to noise losses inherent in multiple layers.

Holographic Data Storage: From Theory to Practical Systems Kevin Curtis, Lisa Dhar, Adrian Hill, William Wilson and Mark Ayres

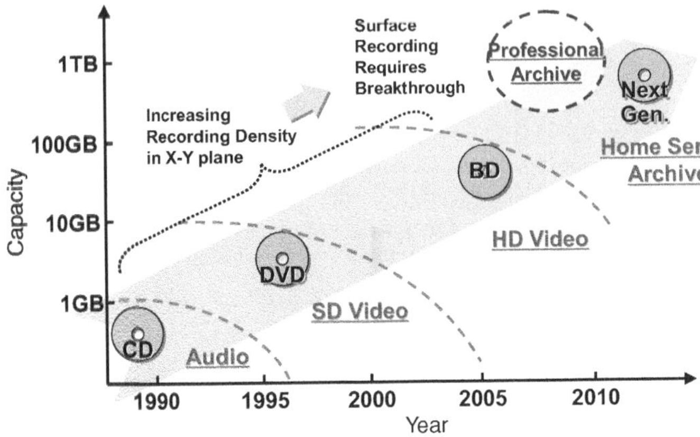

Figure 1.1 *Optical storage technology roadmap*

However, the use of multiple layers does not address the need for increased transfer rates that are required to effectively use higher disk capacities. In fact, the use of multi-layers makes increasing the transfer rate more difficult. Taking all these issues into consideration, the practical limit for the storage capacity of BD is thought to be around 100 GB, with a transfer rate of 15–20 MB s^{-1}.

Figure 1.1 shows the storage capacity of these optical technologies. The increasing difficulty in continuing to provide higher storage density and data transfer rate has triggered a search for the next generation of optical storage.

Alternative optical recording technologies, such as near field [5,6] and super resolution methods [7,8], aim to increase density by creating still smaller data marks. As the name suggests, near field methods record in the near field of the lens or aperture, so that the optical diffraction limit does not apply. Super resolution systems typically use special media structures to shorten the recorded marks. However, neither near field nor super resolution methods has shown compelling improvements over BD.

Another approach that produces multiple layers is two-photon recording in homogeneous media [9–11]. This method uses a first laser wavelength to record by producing a local perturbation in the absorption and fluorescence of the media, which introduces a small, localized index change through the Kramers–Kronig relationship [12]. A second wavelength is used to read out the data by stimulating an incoherent fluorescence at a different wavelength. The amount of fluorescence is used to determine whether a one or zero was recorded at a given location. Many layers of bits are recorded to achieve high density. Unfortunately, two-photon approaches suffer from an inherent trade-off between the cross-section of the virtual or real state (sensitivity) and the lifetime of this state (transfer rate). If the sensitivity is high enough for reasonable data density, then the transfer rate is typically low because of the lifetime of the state. In addition, in at least one example [9], the media is partially erased by each read out. Thus, two-photon techniques face both difficult media development and transfer rate or laser power issues.

With all other optical technologies facing obstacles to significant performance improvements, interest in holographic data storage has dramatically increased in recent years. For example, at the 2008 Joint International Symposium on Optical Memories and Optical Data Storage held in Hawaii, nearly half of the papers were related to holographic systems, media, components, and data channels.

1.2 Holographic Data Storage

Holographic data storage (HDS) breaks through the density limitations of conventional storage technologies by going beyond two-dimensional layered approaches, to write data in three dimensions. Before discussing page-based HDS, which is the focus of this book, we will briefly outline an alternate approach; bitwise holographic storage.

In bitwise holographic storage (also called micro-holographic storage) [13–16], multiple layers of small localized holograms are recorded at the focus of two counter-propagating beams. Each of these holograms represents a single bit that is subsequently read out by monitoring the reflectance of a single focused beam. Tracking the hologram locations through the volume in three dimensions is typically accomplished using a reference surface or part of the holograms themselves [17,18]. Bitwise holographic storage is appealing because the drive technology and components are similar to traditional optical storage, and because the media is homogenous and hence easy to manufacture. However, there are several serious drawbacks. First, it is difficult to achieve fast transfer rates. Also, it requires the invention of a material that is optically nonlinear. The technique also requires a complex servo system because the two recording beams must be dynamically focused into the same volume. Finally, the multiple layers of micro holograms cause distortion in the optical beams, which significantly limits the achievable density [19].

Unlike serial technologies (including bitwise holographic storage) which record one data bit at a time, page-wise holography records and reads over a million bits of data with a single flash of light, enabling transfer rates significantly higher than traditional optical storage devices. Page-wise HDS has demonstrated the highest storage densities (712 Gb in^{-2}) of any removable technology [20], and has a theoretically achievable density of around 40 Tb in^{-2} (see Section 2.6). High storage densities, fast transfer rates and random access, combined with durable, reliable, low cost media, make page-wise holography a compelling choice for next-generation storage and content distribution applications. As shown in Chapters 3 and 15, the flexibility of the technology allows the development of a wide variety of holographic storage products, ranging from handheld devices for consumers to storage products for the enterprise market.

1.2.1 Why Now?

Page-wise holographic storage was heavily researched in the 1960s and 1970s [21–29], but no commercial products came out of these efforts. The research was stymied by significant technical challenges, including poor media performance and a lack of input and output devices such as spatial light modulators and cameras. In the last few years, there has been a resurgence of activity and development in holographic storage, and commercial products are now within sight.

In the mid 1990s, the Defense Advanced Research Program Agency (DARPA) formed a consortium of companies and universities in the United States, led by IBM and Stanford

University, to develop high performance holographic storage systems [30–33]. The goal of the consortium was to demonstrate high density and transfer rate by developing the necessary technology and components, such as custom high speed cameras and spatial light modulators. Research in data channel modulation and detection schemes was also undertaken. Two types of storage systems were developed: one using a large crystal without mechanical motion as a recording medium, and the other using a spinning disk. The recording materials were primarily based on photorefractive crystals and on the then-available photopolymer films originally intended for display holograms [34,35]. These materials allowed basic demonstrations of HDS but did not meet the requirements for a commercial product. The consortium grew to include Polaroid (and later, Aprilis, a company spun out of Polaroid), who started developing photopolymers specifically designed for HDS [36,37]. This addition, together with the efforts of the other members, led to several significant achievements. Stanford University demonstrated high data transfer rates from a spinning disk – up to 1 GB s^{-1} [31], while IBM demonstrated storage densities of 250 Gb in^{-2} in very thick $LiNbO_3$ crystals [38].

Also in the mid 1990s, work in holographic storage began at Bell Laboratories, Lucent Technologies. Aimed at developing a suitable recording media in conjunction with a practically implement-able drive, the program targeted systems that would lead to commercially feasible products. By designing and developing both the media and drive in concert, several important technical milestones were reached: a process allowing for optically flat recording media to be fabricated using standard optical media manufacturing methods (Zerowave®) [39]; the invention of a new class of photopolymer recording material for holography (Tapestry®, two-chemistry materials) enabling both high performance and robust lifetime characteristics; and drive designs that improved signal to noise ratio and simplified servo techniques over previous systems. By 1998, data densities of 49 Gb $in.^{-2}$ were achieved in the two-chemistry materials [40]. With these technology breakthroughs in place, in 2000, Lucent Technologies spun out an independent company, InPhase Technologies®, to commercialize holographic storage systems.

InPhase has primarily focused on the development of a storage system suitable for archival applications in the professional market. The drive's architecture (see Chapter 3) was designed for ease of implementation and operation, minimizing the use of custom-developed components and ensuring environmental robustness. With this strategy, InPhase has demonstrated the highest storage density to date (712 Gb $in.^{-2}$) of any removable storage technology, media interchange between drives for the first time, and operation over a temperature range of 40°C. In addition, InPhase has partnered with some of the leading companies and organizations in the world of optical storage to productize its system, including Bayer Material Science, Hitachi Maxell Corporation, Nichia, Sanyo, Lite-On, Displaytech, Cypress, University of California at San Diego, and Carnegie Mellon University.

Also in the 2000s, companies in Japan and Korea started research into holographic storage drives and media, and several consortiums sponsored by the Japanese government were formed. Companies such as Sony and a small start-up, Optware, focused their efforts on a coaxial or collinear architecture that leverages CD and DVD technologies (this architecture is presented in detail in Chapter 3). Sony has demonstrated a storage density of 415 Gb in^{-2} [41] using collinear geometries. Sony also directed some of their efforts into bitwise holographic storage, developing methods to replicate media for read only memories (ROMs). These ROM replication efforts will be covered in Chapter 15. More recently,

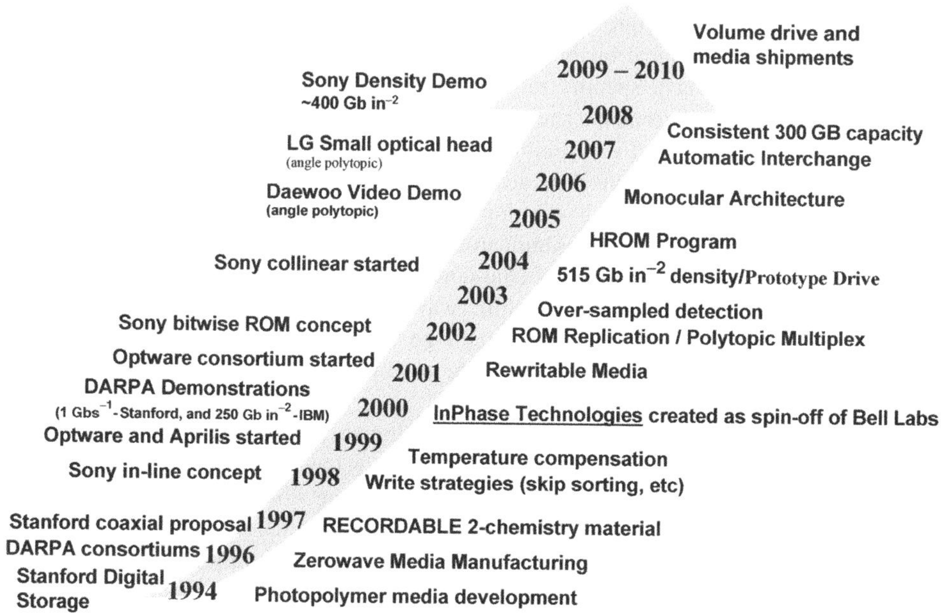

Figure 1.2 *Key holographic technology advancements of the last 15 years*

Lucky Goldstar in Korea has used the InPhase architecture to design and build a miniature optical head [25], and Korea's Daewoo has used the same InPhase architecture to achieve high speed video recording [42] and playback.

Figure 1.2 shows the highlights in holographic storage developments over the last 15 years. The right-hand side of the figure shows technical advances made by Bell Laboratories and InPhase Technologies, while those of other companies and institutions are shown on the left-hand side of the figure.

1.2.2 Focus of the Book

This book aims to present in an integrated manner, the technologies that enable practical holographic storage systems. To this end, the majority of this book will focus on the design, implementation, integration and operation of a drive and media using InPhase's drive architecture. This drive is targeted at professional archival storage applications, which require high capacity and transfer rate, media with a long archival life, and a product roadmap with performance improvements while maintaining backward read compatibility.

Focusing on a single drive architecture allows us to present a complete picture of how the underlying requirements and performance targets for holographic storage dictate the specifications for components and subsystems, and how those subsystems are developed, designed, and integrated into a complete drive.

The key features of the InPhase Architecture are (i) the optical architecture of the drive used to achieve the three-dimensional recording of the holographic data, (ii) the servo systems used to write and read the data, and (iii) the recording media which stores the holographic data. These features govern the system's performance, and sensitivity to environmental and mechanical factors.

While the focus is on a specific implementation, the principles are general – the relationships between requirements and specifications and the trade-offs between different subsystems will be common to all architectures for page-based holographic storage. To illustrate these commonalities, this book also discusses how to build on the basic technology of the professional archival drive to develop consumer products.

The optical architecture of a drive is built around a multiplexing strategy that provides the ability to overlap many holograms within the same volume of the recording medium. Many multiplexing methods such as angle, shift, wavelength, peristrophic (rotational) and correlation techniques have been investigated (see Chapter 3 for a detailed discussion), but no single multiplexing method has been able to achieve both high storage density and a robust implementation.

For example, angle multiplexing is simple to implement, provides high-speed recording and read-out, allows easy media interchange, and exhibits low sensitivity to environmental changes. However, geometrical factors ultimately limit the storage densities achievable with angle multiplexing to less than 140 Gb in^{-2} (see Section 3.3.1).

The InPhase architecture adds a new type of multiplexing, polytopic, onto angle multiplexing to mitigate the geometrical limitations on storage densities. Polytopic multiplexing maintains the speed, media interchange and robustness advantages of angle multiplexing, while allowing a more than 20-fold increase in the storage capacity of a system. In addition, by using a phase conjugate architecture with polytopic multiplexing, all the optics can be placed on one side of the media in the drive, which simplifies the optics compared with other approaches.

The InPhase drive is built around Tapestry®, a two-chemistry photopolymer recording material and media (discussed in detail in Chapter 6). The recording material is based on an interpenetrating network of two polymer systems: a cross-linked polymer that is the majority of the system and acts as the support or matrix, and a second photopolymerizable material which reacts during recording and leads to the formation of the holographic pattern. This material allows independent optimization of media performance metrics such as storage density, data transfer rate, and data lifetimes, to meet the requirements of holographic storage. In addition, the Zerowave® manufacturing process is used to fabricate inexpensive, optically flat media, using plastic substrates. This flatness improves the overall performance and signal to noise ratio (SNR) of page-based holographic systems.

Implementing the optical architecture and the recording media requires a highly interdependent effort. Aspects of the implementation such as the manufacturing of the media, the components used in the drive, the data layout format used during writing, the servo and feedback on the disk during recording and reading, and the error correction strategy, are developed by simultaneously trading off the requirements and capabilities of both the media and the drive. The servo system governs the interface between these two components.

For example, because holography records throughout the volume of the medium and the volume of the polymer-based medium can change with temperature fluctuations, a servo strategy to compensate for thermal effects is necessary. Varying the wavelength of the laser used to read out the hologram can compensate for the effects of temperature changes. The InPhase system is therefore built around a tunable laser: a coated gallium nitride laser diode, in a small, simple, stable, relatively high-power, external cavity. Also, the thermal expansion of the media can be minimized by using plastic substrates rather than glass.

Other examples, which will be expanded upon throughout this book, demonstrate the interdisciplinary development that is essential to achieving a commercially viable system:

- Writing strategies and multiplexing methods for achieving high fidelity and high-density storage in photopolymer systems.
- Parallel data channels that are significantly different from conventional serial data channels, requiring new channel detection schemes, data formatting and the use of advanced error correction codes.
- Servo methods for tracking and finding the data for the key axes such as galvo angles, wavelength, and temperature changes, allowing for fast transfer rates.
- Interchange and servo algorithms, and build processes and tools, which can be implemented in a real-world environment.

1.2.3 Other Examples of System using the InPhase Architecture

The InPhase Architecture, including media, servo, and data channel technologies, can be used to develop consumer products. The path from professional drives to consumer products using holography is similar to the path that was followed in the history of CD development. The first CD-R was a similar size to the InPhase professional drive (approximately 5.25 in × 5.25 in × 25 in), and cost US$15 000 in the 1970s (which was roughly the price of a house in Southern California at the time). Currently, the cost of a higher performance CD-R drive is around US$10, and the drive height is less than 13 mm. The following paragraphs discuss the preliminary development work on two holographic systems that are suitable for consumer markets.

The first concept is a holographic read only memory (HROM) built as a unique, optical card or chip reader that is backwards compatible with solid state memories (SSMs). In this chip reader, the slot for the replicated holographic media chip can also be used to read the SSM. InPhase has developed the process and custom tools that allow full holographic media replication in times similar to those of CD and DVD replications. The key two-step mastering process produces masters that have high diffraction efficiency and high fidelity at the high densities required for use in a fast lens-less replication process. Replicated media is read using a small HROM prototype reader. Chapter 15 describes this concept and implementation in detail.

Working with Hitachi, InPhase has also developed a consumer optical storage system; an implementation of the InPhase Architecture that is backwards compatible with Blu-ray. The system uses a monocular architecture that passes both the data beam and the plane wave reference through the same high numerical aperture lens. The media uses a grating to enable phase conjugate read-out, which allows for a slim height (12.7 mm) using appropriately sized components. With the already demonstrated density of 712 Gb in.$^{-2}$, a 120 mm disk can store >500 GB of user data, with a transfer rate of 100 MB s^{-1} or more. Chapter 3 introduces the monocular concept and Chapter 4 specifies the required components needed to implement an inexpensive, slim height drive.

1.3 Holographic Data Storage Markets

1.3.1 Professional Archival Storage

The first market for HDS is professional archival storage – the long term storage of digital assets. Demand for long term archival storage and fast data access is being driven by regulatory compliance requirements, an increased volume of fixed-content data, surveillance and security systems, and the explosion of rich media applications. Storage for these archive and data distribution markets is primarily based on removable media.

'Long term' archiving means being able to store data for several decades without the need to refresh or migrate them (data migration is typical for tape-based storage). These time periods are considerably longer than the 3–7 years commonly required for transaction data. In 2005, the United States Government Information Preservation Working Group (GIPWoG) surveyed users about their longevity requirements for archival storage. Partial results from the survey are summarized in Figure 1.3. Close to 60% of the 4483 respondents indicated an archival life requirement of over 40 years for their data. Further details are available in the INSIC International Optical Data Storage Roadmap [43].

Regulatory compliance legislation, passed in the US in the early 2000s, has raised the importance of data protection and archiving. The intent of many of the regulations is to protect data that may be of value in litigation. The write once aspect of holographic write once read many (WORM) media is a good fit for this requirement. The legislation also mandates that data must be archived for periods of up to decades. These compliance regulations impact a broad range of industries such as financial services, healthcare,

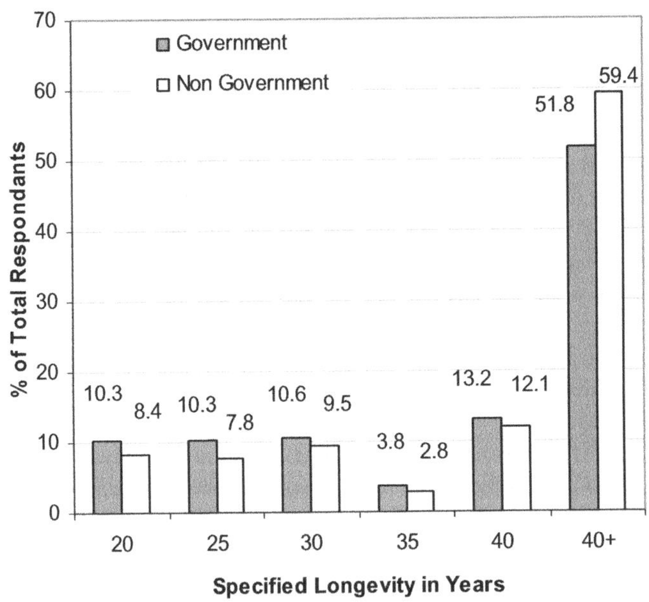

Figure 1.3 Results of a 2005 user survey by the US Government GIPWoG group. There is a strong preference for 40+ year longevity for archival data

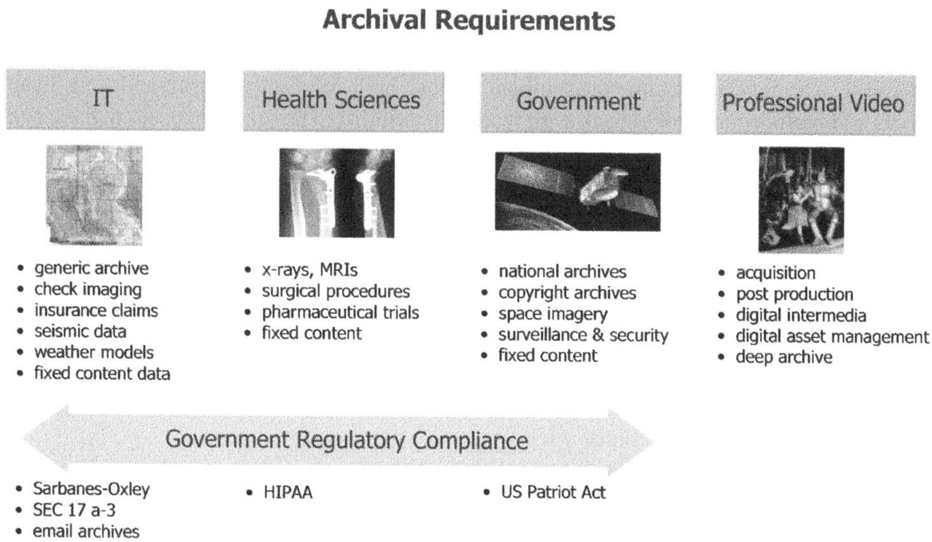

Figure 1.4 *Professional archival markets overview*

pharmaceuticals, and government data, as well as email archives in all industries. Data archiving has gone from being an irritant to becoming a major application, with additional requirements to protect the data from alteration and unauthorized access. Significant fines have been levied against companies that fail to comply. Figure 1.4 summarizes several important archival markets and the regulations that are driving some of them.

Rich media video and audio applications have emerged as another new market that has very long archive requirements. Content in the incumbent analog video and audio technologies is being migrated to digital formats, to leverage the lower cost of managing digital workflow from content acquisition, to post production, distribution, and archiving. The market is also expanding because high definition (HD) formats generate even more data than the older standard definition formats. For example, 1 s of video may generate 12 MB of data. Often over 100 times more video footage may be shot than is actually distributed, so a 30 s commercial or a 2 h movie will generate terabytes of content. The high cost of acquisition and the revenue generating nature of the content, mean that archive expectations are 'forever'.

Historically, magnetic tape has been the predominant technology for back-up and archive applications because of its high capacity, high transfer rate, and low cost media. However, when archiving data for more than a few years, data tapes are often stored in a temperature and humidity controlled environment, which is expensive to construct and maintain. If the stored data are especially valuable, it will be migrated to new tapes anywhere from once a year to once every 7 years, which incurs further labor and media costs. In addition to its reputation as an unreliable data recovery format, tape has long access times to data because of the need to rewind or advance the tape spool. In spite of these problems, until the advent of holographic storage, no other technology has been compelling enough to displace tape.

Conventional optical drives such as magneto-optical drives, and to some extent DVDs, have also been used in the IT sector for digitally archiving items such as medical records, bank check images, and telecommunications logs. The primary need in these applications is the write once feature of optical technology which protects the data from being erased or altered. Customers in this market segment require high reliability, stable and long-lived media and multi-generational backward read compatibility. The random access to data, and low cost unalterable WORM media, also provide advantages over tape. However, the principal challenge to technology is its limited capacity and slow transfer rates, even when the latest generation – Blu-ray – is considered.

Hard disk RAID arrays have dramatically impacted both back-up and high access, short term (1–3 years) archive markets. These arrays have tremendous performance and have benefited from increases in disk drive capacity. However, disk drives have limited lifetimes (3–5 years) if they are kept spinning or are spun up at least every few weeks. Thus, RAID arrays have been limited to use for short term, high access, fixed content and back-up markets. For long term archive (over 7 years) these solutions are very expensive to buy, maintain, power, and keep cooled.

Several companies have implemented disk-to-disk-to-tape systems to address the issues of limited accessibility and recoverability associated with a pure tape back-up and archive strategy. However, the issue of long term archiving remains a problem with these hybrid solutions because hard drives are expensive for storing infrequently accessed data, and tape remains a problematic data recovery medium.

Holographic storage offers a compelling alternative because the capacity and transfer rate are competitive with tape, with the additional benefit of random access in a disk format. The media cost is lower than for the new blue laser optical formats, and has the advantage of a 50 year media archive life in unalterable WORM media.

Holographic technologies offer improvements in the performance and cost curves of storage that make increasingly large amounts of data accessible to users, while reducing the total cost of storing the data.

The value proposition for holographic data storage products includes:

- Highest performance for removable storage, which combines a demonstrated data density of over 712 Gb in^{-2}, random access (around 250 ms), and transfer rates capable of exceeding 120 MB s^{-1}.
- A 50+ year media archive life, requiring no special handling, refreshing or environmental controls; and no wear from media contact with a read/write head.
- Near-line random access to content, making petabytes of data almost instantly accessible.
- Smaller media format with higher density per cubic foot.
- Lowest cost per gigabyte for professional grade media, making archiving affordable for terabytes to exabytes of data.
- Improved data protection with a true (intrinsic) WORM media format that ensures that the data retains its original state.
- Lowest total cost of ownership, resulting from low media costs; reduced frequency of media migration; smaller media size (which reduces data center floor space requirements); and power savings, achieved by decreasing the use of hard disk drives to store infrequently accessed data.

HDS has a sustainable advantage over other technologies, with a roadmap that allows drive functionality to improve over time while maintaining backwards compatibility. The current InPhase Technologies roadmap has the second generation drive (800 GB capacity, with

80 MB s^{-1} transfer rate) appearing 2 years after the 300 GB drive, and the third generation (1.6 TB capacity and 120 MB s^{-1}) appearing a further 2 years later. This is a faster growth curve than for tape, hard disk or SSM technologies.

Magnetic tape, hard disk, and CD/DVD/Blu-ray are the current competitors for HDS in the archive market. Media reliability is the major problem with tape. Performance is the major limitation for traditional optical storage, while lifetime, cost, and power usage are the issues with standard hard disk. Figure 1.5 summarizes the pros and cons for each technology.

According to IDC, the OEM market size for archive drives and media will be US$17.5 billion in 2010.

1.3.2 Consumer Applications

Removable storage for consumer applications is largely dependent on optical disk and solid state memory technologies to satisfy the ever-increasing demands for distribution, recording, and storage. For distribution, archiving, and video recording, traditional optical storage is by far the predominant removable storage technology in use today. Blu-ray has pushed the limits of capacity and transfer rate of surface recording technologies, and next generation removable products require densities and transfer rates that cannot be provided by incremental improvements of these technologies.

 Tape		 Hard Drives		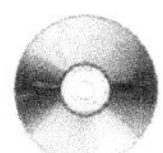 CD/DVD	
Pros	Cons	Pros	Cons	Pros	Cons
• High Capacity • High Transfer Rate • Low Cost Media	• Media Reliability • High Media Maintenance $ • Slow Data Access • Not True WORM	• High Capacity • Low Cost / GB for device • Easy to use • Random Access to Data	• High Power Usage • Device Life 3-5 years • Not Archival Format	• Good Media Archive Life • Low Cost • True WORM Format	• Low Capacity • Low Transfer Rate

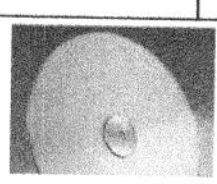 Holographic Technology Benefits	
Pros	Cons
• High Capacity competes with tape and hard drives, 6X more capacity than BD • Media Archive life +50 years vs 3-7 years for tape and hard drives • Random Access to data; milliseconds to file vs minutes for tape • True WORM Format Protects Archive Data • Low $/GB media competitive against tape and existing optical • Transfer rate much higher than existing optical • Low power requirements	• New technology • WORM only format at Introduction

Figure 1.5 Competing technology options for archival storage applications

The initial consumer markets for HDS will likely leverage its removability, inexpensive media and replication, and long archival life, for home archiving and content distribution. As consumers produce more digital content in the form of movies and pictures, the need to archive them effectively will become painfully clear. This market shares many of the requirements of professional or enterprise archiving, where the content is expected to be preserved for a person's lifetime.

InPhase has developed two concepts for consumer markets that leverage the technology developed for the professional drive. The first concept is a holographic ROM with high speed, full media replication for content distribution. The reader and media card are the same size as SSM formats to allow the drive to read both SSM media and holographic media with replicated content. Prototype replication equipment and a small prototype reader have been developed (see Chapter 15). While SSM has significant advantages for mobile applications, it does not have an easy, fast, and inexpensive method to distribute content. The InPhase Technologies reader and low cost replicated media is intended to complement SSM in applications where content distribution is required, for example, for games, software, maps, videos, movies, and so on.

The second concept, called the monocular architecture, implements a slim height, optical drive that is backwards compatible with Blu-ray (see Chapter 3) and would be used for both home archive and content distribution. This uses a Blu-ray like 0.85 numerical aperture lens to introduce both reference and data beams to the media.

For the consumer market, HDS provides:

- Highest performance removable storage **with** highest capacity (500 GB–2 TB per disk), random access (~250 ms), and highest transfer rates (>50 MB s^{-1}).
- Rugged, able to withstand on-the-go use because there is no contact between media and head.
- Low power, and low noise because the media does not rotate.
- Low cost, small size drives and media for use in mobile applications.
- Long archival media life requiring no special handling or environmental controls.
- Random access to content.
- High-speed and low-cost replication for physical content distribution.
- Unique formats with a card reader that is compatible with SSM.
- Low cost media due to plastic substrates and photopolymer.

Drives based on the monocular architecture can be the next generation of optical storage (Figure 1.1) because of the ability to efficiently replicate disks, backward compatibility with Blu-ray (BD), the slim height of the drive, and the advent of inexpensive media. While a version of BD with 50 GB per disk (2 layers) is available in Japan, it is possible that a 100 GB BD will be commercialized eventually. Significant modifications to the Blu-ray drive will be required to compensate for the change in focal depth inside the media, and to increase the SNR and light throughput. HDS will allow the next step up in both capacity and transfer rate for optical technology, which will allow for effective home archiving, as well as distribution of three-dimensional, ultra high resolution, or user-controlled content.

Figure 1.6 compares the prevalent storage technologies for consumer archive and content distribution. Flash or SSM is, and will remain, dominant for mobile applications because of its power, robustness, and size advantages. However, it does not support an inexpensive method to physically distribute content. Hard drives are dominant in computers, but are not

Flash		Hard drives		CD/DVD	
				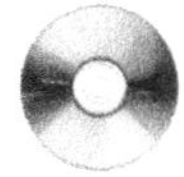	
Pros	**Cons**	**Pros**	**Cons**	**Pros**	**Cons**
• Small size • Robust • Ease of use	• Low capacity • High $/GB • Slow replication	• High capacity • Low $/GB	• High power usage • Shock sensitive	• Low $/GB • Low cost replication • Standard for distribution	• 120 mm • Large device

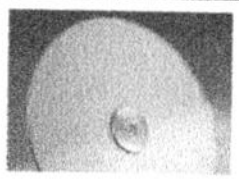

Holographic technology potential

Pros	**Cons**
• High data density • Flexible formats • Low $/GB media • Low cost / high speed replication • Low power usage	• New technology • Need new standard

Figure 1.6 *Advantages and disadvantages of various technologies for consumers*

appropriate for these two initial applications. BD/DVD/CD are dominant in content distribution but have reached, or nearly reached, the end of their technology roadmaps, and cannot supply the next increase in performance.

IDC estimates that the consumer market for drives and media in these technologies will be around US$32 billion in 2010. The fraction of the market for distribution and archival (all optical and some SSM) applications is estimated from IDC numbers at around US$18 billion for drives and media. Clearly, the current market potential is very large, and expected to grow significantly.

Advances in network technology are expected to significantly affect the content distribution market. As connection speeds increase, Internet or pay for view services will continue to be a force in content distribution to the home. While this trend is likely to make the home archival market segment even better for holographic technology (because this content is not delivered in a form that is already archived like an optical disk), it does represent competition to physical distribution. Even if the content is not user generated, if the data will be owned rather than rented, some archive storage must be used for storing the downloaded material.

Physical distribution has some advantages over network distribution in that the content is already archived, can be taken anywhere, can be more securely distributed, and large content can be distributed easily. However, renting content over the network is easier than renting physical media, and you do not need to worry about keeping it. The market will most likely see both physical and network distribution thrive, and even as distribution to mobile devices becomes available, physical distribution will continue to be an important part of the huge and growing market.

In Chapter 16, we will return to this topic and speculate on how the technology may evolve, and which other markets may then open up for holographic technology.

1.4 Summary

This book coincides with the commercialization of the first HDS product. The drive has 300 GB capacity, fast transfer rates through standard computer interfaces and operates over wide environmental conditions. However, this is the just the first step for the technology, and much more can be done. The performance of both media and drive can be dramatically improved. New technology can be developed that simplifies the drive construction, improves performance, and lowers cost. The technology will also be applied to the consumer market.

The last book published on HDS is now 10 years old, and out of print [44]. While it was an excellent introduction, it was a snapshot of the status of a number of research groups at that time. This book describes the advances made since that time and details the technology required to make data storage products using holography. It is hoped that by understanding this technology, others will be able to use it to develop their own holographic storage products.

Acknowledgements

We wish to thank the many people that have helped us understand the storage market, including Tom Burniece, Mark Cardillo, Hiroshi Kamada,Terry Loseke, Art Rancis, Rusty Rosenberger, Ed Schlesinger, Wolfgang Schlichting and Stephanie Soule. In addition, we deeply appreciate Maxell, Turner Broadcasting, Ikegami, Paramount, Disney, USGS, National Recognizance Organization, SAIC, and several other organizations for contributing to our understanding of their archival needs. Thanks also to Clyde Smith of Turner Broadcasting, Garrett Smith of Paramount, and Naoki Kashimura of Ikegami, for their support and patience over the years. Thanks also to Bart Stuck and Stephen Socolof and the rest of the investors of InPhase for their confidence in investing in InPhase. Finally, our sincere thanks to two of the original founders of InPhase, Michael Tackitt and Melinda Schnoes, who made significant contributions to this work at both Bell Laboratories and InPhase Technologies.

References

1. A. Mitsumori, *et al.*, Multi-layer 400GB optical disk, *Joint Int. Symp. on Opt. Memories and Opt. Data Storage*, Waikoloa, Hawaii, July (2008), paper MB01.
2. H. Mikami, *et al.*, Read-out signal amplification by homodyne detection scheme, *Joint Int. Symp. on Opt. Memories and Opt. Data Storage*, Waikoloa, Hawaii, July (2008), paper TuA01.

3. S. Aoki, *et al.*, A novel deformable mirror for spherical aberration compensation, *Joint Int. Symp. on Opt. Memories and Opt. Data Storage*, Waikoloa, Hawaii, July (2008), paper TuB02.
4. A.M. van der Lee and E. Altewischer, Drive considerations for multi-layer discs, *Int. Symp. on Opt. Memories,* Takamatsu, Japan, October (2006), paper Mo-C-05.
5. D. Bruls, *et al.*, Practical and robust near field optical recording systems, *Int. Symp. on Opt. Memories,* Takamatsu, Japan, October (2006), paper Mo-C-01.
6. J.M.A. van den Eerenbeemd, *et al.*, Towards a multi-layer near field recording system, dual layer recording results, *Int. Symp. on Opt.l Memories*, Takamatsu, Japan, October (2006), paper Tu-F-03.
7. J. Kim, *et al.*, The error rate improvement of Super-RENS Disc, *Int. Symp. on Opt. Memories*, Takamatsu, Japan, October (2006), paper Mo-B-01.
8. J. Tominaga and T. Nakano, *Optical Near-Field Recording - Science and Technology*, Springer-Verlag, New York, 2005.
9. E. P. Walker, *et al.*, Terabyte recorded in a two-photon 3D disc, *Joint Int. Symp. on Opt. Memories and Opt. Data Storage*, Waikoloa, Hawaii, July (2008), paper MB01.
10. A. N. Shipway, *et al.*, A new media for two-photon volumetric data recording and playback, *Joint Int. Symp. on Opt. Memories and Opt. Data Storage*, Honolulu, Hawaii, July (2005), paper MC6.
11. M. Akselrod, *et al.*, Progress in bit-wise volumetric optical storage using alumina-based materials, *Opt. Data Storage Conf.*, Portland, Oregon, May (2007), paper MA2.
12. A. Yariv, *Optical Electronics*, Holt, Rinehart and Winston, New York, 1985.
13. S. Orlic, *et al.*, Microholographic data storage towards dynamic disk recording, *Joint Int. Symp. on Opt. Memories and Opt. Data Storage,* Waikoloa, Hawaii, July (2008), paper MB05.
14. R.R. McLeod, *et al.*, Micro-holographic multi-layer optical disk data storage, *Joint Int. Symp. on Opt. Memories and Opt. Data Storage*, Honolulu, Hawaii, July (2005), paper MB03.
15. T. Horigome, *et al.*, Drive system for micro-reflector recording employing blue laser diode, *Int. Symp. on Opt. Memories*, Takamatsu, Japan, October (2006), paper Mo-D-02.
16. B. L. Lawrence, Micro-holographic storage and threshold holographic recording materials, *Joint Int. Symp. on Opt. Memories and Opt. Data Storage*, Waikoloa, Hawaii, July (2008), paper MB03.
17. K. Saito, *et al.*, Drive system and readout characteristics of micro reflector optical disc, *Opt. Data Storage Conf.*, Portland, Oregon, May (2008), paper MB1.
18. H. Miyamoto, *et al.*, Direct servo error signal detection method for recorded micro-reflectors, *Joint Int. Symp. on Opt. Memories and Opt. Data Storage*, Waikoloa, Hawaii, July (2008), paper MB04.
19. R. R. McLeod, Impact of phase aberrations on three-dimensional optical data storage in homogeneous media, *J. Opt. Soc. Am. B*, Vol. 26, pp. 308–317 (2009).
20. K. Shimada, *et al.*, High density recording using Monocular architecture for 500 GB consumer system, *Opt. Data Storage Conf.*, Buena Vista, Florida, May (2009), paper TuC2.
21. P.J. van Heerden, Theory of optical information storage in solids, *Appl. Opt.*, Vol. 2, pp. 393–400 (1963).
22. D.L. Staebler, *et al.*, Multiple storage and erasure of fixed holograms in Fe-doped $LiNbO_3$, *Appl. Phys. Lett.*, Vol. 26, p. 182 (1975).
23. H. Fleisher, *et al.*, An optically accessed memory using Lippmann process for information storage, in *Optical and Electro-optical Information Processing*, J. Tippett, *et al.*, eds. MIT Press, Cambridge, Massachusetts, pp. 1–30, 1965.
24. E.N. Leith, *et al.*, Holographic data storage in three dimensional media, *Appl. Opt.*, Vol. 5, No. 8, pp. 1303–1311 (1966).
25. L.K. Anderson, Holographic Optical Memory for bulk data storage, *Bell Laboratories Record*, Vol. 45, pp. 319–326 (1968).
26. L. d'Auria, *et al.*, Experimental holographic read-write holographic storage system, *Appl. Opt.*, Vol. 13, No. 4, pp. 808–818 (1974).
27. N. Nishida, *et al.*, Holographic coding plate: a new application of holographic memory, *Appl. Opt.*, Vol. 12, No. 7, pp. 1663–1674 (1973).

28. Y. Tsunoda, *et al.*, Holographic videodisc: an alternative approach to optical videodisks, *Appl. Opt.*, Vol. 15, No. 6, pp. 1398–1403 (1976).
29. K. Kubota, *et al.*, Holographic disk with high data transfer rate: its application to an audio response memory, *Appl. Opt.*, Vol. 19, No. 60, pp. 944–951 (1980).
30. J. Heanue, M. Bashaw and L. Hesselink, Volume holographic storage and retrieval of digital data, *Science*, Vol. 265, pp. 749–752 (1994).
31. S.S. Orlov, *et al.*, High transfer rate (1 Gbit/sec) high-capacity holographic disk digital data storage system, *Conf. on Lasers and Electro-Optics (CLEO 2000)*, Vol. 39, San Francisco, CA, May (2000), paper TuC4.
32. L. Hesselink, Digital holographic demonstration systems by Stanford University and Siros Technologies, in *Holographic Data Storage*, H. J. Coufal, D. Psaltis, and G. Sincerbox,eds. Springer-Verlag, New York, pp. 383–397, 2000.
33. C.M. Jefferson, G.W. Burr and J.A. Hoffnagle, "IBM holographic digital data storage test platforms," in *Holographic Data Storage*, H. J. Coufal, D. Psaltis and G. Sincerbox, eds. Springer-Verlag, New York, pp. 369–381, 2000.
34. S. Redfield, Tamarack optical head holographic storage, in *Holographic Data Storage*, H. J. Coufal, D. Psaltis and G. Sincerbox, eds. Springer-Verlag, New York, pp. 343–357, 2000.
35. J. Ma, T. Chang, S. Choi, and J. Hong, Digital holographic data storage with fast access, in *Holographic Data Storage*, H. J. Coufal, D. Psaltis, and G. Sincerbox, eds. Springer-Verlag, New York, pp. 409–418, 2000.
36. D.A. Waldman, *et al.*, Cationic ring opening photo-polymerization methods for holography, *Proc. SPIE*, Vol. 2689, pp. 127–141 (1996).
37. R.T. Ingwall and D. Waldman, Photopolymer systems, in *Holographic Data Storage*, H. J. Coufal, D. Psaltis, and G. Sincerbox,eds. Springer-Verlag, New York, pp. 171–197, 2000.
38. G. Burr, *et al.* Volume holographic data storage at an areal density of 250 gigapixels/in.^{2}, *Opt. Lett.*, Vol. 26, No. 7, pp. 444–446 (2001).
39. S. Campbell, *et al.*, Method for fabricating a multilayer optical article, US Patent 5,932,045, August 3, 1999.
40. W. L. Wilson, *et al.*, High density, high performance data storage via volume holography, *Int. Phot. Conf. 98*, Taiwan, December (1998), paper We2.
41. K. Tanaka, *et al.*, 415 Gbit/in^2 recording in coaxial holographic storage using low-density parity-check codes, *Opt. Data Storage Conf.*, Buena Vista, Florida (2009), paper TuC3.
42. E. Hwang, *et al.*, Real-time video demonstration of holographic disk data storage system, *Proc. SPIE*, Vol. 6282, pp. 6282–6285 (2006).
43. Information Storage Industry Consortium (INSIC) Optical Disk Storage Roadmap, August 2006.
44. H. J. Coufal, D. Psaltis, and G. Sincerbox,eds. *Holographic Data Storage*, Springer-Verlag, New York, 2000.

2

Introduction to Holographic Data Recording

William Wilson, Alan Hoskins, Mark Ayres, Adrian Hill and Kevin Curtis

2.1 Introduction

From the earliest drawings on cave walls to the latest Blu-ray disk technology, humans have employed a wide range of optical recording methods to store and disseminate information. Optical technologies have proven remarkably versatile and have used a variety of recording surfaces. The first recordings were readable only by eye, with later recordings requiring sophisticated machines. All these storage techniques employ the same basic principle – that of inscribing data marks on a two-dimensional (2D) medium. In some cases, the optical component manifests only in the read function; in others the inscribing method is also optical. In either case, upper limits for storage speed and density are determined by optical physics.

For millennia, the capabilities of light and its interactions with matter far exceeded the meager requirements of these applications. More recently, however, the dimensions of the data marks have begun to approach the fundamental physical resolution afforded by optics. Blu-ray data pit dimensions, for example, already approach the Rayleigh resolution limit for its aggressively designed optical path. From this perspective, traditional optical storage, with its long march of advancement and utility, is fast approaching a dead end.

There remains, however, a glimmer of hope in the prospect of stacking many optically recorded surfaces in the third dimension. This technique has been employed in some fashion ever since pages have been bound into books. More recently, DVD and Blu-ray disks have been produced with two separate recording layers, and other multi-layer technologies are

Holographic Data Storage: From Theory to Practical Systems Kevin Curtis, Lisa Dhar, Adrian Hill, William Wilson and Mark Ayres

being pursued in the laboratory. Optical technologies are uniquely poised to reap the benefits of volumetric storage, for electromagnetic radiation can probe the depths of a solid where no magnetic recording head can fly, and no atomic force stylus can penetrate. However, true three-dimensional (3D) recording systems, with a bit density in depth that rivals that of the other two dimensions, have remained elusive. The probe beam used by these modern devices for both reading and writing must penetrate all the other recording layers without significant loss or aberration in order to form a diffraction-limited focal point, yet interact with the desired layer strongly enough to write and read the data. It seems that multi-layer recording is trapped in a vicious cycle wherein the more densely the medium is modulated, the less resolution is available to probe it – the technique just does not scale well.

What is needed is a method to access each of the myriad resolvable bits within a volume of solid matter in a completely independent manner. Enter holographic data recording. The Bragg effect allows a vast number of diffraction gratings to coexist within a volume, with pristine mathematical independence. The number of independent gratings so permitted can be shown to be precisely equal to the number of bits that would be obtained by a hypothetical 3D imaging system with the same optical parameters. Only now, each of these transformed volumetrically stored bits can – in theory, at least – coexist happily in a medium that has been modulated to its very fullest extent: *holographic data recording scales.*

The application of the principle of wavefront capture (pioneered by Gabor in 1948) to data storage was first suggested by van Heerden [1] in 1963. The appeal was obvious: the notion of storage densities approaching $1/\lambda^3$ made holographic data recording a very compelling storage technology. The transition from potential to product has taken nearly 50 years and is the subject of this book. While a simple understanding of the physics involved was common knowledge in the late 1960s, development of the materials, components and supporting technologies needed to build viable systems have taken over four decades to mature. In this chapter we will briefly cover the history of holographic data storage and review the basic physical principles and methods used to store data into a volume of material. Finally, we will discuss achievable densities for volume holographic storage.

2.2 Brief History of Holography

Holography is in essence a technique for recording optical wavefronts. In 1948, Hungarian scientist Dennis Gabor first conceived of the notion of 'two-step optical imagery' [2] through his work on electron microscopy. The basic idea was suggested to Gabor by Bragg's X-ray microscopy research [3]. Bragg was able to form crystal lattice images by diffraction from a photographic plate. The fundamental concept here is simple double refraction. Given that the field diffracted from an object can be represented as a Fourier transform of the light distribution at the object, the secondary diffraction becomes a Fourier transform of the Fourier transform, that is, the object field itself. This is provided that all the amplitude and phase information of both field diffractions are preserved. Although Gabor conceived of this strategy to improve the resolution of the electron microscope, he first demonstrated the feasibility of the holographic approach in 1948 using photons. Using filtered light from a mercury arc lamp (imaged through a pinhole to acquire a degree of spatial coherence), he was able to record holograms of a diffracted electron beam. Although the method never solved the problem he was pursuing, the concept was born. Holographic recording for the most part lay dormant as a research subject until the early 1960s when it was revived by Leith and Upatnieks [4]. This resurrection was energized by the introduction of a new coherent light source, the laser. Leith and Upatnieks pioneered the technique of off-axis holography, and recast holographic theory in a communications framework.

As mentioned previously, the notion of using holography for data storage was first suggested by P. J. van Heerden in 1963. He suggested bleaching alkali halide color centers in a medium as a method to record data-bearing holograms. He described both wavelength and angle multiplexing and even derived a capacity of 10^{13} bits in a 1 cm^3 crystal with 10^{15} color centers ($\sim$25% absorption). While the derived capacity was purely theoretical (he did not consider actual system geometries) he was the first to suggest what might be possible. Soon after this work, others [5,6] began to expand and refine the system and theoretical concepts needed for holographic data storage. A large effort involving many companies in the United States in the early 1970s explored holographic storage, mostly using photorefractive crystals as a recording medium. In the 1980s, many Japanese companies tried to develop optical disk versions of the technology. In the mid 1990s, with extensive funding from DARPA, a number of US companies again started developing holographic systems and materials, as the drive components (modulators, detectors, and lasers) became commercially available. This momentum carried over into the 2000s, when many companies and universities worldwide are again researching holographic data storage. Chapter 3 provides more details on the two prominent approaches to drive developments of the last 10 years. (For a very detailed account of the earlier developments, see the review chapter authored by Sincerbox [7].)

2.3 Holographic Basics

2.3.1 Introduction

The simplest hologram is recorded using collimated light (plane waves) for both the object and reference beams (Figure 2.1). The two beams of wavelength λ are derived from the same monochromatic laser source, and are thus coherent with each other. Both beams are incident on a recording medium, with an angle θ between them [Figure 2.1(a)]. The interaction of the two optical fields results in a stationary interference pattern at the media, consisting of bright and dark fringes. The fringes are recorded throughout the media volume via modulation of the dielectric constant of the material system [Figure 2.1(b)]. (Chapter 6 describes how holographic materials convert this intensity pattern into a refractive index perturbation.) To reconstruct the hologram, a replica of either one of the recording beams is used to illuminate the medium, and diffraction from the hologram reconstructs the other beam [Figure 2.1(c)].

One parameter that characterizes the diffraction efficiency of the hologram is the contrast ratio, or *modulation depth*, of the fringes. Since all wavefronts can be decomposed into a sum of plane waves, understanding the nature of a hologram formed by two plane waves is very useful. It in essence provides all the information required to record any complex wavefront. A hologram of two plane waves comprises a sinusoid refractive index pattern, that is, a *grating*. The modulation depth of a grating is a function of beam intensity, as shown in Figure 2.2. This quantity is optimized when the two beams have equal intensity. As the

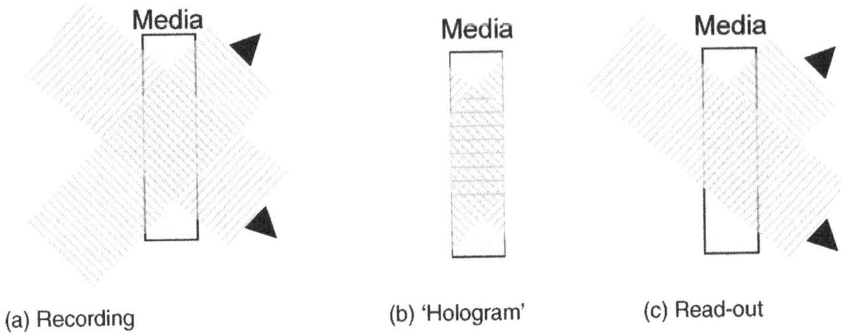

Figure 2.1 *Diagram of hologram formation and read-out*

modulation depth (m) is decreased, the DC component of the resulting interference pattern is increased. Since diffraction arises from the interference modulation and not the DC portion, an excessive DC portion wastes the media dynamic range, resulting in lower diffraction efficiency for a given exposure energy.

The hologram diffraction efficiency, η, is defined as the diffracted optical power divided by the incident optical power. The diffraction efficiency of a hologram varies with the square root of the modulation depth. Holograms are recorded via changes of the real and/or imaginary parts of the dielectric constant (which determine the index of refraction and absorption/emission, respectively) within the medium. These two parts of the dielectric constant produce gratings of dramatically different character. The lossy nature of absorptive gratings (pure amplitude) limits their maximum diffraction efficiency to less than 4%. Refractive index gratings (pure phase gratings) on the other hand, can achieve diffraction efficiencies of 100% and support many small holograms whose cumulative grating strength would be equivalent to a diffraction efficiency of greater than 100% (of course, only one of these gratings may be read out at a time – see the Chapter 6 discussion on *M/#*.) Thus for data storage, refractive index gratings are the most important type, and the discussion below is restricted to that form.

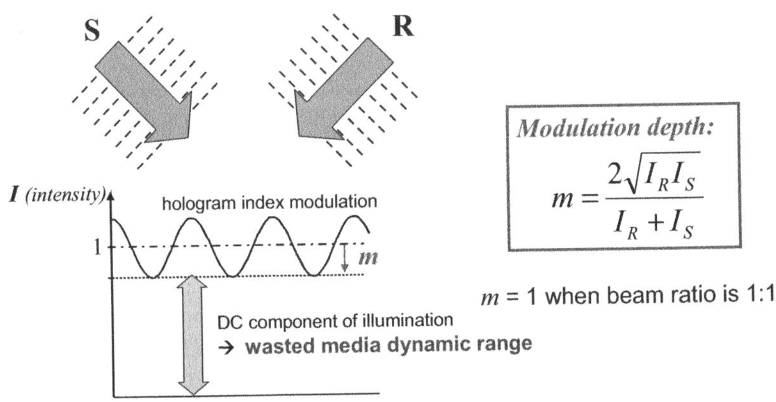

Figure 2.2 *Modulation ratio definition and schematic*

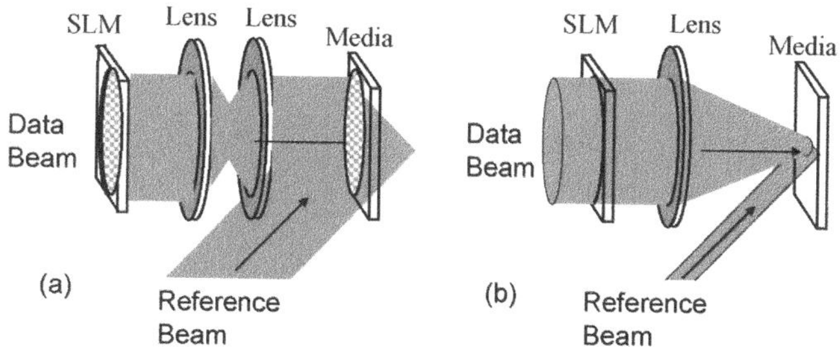

Figure 2.3 *(a) Image plane and (b) Fourier plane recording geometries for data storage*

More complex object waves can be recorded. For example, the optical field scattered by any illuminated object can be decomposed into a Fourier sum of plane waves of varying angles (spatial frequencies). The complex interference pattern resulting from the superposition of a complex object and a reference generates a spectrum of gratings in the material's volume. Diffraction from this complex grating structure reconstructs the complex field recorded.

While light diffracted by an object could be recorded at any propagation distance, two configurations are most common. In the first, the optical system is set up such that an image of the object being recorded is projected into the medium – such an *image plane geometry* is illustrated in Figure 2.3(a). An image plane system can achieve moderately high density because the image plane is located at a beam waist, and so results in relatively small holograms. The image plane geometry is commonly used for recording objects for holographic display. In this application, the object is a two-dimensional block or page of data generated by a pixellated spatial light modulator (SLM) that modulates the beam. This beam is called the signal or data beam.

However, for data storage and optical processing applications, the *Fourier plane geometry* is more common. Here the optical system is configured such that a Fourier plane of the SLM is located within or near the media, as shown in Figure 2.3(b). Given that, for these applications, the wavefront being recorded has little or no depth information (that is, the data page lies in a single plane), Fourier recording offers many advantages. The Fourier plane is also located at a beam waist, and, for typical pixel sizes and lenses, the Fourier geometry minimizes the hologram size. Indeed, it is because the Fourier geometry maps the spatial frequencies of the object onto the transverse dimensions of the media that the hologram size may be reduced to record only the essential information-preserving components as determined by Shannon's sampling theorem. Additionally, the Fourier geometry allows for more fault tolerant operation (through, for example, shift invariance), and could enable sophisticated optical filtering, potentially leading to direct optical processing of holographic information (see Chapter 16).

The descriptions used thus far suggest a recording geometry where the reconstructing light is transmitted through the medium, in other words, a transmissive geometry [Figure 2.4(a)]. Angle multiplexing has higher selectivity for that geometry (as well as for a 90° geometry [8]), and therefore angle-multiplexed holograms are usually recorded in transmission.

However, one can also record holograms in a reflective geometry, as shown in Figure 2.4(b). This method of holography was suggested by Denisyuk in the early

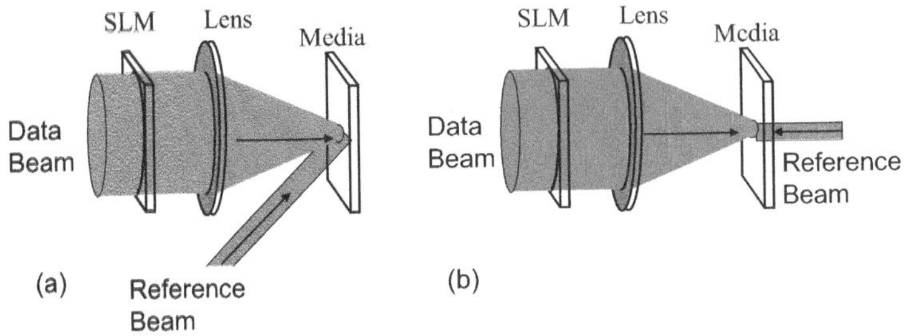

Figure 2.4 *(a) Transmissive, and (b) reflective holographic recording geometries*

1960s [9]. To record in a reflective geometry, the signal and reference beams enter the medium from opposite surfaces (in a pure reflective geometry the beams are counter-propagating), and thus the diffracted beam exits during reconstruction from the same face that the probe beam (reference) enters during hologram formation. Since reflective gratings have the highest spatial frequencies (with fringe period approximating one half the recording wavelength), they are more sensitive to wavelength and anisotropic media dimensional changes due to temperature variations than their transmissive counterparts. However, because of this higher selectivity for wavelength changes, wavelength-multiplexing systems typically employ a reflective geometry.

2.3.2 Using Holography for Data Storage

Holographic data storage takes advantage of the fact that a known, reproducible set of reference beams may be used to record and recall a set of unknown data beams that encode digital data. During recording, light from a laser is split into two beams: a data or signal beam (i.e. a digitally encoded object beam) and a reference beam. A hologram is formed where these two beams interfere in the recording medium. While there are many methods one might devise to digitally encode a light beam, a page-wise system typically employs a device called a *spatial light modulator* (SLM). The SLM translates the electronic data of 0s and 1s into a pixel pattern of optically distinguishable states – typically light and dark pixels. The digital data stream is arranged in a 2D image, or *page*, typically comprising over 1 million bits. The exact number of bits is determined by the SLM's pixel count, in conjunction with the optical resolution of the data path. Each data page is formatted with encoded user data, control headers, blocks of known bits, and error correction information. Figure 2.5 shows a schematic of the write process. More information on the data channel is given in Chapter 11.

The optical energy of the recording beams induces a material change in the recording medium. In an ideal photopolymer medium, chemical reactions and chemical diffusion render optical intensity linearly as a refractive index perturbation. A linear, high resolution recording of the interference pattern leads to a high fidelity reconstruction of the signal beam provided the diffraction efficiency is in the weak linear (Born approximation) regime.

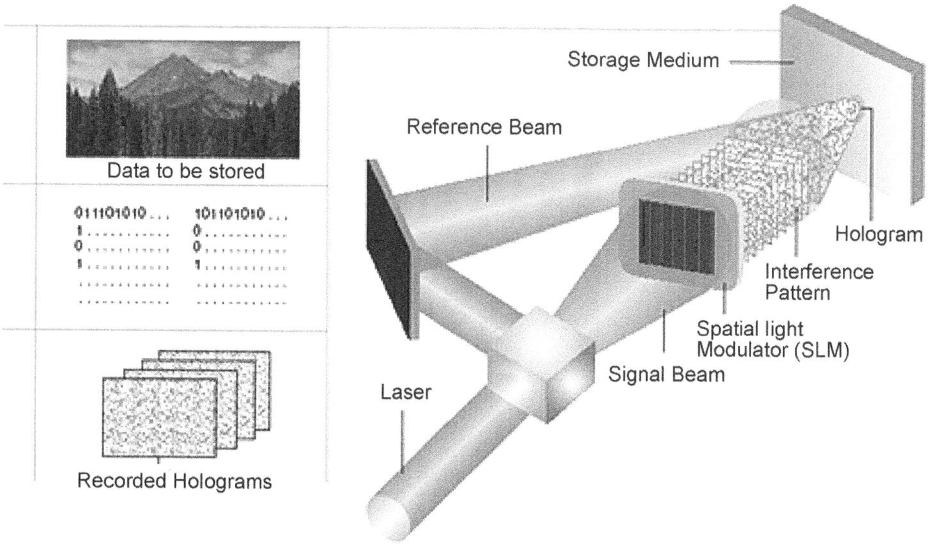

Figure 2.5 *Writing data*

A single hologram in a well-designed HDS system achieves an areal storage density approximately determined by the resolvable spot size of its object path. This density approaches that of a single layer bitwise technology (such as DVD), but will be somewhat lower since the holographic device must share its available numerical aperture (and hence its resolving power) with a reference beam. Thus, an HDS system might achieve about a gigabyte of data on a DVD-sized disk using nonoverlapping holograms. The true power of holographic recording is realized by overlapping many holograms in the same media volume by the process of multiplexing.

Many multiplexing techniques have been proposed, and some are discussed in Section 2.5. InPhase's primary multiplexing method is angle multiplexing, which allows data pages to be superimposed by varying the reference beam angle. For each reference beam angle, a different data page is stored. In this way, hundreds of unique holograms are recorded within the same volume of material. This group of angle-multiplexed holograms is referred to as a *book* (or sometimes, a *stack*). To achieve high capacity, fully multiplexed books may be recorded at every location on the medium, thus multiplying the 2D resolution-limited capacity by a factor of several hundred.

For data recovery, a *probe* or *reference beam* (which is nominally a replica of the recording reference beam) is used to illuminate the medium. The hologram diffracts the probe beam, thus reconstructing a replica of the encoded signal beam. This *diffracted beam* is then projected onto a pixellated photodetector that captures the entire data page of over 1 million bits at once. This parallel read-out provides holography with its fast transfer rates. Figure 2.6 illustrates a basic geometry for reading holograms. The detected data page images are then processed and decoded in order to recover the stored information.

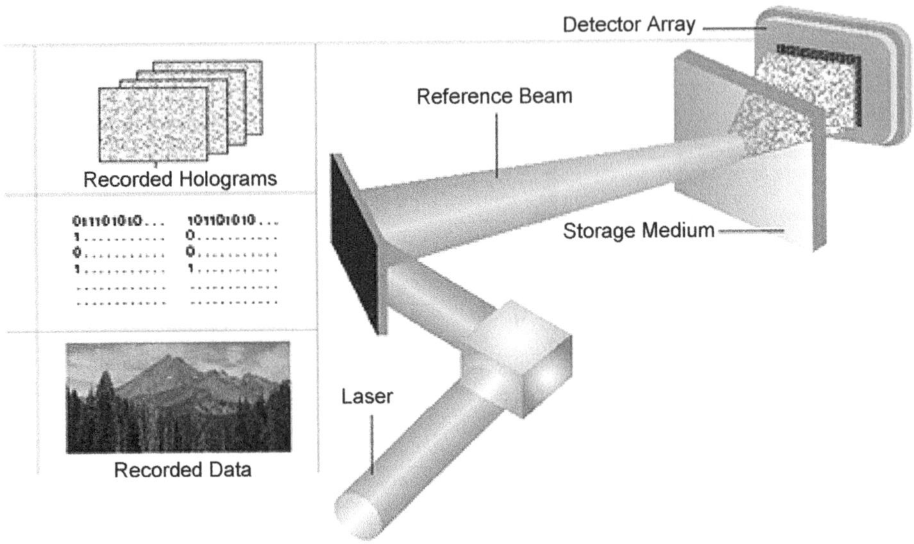

Figure 2.6 *Reading data*

2.4 Volume Holograms

2.4.1 Introduction

Holograms may be written in either the thick or thin regime. Thin holograms, such as those written on ordinary photographic film, effectively confine their diffractive interaction to a single plane and are not suitable for dense multiplexing. One basic criteria for defining holograms as thin or thick is the Q parameter suggested by Klein and Cook, and a grating strength parameter [10,11]. These parameters determine whether the grating is in the Bragg (a single order is diffracted) or Raman–Nath (multiple orders are diffracted) diffraction regimes. The Q parameter is given by:

$$Q = 2\pi\lambda L / n_0 \Lambda^2 \tag{2.1}$$

where λ is the wavelength of the light in a vacuum, L is the thickness of the recording layer, n_0 is the nominal refractive index of the medium, and Λ is the grating period. If $Q < 1$ and the grating has a large enough index change (equivalent to a few percent diffraction efficiency in thin media) then the gratings are considered in the thin media or Raman–Nath regime. For values of $Q > 1$, the gratings are considered to be thick, volume gratings in the Bragg diffraction regime. The presence of many grating planes or grating periods supports Bragg diffraction from these structures.

For storage applications, the recording medium will have a thickness between 500 μm and a couple of millimeters, with grating periods of less than 1 μm. Thus, for storage applications, the holograms are firmly in the Bragg regime. In fact, as discussed later, Bragg selectivity is the foundation of high density storage. The holograms manifest as a 3D structure with a quasi-periodic modulation of refractive index. The periodicity results in a maximum in the diffracted amplitude when the Bragg condition (correct angle and wavelength) is satisfied, and diffraction efficiency tending strongly to zero when it is not. Thus, the desired hologram can be reconstructed with the correct wavelength and angle while other holograms within the same volume cause negligible diffraction. In this manner, Bragg selectivity is used to achieve high density storage. More complex reference beams (such as spherical beams) can be analyzed by decomposing them into plane wave components.

In the Fourier plane geometry, each data pixel is transformed into a unique plane wave within the signal beam, and is therefore recorded as a unique sinusoidal grating by each plane wave component of the reference beam. As a result, in storage applications with hundreds of holograms each consisting of millions of gratings, the diffraction efficiency of the individual gratings is very weak indeed. Thus the Born approximation can be used to solve or model the diffraction of light [12], rather than the more complex coupled wave equations [13]. Both models will be briefly summarized below.

2.4.2 Kogelnik's Coupled Wave Equations

Even though the Born approximation is typically used to analyze data storage systems, measuring the basic parameters of materials (described further in Chapter 8) uses plane wave holograms. Due to the higher diffraction efficiency involved in these tests, the results are analyzed using Kogelnik's famous equations [13] for plane wave holograms.

For a pure index grating in transmission geometry the diffraction efficiency is given by:

$$\eta = \sin^2\left(\sqrt{\nu^2+\xi^2}\right) \Big/ \left(1+\frac{\xi^2}{\nu^2}\right) \tag{2.2}$$

where the parameters ν and ξ are given by:

$$\nu = \frac{\pi n_1 L}{\lambda\left(\cos^2\theta - \dfrac{K\cos\theta\cos\phi}{\beta}\right)^{1/2}} \tag{2.3}$$

and

$$\xi = \frac{[\Delta\theta\cdot K\sin(\phi-\theta_o) - \Delta\lambda K^2/4\pi n_0]d}{2\left(\cos\theta - \dfrac{K\cos\phi}{\beta}\right)} \tag{2.4}$$

In these equations, η is the diffraction efficiency, n_1 is the index perturbation, L is the thickness of the recording layer, λ is the wavelength of the light in a vacuum, θ is the angle of the reference beam outside the media, measured from media normal, on read-out, K is the grating number $(2\pi/\Lambda)$, where Λ is the grating period, n_0 is the bulk index of the material, ϕ is the grating slant angle inside the material measured from normal, β is $2\pi(\varepsilon_o)^{1/2}/\lambda$, where ε_o is the bulk dielectric constant, $\Delta\lambda$ is the deviation from the Bragg condition for wavelength and $\Delta\theta$ is the deviation from the Bragg condition for angle.

The functional form for reflection mode holograms is very different. Typically, reflection holograms are not used in data storage as they impose very stringent requirements for isotropic thermal expansion and contraction in order to fully compensate for thermal changes (see Chapter 15 for details). However, the Kogelnik equations for a pure index grating in reflection are given by:

$$\eta = 1/\left[1 + (1-\xi^2/\nu^2)/\sinh^2(\nu^2-\xi^2)^{1/2}\right] \tag{2.5}$$

where ν is given by Equation (2.3) but multiplied by j, and ζ is equal to the negative of Equation (2.4), because the angle of the signal is negative in this geometry.

2.4.3 *k*-Space Formalism

Kogelnik's analysis is a rigorous solution for diffraction in volume holograms. However, for many applications, coupled-mode analysis can be computationally difficult and may not yield an intuitive understanding of the physical processes. Fortunately, holographic memories operate in a regime where several simplifying approximations are applicable, and a powerful, intuitive formalism may be developed. References [12] and [14] demonstrate a very general formulation in which each spatially varying complex scalar optical field component is represented by its wave vector, $\vec{k}$, in a Fourier transform space, or *k-space* (sometimes referred to as *momentum space*). Other important spatially varying quantities – such as the refractive index distribution and the polarization distribution of the medium when illuminated by a probe beam – are similarly represented by 3D Fourier transforms of their spatial distributions. Many important diffraction problems can be solved by simple graphical manipulation of these distributions in k-space.

In scalar wave optics, a monochromatic plane wave of infinite extent may be represented by acomplex amplitude, for example:

$$E_R(\vec{r}) = E_{R0}\exp\left(j\vec{k}_R \cdot \vec{r}\right) \tag{2.6}$$

where $\vec{k}_R$ is the wave vector of the light, $\vec{r} = \{x, y, z\}$ is the spatial coordinate vector, and $\cdot$ is the dot-product operator. E_{R0} is a complex scalar denoting the field strength, which will be subsequently omitted for brevity. The complex amplitude representation likewise omits time dependencies, which would take the form of a common $\exp(j\omega t)$ factor in all the monochromatic optical fields. As long as the plane wave is propagating in an isotropic and weakly inhomogeneous medium (such as a holographic recording medium), the length of the wave vector will be effectively constant at $|\vec{k}_R| = 2\pi n_0/\lambda \equiv k_n$. Defining $F(\vec{k}) = \Im\left[F(\vec{r})\right] \equiv \iiint F(\vec{r})e^{-j\vec{k}\cdot\vec{r}}\,\mathrm{d}\vec{r}$ as the 3D Fourier transform, the k-space representation of the plane wave becomes:

$$E_R(\vec{k}) = \Im\left[E_R(\vec{r})\right] = \delta\left(\vec{k} - \vec{k}_R\right) \tag{2.7}$$

where $\vec{k} = \left[k_x, k_y, k_z\right]$ is the 3D spatial frequency coordinate vector. In other words, the k-space representation of an infinite plane wave is a delta function. The delta function takes a nonzero value only at an infinitesimal point, so it is convenient to indicate its location graphically by plotting the $\vec{k}_R$ wave vector with an arrow, as shown in Figure 2.7(b). Since all of the light used derives from a single monochromatic source, all of the k-vectors must lie on a sphere of radius k_{n} referred to as the *k-sphere* (or *Ewald* sphere).

The refractive index perturbations constituting the holograms may also be represented in k-space. More generally, the spatial *dielectric constant* (or *relative permittivity*) distribution, $\varepsilon(\vec{r})$, may be transformed to yield a k-space representation, $\varepsilon(\vec{k})$. The real

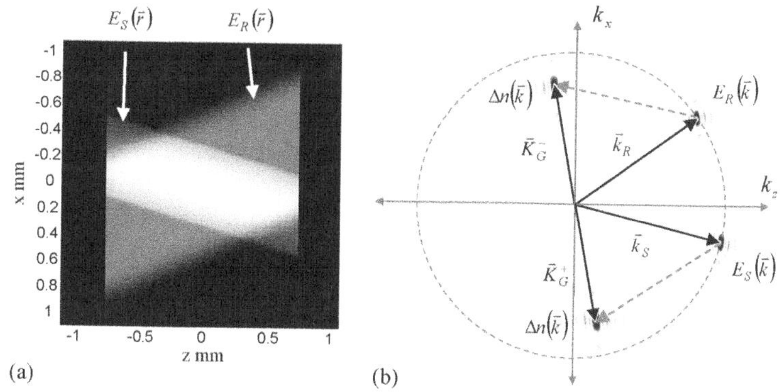

Figure 2.7 *(a) Cross-section of recording beam complex amplitudes within a thick recording layer in real space. (b) k-space representation showing uncertainty envelopes of the optical and grating distributions (sizes greatly exaggerated), along with simplified vectorial representation*

part of the dielectric constant corresponds to the square of the refractive index; the imaginary part corresponds to absorption/emission, which is usually assumed to be negligible for holographic media. Alternatively, in a linear recording medium the $\varepsilon(\vec{k})$ distribution may be determined directly from the recording beams, as the following plane-wave example shows.

Consider a hologram written with two plane waves, $E_R(\vec{r}) = \exp(j\vec{k}_R \cdot \vec{r})$ and $E_S(\vec{r}) = \exp(j\vec{k}_S \cdot \vec{r})$, as illustrated in Figure 2.7(a). If the medium is linear, the refractive index modulation is locally proportional to the intensity of the writing beams:

$$\begin{aligned}\Delta n(\vec{r}) &= STI(\vec{r}) \\ &= \frac{\varepsilon_0 c_0 n_0}{2} ST\left[\left|E_R(\vec{r})\right|^2 + \left|E_S(\vec{r})\right|^2 + E_R^*(\vec{r})E_S(\vec{r}) + E_R(\vec{r})E_S^*(\vec{r})\right]\end{aligned} \tag{2.8}$$

where $\Delta n(\vec{r})$ is the optically induced refractive index modulation, $I(\vec{r})$ is the time-invariant optical recording intensity, and * is the unary complex-conjugation operator. T is the exposure time, S is the medium sensitivity in units of $m^2 J^{-1}$, and ε_0 and c_0 are the electric permittivity and speed of light in a vacuum, respectively. The third and fourth terms in the intensity expansion represent the holographic grating, whereas the first and second terms are merely the constant intensity of the reference and signal beams in isolation, and will henceforth be omitted. No optical polarization term appears in the simple scalar model, so the beams are effectively assumed to exist in the same polarization state.

The products in the spatial distribution, $E_R^*(\vec{r})E_S(\vec{r}) + E_R(\vec{r})E_S^*(\vec{r})$, become *cross-correlations* in k-space (by the Wiener–Khinchin theorem), that is:

$$\Delta n(\vec{k}) \propto E_R(\vec{k}) \otimes E_S(\vec{k}) + E_S(\vec{k}) \otimes E_R(\vec{k}) \tag{2.9}$$

where $\otimes$ is the 3D cross-correlation operator. Thus, the refractive index distribution comprising a hologram can be computed simply and directly from the complex amplitudes of the writing beams. In the case of infinite plane waves, the procedure is even simpler since the cross-correlations may be graphically evaluated by vector subtraction – the two sidebands of the grating vector written by two plane waves are given by $\vec{K}_G^+ = \vec{k}_S - \vec{k}_R$

and $\vec{K}_{\bar{G}} = \vec{k}_R - \vec{k}_S$, as illustrated in Figure 2.7(b). These two opposing vectors represent the two conjugate sidebands of the sinusoidal refractive index distribution, which is real-valued in the spatial domain.

In practice, all holograms have finite size, determined by the thickness of the medium and the shape of the overlapping region of the recording beams. In this case, the cross-correlation method is still valid, but the k-space distribution will not simplify to an infinitesimal delta function. However, for gratings written by individual collimated beam components – such as the light from an individual signal pixel interfering with a plane wave reference beam in a Fourier geometry – the true k-space distributions of both the finite beams and the localized gratings will consist of a small uncertainty envelope centered on the delta function determined by the infinite plane wave approximation, as shown in Figure 2.7. In Figure 2.7(a), the location and region of overlap between two collimated recording beams is indicated in real space. In Figure 2.7(b), the corresponding k-space distributions are illustrated, with the size of the uncertainty envelopes greatly exaggerated for clarity. The uncertainty envelope of each element is determined by the 3D Fourier transform of its spatial envelope. Of particular significance is the thickness of the recording layer, which effectively imparts a common multiplicative rect function on all spatial distributions, and thus convolves all k-space components by a sinc which is the transform of the rect, that is:

$$\mathrm{rect}\left(\frac{z}{L}\right)F(\vec{r}) \xrightarrow{\Im} \frac{L}{\pi}\mathrm{sinc}\left(\frac{L}{2}k_z\right) * F(\vec{k}) \tag{2.10}$$

where $\mathrm{sinc}(x) \equiv \sin(x)/x$. This can be used to quickly estimate Bragg selectivity in the k-vector based simplification without requiring full evaluation of the cross-correlation integrals, as will be elaborated below.

The k-space model can be applied to both data recovery and recording processes. It yields accurate results so long as the holograms satisfy the conditions for the Born approximation, where diffraction is weak and linear with dielectric modulation. To begin, assume that the electric permittivity distribution of the medium interacts with a propagating optical probe beam, $E_P(\vec{r})$:

$$D(\vec{r}) = \varepsilon(\vec{r})E_P(\vec{r}) = \varepsilon_0\varepsilon_r E_P(\vec{r}) + \varepsilon_0\Delta\varepsilon(\vec{r})E_P(\vec{r}) \tag{2.11}$$

where $D(\vec{r})$ is the complex amplitude of the induced electric flux density, and $\varepsilon(\vec{r})$ is the electric permittivity distribution. The permittivity is then separated into nonspatially varying and spatially varying (homogeneous and inhomogeneous) dielectric components, ε_r and $\Delta\varepsilon(\vec{r})$, respectively. [Note that in the weakly inhomogeneous regime the modulated distribution is related to the refractive index modulation by $\Delta\varepsilon(\vec{r}) \cong 2n_0\Delta n(\vec{r})$]. The second term of Equation (2.11) constitutes an inhomogeneous polarization density, $P_1(\vec{r})$, which gives rise to the diffraction. Since $P_1(\vec{r})$ is a product in real space, it becomes a convolution in k-space:

$$P_1(\vec{r}) = \varepsilon_0\Delta\varepsilon(\vec{r})E_P(\vec{r}) \xrightarrow{\Im} P_1(\vec{k}) = \varepsilon_0\Delta\varepsilon(\vec{k}) * E_P(\vec{k}) \tag{2.12}$$

where $*$ is the 3D convolution operator. Physically, the polarization density represents atomic dipoles of the medium oscillating at the optical frequency in the spatial frequency distribution of Equation (2.12). Diffraction occurs where these dipoles can radiate into

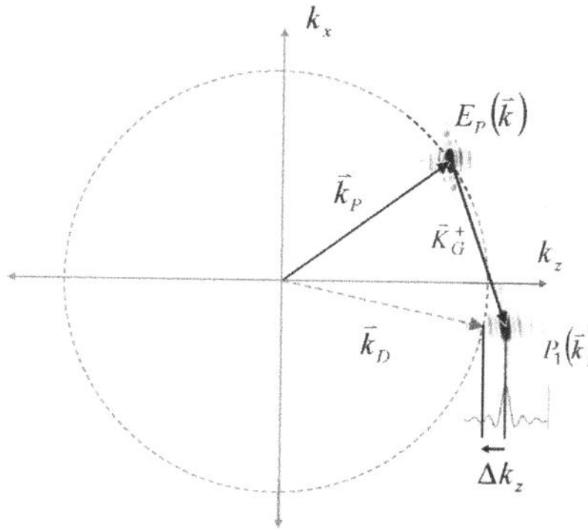

Figure 2.8 *Cross-section of recovery in k-space showing uncertainty envelopes of the probe beam and polarization density distributions (sizes greatly exaggerated). The vectorial representation is superimposed, including evaluation of the Δk_z Bragg-mismatch vector*

propagating optical modes; that is, those components of $P_1(\vec{k})$ that lie on the k-sphere. Since diffraction is linear in the Born regime, the diffracted field can only have the form:

$$E_D(\vec{k}) = \kappa\Delta\varepsilon(\vec{k}) * E_P(\vec{k})|_{|\vec{k}|=k_n} \tag{2.13}$$

where $E_D(\vec{k})$ is the propagating diffracted field and κ incorporates all linear scaling factors.[1] The notation '$(\,)|_{|\vec{k}|=k_n}$' indicates that the referenced expression is evaluated at $|\vec{k}| = k_n$.

As with recording, reconstruction can be computed graphically for infinite plane waves and gratings. In this case, it is the convolution of Equation (2.12) that may be performed with vector addition. Figure 2.8 illustrates such a reconstruction in k-space including a Bragg mismatch. Because the vector addition $\vec{k}_P + \vec{K}_G^+$ falls to the right of the k-sphere, the reconstruction does not reach its full, Bragg-matched diffraction efficiency. (The $\vec{K}_G^-$ sideband of the reconstruction is highly mismatched and is omitted for clarity.) The diffracted wave vector, $\vec{k}_D$, may be estimated by projecting the tip of the vector sum back onto the k-sphere. The resulting Bragg-mismatch vector, Δk_z, can be used to determine the diffracted field strength by evaluating the thickness selectivity sinc function of Equation (2.10) at Δk_z. This method is accurate provided the transverse dimensions of the hologram are large compared with its thickness; in other cases, a full evaluation of Equation (2.12) may be performed either analytically or numerically.

Data storage systems typically use high numerical aperture (NA) signal beams with a large number of pixels in a Fourier plane geometry. The analysis of this geometry can be performed by treating each data pixel as a point source, or equivalently, treating the field of each pixel as a plane wave in the holographic media. The reference and probe beams can likewise be decomposed into plane components (with each probe beam being composed of a single component in the angle multiplexing case). The k-space formalism developed above can then be applied to the resulting complex beams and holographic gratings, and the vectorial simplification can be applied to each individual grating component. An entire data

[1] It can be shown that $\kappa = jk_0^2 / \left[(2\pi)^3 2k_{Dz}\right]$, where k_{Dz} is the z-component of the diffracted wave vector.

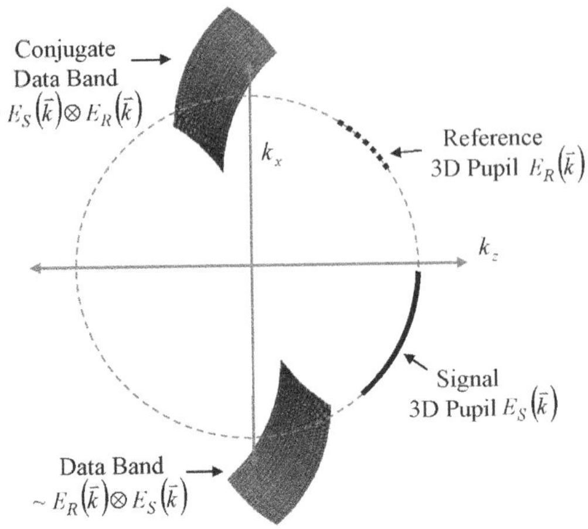

Figure 2.9 *Cross-section of angle-multiplexed book of holograms in k-space*

page is stored as a hologram manifold, and an entire book is stored as a book manifold. When a Bragg-matched probe beam is used to read out a hologram, the resulting polarization density of the entire hologram manifold lies on the *k*-sphere and the entire data page is reconstructed at once.

Figure 2.9 illustrates the layout of a book manifold in *k*-space for an angle-multiplexed system. The *3D signal pupil* illustrates the locus of all the signal beam uncertainty envelopes. In the 2D cross-section it appears as an arc spanning the angular aperture of the signal beam, but it actually subtends a page-shaped patch on the surface of the *k*-sphere extending out of the plane of the figure. The *3D reference pupil*, by contrast, is an arc representing the range of the reference beam distributions, and remains confined to the vicinity of the k_x, k_z plane. The locus of the book manifold, which constitutes the data-bearing refractive index distribution for the whole book of holograms, may be determined from the two optical pupils using the cross-correlation rule. This yields the two conjugate sidebands of the book manifold, denoted as the *data band* and the *conjugate data band* in Figure 2.9.

As an example of the *k*-space formalism applied to a data recovery process, consider what happens with a temperature change, as shown in Figure 2.10. When temperature is varied, the bulk index of refraction, n_0, changes, effectively changing the radius of the *k*-sphere. The individual grating components may also tilt if the expansion or contraction of the media is anisotropic, imparting a slight ellipticity to the shape of the hologram manifold. Because of these effects, the hologram manifold no longer fits precisely onto the spherical surface. With a full data page, the fraction of the grating manifold that Bragg matches will at best form a narrow band in the recovered image, as shown in Figure 2.10(b). This condition can be compensated by changing the probe beam wavelength and angle, and thus adjusting the size of the *k*-sphere to best match the hologram manifold. The method does result in a magnification of the recovered data page, necessitating the use of oversampled detection or a zoom lens in the recovery imaging system.

The *k*-space formalism is very useful for developing an intuitive understanding of holographic systems, as well as for simulating system effects, such as the thermal example

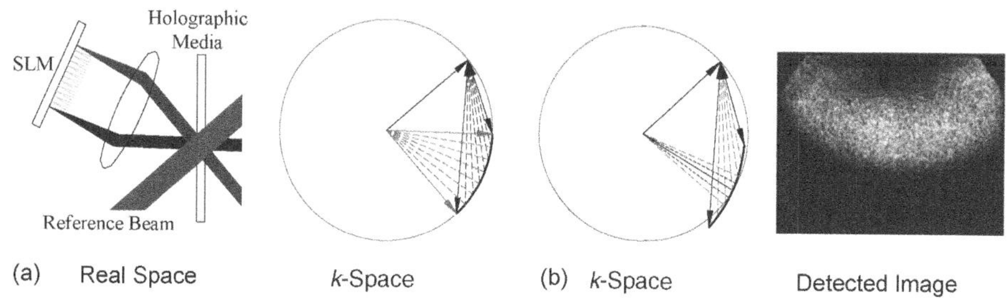

Figure 2.10 *(a) Holographic system employing a Fourier plane geometry and its representation in k-space. Each data pixel is represented by a delta function on the surface of the k-sphere creating a manifold of grating vectors. (b) The effects of temperature on a stored data page cause only a narrow band of the manifold of grating vectors to Bragg match. The k-space representation shows that only a fraction of the grating manifold can Bragg match at a given reference beam angle*

just described. References [15–19] use the method to estimate multiplexing cross-talk noise in Bragg-based holographic memories (that is, noise from the other holograms that are multiplexed in the same volume). Because the method uses plane waves, computer versions of the model can be integrated with ray tracing software and used to model the optical and holographic systems together. Chapter 14 gives examples of the *k*-space formalism applied to alignment problems, thermal effects, and tolerance analysis.

2.5 Multiplexing Techniques

2.5.1 Introduction

In order to achieve high storage densities, many holograms are superimposed in the same volume. '*Multiplexing*' is the term used for techniques designed to superimpose holograms in the same or nearly the same volume. Optical physics affords a great variety of multiplexing methods, and their elucidation occupies a large segment of the optical data storage literature. A selection of possible options is presented in Figure 2.11. The strategies fall roughly into three categories, driven by the dominant physical contribution: Bragg-based, momentum-based, and correlation-based methods. Each exploits a particular attribute of holography. Bragg-based methods depend on the multi-scatter properties of holograms.

Bragg Based	Momentum Based	Correlation Based
• Angle • Wavelength • Phase Code • Peristrophic • Shift • Correlation	• Fractal • Peristrophic • Aperture • Polytopic	• Correlation • Correlation Distance

Figure 2.11 *Summary of multiplexing options*

Momentum-based approaches depend on the directionality of the diffraction, and finally correlation methods depend on the specific spatial and polarization structure of the reference fields employed. (Some methods, such as peristrophic multiplexing can be dominated by either Bragg effects or conservation of momentum, depending on the exact geometry and system configuration used.) Correlation-based techniques typically exploit the Bragg effect in some subset of their grating distributions. Often many of the multiplexing methods are used together to optimize the geometry used.

2.5.2 Bragg-Based Techniques

These methods are based on the physics of the Bragg effect. Individual hologram reconstruction is dominated by diffraction from fringes that are highly periodic in depth, thus constituting index gratings. Multiple holograms may be superimposed in the same volume because the structured destructive interference of Bragg mismatched holograms serves to suppress multiplexing cross-talk. The Bragg selectivity of a hologram is determined by the thickness of the recording medium, the wavelength of the light, and the recording geometry. Angle multiplexing is the most common of these techniques [20–22], but wavelength [23,24], phase code [25–28], peristrophic [29], shift [30,31] and correlation multiplexing [32–34] all to some extent have Bragg constructive and/or destructive interference contributions to their selectivity.

Angle multiplexing, as described earlier, uses a discrete, angularly distinct set of plane waves as reference beams for storing the data holograms. A different hologram is stored at each angle by interfering its signal beam with a plane wave reference beam at the given angle of incidence. Another data page is then composed and stored using a reference beam at a different angle. The angular selectivity, that is, the spacing between the diffraction maximum and the first null, is given by the expression:

$$\Delta\theta = \frac{\lambda}{L}\frac{\cos\theta_s}{\sin(\theta_r + \theta_s)} \tag{2.14}$$

where L is the media thickness, and θ_r and θ_s are the reference and signal angles, respectively. For storage with no multiplexing crosstalk, data pages are placed so that each has its Bragg peak at a 'null' position in the sinc^2-shaped diffraction efficiency functions of every other page, potentially allowing for data recovery with a very high signal to noise ratio (SNR). This property means that, in principle at least, Bragg-based multiplexing crosstalk does not scale with recording density, a desirable property that may be hard to obtain with correlation or non-Fourier (for example, bitwise) architectures.

Equation (2.14) applies to plane wave holograms. For a signal beam that subtends a high NA, a slightly different situation applies. Since each data page generally spans many angles, the selectivity given by Equation (2.14) varies across the page. Thus the angular diffraction efficiency function of a high NA page resembles a superposition of sinc^2 functions of varying widths, resulting in a composite $sinc^2$-like function, with minima that (instead of nulls) do not go all the way to zero. A rough estimate of the position of the first minimum may be calculated from the selectivity of a plane wave component that is two-thirds between the center pixel and the worst case pixel (the one nearest the reference beam). The exact calculation can be achieved by summing the contributions for all the pixels in the page as a function of angle change. Because these minima are not actually nulls, high NA recording (in principle) suffers some multiplexing crosstalk. However, the overall $1/x^2$ envelope of the sinc^2 diffraction efficiency functions still bounds multiplexing crosstalk in a favorable manner.

Wavelength multiplexing uses the same physics as angle multiplexing. Wavelength Bragg selectivity is defined as the change in wavelength needed to reach the first null of the sinc^2 diffraction efficiency function. The wavelength selectivity for a reflection hologram is given by:

$$(\Delta\lambda)_{\mathrm{Bragg}} = \frac{\lambda^2 \cos\theta_s}{2L\sin^2[(\theta_r + \theta_s)/2]} \tag{2.15}$$

Phase-code multiplexing is effectively angle multiplexing with each page recorded with a set of reference beam components of varying angle, spaced by some multiple of their mutual Bragg selectivity. Additionally, each component is assigned its own phase. The read-out condition requires simultaneously satisfying the Bragg condition for each of the components *and* presenting each with its correct phase. The phase pattern of the components together comprises the phase code of the hologram. The number of holograms that can be stored is the same as for an angle multiplexed system of the same media thickness, wavelength, and geometry. The performance is determined by the orthogonality of the phase code set chosen, and, more importantly, how accurately the phase modulation is realized. The best choices of codes are orthogonal codes such as Walsh–Hadamard (though one of these code words needs to be removed [18]). In principle, phase code multiplexing should have excellent crosstalk performance – but in practice the difficulty of achieving accurate phase modulation of the reference beam makes this technique problematic, even at a fixed temperature.

Shift multiplexing uses a focusing reference beam such as a spherical or cylindrical wave. These references can be treated as the sum of a spectrum of plane waves. Usually, the reference beam is a spherical wave created by a focused spot near the media surface. Each location on the spherical wavefront of the reference beam can be modeled as a plane wave orthogonal to the wavefront at that location. If the focus point is moved relative to the media, any point on the media receives light from a different part of the reference wavefront, which means that locally the plane wave's angle has changed. This change in the angle will have the same Bragg selectivity as angle multiplexing; however, the overall selectivity has an additional term that is the uncertainty in the focus spot's location. The shift selectivity of a plane wave signal beam and a spherical reference beam in the Bragg direction (in the Bragg plane, the plane defined by the optical axes of the reference and signal beams) is given by:

$$\Delta x = \frac{\lambda z_o}{L\tan\theta_s} + \frac{\lambda}{2(NA)} \tag{2.16}$$

where L is the thickness of the media, λ is the wavelength, z_o is the distance from the center of the material to the focal point of the reference beam, θ_s is the angle of incidence of the signal beam measured from media normal, and NA is the numerical aperture of the reference beam. The first term is the normal Bragg angle selectivity due to a shift, and the second term is the uncertainty in the focal spot location. In the shift direction perpendicular to the Bragg plane along the media surface, the selectivity is typically much lower. In some configurations (as described in Chapter 3), the signal beam optical axis is close to the medium normal, and the reference components lie at higher angles of incidence in all directions, so there is no well-defined Bragg plane. In this case, selectivity may be less than indicated by Equation (2.16), but it may be equal in both shift directions, permitting dense 2D shift multiplexing. Using a reference comprised of a finite number of plane waves separated by their Bragg spacing also results in shift selectivity [30].

Peristrophic (Greek for 'turn about') *multiplexing* involves rotating the medium relative to the recording beams in order to multiplex holograms. Peristrophic multiplexing can be momentum based or Bragg based, depending on material thickness and geometry. For practical media thicknesses, peristrophic multiplexing is dominated by Bragg selectivity. In a transmission geometry, the Bragg selectivity can be expressed as the required rotation about the hologram center to the first null, and is given by:

$$\Delta\theta = \left\{\frac{2\pi}{L}\left[\frac{\cos\theta_r}{\sin\theta_r(\sin\theta_s+\sin\theta_r)}\right]\right\}^{1/2} \tag{2.17}$$

where the reference plane wave is incident at θ_r measured from media normal, and the center pixel of the image is incident at θ_s measured from media normal. Bragg selectivity is the dominant effect, if the inequality in Equation (2.18) is satisfied.

$$\frac{d}{F} > \left[\frac{2\lambda\cos\theta_s(\sin\theta_s+\sin\theta_r)}{L\sin\theta_r}\right]^{1/2} \tag{2.18}$$

In this equation, d is the width of the image (data page) in the direction of the peristropic rotation, and F is the focal length of the objective lens (assuming Fourier plane storage). When momentum change is the dominant effect, the peristrophic selectivity is simply the amount of rotation required to rotate the diffracted beam entirely out of the data path, thereby translating the reconstructed data page (image) off the detector. This rotation is given approximately by:

$$\Delta\theta \geq \frac{d/F}{\sin\theta_s+\sin\theta_r} \tag{2.19}$$

For image plane holograms when momentum change is the dominant effect, the reconstruction has to be filtered in the Fourier plane as that is where the reconstruction shifts with image rotation. The expression for the required rotation for the hologram to be filtered in the Fourier plane is given by:

$$\Delta\theta \geq \frac{2\lambda/\delta}{\sin\theta_s+\sin\theta_r} \tag{2.20}$$

where $1/\delta$ is the highest spatial frequency in the signal beam.

2.5.3 Momentum-Based Techniques

When a hologram is reconstructed with a small Bragg mismatch (within the applicable Bragg selectivity function), then the hologram is still reconstructed, but the angle of the reconstruction may be changed because of conservation of momentum. The change in the reconstruction direction means that at a Fourier or image plane the reconstruction is shifted off the optical axis of the data path. This shift in the reconstruction allows holograms to be multiplexed in that dimension, even when the spacing is less than the inherent Bragg selectivity. Under these conditions, momentum conservation may be used to intentionally separate desired and undesired reconstructions. Polytopic multiplexing does not result in a change in read-out angle but change in actual location and is described below. Undesired reconstructions may be prevented from reaching the detector with a spatial filter. This is a fundamentally different approach than Bragg selectivity, where instead the undesired reconstructions have very little diffraction strength. In momentum-based cases, the undesired holograms may have diffraction efficiencies similar to the desired hologram. The most important momentum-based method is polytopic multiplexing [35]. Other methods that may have momentum-based components are peristrophic [29], fractal [36,37], and aperture multiplexing [38].

Polytopic (Greek for 'many places') *multiplexing* is illustrated conceptually in Figure 2.12. Figure 2.12(a) shows a traditional angle-multiplexing approach without polytopic multiplexing in which the books (angle-multiplexed holograms at one location) must be spatially separated in order to prevent crosstalk from the neighboring books. The effective area occupied by each book is much larger than the Fourier plane located at the signal beam waist because the beam is much wider everywhere else through the media thickness, and because the reference beam path must also be separated from the data-bearing fringes of neighboring books. By contrast, Figure 2.12(b) shows a combination polytopic-angle multiplexing scheme, where the books overlap within the media. The book spacing is chosen so that the signal beam waists (here shown outside the medium) do not overlap.

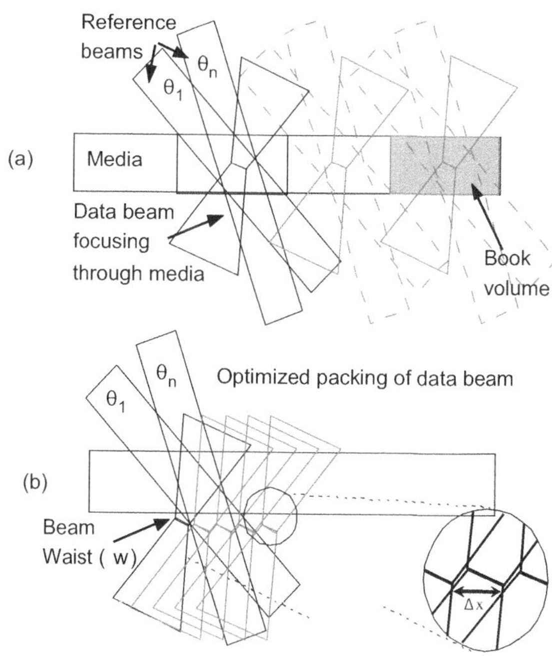

Figure 2.12 (a) Traditional angle multiplexing without book overlap. (b) Polytopic overlap of books but without overlapping the beam waists

Since the waists do not overlap, undesired reconstructions from neighboring books can be filtered out by introducing an aperture that passes only the desired beam waist. The writing process is the same as for plain angle multiplexing, except that the aperture is present and the book spacing is determined by the polytopic criterion. The dynamic range of the medium must also be budgeted to account for the increased hologram overlap. Neighboring books of holograms are subsequently recorded by simply moving the media by an amount equal to:

$$\Delta x = d\cos(\theta) + d\sin(\theta)\tan(\theta + \alpha) \tag{2.21}$$

where d is the Fourier plane aperture size, θ is the data beam center angle, and α is equal to the arcsine of the numerical aperture of the Fourier transform lens. Moving by Δx is required to ensure that the Fourier planes of neighboring holograms do not overlap. In the orthogonal direction, the required movement is simply:

$$\Delta y = d\tan(\alpha) \tag{2.22}$$

On read-out, the desired page and several of its neighbors will be simultaneously reconstructed since the probe beam illuminates data-bearing fringes from several books at once. However, the physical aperture blocks the components diffracted from the neighboring books, which allows the books to be separated by the Fourier plane width of the signal beam as opposed to the maximum width of both beams anywhere in the medium. This smaller required separation is particularly important when, as is required for high-density storage, a high NA and a thick medium are used.

Figure 2.13(a) shows a typical arrangement for polytopic multiplexing. For recording, the signal beam is modulated by the SLM and then Fourier-transformed by a lens. The

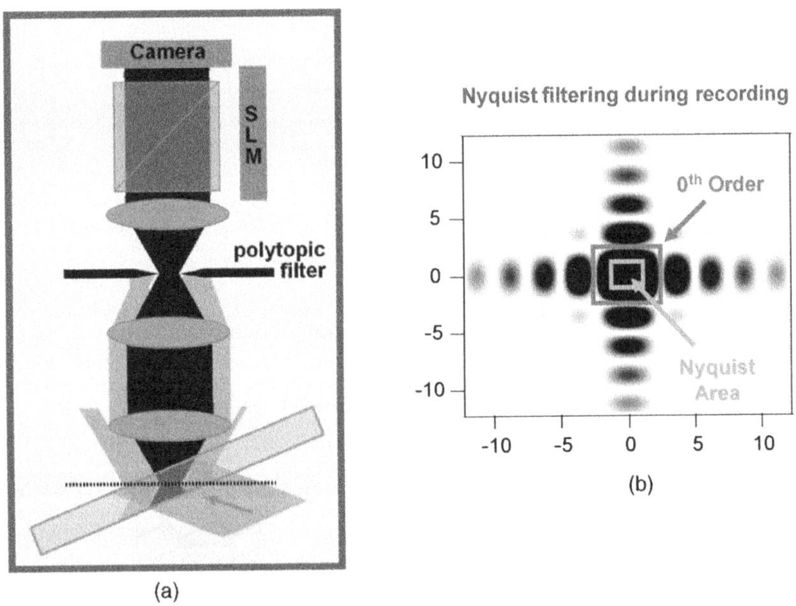

Figure 2.13 *(a) Arrangement for using polytopic multiplexing and filter size in Fourier plane of lens. (b) Fourier transform of SLM and appropriate filtering*

polytopic filter, or aperture, is placed in the Fourier plane of this first lens to block higher diffraction orders, making the holograms as small as possible. This aperture is sized somewhat larger than the Nyquist area, $(\lambda f/\Delta)^2$, of the data pattern within the signal beam, where f is the focal length of the lens and Δ is the pixel spacing of the SLM. The Nyquist area occupies approximately the central quarter of the zeroth diffraction order produced by the SLM, as illustrated in Figure 2.13(b). The Nyquist condition ensures that only data-bearing optical components pass through the aperture, thus optimizing SNR performance and density. The apertured signal beam is then imaged into the medium with a $4f$ lens relay, which produces another copy of the filtered Fourier plane in the middle of the recording layer.

On recovery, the desired page is read out by a phase conjugating reconstruction process – see Chapter 3 for more details. Phase-conjugate read-out uses a plane wave probe beam that is counter propagating with respect to the recording reference beam [as shown with an arrow in Figure 2.13(a)], causing the reconstructions to propagate back towards the camera. Only the desired page passes through the polytopic filter and impinges on the detector (camera). Undesired holograms from neighboring books (shown in gray in the figure) are also reconstructed, but are blocked by the polytopic filter. Thus, the same polytopic filter serves both to minimize hologram size during recording and to filter out book crosstalk during reconstruction. Additionally, the total overlap of the books is minimized by placing the Fourier plane in the middle of the recording layer, which in turn minimizes the dynamic range requirements for the recording medium.

Polytopic multiplexing is important because it allows for the full address space of the medium to be accessed without incurring a penalty for using a thick medium. Without polytopic multiplexing, the spacing of the books must increase with media thickness so that they do not overlap. This increased spacing causes the achievable density to saturate at a low value as the media thickness is increased because the book pitch (spacing) increases faster than the number of angle multiplexed pages in thicker media, and significant portions of media volume that lie between books are simply wasted (see Section 3.3.1). By making book pitch independent of media thickness using a polytopic filter, storage capacity can be made to scale linearly with media thickness, which is after all, the main promise of holographic recording.

Fractal multiplexing uses plane wave reference beams much as angle multiplexing does. However, if the whole angular bandwidth is used for the data page, then the angles for angle multiplexing only can use one line (or sweep of reference angles). If more holograms are to be recorded, the reference angle must be changed to be outside the angular bandwidth of the data page before another set of angle multiplexed holograms can be recorded. This second set of angle multiplexed holograms is fractally multiplexed relative to the first set. This technique gets its name from considering the gratings as interconnections between the data and reference planes, with each set of points Fourier transformed in to the media. Unfortunately 3D media can only implement a 3D interconnection matrix, and to fully interconnect two 2D planes of points requires a four-dimensional (4D) interconnection matrix. Thus only various 3D interconnections – like a 2D plane for the data and a one-dimensional (1D) line of points for the reference can uniquely be supported. However, more sparse distributions of points can be connected – for example a 1.5D plane of points can be connect to a second 1.5D plane of points. Thus, the name *fractal multiplexing,* and the term fractal interconnection matrix are used.

Shift multiplexing with a spherical reference beam produces strong Bragg selectivity for the whole data page only when shifting in the Bragg plane (x direction). In the perpendicular direction (y direction), Bragg selectivity is relatively coarse. However, this motion may still cause the reconstruction to come out in a different direction, a feature, which can be used to separate, and thus multiplex holograms in this direction. This is called *aperture multiplexing*. If the holograms are stored in an image plane, then the filtering is done in a Fourier plane; whereas if the holograms are stored in a Fourier plane then the filtering is done in an image plane. The required shift is:

$$\Delta y > z_o d / F \tag{2.23}$$

where z_o is the distance from the focal point of the spherical reference beam to the center of the medium, d is the width of the holographic image in the y direction, and F is the focal length of the Fourier transform lens.

2.5.4 Correlation-Based Techniques

Correlation multiplexing exploits both the amplitude and phase structure of a complex reference beam to introduce selectivity. A hologram is recorded with a complex reference beam wavefront, and then the media is moved relative to the recording location in a manner similar to shift multiplexing. As in shift multiplexing, there will be some component of Bragg selectivity that helps to reduce multiplexing crosstalk. However, unlike shift multiplexing, a sharply correlated reference wavefront can have shift selectivity that is greater than the equivalent Bragg selectivity. This enhanced selectivity is possible because the autocorrelation peak width of a complex reference beam can be designed to be smaller than the Bragg selectivity. The strategy depends upon the introduction of specific *complexity* into the reference beam by modulating both its phase and amplitude. This designed-in complexity can be induced by lenslet arrays, multimode fibers, and/or more controlled optical elements such as a filtered phase mask.

A simplified mathematical description illustrating how correlation shift selectivity is achieved is as follows. The arrangement of the reference and signal beams and reconstructed image beam is shown in Figure 2.14. The data input images $S(r;\delta_j)$ [shown as $S(x,y)$ in Figure 2.14], and reconstructed images, $S'(r;\delta_j)$, are indexed by the shifts δ_j (spatial locations), at which the holograms were written. The stored holograms are conventionally written by interfering the signal beam (which contains the data to be stored) with a reference beam in the recording medium. In the Fourier plane recording geometry shown, the data have been imaged onto a phase mask to make the Fourier transform of the data more spatially homogeneous in intensity for better recording. The reference beam starts as a plane wave illumination of a phase mask, which can be comprised of random phase. This reference beam is Fourier transformed and the zeroth diffraction order is blocked. This high pass filtered version is then Fourier transformed back into the media as the reference. This reference beam has very sharp amplitude and phase features that provide the correlation function. To reconstruct a stored image at position δ_j, the medium is illuminated with the same read-out reference beam as was used to store it. The reconstructed image in the focal plane of output lens is:

$$S'(\vec{r};\vec{\delta}_j) \propto \sum_i \int d\vec{r}' S(\vec{r}-\vec{r}';\vec{\delta}_j-\vec{\delta}_i) g(\vec{r}',\vec{\delta}_i) \tag{2.24}$$

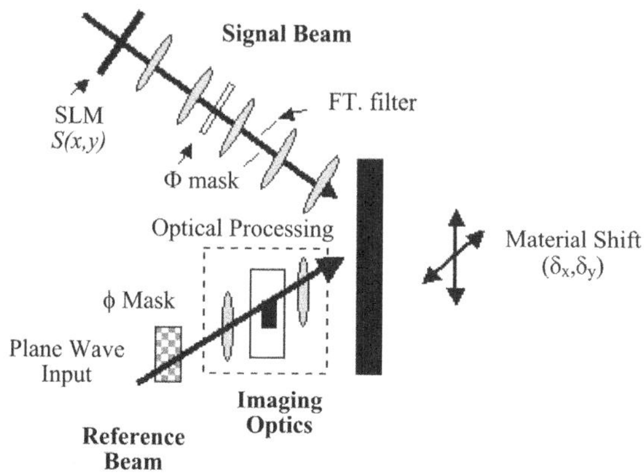

Figure 2.14 *Configuration to introduce complex reference structure*

where

$$g(\vec{r};\vec{\delta}) = f_{\text{Bragg}}(x) g_{2D}(\vec{r};\vec{\delta}) \tag{2.25}$$

and g_{2D} is given by:

$$g_{2D}(\vec{r};\vec{\delta}) = \int d\vec{r}'' R^*(\vec{r}'') R(\vec{r}'' + \vec{\delta}) \exp\left(\frac{2\pi i}{\lambda f} \vec{r}.\vec{r}''\right) \tag{2.26}$$

The Bragg effect is incorporated by using $f_{\text{Bragg}}(x) = d\text{sinc}(\pi x d/\lambda f)$, where d is the effective optical thickness of the medium, λ is the wavelength of the light, and f is the focal length of the lens.

The relationship between the reconstructed signal (S') and the stored original signal (S) is a convolution in two spatial variables with a kernel $g(r,\delta)$. g is dominated by the spatial structure of the recording and read-out reference beams. In conventional holography with plane wave references, this kernel is a constant. By manipulating $g(r,\delta)$, the reconstruction of the recorded image can be controlled.

For example, if $R(r)$ is a pseudo-random high-pass filtered amplitude and phase function, $g(r,\delta)$ will behave like a delta function. In this case, a small shift (a few micrometers) of the medium relative to the recording position will dramatically reduce the diffracted signal. While arbitrary manipulation of $g(r,\delta)$ is not possible, there is considerable freedom to distribute the weight in g to enhance the performance of the system. Experiments show that a simple random phase reference is a poor choice because the amplitude portion is critical. Phase masks with a systematic change (or ramp) in pixel (feature) size across the illuminated portion provide correlation properties that result in sharp selectivity and with improved SNR.

Using a highly complex reference wavefront results in high selectivity, which is independent of the media thickness and dramatically affects the diffractive properties of the recorded image. Figure 2.15 illustrates this *thickness independence* with the hologram shift selectivity curve (diffracted energy versus position of the media relative to the beams) in a 4 mm sample of $LiNbO_3$ and a 0.7 μm sample of holographic thermoplastic. Although

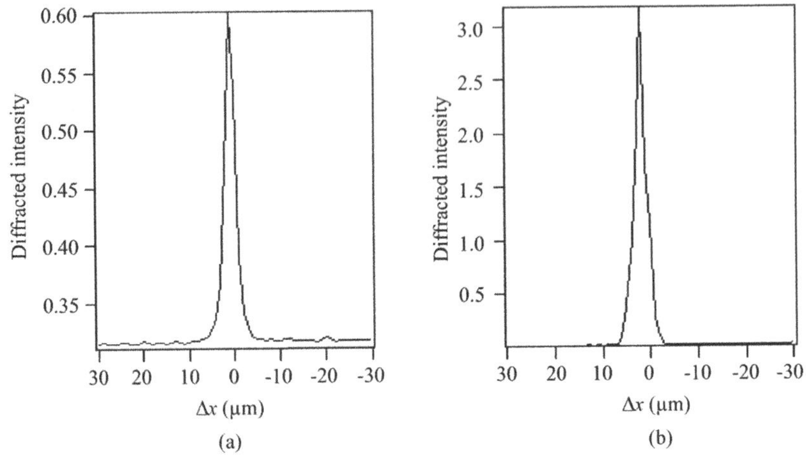

Figure 2.15 *Hologram shift selectivity for two different thickness media: (a) 0.7 μm thermoplastic film and (b) 4 mm $LiNbO_3$ crystal. Structure drives selectivity not media thickness*

the thickness difference is greater than 1000:1 the selectivity observed is essentially identical.

While the selectivity is independent of media thickness, the noise build-up is not. A thicker media may have higher capacity because the Bragg selectivity suppresses some of the multiplexing crosstalk noise. However, noise ultimately increases with the number of superimposed holograms, limiting the potential to increase capacity by increasing thickness. In addition, it is more difficult to achieve interchange and compensate for thermal effects with a complex reference wavefront.

In correlation multiplexing, the reference structure effectively encodes the data on or in the media. This method can be used as form of optical data encryption. Data written by different devices need to have identical reference beams for read-out. As shown in Figure 2.15, selectivity can be made extraordinarily high leading to the potential of high capacity. However, these systems typically have more noise at a given density than a pure Bragg-based system.

The efficacy of correlation multiplexing was demonstrated in an experiment conducted at Bell Laboratories in 1993: 17 000 holograms were stored in a 4 mm thick $LiNbO_3$ cube. The megabit data pages were roughly 2×2 mm, spaced by 25 μm by 25 μm in x and y. A raw channel density in excess of 700 bits μm^{-2}, with an average SNR of 1–1.5 dB was achieved. Figure 2.16 shows a local scan of intensities of a small part of the array and an individual recovered data page.

2.5.5 Combinations of Multiplexing Methods

Since many of the methods use plane wave references, the methods can be combined to increase storage density. The InPhase design uses both angle multiplexing and polytopic multiplexing. Fractal or peristrophic multiplexing methods could also be added [39]. Shift and aperture multiplexing with a spherical reference beam are also simple to combine.

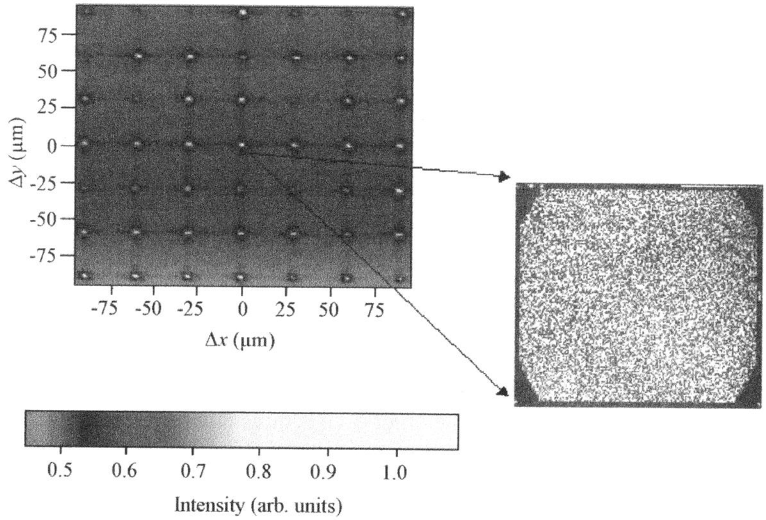

Figure 2.16 *Two-dimensional array of correlation multiplexed data pages, showing a local intensity scan and a recovered data page. Density is ~700 bits um^{-2}*

These combinations are key to accessing all the available angular bandwidth that the media can store, and so yield the highest possible data densities.

2.6 Address Space Limitations on Holographic Densities

A simple degrees-of-freedom argument can be used to establish an upper limit for holographic storage density. Assuming an isotropic media with refractive index n_0 and a recording wavelength of λ, the highest frequency grating that can be recorded is $\lambda/(2n_0)$. The number (n) of orthogonal Fourier components that can be stored in a 3D cube of size L and volume V is given by:

$$n = \left(\frac{L}{\lambda/2n_0} + 1\right)^3 \cong 8n_0^3\frac{V}{\lambda^3} \tag{2.27}$$

In a practical design using a disk or a card-shaped medium, much of the angular space cannot be accessed. Polytopic-angle and correlation multiplexing might allow systems to approach this density, but they do not fundamentally change this limit.

For example, for a medium 1.5 mm thick and a bulk index of refraction of 1.5, and a recording wavelength of 0.4 μm, the addressable limit for the system is approximately 19 kbits μm^{-2}, or 12 Tb in.$^{-2}$. By increasing the material thickness, and index of refraction, it is possible to build devices and media with addressable densities of 40 Tb in.$^{-2}$ with the technology of the professional drive discussed in Chapter 4. However, using quadrature multiplexing (see Chapter 11) does double or triple the achievable addressable space limit.

2.7 Summary

Key points of this chapter include:

- Holographic recording breaks through the density limitations of conventional forms of storage by recording throughout the full depth of a medium rather than in just one or a few surfaces.
- Multiplexing techniques are used to superimpose large numbers of holograms within a common recording volume. Multiplexing can be based on Bragg selectivity, conservation of momentum, correlation, or combinations of these.
- Unlike serial technologies that record one data bit at a time, holography can record and recover over a million bits of data in parallel, theoretically enabling transfer rates significantly higher than tradition storage devices
- Very high densities are possible, on the order of the equivalent of 40 Tb in.$^{-2}$ in a medium around 1 mm thick.
- The k-space formalism is a powerful tool for analysis of HDS systems.

References

1. P. J. van Heerden, Theory of optical information storage in solids, *Appl. Opt.*, Vol. 2, pp. 393–400 (1963).
2. D. Gabor, A new microscope principle, *Nature*, Vol. 161, p. 777 (1948).
3. W. L. Bragg, A new type of X-ray microscope, *Nature*, Vol. 143, pp. 678–678 (1939).
4. E. N. Leith and J. Upatnieks, Reconstructed wavefronts and communication theory, *J. Opt. Soc. Am.*, Vol. 52, pp. 1123–1128 (1962).
5. H Fleisher, *et al.*, An optically accessed memory using the Lippmann process for information storage, in *Optical and Electro-Optical Information Processing*. J. Tippett *et al.*, eds. MIT Press, Cambridge, Massachusetts, pp. 1–30 1965.
6. E. N. Leith, A. Kozma, J. Upatnieks, J. Marks, and N. Massey, Holographic data storage in three-dimensional media, *Appl. Opt.*, Vol. 5, pp. 1303–1311 (1966).
7. G. T. Sincerbox, History and Physical Principles, in *Holographic Data Storage*, H. J. Coufal, D. Psaltis, and G. Sincerbox,eds. Springer-Verlag, New York, pp. 3–20, 2000.
8. G. Burr,Volume holographic storage using the 90° geometry. PhD Thesis, California Institute of Technology, 1996.
9. Y. N. Denisyuk, On the reflection of optical properties of an object in a wave field of light scattered by it, *Dokl. Akad. Nauk SSSR*, Vol. 144, pp. 1275–1276 (1962).
10. W.R. Klein and B.D. Cook, Unified approach to ultrasonic light diffraction, *IEEE Trans. Sonics Ultrasonics*, Vol. SU-14, pp. 123–134 (1967).
11. M. G. Moharam and L. Young, Criterion for Bragg and Raman–Nath diffraction regimes, *Appl. Opt.*, Vol. 17, pp. 1757–1759 (1978).
12. H.S. Li, Photorefractive 3-D Disks for optical data storage and artificial neural networks, Chapter 2. PhD Thesis, California Institute of Technology, 1994.
13. H. Kogelnik, Coupled wave theory for thick hologram gratings, *Bell System Tech. J.*, Vol. 48, No. 9, pp. 2909–2947 (1969).
14. G. Barbastathis and D. Psaltis, Volume holographic multiplexing methods, in *Holographic Data Storage*, H. J. Coufal, D. Psaltis, and G. Sincerbox,eds. Springer-Verlag, New York, pp. 21–62, 2000.
15. X.-G. Gu, *et al.*, Cross-talk-limited storage capacity of volume holographic memory, *J. Opt. Soc. Am. A*, Vol. 9, No. 11, pp. 1–6 (1992).

16. K. Curtis, X.-G. Gu and D. Psaltis, Crosstalk in wavelength multiplexed holographic memories, *Opt. Lett.*, Vol. 18, No. 12, pp. 1001–1003 (1993).
17. K. Curtis and D. Psaltis, Cross-talk for angle and wavelength multiplexed image plane holograms, *Opt. Lett.*, Vol. 19, No. 21, pp. 1774–1776 (1994).
18. K. Curtis and D. Psaltis, Crosstalk in phase coded holographic memories, *J. Opt. Soc. Am. A*, Vol. 10, No. 12, pp. 2547–2550 (1993).
19. X. Yi, P. Yeh and X.-G. Gu, Cross-talk noise in volume holographic memory with spherical reference beams, *Opt Lett.*, Vol. 20, No. 17, pp. 1812–1814 (1995).
20. F. H. Mok, Angle-multiplexed storage of 5000 holograms in lithium niobate, *Opt. Lett.*, Vol. 18, No. 11, pp. 915–917 (1993).
21. D. L. Staebler, *et al.*, Multiple storage and erasure of fixed holograms in Fe-doped $LiNbO_3$, *Appl. Phys. Lett.*, Vol. 26, p. 182 (1975).
22. J. F. Heanue, M.C. Bashaw and L. Hesselink, Volume holographic storage and retrieval of digital data, *Science*, Vol. 265, p. 749 (1994).
23. G. A. Rakuljic, *et al.*, Optical data storage using orthogonal wavelength multiplexed volume holograms, *Opt. Lett.*, Vol. 17, No. 20, pp. 1471–1473 (1992).
24. S. Campbell, *et al.*, Hybrid sparse-wavelength angle-multiplexed optical data storage system, *Opt. Lett.*, Vol. 19, No. 24, pp. 2161–2163 (1994).
25. V.N. Morozov, Theory of holograms formed using coded reference beams, *Sov. J. Quantum Electron.*, Vol. 7, No. 8, pp. 961–965 (1977).
26. C. Denz, *et al.*, Volume hologram multiplexing using a deterministic phase encoding method, *Opt. Commun.*, Vol. 85, pp. 171–176 (1991).
27. J. F. Heanue, M. C. Bashaw and L. Hesselink, Recall of linear combinations of stored data pages based on phase-code multiplexing in volume holography, *Opt. Lett.*, Vol. 19, No. 14, pp. 1079–1081 (1994).
28. C. Denz, G. Pauliat, G. Roosen and T. Tschudi, Potentialities and limitations of hologram multiplexing by using the phase-encoding technique, *Appl. Opt.*, Vol. 31, pp. 5700–5705 (1992).
29. K. Curtis, A. Pu and D. Psaltis, Method for holographic storage using peristrophic multiplexing, *Opt. Lett.*, Vol. 19, No. 13, pp. 993–994 (1994).
30. D. Psaltis, A. Pu, M. Levene, K. Curtis and G. Barbastathis, Holographic storage using shift multiplexing, *Opt. Lett.*, Vol. 20, No. 7, pp. 782–784 (1995).
31. A. P. Yakimovich, Selective properties of 3-D holographic gratings using spherical wave fronts, *Opt. Spectrosc.*, Vol. 47, No. 5, pp. 530–535 (1979).
32. K. Curtis and W.L. Wilson,Phase correlation multiplex holography, US Patent 5,719,691, February 17, 1998.
33. K. Curtis,Method and apparatus for a phase correlated holographic drive, US Patent 6,909,529, June 21, 2005.
34. K. Curtis, *et al.*, High density, high performance data storage via volume holography: the Lucent Technologies Hardware Platform, in *Holographic Data Storage*, H. J. Coufal, D. Psaltis and G. Sincerbox,eds. Springer-Verlag, New York, pp. 359–368, 2000.
35. K. Anderson and K. Curtis, Polytopic multiplexing, *Opt. Lett.*, Vol. 29, No. 12, pp. 1402–1404 (2004).
36. D. Psaltis, X.-G. Gu and D. Brady, Fractal sampling grids for holographic interconnections, *Proc. SPIE*, Vol. 963, pp. 468–474 (1988).
37. H. Lee, X.-G. Gu and D. Psaltis, Volume holographic interconnections with maximal capacity and minimal cross-talk, *J. Appl. Phys.*, Vol. 65, No. 6, pp. 2191–2194 (1989).
38. K. Curtis and W.L. Wilson,Multiplex holography, US Patent 5,892,601, April 6, 1999.
39. J.-S. Jang,*et al.*, Holographic data storage by combined use of peristrophic, angular, and spatial multiplexing, *Opt. Eng.*, Vol. 31, No. 11, pp. 2975–2981 (2000).

3

Drive Architectures

Kevin Curtis, Adrian Hill and Mark Ayres

3.1 Introduction

As described in the previous chapter, there are several multiplexing methods that can be used, either singly or in combination. The primary architectural decision required in designing a holographic data storage system is the selection of the appropriate multiplexing scheme. This chapter focuses on the primary architectures that have garnered the most attention over the last few years.

The book, *Holographic Data Storage* [1] describes the wide range of architectures that were investigated until the late 1990s. More recently, Yamatsu *et al.* [2] made a comprehensive comparison of many multiplexing techniques and system geometries. Their analysis provided a platform for making the key engineering decisions.

Over the last few years, the storage industry has settled on two basic architecture paths: the collinear or coaxial architecture, and the angle-polytopic architecture. These architectures are described in detail in the rest of this chapter. For the angle-polytopic geometry, two systems developed at InPhase Technologies are described – the current drive ('Tapestry®') and a consumer version. The holographic read only memory (HROM) implementation of the InPhase architecture [3] is presented in Chapter 15.

3.2 Collinear/Coaxial Architecture

3.2.1 Introduction

In 1996, a team at Stanford University proposed using a storage lens with a numerical aperture (NA) large enough to allow both the signal beam and reference beam to pass

Holographic Data Storage: From Theory to Practical Systems Kevin Curtis, Lisa Dhar, Adrian Hill, William Wilson and Mark Ayres

through it [4,5]. This idea was the genesis of the collinear and coaxial architectures. The Stanford team used correlation multiplexing and a pulsed laser system to record holograms on a continuously spinning disk. In 1998, Sony Corporation presented a similar concept using shift multiplexing [6]. Following this, Optware Corporation and a number of others in the industry began developing systems based on these concepts. Optware called their system 'coaxial', while Sony called their version 'collinear'. Other variations have been proposed [7,8], but these two are the dominant variations so far. Both are presented below.

3.2.2 Coaxial Architecture

In the early 2000s, Optware Corporation [9–12] developed systems using the coaxial architecture. Figure 3.1 illustrates the coaxial architecture [13]. The SLM for the system displays the page [shown in Figure 3.1(b)] that has the data and modulation marks located in the center area. The reference beam is shown as lines of 'on' pixels around the perimeter of the SLM. Lines, random pixels, and other patterns have all been used as reference patterns in

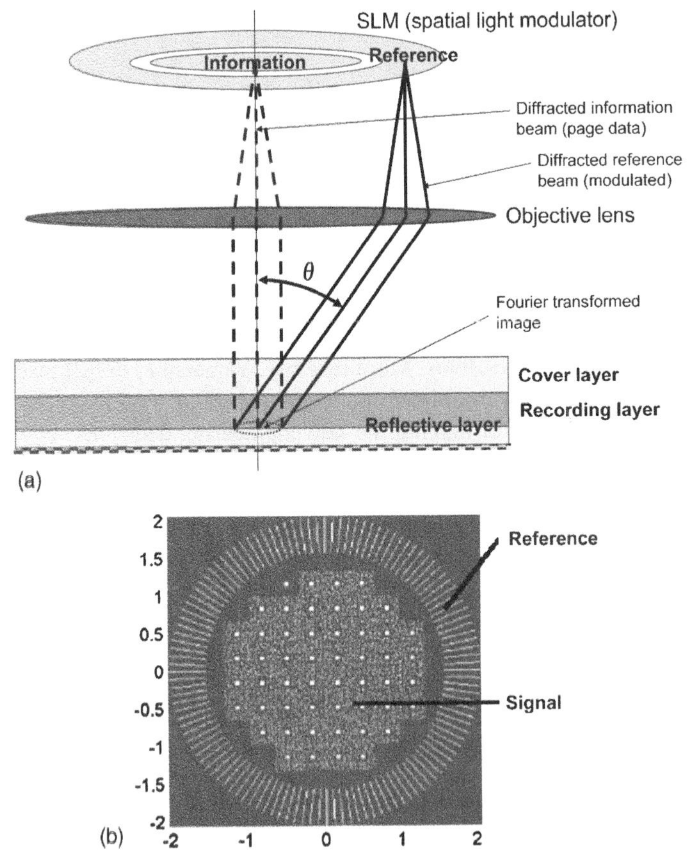

Figure 3.1 (a) Schematic of collinear/coaxial concept. (b) Example page with data and reference marks [13]. Reprinted with permission from Applied Optics, Collinear Holography by H. Horimai, X. Tan and J. Li, 44, 2575–2579 Copyright (2005) Optical Society of America

different versions of this system. The storage lens converts a pixel on the SLM to a plane wave traveling at a unique angle corresponding to that pixel's location. During recording, every reference 'on' pixel's corresponding plane wave intersects in the media to form a grating with every other 'on' pixel on the SLM display (both data and reference) as shown in Figure 3.1(a). On read-out, only the reference pattern is turned 'on' and these components are introduced to the media. The reference components are diffracted by the hologram and reconstruct the image back at the detector array (not shown).

The media for this particular implementation of collinear/coaxial architecture has a number of layers. Starting from the objective lens side of the media, there is a cover substrate with an anti-reflective coating, followed by the recording material, a gap layer, a reflective layer, and finally the bottom substrate. The gap layer creates some space between the Fourier transform (FT) plane and the recording layer so that the FT plane's hot spot does not saturate the media. The recording layer records the hologram where the components of the light overlap. The reflective layer is a diachronic layer, reflecting signal light back to the detector while passing co-propagating red laser light to a second reflecting layer embossed with servo marks (not shown). These servo marks allow the red laser to track the disk with the same servo concepts used in CD or DVD drives. (Both transmission and reflection holograms are recorded because of the presence of the reflective layer.)

Early systems used shift multiplexing, but more recent systems have switched to phase correlation multiplexing. The phase structure is either embedded into the SLM, or imaged onto it [14,15], which homogenizes the FT of the data beam in addition to providing the phase information needed for phase correlation. Optware Corporation developed a number of these aspects, including the reflective media structure and the servo system for tracking the disk position using CD and DVD techniques. Optware also developed a servo system that lowers the power required to read and write on a moving disk [16]. This servo system, called an 'asking servo', shifts the objective head along the track while steering the SLM encoded beam to match the objective's motion. This allows the beam to expose the same location on the media for a longer period, thus lowering the intensity (by increasing the exposure time) needed to record the hologram while the disk spins. This system requires a fast beam scanner and requires part of the lens NA to be sacrificed for the necessary angular space. (The 'asking servo' was first implemented in an experiment by a team at Sony [17,18].)

In addition to a recordable system concept, Optware proposed a ROM strategy with an associated replication process [19]. This scheme, shown in Figure 3.2, introduces a third, master reference beam from outside the objective, and records a separate set of holograms, as seen in Figure 3.2(a). The master reference plane wave would be used to read out only a small section (a few holograms) of the disk since the disk rotates the angle changes quickly. These reconstructed holograms would be imaged into another blank disk by a 4F imaging system, as seen in Figure 3.2(b). Since the reconstructions contain both reference and signal components, the holograms would be copied into this blank media and could therefore be read in a coaxial drive, as shown in Figure 3.2(c).

While no fully functioning system came out of this work, it represented a tremendous step forward in the optical storage industry's understanding and participation in holographic storage, and stimulated advances aimed at leveraging traditional optical storage technologies in holography. In addition, an industry standard for the coaxial architecture, ECMA (European Computer Manufacturers' Association) International standard ECMA/TC44/2006/0XX [20–22], was granted to Optware Corporation. This activity also fostered

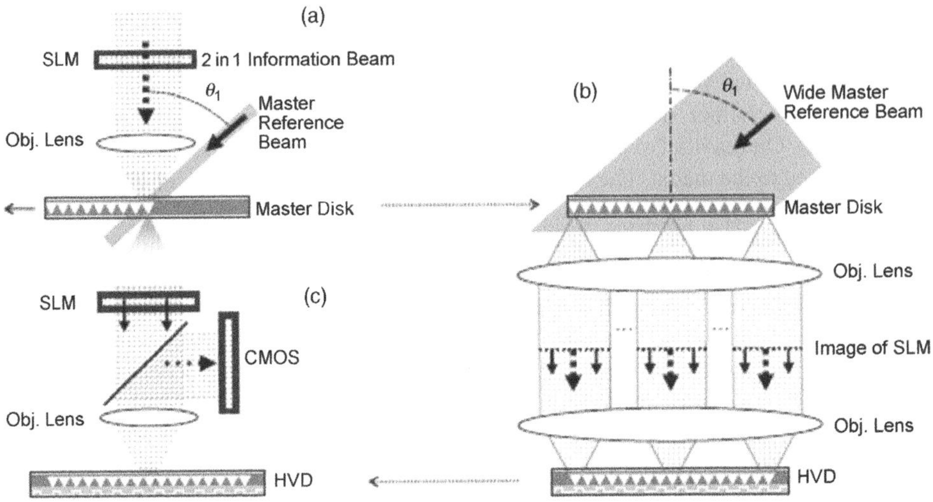

Figure 3.2 Conceptual replication proposed for collinear/coaxial architecture. HVD, Holographic versatile disk; CMOS, detector array for detecting the data. Reprinted with permission from Technical digest of Internaional Symposium on Optical Memory and Optical Data Storage – ISOM/ODS '05, Duplication Technology for Secured Read-Only Holographic Versatile Disc by Hideyoshi Horimai and Xiaodi Tan, Honolulu, Hawaii, USA 2005, MB7 Copyright (2005) SPIE

collaboration between several companies; as an example, Pulstec Corporation produces digital media testers based on the coaxial architecture.

3.2.3 Collinear Architecture

A team at Sony Corporation has also designed recording platforms using this type of architecture, and its work has led to several implementation improvements [23–25]. Using correlation multiplexing, the team has demonstrated channel densities around 400 Gb in^{-2} [26,27]. Figure 3.3 shows a diagram of a typical Sony 'collinear' implementation. The blue laser illuminates the SLM containing both reference and signal components as for the coaxial architecture. Typically, the SLM would have a phase mask integrated into it, or the SLM would be imaged onto a phase mask; Figure 3.3 shows the SLM imaged onto the phase mask. In this case, the relay lenses accommodate an FT plane filter that filters out the higher orders of the SLM. There is also a polarizing diffractive filter before the storage lenses. The storage lens is a high NA (0.85) objective lens which Fourier transforms the SLM pattern into the media. On reconstruction, only the reference pattern is directed into the media to reconstruct the hologram. The polarizing filter eliminates the reference beam pattern and high angular frequency noise, in addition to changing the polarization of the signal beam so that it can be routed (imaged) back to the detector array.

Correlation multiplexing is used to multiplex holograms by shifting the media. The phase mask is responsible for the phase modulation of both the signal and reference beams, which allows holograms to be multiplexed in both translation directions with media motion of only

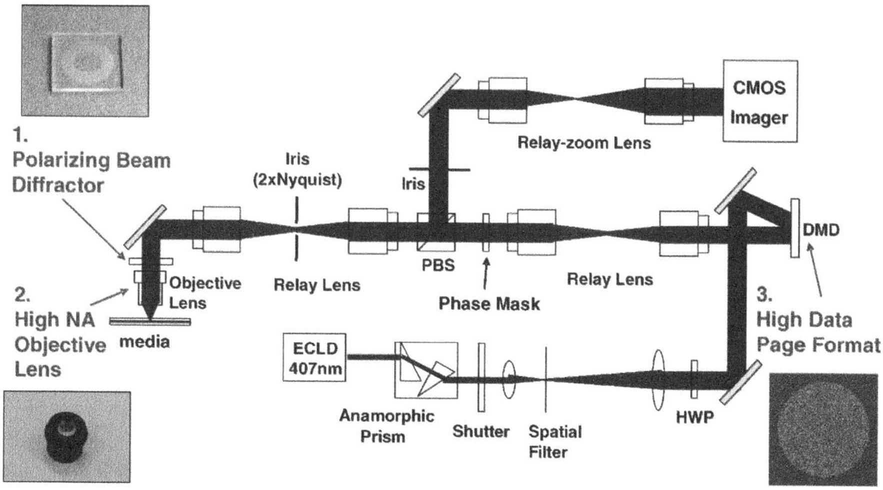

Figure 3.3 Diagram of Sony's collinear system [28]. Reprinted with permission from Improvements in Multiplexed Recording Performance in Coaxial Holographic Data Storage by Kenji Tanaka, ISOM Conference, Takamatsu, Japan Copyright (2006) Sony

a few micrometers. These systems use a similar basic media construction and servo strategy as the coaxial system, with the exception that the gap layer in the media is not required because a phase mask is used. Coherent addition (homodyne detection) has been implemented in a collinear system to increase the signal to noise ratio (SNR) by up to 3.5 dB, but results in a system with very tight tilt and focus tolerances [29]. Recently, Sony [30] demonstrated a system to compensate for thermal changes in collinear architectures using a zoom lens and a tunable laser to change the angle and wavelength as a function of media temperature. Good results were shown for a greater than 6 °C temperature change, using achromatic zoom and objective lenses.

3.3 InPhase Architecture

3.3.1 Introduction

The group at InPhase Technologies took a more conventional approach to holographic drive development. Systems which use a simple plane wave reference beam offer three distinct advantages: (i) it is possible to compensate for media temperature changes (which is extremely important for polymer media); (ii) the system can be modeled in a tractable manner; and (iii) these systems offer improved SNR when compared with systems employing high bandwidth reference beams.

In 1998, a team at Bell Laboratories invented a method for compensating for temperature changes in the media by simply changing the angle and wavelength of the read-out beam. In addition, the team reduced thermal changes in the media by using plastic substrates or substrates that better matched the thermal expansion of the recording material [31,32]. The compensation method, while effective for the plane wave reference beam used in angle multiplexing, has limited utility in systems using shift or correlation multiplexing where the reference beams contain many plane wave spectral components which with a thick lens require a slight nonlinear correction of angle versus recording angle for each component.

A detailed model of the system's optical path is a key prerequisite to developing a commercial drive. With a single plane wave reference beam, the analysis must incorporate the effects of the approximately 1 million gratings formed from a single high bandwidth data page. This analysis is tractable, and InPhase engineers have developed models that can be incorporated into standard ray tracing software to allow detailed modeling of systems for many different platforms and media perturbations.

However, detailed models are intractable for shift multiplexing systems, where the combination of high bandwidth reference and signal beams results in the formation of around 1 trillion gratings for a single data page. With correlation, the phase and local amplitude of these 1 trillion gratings must also be calculated throughout the media volume. The number of degrees of freedom makes quantitative models problematic. While some very simple models can predict qualitative measures [33–36], detailed system perturbation calculations require the type of k-space modeling discussed in the previous chapter.

Early work at Bell Laboratories showed that when many holograms are recorded in the same volume, shift and correlation multiplexed holograms have lower SNR than angle multiplexed holograms. This is because Bragg selectivity with a plane wave reference beam is effectively more 'orthogonal' than for correlation multiplexing, which suffers from statistical root mean square crosstalk [36,37]. Also, recording at high density results in greater distortion of the overlapping holograms. With shift and correlation multiplexing, modest numbers of holograms can be written and recovered with high SNR but the SNR drops significantly as the storage density is increased. This SNR drop is significantly smaller for angle multiplexed holograms.

While angle multiplexing has these advantages, it suffers from a limited achievable storage density [38]. This is illustrated in Figure 3.4, which shows the dependence of the addressable user data capacity on media thickness for a 130 mm disk, calculated assuming a

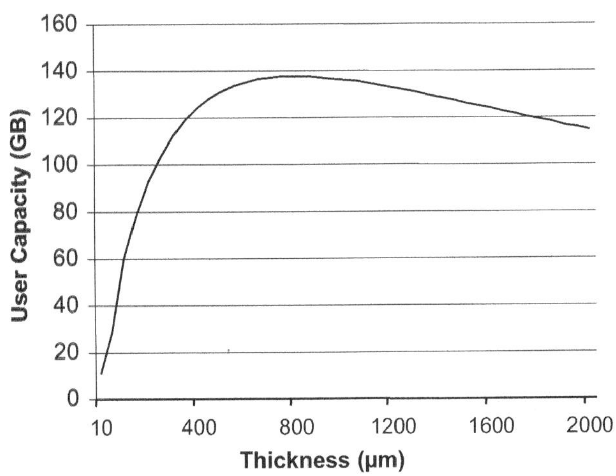

Figure 3.4 *Capacity versus media thickness for an angle multiplexed disk*

405 nm blue laser, an object beam NA of around 0.6, a 1000 × 1000 pixel data page, and a combined code rate of 0.488 (which allows for ECC and modulation coding). The graph shows that the achievable user capacity (considering only the optical system limits) reaches a maximum of around 140 GB for a media thickness of about 800 μm. The capacity curve saturates because although the capacity within a single book increases in proportion to the media thickness, the signal bandwidth of the data pages also requires that the book pitch (spacing) increases rapidly as the media thickness increases, so that the books do not overlap in the media.

Polytopic multiplexing (described next) overcomes this angle multiplexing capacity limit while retaining the advantages of thermal compensation, modeling tractability, and SNR enhancement afforded by plane wave reference beams.

3.3.2 Angle-Polytopic, Phase Conjugate Architecture (InPhase Architecture)

The invention of polytopic multiplexing [39] allowed InPhase Technologies to achieve high density storage beyond the limits imposed by angle multiplexing alone. As discussed in the previous chapter, combining polytopic multiplexing with angle multiplexing allows books to be overlapped within the media, which increases the storage density (see Figure 3.5). If the set of angle multiplexed holograms in one book overlaps with those of another book, then introducing the reference beam with the angle required to read a particular hologram (page) in one book will lead to the reconstruction of that required page from *all* books that overlap at that location in the media, since those pages share the same reference beam angle. However, the undesired reconstructions can be rejected if the books are spaced at least one signal beam waist width apart. The 'polytopic' filter can be used to pass only the required page reconstruction, while blocking the unwanted page reconstructions. Polytopic multiplexing removes the dependence of book pitch on media thickness, which in turn means that the capacity no longer saturates with increasing media thickness. (Chapter 4 describes polytopic filters in detail.)

Architecting the drive to incorporate phase conjugation with angle-polytopic multiplexing was another significant innovation, which simplified the optical path by placing all

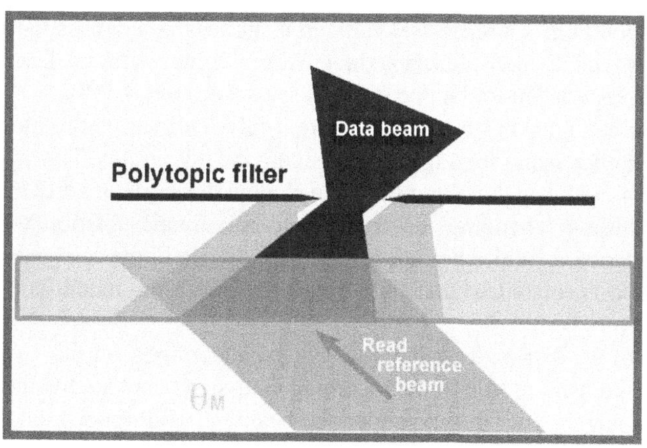

Figure 3.5 Polytopic filtering

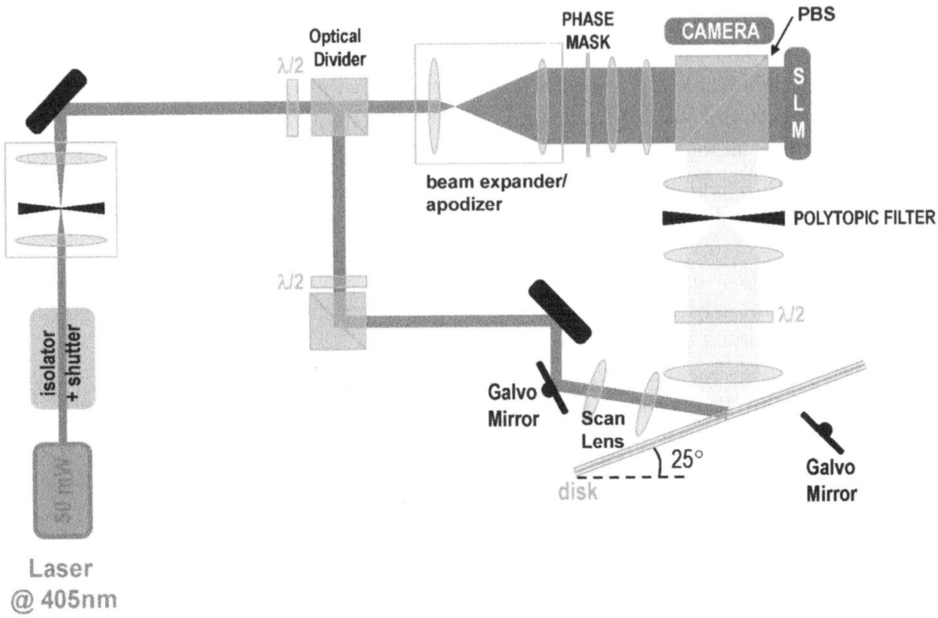

Figure 3.6 *Diagram of light path during write operation*

the optical elements on one side of the media, and allowed simpler lenses to be used because of the inherent aberration correction. Strictly, the drive does not read out using pure phase conjugation, which typically requires a nonlinear crystal [40,41]. Instead, since the reference beams are plane waves, and if their optical quality is adequate, a simple retro reflection provides a practical means of achieving quasi phase conjugation. The basic phase conjugate architecture of the Tapestry drive [42,43] is shown in Figures 3.6 and 3.7.

Figure 3.6 shows the light path through the system during a write operation. Light from the tunable 405 nm laser source passes through a shutter that controls the write exposure time. Next, the optical divider splits the light into the reference path and data path. Typically, the ratio of reference beam to data beam power is set around 4.5 : 1.

The reference beam passes to a galvanometer (galvo) mirror that rotates to control the angle of the reference beam for angle multiplexing the holograms. This mirror is imaged into the center of media so that the mirror can change the angle of incidence of the beam on the media without changing the (radial and rotational) position where the beam impinges on the media. While holograms are being written, the read galvo on the other side of the media is rotated so that light passing through the media is deflected into a beam dump.

The first part of the data path contains a beam expander/apodizer that expands the beam and flattens the intensity profile of the incoming Gaussian beam so that the entire SLM is uniformly illuminated. After this expander, the beam passes through a phase mask which will make the intensity of the Fourier transform of the SLM more uniform at the media. (More information about the phase mask is presented in Chapter 13.) The phase mask is

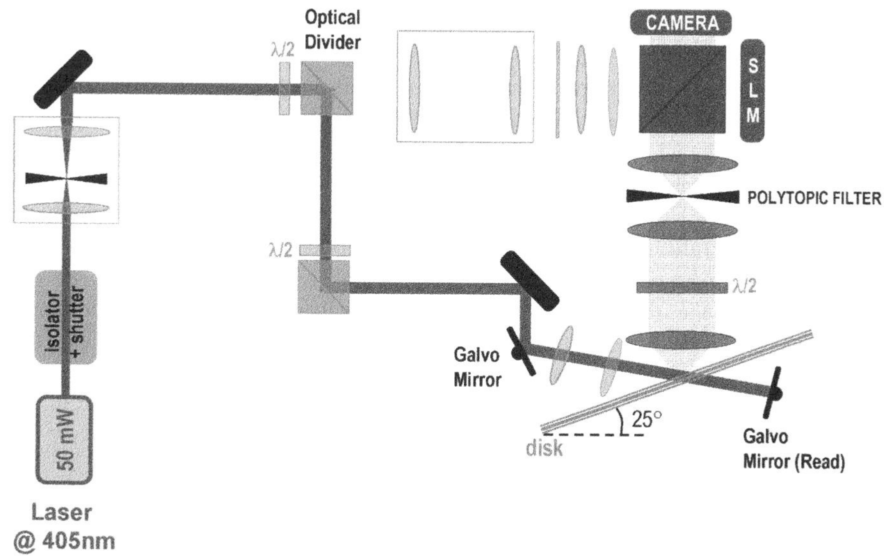

Figure 3.7 *Diagram of light path during read operation*

imaged onto the SLM with a 4F imaging system to place the phase modulation at the image plane. A 4F imaging system consists of two lenses separated by two focal lengths, with two image planes that are each one focal length outside of the lens system (so, the total length is four focal lengths, hence '4F'). Since the phase modulation is in the image plane, it does not impact the intensity detected by the camera on recovery. The SLM amplitude modulates the light beam with the data pattern. The light from the SLM is Fourier transformed by the first relay lens, and then spatially filtered by a filter at the Fourier plane. This filter removes the redundant higher diffraction orders from the SLM, and limits the zeroth order to the minimum size. On recovery, this filter also serves as the polytopic filter.

Typically the filter size is about 1.2 times the Nyquist area. The Nyquist area represents the minimum aperture size required to resolve the SLM pixels according to the sampling theorem, and is one quarter of the area of the zeroth order of the SLM centered on the optical axis. The next relay lens converts the data page back into an image, which is finally transformed into the center of the media by the high NA storage lens This high (around 0.65) NA lens decreases the volume of the hologram in the media, which minimizes the overlap of the books. This in turn reduces the *M/#* requirement of the media. The reference and data beam intersect in the media and form the hologram.

Angle multiplexing is implemented by simply rotating the galvo mirror between each data page exposure. By this means, a complete book of holograms is recorded at the same physical location in the media. To record other books, the disk media is either rotated, or translated along a radius relative to the optics.

Figure 3.7 shows the path of the light during a read operation. All the light from the laser is directed into the reference path. To read out a particular hologram, the disk is first positioned at the correct radius and rotation angle to read out the holograms from the corresponding book. The galvo mirror is then adjusted to the correct angle to read out the hologram.

The reference beam passes through the media and is retro-reflected by the read galvo. The read galvo consists of a mirror with an integrated quarter wave structure. The read galvo is synchronized with the write galvo to retro-reflect the light with a 90° polarization change. The hologram is reconstructed with this retro-reflected beam. The polytopic filter eliminates the crosstalk from adjacent books. Because of the polarization change, the reconstruction passes directly through the polarizing beam splitter (PBS) and is detected on the CMOS active pixel detector array (camera).

To read another page from the same book, only the angle of the galvo mirror is changed. To recover data from other books, the disk is moved (radially and/or rotationally) relative to the optics.

The majority of this book is devoted to discussing the functions of this drive in more detail. Some key attributes of this drive are:

- Wide temperature range operation (10–45 °C media temperatures).
- Ability to interchange media between different drives.
- Demonstrated high density storage ($>$600 Gb in.$^{-2}$ demonstrated in actual drive).
- Fast transfer rate.
- Large capacity.

The first product has a capacity of 300 GB and a transfer rate of 20 MB s^{-1} for both recording and reading. A technology roadmap to 1.6 TB on a disk and 120 MB s^{-1} transfer rate, with backward read compatibility, has been identified for this architecture, and is shown in Figure 3.8. The two shaded columns, labeled 2005 and 2008, describe the actual demonstrated performance and the hardware that was used. The 2005 transfer rates were demonstrated for a series of holograms within a single book, with a software implementation of the ECC decoder. Chapter 9 presents the details of the initial implementation of the Tapestry drive.

Drive tolerances are a key aspect of the viability of any drive architecture, and are discussed in detail in the Chapter 13.

While the drive architecture just described is for a recordable drive, HROM versions of this architecture are also possible. Chapter 15 describes a small HROM reader, and the critical two-step mastering process for high speed replication at high density.

3.4 Monocular Architecture

3.4.1 Introduction

The angle-polytopic drive architecture described in the previous section was developed to fill the need for a high performance archive drive for professional archive applications. The experience gained during the development of that drive, and the HROM consumer product, is being applied to developing a drive aimed at the consumer market. This section describes the implementation and early results for the consumer product based on the InPhase architecture. This work has been a partnership between InPhase Technologies and Hitachi Corporation.

	2005	**2008** (DVT)	**2009** (DVT)	**2010**	**2011**
Specs	515 Gb in^{-2} 160 Mbits s^{-1}	440 Gb in^{-2} (592) 160 Mbits s^{-1}	592 Gb in^{-2} 180 Mbits s^{-1}	970 Gb in^{-2} 640 Mbits s^{-1}	1900 Gb in^{-2} 960 Mbits s^{-1}
Disk capacity and format type	NA	360 GB/ isolated bookcases	480 GB/ **wave format**	800 GB/ wave format	1600 GB/ wave format+
# of pages per book	340	320	320	**385**	**730**
Angle spacing	0.075	0.075	0.075	**0.07**	**0.045 (ave)**
Ref. beam sweep (°)	25	24	24	**27**	**33**
Book pitch (q, r) (mm)	0.82, 0.6	0.86, 0.575	0.86, 0.575	**0.63**, 0.575	0.63, 0.575
NA of object beam	0.65	0.65	0.65	0.65	0.65
User data/page (kB)	77	69	69	**70**	**72**
SLM pixels	1280x1024	1200x1200	1200x1200	1200x1200	1200x1200
Camera pixels	1280x1024	1696x1710	1696x1710	1696x1710	1696x1710
Camera sensitivity (counts/J m^{-2})	176 000	350 000	350 000	700 000	1 400 000
Laser power (mW)	30	40	60	80	80
Wavelength (nm)	405	405	405	405	405
Material thickness (mm)	1.5	1.5	1.5	1.5	1.5
Media sensitivity dn J^{-1}	10^{-5}	$6x10^{-6}$	$6x10^{-6}$	**$8x10^{-6}$**	**10^{-5}**
M# of media @1.5mm	50	30	30	**50**	**100**

Figure 3.8 *Technology roadmap for improvement to first product. DVT, 'Design, Verification and Test' unit*

The primary goal was to develop a slim height (12.7 mm) consumer drive that has a capacity of 500 GB, and a transfer rate of 100 MB s^{-1}. For the consumer market, the drive must be inexpensive, and also be backward compatible with Blu-ray media. Lastly, the drive must support mastered and replicated media for content distribution, in the same manner as the HROM drive described in Chapter 15.

The consumer drive employs the same basic technology as the Tapestry archive drive, which allows most of the servo, data channel, data formatting, firmware, software, and electronics to be reused. The custom components specifications needed to make a consumer product for the laser, SLM and camera are given in Chapter 4.

3.4.2 Monocular Implementation

The Monocular architecture [44–46] is promising because it reduces the drive size and complexity by placing both the data and reference beams through a single objective lens, as shown in Figure 3.9(a). The laser's output is collimated and circularized, and then split into reference and data paths. The splitting ratio is dynamically controlled with a liquid crystal plate and a PBS. The ability to dynamically control the reference and data beam ratio results in enhanced optical throughput both when writing and reading.

The objective lens collimates the reference beam while overlapping it with the object (data) beam in the media to record holograms. The objective lens's curved surface contains a thin film angular filter for implementing the polytopic and Nyquist filtering, as explained in Chapter 4.

For cost and size reasons, all the components are small. The SLM and camera have pixel pitches of 4 and 3 μm respectively. The objective lens has a focal length of 2.5 mm and an NA of 0.85 similar to the specifications for a Blu-ray disk objective lens.

Angle multiplexing is accomplished by changing the mirror angle or by shifting the objective lens position [Figure 3.9(b) and (c)]. Transfer rate is maximized by using a fast mirror actuator for multiplexing, and only moving the objective lens for radial tilt

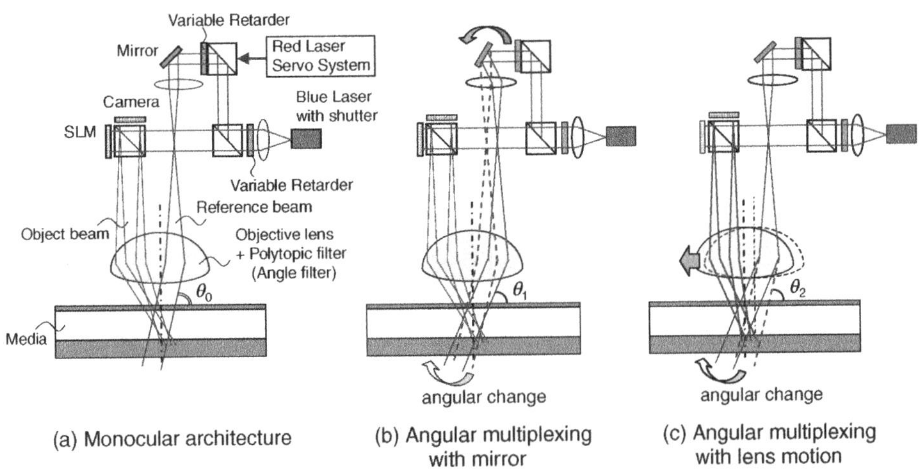

Figure 3.9 *Monocular concept with multiplexing*

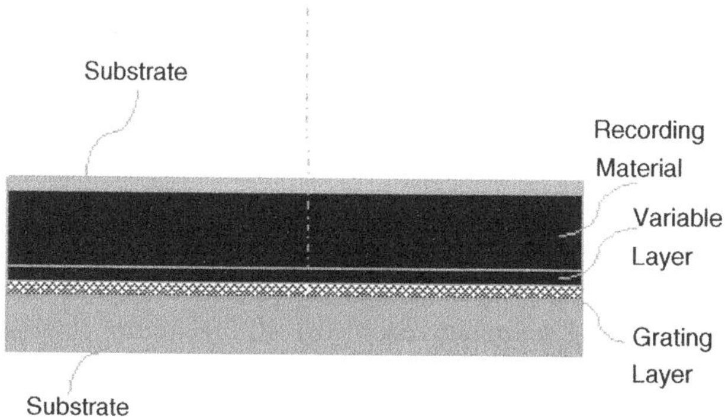

Figure 3.10 *Schematic of media structure*

corrections. As in the recordable drive, books of holograms are recorded by angle multiplexing, and multiple books are recorded by rotating the disk and/or moving the head radially across the disk. The spindle motor for rotating the disk can be a standard brushless motor, so that it can step under servo control or spin at high speed for use with a Blu-ray disk.

The media structure for the monocular drive is shown in Figure 3.10. The media consists of a 100 μm thick top cover substrate with an anti-reflection coating, on top of a 1.55 mm thick recording layer. Under the recording layer there is a variable layer and grating layer (both described next), and finally a 1 mm thick polycarbonate substrate that provides mechanical support and the attachment to the drive. The focal point of the objective lens is set to the middle of the recording layer.

The variable layer is switched to absorb light during recording to prevent reflection holograms from being recorded at the same time as the intended transmission holograms. During read out, the variable layer is switched to allow light to pass through it. A polarizer with as little as a 20 : 1 contrast ratio is sufficient for a variable layer. A suitable layer could be fabricated with a radial symmetric polarizer formed by nano-imprint lithography [47].

On read out the polarization of the reference beam is changed so that the beam will pass through the polarizer. This change of polarization also allows the reconstructed hologram to be directed to the detector by the PBS.

The grating layer is a reflection grating with a period of approximately 315 nm that is chirped or blazed to reduce the diffraction efficiency of the unwanted higher orders. Initial designs indicate that the polarizer and blazed grating combination can achieve 40 : 1 contrast ratio for both polarizations. On read out, the combination can achieve more than 40% reflectivity of the desired order. A servo pattern like the standard push-pull servo of a CD/DVD can be placed in this layer to be read by a red laser. Both the grating and the servo pattern can be stamped into the bottom substrate and then coated with metal for reflection.

On read out, all the laser light power is directed into the reference beam. The polarization is switched to the read-out state, and the reference mirror actuator set to the correct angle to read out the desired hologram. The read-out beam goes through the recording layer and the variable layer, and is reflected by the grating layer. The reflected reference beam goes back through the recording layer, where it causes phase conjugate read-out of the hologram. The reconstructed hologram image passes through the polytopic filter on the storage lens, which filters out the undesired reconstructions and noise. The data page is then steered to the camera by the PBS, and detected.

The fixed grating layer is used in conjunction with the reference beam angle to read a given hologram within a stack. The mechanism is most easily explained with an example. Consider a stack of holograms written with reference beam angles of 36, 38, 40, 42 and 44°, where the angles are measured relative to the media normal. The grating would be configured to retro-reflect the center angle of the reference beam's scan range, in this example, 40°. To read out the hologram recorded at 40°, the mirror actuator would set the reference beam angle to the same 40° position. This beam would pass through the media and variable layer, and be retro-reflected by the grating to cause the phase conjugate reconstruction of the hologram. The phase conjugate reconstruction would then detected by the camera and decoded.

To read the hologram recorded at 36°, the mirror actuator would be set so that the reference beam is incident at 44°. Since the grating conserves the momentum of the beam, the reflection from the grating enters the media at an angle of 36°, which causes the hologram originally recorded at 36° to be read out. In this way, with a single fixed grating, all holograms in the angle scan can be read out in with a phase conjugate beam.

Chapter 15 on HROM describes how to replicate disks that are angle-polytopic multiplexed using a high speed, disk at a time, process. This uses a two-step mastering process to produce high efficiency masters at high density.

3.4.3 Experimental System

The experimental system used to explore the feasibility of the monocular architecture is shown in Figure 3.11. The blue (405 nm) laser diode unit is a modular laser that consists of an external cavity tunable laser diode, isolator, and beam expander. The beam from this unit is split into a data beam and a reference beam. The amplitude of the data beam is spatially modulated using a 1200×820 pixel data page on an SLM with a 10.7 μm pixel pitch.

The objective lens is a custom, high NA lens (NA = 0.85, focal length = 7.4 mm) with six spherical elements. The objective lens allows the reference beam to sweep through a range of approximately 29°. Figure 3.12 shows the angles used by the reference and data beams, as well as the angles separating the two beams. Angles at either end of the range are reserved for system alignment and pitch correction using lens shift. Both the object and reference beams pass through the objective lens so that the focal point of the reference beam is at the back focal plane of the lens. This alignment produces a collimated reference beam at the media. The media is a custom transparent disk with a 1.5 mm recording layer (photopolymer) sandwiched between a 0.1 mm first substrate and a 1.0 mm second substrate. The hologram size at the Fourier plane in the medium is 0.308 mm. The raw data page capacity is 100 676 bytes.

In the experimental system in Figure 3.11, phase conjugate read-out is used to recover the holograms. The phase conjugate read-out is implemented with one stationary mirror, and two galvo mirrors (Galvo2, Galvo3 in Figure 3.11) on the back side of the media to retro-reflect the reference beam. Recovered holograms are imaged onto a 1696×1710 pixel camera that has an 8.0 μm pixel pitch.

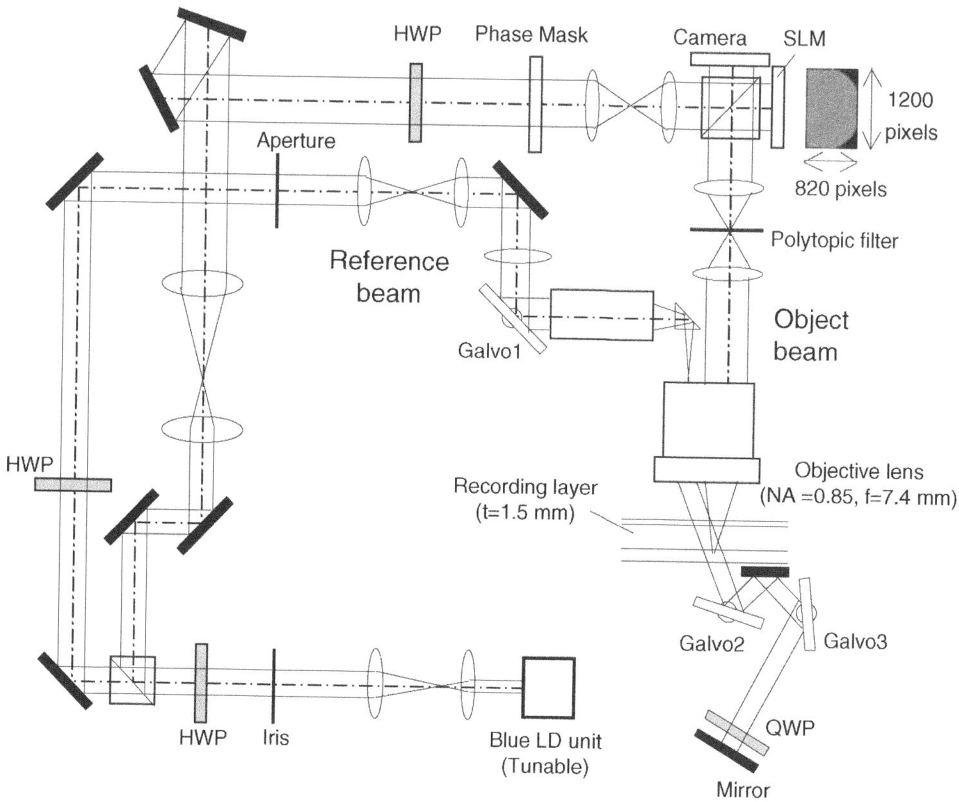

Figure 3.11 *Schematic diagram of experimental system. Reprinted with permission from Optical Letters, Polytopic multiplexing by K. Anderson and K. Curtis,* ***29****, 1402–1404 Copyright (2004) Optical Society of America*

Additional components not shown in the diagram include a media curing system, used to fully and quickly polymerize and fix a region of the medium after all recording is complete for that region.

3.4.4 Preliminary Experimental Results

Figure 3.13 shows the SNR and relative intensity of 130 holograms that were angle multiplexed by changing the 'Galvo1' mirror angle. Each hologram was offset from the next by an angle step that varied from 0.14 to 0.39° over the set of holograms.

Figure 3.14 shows the relationship between SNR and Bit Error Rate (BER) for the error correction code used in this system. This 'waterfall' curve is typical of a Low Density Parity Check (LDPC) code with a rate of 0.5, a code length of 16 384 data bits, and fewer than 40 iterations. The code, encoder, and decoder were developed by Icoding Corporation. BER values of 0 are plotted as values of 10^{-10} on the graph, to allow the standard log axis scale to be used.

The SNR graph shows that the fidelity of the 130 holograms was sufficient for error-free recovery because the average SNR for all holograms was approximately 2.8 dB, without

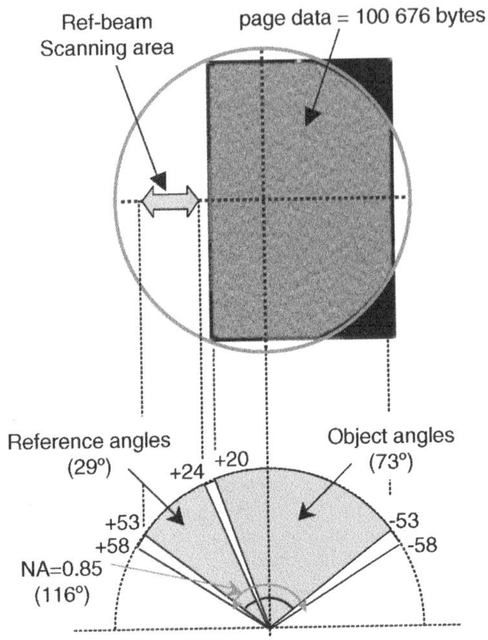

Figure 3.12 *Spot diagram at the pupil of the objective lens*

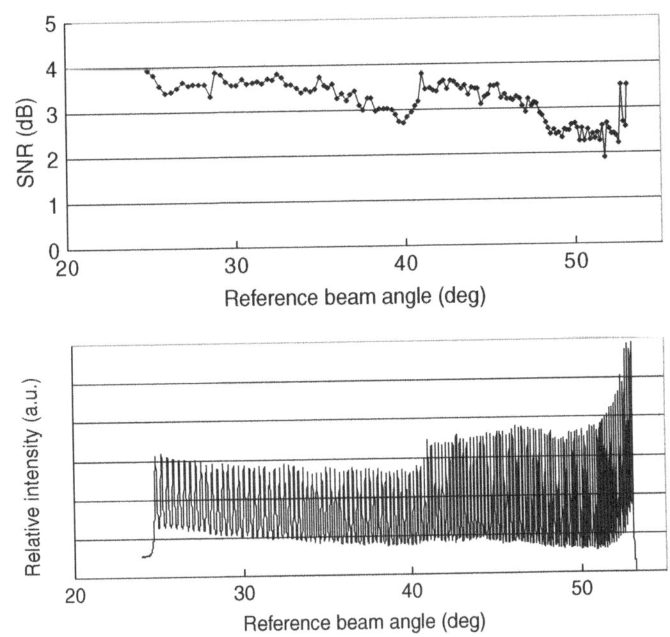

Figure 3.13 *Recovered multiplexed data page's SNR and relative intensity*

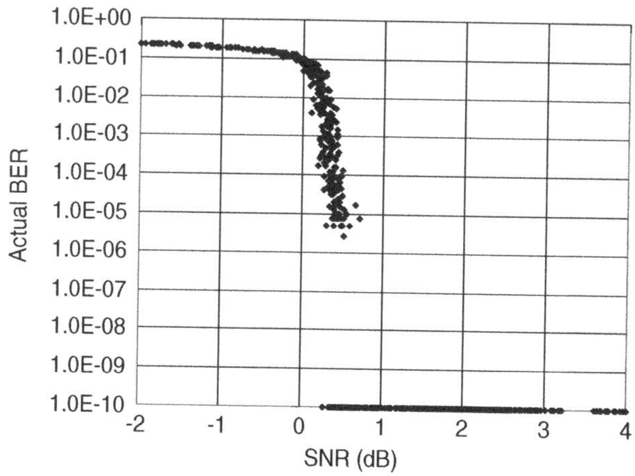

Figure 3.14 *Relationship between BER and SNR*

correcting for pitch errors. (With pitch correction, the average SNR increased to 3.9 dB.) There were no data errors on recovery.

The raw recoding density in this system was 712 Gb in.$^{-2}$, which would correspond to approximately 590 GB of user data on a standard 120 mm disk:

$$\text{Raw data recording density} = \frac{100\,676\text{ bytes}}{0.308\text{ mm} \times 0.308\text{ mm}} \times 130\text{ holograms} = 712\text{ Gbits in.}^{-2} \tag{3.1}$$

The current monocular test-bed uses the Tapestry drive's SLM, camera, and laser except for the custom 0.85 NA objective lenses design to work with those parts. Measurements of the system tolerances match the expected tolerances obtained by scaling the Tapestry drive's tolerances by the ratio of their respective focal lengths [48]. (See Chapter 14 for a detailed description of the drive tolerances.)

Next steps for the monocular drive include fabricating a custom SLM and camera, and producing custom media substrates, in order to shrink the system and start the optomechanical design of the slim height version.

Acknowledgements

Many people at InPhase contributed to this work including Ken Anderson, Ernest Chuang, Alan Hoskins, Steve Hughes, Martin Pane, Ian Redmond and Brian Riley.

Development of the monocular architecture was the result of joint work between InPhase Technologies and Hitachi. We gratefully acknowledge the contributions of Tatsuro Ide, Toshiki Ishii, Ken-ichi Shimada, Takeshi Shimano, Yukinobu Tada and Masahiko Takahashi.

References

1. H. J. Coufal, D. Psaltis and G. Sincerbox, eds. *Holographic Data Storage*, Springer-Verlag, New York, 2000.
2. H. Yamatsu, *et al.*, Study on Multiplexing methods for volume holographic memory, *Joint Int. Symp. on Opt. Memories and Opt. Data Storage*, Honolulu, Hawaii, July (2005), paper ThE1.
3. E. Chuang, *et al.*, Demonstration of holographic ROM mastering, replication, and playback with a compact reader, *Jpn. J. Appl. Phys.*, Vol. 47, No. 7, pp. 5909–5911 (2008).
4. S.S. Orlov, *et al.*, High transfer rate (1 Gbit/sec) high-capacity holographic disk digital data storage system, *Conf. on Lasers and Electro-Optics (CLEO 2000)*, Vol. 39, San Francisco, CA, (2000), post deadline.
5. L. Hesselink, Digital holographic demonstration systems by Stanford University and Siros Technologies, in *Holographic Data Storage*, H. J. Coufal, D. Psaltis and G. Sincerbox, eds. Springer-Verlag, New York, pp. 383–398, 2000.
6. K. Saito and H. Hormai, Holographic 3-D disk using in-line face-to-face recording, Opt. Data Storage Conf., Aspen, Colorado (1998), pp. 162–164.
7. J. Knittel, A reflective counter-propagating holographic setup, *Joint Int. Symp. on Opt. Memories and Opt. Data Storage*, Waikoloa, Hawaii, July (2008), paper ThB04.
8. S. Solanki, *et al.*, Polarization and random phase modulated reference beam for high-density holographic recording with 2D shift-multiplexing, *Joint Int. Symp. on Opt. Memories and Opt. Data Storage*, Waikoloa, Hawaii, July (2008), paper MP06.
9. Y. Kaneko, *et al.*, High density recording of Collinear Holographic Data Storage, *Joint Int. Symp. on Opt. Memories and Opt. Data Storage*, Honolulu, Hawaii, July (2005), paper MP22.
10. H. Horimai, *et al.*, Holographic Versatile Disc (HVD), *Joint Int. Symp. on Opt. Memories and Opt. Data Storage*, Honolulu, Hawaii, July (2005), paper ThE6.
11. K. Curtis, *et al.*, Volumetric Storage Technologies, *National Storage Industry Consortium (NSIC) Optical Disk Storage Roadmap*, April 2006.
12. H. Horimai and J. Li, A novel collinear optical setup for holographic data storage system, ISOM, Jeju Island, Korea (2004), post deadline.
13. H. Horimai, X. Tan and J. Li, Collinear holography, *Appl. Opt.*, Vol. 44, pp. 2575–2579 (2005).
14. M.J. O'Callaghan, *et al.*, Spatial light modulators with integrated phase masks for holographic data storage, *Opt. Data Storage Conf.*, Montreal, Canada (2006), pp. 23–25.
15. K. Ishioka, *et al.*, Optical collinear holographic recording system using a blue laser and a random phase mask, *Joint Int. Symp. on Opt. Memories and Opt. Data Storage*, Honolulu, Hawaii, July (2005), paper ThE3.
16. H. Hideyoshi, Method and device for recording optical information, Japanese Patent Application 2001-278678, Publication number 2003-085768.
17. K. Takasaki, *et al.*, Optical system designed for coaxial holographic recording on continuously rotating disc, *Opt. Data Storage Conf.*, Montreal, Canada (2006), pp. 159–161.
18. K. Hirooka, *et al.*, A new coaxial-type holographic storage evaluation system, capable of writing and reading 500 consecutive fps, *Opt. Data Storage Conf.*, Montreal, Canada (2006), pp. 12–14.
19. H. Horimai and X. Tan, Duplication technology for secured read-only holographic versatile disc, *Joint Int. Symp. on Opt. Memories and Opt. Data Storage*, Honolulu, Hawaii, July (2005), paper MB7.
20. See http://www.ecma-international.org/publications/standards/Ecma-378.htm.
21. See http://www.ecma-international.org/publications/standards/Ecma-377.htm.
22. See http://www.ecma-international.org/publications/standards/Ecma-375.htm.
23. A. Fukumoto, Development of a coaxial holographic data recording system, *Joint Int. Symp. on Opt. Memories and Opt. Data Storage*, Waikoloa, Hawaii, July (2008), paper ThB03.
24. N.Tanabe, *et al.*, "Experimental research on hologram number criterion for evaluating bit error rates of shift multiplexed holograms," *Int. Sym. on Opt. Memories,* Jeju, Korea (2004), paper Th-PP-05.
25. K. Sako, *et al.*, Line segment hologram writer utilizing 1-D spatial light modulator, *Jpn. J. Appl. Phys.*, Vol. 44, pp. 6090–6091 (2005).

26. K. Tanaka, *et al.*, 415 Gbit/in.^{2} recording in coaxial holographic storage using low-density parity-check codes, *Opt. Data Storage Conf.*, Buena Vista, Florida (2009), paper TuC3.
27. K. Kimura, Improvement of the optical signal-to-noise ratio in common-path holographic storage by use of a polarization-controlling media structure, *Opt. Lett.*, Vol. 30, pp. 878–880 (2005).
28. K. Tanaka, Improvements in multiplexed recording performance in coaxial holographic data storage, *Int. Sym. on Opt. Memories*, Takamatsu, Japan (2006), paper Th-02.
29. K. Tanaka, *et al.*, Experimental verification of coherent addition technique for coaxial holographic data storage, *Opt. Data Storage Conf.*, Buena Vista, Florida (2009), post deadline.
30. T. Ito, *et al.*, Improvement in temperature tolerance of coaxial holographic data storage, *Opt. Data Storage Conf.*, Buena Vista, Florida (2009), paper WB3.
31. K. Curtis, *et al.*, Holographic storage medium having enhanced temperature range and method of manufacturing the same, US Patent 6,348,983, February 19, 2002.
32. K. Curtis, *et al.*, Holographic storage medium having enhanced temperature range and method of manufacturing the same, US Patent 6,650,447, November 18, 2003.
33. T. Shimura, *et al.*, Calculation of the pixel spread function with a simple numerical model for the collinear holographic storage system, *Joint Int. Symp. on Opt. Memories and Opt. Data Storage*, Honolulu, Hawaii, July (2005), post deadline.
34. S.R. Lambourdiere, *et al.*, Holographic data storage simulator for collinear optical system using shift correlation multiplexing," *Joint Int. Symp. on Opt. Memories and Opt. Data Storage*, Honolulu, Hawaii, July (2005), paper MB4.
35. T. Shimura, *et al.*, A calculation model for the collinear holographic systems, *ISO/IEC JTC 1/SC 23, Holographic Storage 12th SC 23 Plenary Meeting*, Kyoto, Japan (2005).
36. T. Shimura, *et al.*, Inter-page cross-talk noise in collinear holographic memory, Joint Int. Symp. on Opt. Memories and Opt. Data Storage, Waikoloa, Hawaii, July (2008), paper TuPO4.
37. T. Shimura, *et al.*, Analysis of a collinear holographic storage system: introduction of pixel spread function, *Opt. Lett.*, Vol. 31, No. 9, pp. 1208–1210 (2006).
38. H.Y.S. Li and D. Psaltis, Three-dimensional holographic disks, *Appl. Opt.*, Vol. 33, pp. 3764–3774 (1994).
39. K. Anderson and K. Curtis, Polytopic multiplexing, *Opt. Lett.*, Vol. 29, No. 12, pp. 1402–1404 (2004).
40. G.W. Burr and R.M. Shelby, Pixel-matched phase-conjugate holographic data storage, *SPIE Holography Newsletter*, p. 8 (1999).
41. F. Zhao and K. Sayano, Compact Read-Only Memory with lensless phase-conjugate holograms, *Opt. Lett.*, Vol. 21, pp. 1295–1297 (1996).
42. K. Curtis and W.L. Wilson, Architecture and function of InPhase's holographic drive, *Asia-Pacific Data Storage Conf.*, Hsinchu, Taiwan (2006), paper MB1.
43. K. Anderson, *et al.*, High speed holographic data storage at 500 Gb/in^2, *SMPTE Motion Imaging J.*, Vol. May/June, pp. 200–203 (2006).
44. A. Hoskins, *et al.*, Monocular architecture, *Jpn. J. Appl. Phys.*, Vol. 47, No. 7, pp. 5912–5914 (2008).
45. A. Hoskins, *et al.*, "Monocular architecture," *Int. Workshop on Holographic Memory,* Penang, Malaysia (2007).
46. K. Shimada, *et al.*, High density recording using Monocular architecture for 500 GB consumer system, *Opt. Data Storage Conf.*, Buena Vista, Florida (2009), paper TuC2.
47. S.W. Ahn, *et al.*, Fabrication of a 50 nm half-pitch wire grid polarizer using nano-imprint lithography, *Nanotechnology*, Vol. 16, No. 9, pp. 1874–1877 (2005).
48. T. Ishii, *et al.*, Tolerance analysis of monocular system, *Opt. Data Storage Conf.*, Buena Vista, Florida (2009), post deadline.

4
Drive Components

Kevin Curtis and Brad Sissom

4.1 Introduction

A holographic data storage (HDS) drive requires several critical components that are either absent in traditional optical drives or are fundamentally different due to the unique requirements of the technology. The most obvious components in this category include the spatial light modulator (SLM) which presents the data to the optical system as a page of information, and the camera, which detects the recovered data page, digitizes it, and quickly transfers it to the data channel for processing. The high numerical aperture (NA) storage lens must allow for large drive tolerances and easy assembly when used for phase conjugate reconstruction and recording. The polytopic filter must be aligned to low-pass filter the signal during recording and then filter out the undesired reconstructions on read-out. In addition, for consumer storage, this filtering should be done without using relay lenses to simplify the system for size and cost. And finally, the laser must have sufficient coherence length to record holograms and yet be able to tune the wavelength in order to compensate for thermal and humidity changes to the media on read-out. An external cavity laser diode (ECLD) was developed for this purpose. These special components and their requirements are presented in this chapter.

4.2 Laser

The laser is the most critical component in traditional optical data storage systems. This axiom holds true for HDS systems as well. However, for HDS, several laser requirements are

Holographic Data Storage: From Theory to Practical Systems Kevin Curtis, Lisa Dhar, Adrian Hill, William Wilson and Mark Ayres

significantly different than those for traditional optical storage. In HDS systems, when the holographic medium expands or contracts due to temperature or humidity changes, the wavelength of the read-out light must be changed to compensate. Thus, a practical drive requires the ability to tune the wavelength of the laser. The details of this compensation are presented in Chapter 14. In addition to tunability, the laser must have sufficient coherence in order to form the interference pattern necessary for good holograms. In practice, a coherence length of at least 1–2 mm is required but significantly more may be needed depending on optical path length differences and manufacturing tolerances of the drive. In practice, this coherence requirement means the laser must operate with a single longitudinal mode. If the laser does not achieve single mode operation during recording, then the hologram will suffer decreased signal to noise ratio (SNR). However, the wide tuning range and operating conditions required of the laser dictate that many lasing modes be achievable, often with several present simultaneously. If the laser falls into multimode operation during recording, then the drive must detect the condition and mark the hologram (or whole book) as bad for rewriting. Thus, the laser or the drive needs to be able to measure the laser coherence. InPhase Technologies and Nichia Corporation have co-developed a tunable external cavity laser diode (ECLD) that performs reliably under practical conditions and monitors the coherence and power of the laser.

In this section, the specification for the current professional drive's (Tapestry® drive) laser is presented, followed by the details of the current ECLD hardware, sensors and control. The laser specification for the monocular consumer drive is also presented, and a discussion of future options and improvements for the laser is given.

4.2.1 Initial Tapestry Drive Laser Specification

The laser for the Tapestry drive has a number of important specifications shown in Table 4.1. The required power is typically determined by the desired write or read transfer rate. From a transfer rate model, a required average exposure time is derived. The laser power needed to achieve this exposure time is determined by the transfer rate, the media sensitivity, and the power losses throughout the system. The current Tapestry drive has not been optimized for throughput, thus it introduces many optical losses that can be reduced in future versions. With the ECLD producing ~40 mW, the drive has enough optical power

Table 4.1 *First generation Tapestry drive laser specification*

Parameter	Specification
Output power	40 mW, less than 110 mA out of 120 mA
Spectral mode	Single longitudinal mode (SLM), except when tuning
Transverse mode	TEM00
Spectral side mode	<100:1 from main peak, typical line width is 15 MHz
Beam polarization	Linear, >80:1
Beam point stability	±2.5 mrad over temperature and time
Wavelength range	402–408 nm
Wavelength resolution	150 pm, accuracy of motion
Wavefront focus error	<0.1 waves rms, dominant error
Operating temperature	15–45 °C
Replaceable	Field replaceable unit (FRU), simple kinematic mount

to support a 20 MB s^{-1} transfer rate during both recording and recovery. The 40 mW output of the ECLD results in about 2.8 mW illuminating the media. Holography requires the laser to be polarized for efficient interference pattern to be formed; and polarization is used to control the power ratio between the signal and reference beams. This beam ratio must be held constant to within about 10%, which requires good polarization control in the drive and the laser (see Chapter 13). The spectral mode and side mode suppression specifications result in very high coherence, with a coherence length typically exceeding several meters. The stringent beam pointing stability specification is due to the long optical path length through the drive and to the media. In part, this is also due to the requirement that the laser be field replaceable (hence 'FRU' for field replaceable unit). Thus, the ECLD is part of a FRU that is kinematically mounted within the rest of the drive. This requires the laser to be optically imaged into the drive for both the reference and object paths so that the tilt and positional tolerances are reasonable. The laser needs to be field replaceable due to the lifetime requirements for a professional storage product (typically 100 000 h). Current blue laser diodes cannot reliably support this operating life, so the drive has been designed to allow field replacement to minimize down time for the customer. The drive can detect if the laser is starting to fail by monitoring the current and voltage increase required to maintain constant power at the end of life. The wavelength tuning range spec is determined by operating temperature range and media coefficient of thermal expansion as well as humidity changes.

4.2.2 Optical System Configuration

The tunable ECLD, shown in Figure 4.1, consists of a blue laser diode [1], optics, stepper motors, sensors, and a field programmable gate array (FPGA) on a small electrical board. The base plate that the ECLD is built on is the FRU's base plate with slots at either end for kinematic mounting. To complete the FRU, a beam expander (for setting the exiting beam size) and two more mirrors are added. Figure 4.2 shows the schematic diagram of the tunable ECLD. The laser cavity consists of a laser diode with one facet having an anti-reflection (AR) coating, a collimating lens, and a transmission grating that controls the wavelength and also acts as the output coupler. The stepper motor adjusts the ECLD wavelength by rotating the grating and changing which wavelength feeds back into the diode according to the standard grating equation. Outside the cavity, anamorphic prisms convert the elliptical beam from the cavity into a circularly symmetric beam.

Figure 4.1 *Photograph of tunable ECLD*

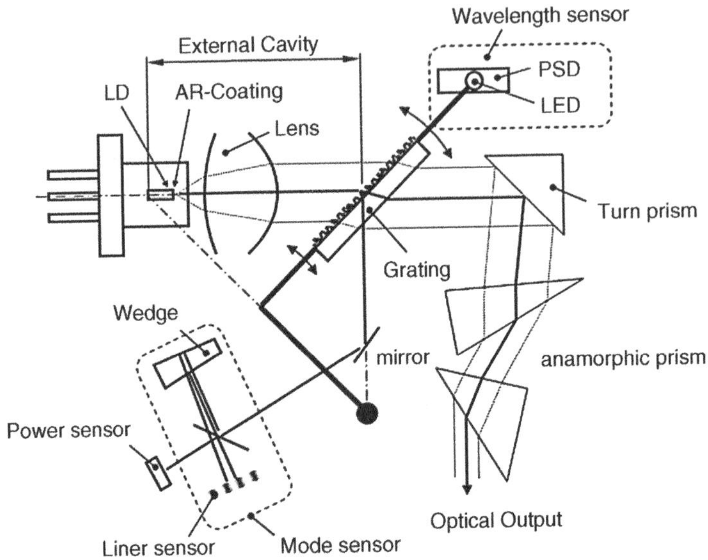

Figure 4.2 *Diagram of ECLD. Reprinted with permission from Joint Meeting of the Optical Data Storage and International Symposium on Optical Memories, Tunable External Cavity Blue Laser Diode for Holographic Data Storage by Masaki Omori et al. Copyright (2008) SPIE*

The ECLD is designed to maximize output power, single-mode tuning range and pointing stability. The Littrow configuration of the cavity, combined with a transmission grating, enables high output power with stable beam pointing [2,3]. The location for the axis of rotation of the grating minimizes the number of cavity mode hops [4,5]. The location for this pivot point is sensitive in the direction across the face of the diode. Tenth of millimeter scale errors in this direction significantly increases the number of mode hops the cavity sees but the pivot point location is rather insensitive to misalignment along the optical axis. Due to the high NA of the collimation lens, defocus on the order of 1 μm can affect laser performance. Thus, the collimation lens is designed to be highly achromatic from 400 to 410 nm, and the mechanism must be very stable over the operating temperature range. The other critical mechanical tolerance is the grating tilt. Tilts as small as 15 mdeg can significantly affect the laser performance in both power and cavity stability. This specification must be held over the shock and vibration range expected. Depending on the cavity design, automatic grating tilt control for long term stability maybe required. The grating tilt is automated by tilting the wing pivot that supports the grating arm with a small motor. The adjustment is done by maximizing the laser power, and is done at start-up of the drive and for large temperature changes. In general, the smaller the cavity can be made, the more stable the cavity is. Cavity stability is also affected by the quality of the laser diode's AR coating. In general, the lower the reflectivity of the laser's facet, the more stable the ECLD will be with regards to mode hops. The coating performance can be measured by measuring the change in threshold current for the bare laser diode when coated compared with uncoated. A reasonable AR coating will roughly double the laser diode's threshold current when not in an external cavity.

4.2.3 Electronics

The ECLD is monitored and controlled by internal sensors and an on-board FPGA. The FRU electronics control the diode current, the sensors, and the actuators. The sensors measure the wavelength, the longitudinal mode, the temperature, and the output power of the ECLD. The FRU electronics are responsible for the laser, and interact with the drive electronics through very high level, low bandwidth, commands for setting, e.g. laser power and wavelength. The FRU electronics can interrupt the drive control electronics if the laser falls out of single mode operation, or if it can not set the laser wavelength to the commanded value.

4.2.4 Mode Sensor

A portion of the output from the cavity is used for monitoring the power and coherence of the laser. The mode sensor consists of a beam splitter, an optical wedge and a linear sensor array. Figure 4.3 shows a detailed schematic of the mode sensor. A part of the incident light is reflected at the first surface of the beam splitter. The reflected beam is incident on the wedge, which creates two overlapping reflections from the first and second surface of the wedge. The wedge creates a spatially varying phase difference between the reflections. After transmission through the beam splitter, the wedge reflections interfere to create a fringe pattern that is detected by the linear detector array as shown in Figure 4.4.

In this configuration, the light creates a high contrast fringe pattern when the laser is running single-mode. When the laser is running multi-mode, the light creates a fringe pattern with low contrast. When using a Fabry Perot etalon, there exists a strong correlation between the number of laser modes and the contrast ratio of the fringes. Further, there is a correlation between the fringe contrast and the SNR of holograms in the HDS drive. When the contrast ratio falls below approximately 0.55, the hologram SNR is decreased below the drive's threshold and the holograms must be rewritten. Data from the linear array is processed by the FPGA to calculate the fringe contrast ratio (CR). The CR can be calculated in a number of ways. One way is to take the ratio of the integral of the peaks divided by the integral of the valleys of the interference pattern. Another method is to take the fast Fourier

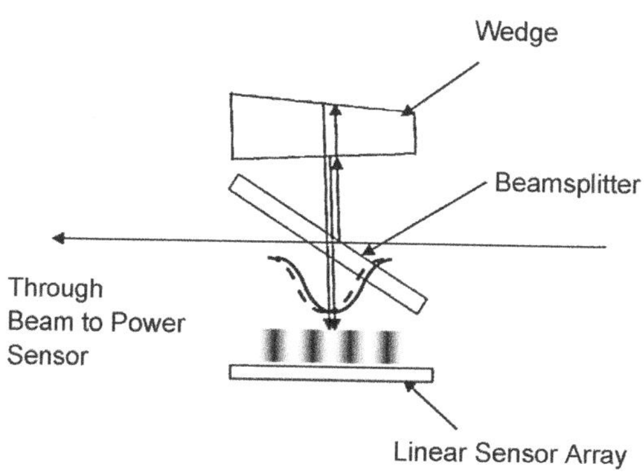

Figure 4.3 Diagram of mode detection mechanism. Reprinted with permission from Joint Meeting of the Optical Data Storage and International Symposium on Optical Memories, Tunable External Cavity Blue Laser Diode for Holographic Data Storage by Masaki Omori et al. Copyright (2008) SPIE

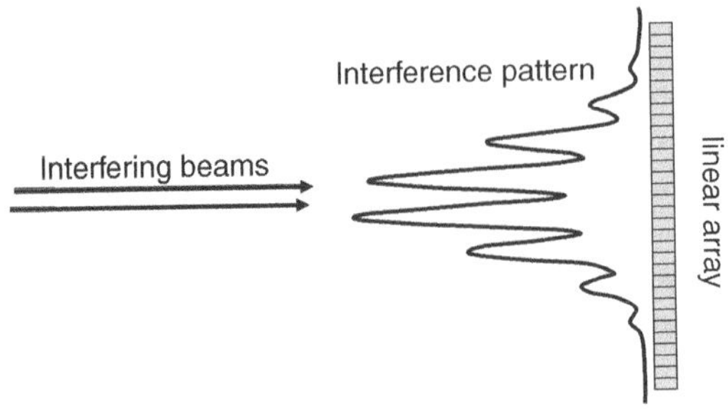

Figure 4.4 *Interference pattern on linear array. Reprinted with permission from Joint Meeting of the Optical Data Storage and International Symposium on Optical Memories, Tunable External Cavity Blue Laser Diode for Holographic Data Storage by Masaki Omori et al. Copyright (2008) SPIE*

transform (FFT) and define the contrast ratio as two times the peak of the side lobes of the FFT divided by the DC term of the FFT. One interpretation of the interference pattern on the detector array is to view it as the product of two functions. The first function is the Gaussian intensity envelope, $e^{-\pi a^2 x^2}$, of the beam, and the second function represents the interference pattern of the two beams, $\beta + \alpha\cos(\omega x)$. The CR is then defined as: $\mathrm{CR} = \alpha/\beta$. Both calculation methods above yield approximately the same answer for CR; however, the FFT method can be made more robust to noise.

4.2.5 Power Sensor

The light transmitted through the mode sensor is incident on a photodiode (see Figure 4.2). The light detected by this photodiode is proportional to the output of the ECLD. During ECLD assembly the photocurrent of the diode as sensed by the photodiode is calibrated using an external power meter. There are small corrections for wavelength to this current measurement as well. The ECLD stores the power sensor calibration in its internal memory. The FPGA recalls and applies the calibration to provide the correct ECLD output power to the drive.

4.2.6 Wavelength Sensor

In the ECLD, the grating angle determines the wavelength of the laser. The wavelength sensor measures the wavelength by detecting the linear position of the end of the grating mount. Figure 4.5 shows the wavelength sensing system. The sensor consists of an LED, mounted to the underside of the grating mount, and a position sensitive detector (PSD) mounted to the mechanical base plate. The LED light is incident on the PSD, which generates a signal that is proportional to the position of LED light. During ECLD assembly, the LED position is calibrated to an external wavelength meter. The calibration coefficients are stored in the memory of the ECLD, which the FPGA can recall for future wavelength measurements. The accuracy of the wavelength sensor is +/–0.07 nm. In the future, a sensor measuring both power and wavelength may be added. These commercially available sensors measure wavelength by the penetration of the light into the sensor's substrate material. With these sensors, power and wavelength sensors can be replaced by a single sensor and the PSD and LED removed.

Figure 4.5 Photograph of wavelength sensor

4.2.7 Characteristics of Optical Power and Tuning Range

Figure 4.6(a) shows the output power versus laser diode (LD) forward current of a tunable ECLD. The threshold current at 405 nm is 38 mA. The output power at 405 nm is 55 mW at 90 mA. The LD used nominally is a 65 mW laser diode without the AR coating. Figure 4.6 (b) shows the output power versus wavelength at 90 mA LD forward current. The maximum output power is 53 mW at 405.7 nm. The minimum wavelength where the ECLD remains in a single longitudinal mode is 400.3 nm and the maximum wavelength is 410 nm. The tuning

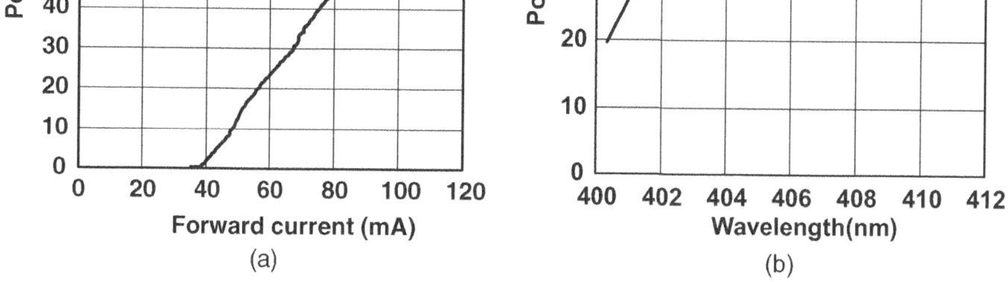

Figure 4.6 (a) Forward current versus output power of the ECLD. (b) Wavelength versus output power of the ECLD at constant current. Reprinted with permission from Joint Meeting of the Optical Data Storage and International Symposium on Optical Memories, Tunable External Cavity Blue Laser Diode for Holographic Data Storage by Masaki Omori et al. Copyright (2008) SPIE

range of the ECLD is 9.7 nm. If the required minimum output power is 35 mW, then the tuning range is 8 nm. Because the central wavelength of the ECLDs may vary (from unit to unit) by +/– 1 nm, the tuning range is specified as 6 nm to give a reasonable yield.

4.2.8 Probability of Single-mode Operation

The laser must operate in a single longitudinal mode ('single-mode') in order to record sufficiently strong holograms. However, the ECLD cavity is subject to mode hops, and single-mode behavior is not guaranteed for all tuning parameters. The mode sensor detects when the laser is multi-mode, and then FRU electronics adjust the LD current, ECLD cavity length, or both in order to return the ECLD to single-mode operation. The absolute recording wavelength is determined by the temperature of the holographic media, so the electronics can only make small adjustments to this parameter. The LD current is instead changed first, which usually results in single-mode operation at the exact commanded wavelength. Figure 4.7(a) shows a map of the mode sensor CR versus laser current and wavelength. In this measurement, the data were taken by stepping the laser's current by 0.2 mA and the wavelength by 0.2 nm. As mentioned above, when the CR exceeds 0.55, the laser is sufficiently coherent for recording operation. The light areas in Figure 4.7(a) correspond to high-contrast (single-mode) operation, while the dark areas correspond to multi-mode operation. Figure 4.7(b) also shows the distribution of the CRs in a histogram format. For reliable operation, the HDS drive requires that the laser achieve single-mode operation over 80% of the wavelength and current tuning space. In the example of Figure 4.7, the laser achieves single-mode operation in 94% of the tuning space. It is very helpful to have a single

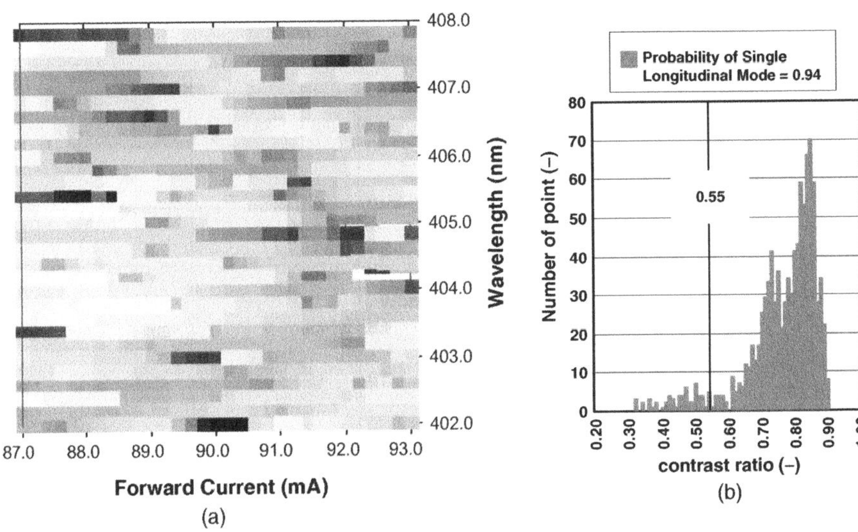

Figure 4.7 (a) Map of the probability of single-mode operation. (b) Plot of CR as measured by the mode sensor versus number of occurrences as the laser was scanned to produce the mode map. Reprinted with permission from Joint Meeting of the Optical Data Storage and International Symposium on Optical Memories, Tunable External Cavity Blue Laser Diode for Holographic Data Storage by Masaki Omori et al. Copyright (2008) SPIE

metric that indicates how good a laser is for holographic storage. The fraction of the current and wavelength space over which a laser exceeds a CR of 0.55 is called the laser's PLG (pronounced 'plug') number – hence the laser of Figure 4.7 has a PLG number of 0.94. (PLG stands for Paul's Laser Goodness after its inventor, Paul Smith.)

4.2.9 Laser Mode Servo

As is evident from Figure 4.7, these ECLDs do sometimes exhibit multi-mode behavior. Typically, ECLDs are temperature controlled so that the effective cavity length does not drift with temperature changes and produce mode hops. In order to not have to temperature control the laser, a laser servo was developed that dramatically reduces mode hops in normal drive environments without having to actively control the temperature of the LD.

The laser servo intentionally wobbles the laser current by very small amounts (tens of microamps) as it measures the CR with the mode sensor. In this way, the more stable current locations can be determined and the laser current slowly adjusted in the direction that produces a higher CR. This effectively tracks the laser's local stable operating point even if the temperature is changing. There are some limits to this technique as the laser is only allowed to change a few milliwatts in power. However, adjustments of fractions of a milliamp typically are sufficient. Wavelength may also be incorporated into the wobble as needed to enlarge the range of the servo.

Figure 4.8(a) shows a plot of CR versus time with a constant current (servo not engaged). Over only 50 s, the laser was multi-mode about 50% of the time even though the measurement was started when the laser was in a stable operating point in current and wavelength. Figure 4.8(b) shows a plot of CR versus measurements taken every 200 ms over 70 h of operation with the laser current wobble servo used. Over 70 h there are only nine individual samples where the CR is below the threshold, indicating that the laser was in single-mode operation 99.9999% of the time.

4.2.10 Lifetime of AR Coated Laser Diode

Figure 4.9 shows the result of a life test of several ECLDs under automatic power control at 45 mW, 25 °C ambient temperature and at 405 nm. The life test was performed with ECLDs in the configuration shown in Figure 4.1. The criterion for end-of-life is if either the LD forward current or voltage increases by 30% from its initial value. After 6000 h the ECLD's single-mode tuning with characteristics similar to the start of the test were confirmed. Based on data, the estimated ECLD lifetime is approximately 4700 h for a mean time to failure (MTTF) without grating tilt compensation, which increases life expectancy.

4.2.11 Future Developments

The lasers needed by future HDS systems will require similar specifications to those given in Table 4.1. For example, a monocular architecture drive as presented in Chapter 3 for a consumer product with 100 MB s^{-1} transfer rate requires approximately 40 mW (with current media) as it benefits from a smaller hologram size. There are a number of methods for decreasing media thermal expansion which will make the required wavelength tuning range drop significantly – approximately two times less than presented. Future generations of the Tapestry drive will require higher powers for faster transfer rates. This maybe

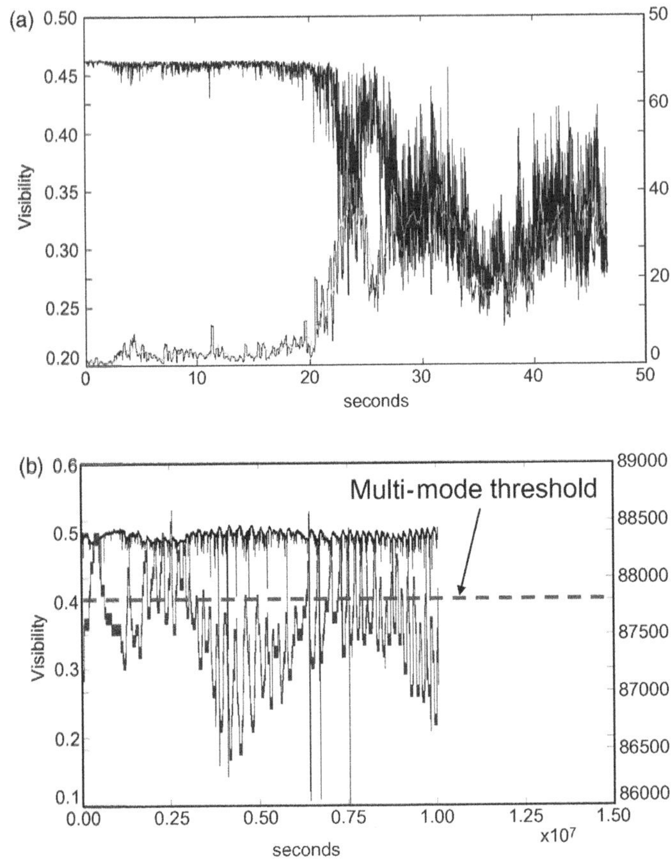

Figure 4.8 *(a) Laser CR (top line on the left side of the graph) versus time in seconds with constant current, 50% multi-mode is shown. (b) CR (top line on the left side of the graph) versus measurement number (taken over 70 h) with laser current wobble servo used. The bottom line on the left side of the graph is the laser current. This shows nine laser errors in 70 h of operation*

achieved in part by pulsing these lasers during recording. By using a technique called idling where the laser is not completely turned off, the coherence of the laser can be maintained even in pulsed mode. Scheduling could be done by adjusting the laser current pulse profile for each hologram. Pulsing for read-out and recording would significantly improve the power available and decrease the exposure times to make the drive very robust to any vibration. Finally, future improvements to media sensitivity and system technology (e.g. homodyne detection) are likely to reduce laser power demand altogether.

A big focus in the future will be to make the lasers simpler and cheaper. There are a number of ways to make an ECLD tunable without moving parts by using liquid crystals or magneto-optical materials in conjunction with grating structures, or incorporated in etalons. In addition to simplifying the ECLD to lower cost, size, and to increase reliability, there is a longer term potential for tunable distributed feedback LDs or tunable distributed

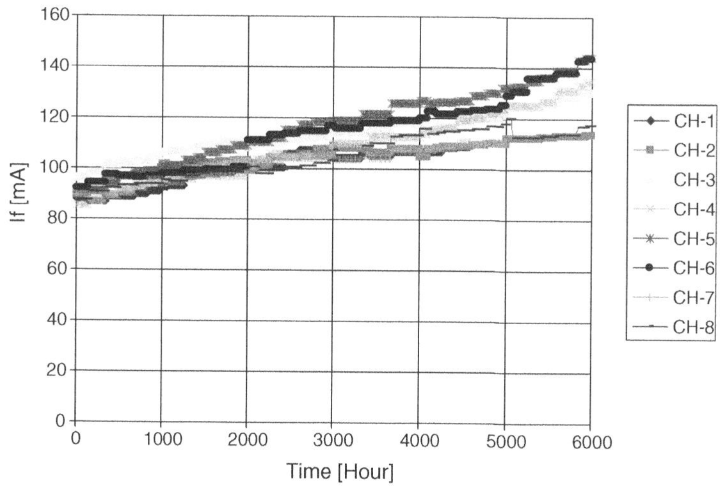

Figure 4.9 *Life test result*

Bragg reflectors that would put the complex tuning function directly into the LD. This would allow for greatly reduced size, cost, and greatly increase reliability.

4.3 SLM

4.3.1 Introduction

The SLM is a critical component that is used to modulate the signal beam, encoding the light with the data to be stored [6]. In the past, these types of optical modulators were developed for the projection display market. This section provides a brief description of the available SLM technologies, followed by the specifications for the SLM used in the first Tapestry drive, and concludes with SLM specifications for an inexpensive consumer drive using the monocular architecture.

SLMs modulate the light to encode it with the data pattern. In the drive, the SLM modulates the light beam's amplitude. However, as explained in Chapter 13, an amplitude SLM must be used in conjunction with a phase mask in order to minimize the DC component in the Fourier transform [7,8]. In addition, it has been determined that moving the phase pattern with each hologram gives improved SNR performance as the motion decorrelates the intra-signal grating patterns within the medium, thus resulting in lower crosstalk [9]. The dynamic phase pattern likely serves to improve SNR by other mechanisms as well. Future HDS devices employing amplitude and phase modulation would greatly simplify the drive by incorporating this phase mask function directly into the SLM. In addition, all data storage applications require very fast data transfer rates, which in turn require faster SLM frame rates than display applications. In practice, sustaining a given transfer rate requires an SLM capable of peak frame rates considerably higher than the average rate. This is due to the variable recording exposure schedule that equalizes hologram strength despite the diminishing sensitivity of the media. The maximum frame rate must be roughly two times the average frame required for a given data transfer rate due to this scheduling.

4.3.2 Available SLM Technologies

The current Tapestry drive uses the ferroelectric liquid crystal (FLC) based SLM shown in Figure 4.10 produced by Displaytech. Most SLMs used in projectors employ twisted nematic liquid crystals; however, these are not fast enough for use in HDS systems. Most liquid crystal based devices work by changing (rotating) the polarization of light in response to the amount of voltage applied to a pixel. A pixel rotates the polarization of the light, and an analyzer elsewhere in the system converts that into the intensity-modulated signal. FLCs are very fast and are normally used in a binary on–off mode. However, it is possible to make analog FLC materials and devices that are nominally off, but can rotate polarization in either direction resulting in 0° or 180° phase for the 'on' pixels [10–12]. Such a SLM could also be used as a phase-only modulator for homodyne systems. The CR of conventional on–off FLC SLMs varies with temperature, so they must have very good CRs at nominal temperatures in order to provide margin for degradation over temperature and aging. CRs as low as 60:1 do not seem to have adverse effects on the HDS system. In addition, the speed and optical throughput can trade-off against each other. For faster modulation, the gap filled by the FLC should be as thin as possible, yet making the gap too small can reduce the optical throughput for the same FLC material.

Another type of modulator is the magneto-optical modulator [13,14]. These modulators operate on the same principle as liquid crystal devices, i.e. rotating the polarization of the light. They have the potential to be very fast as the material response times can be in the ~10 ns range. However, currently these devices have substantial absorption in blue wavelengths and are in a very early stage of development. It is possible with these devices to have ternary modulation (i.e. 1, –1, 0) which is desirable for HDS [15].

Using microelectromechanical systems (MEMs) technology to make SLMs for data storage is a very promising and well-developed approach [16–20]. Typically, amplitude modulation is achieved by a tilting pixel structure, or by forming a switchable diffraction grating and using a low pass filter later in the system to block the unwanted light. Phase

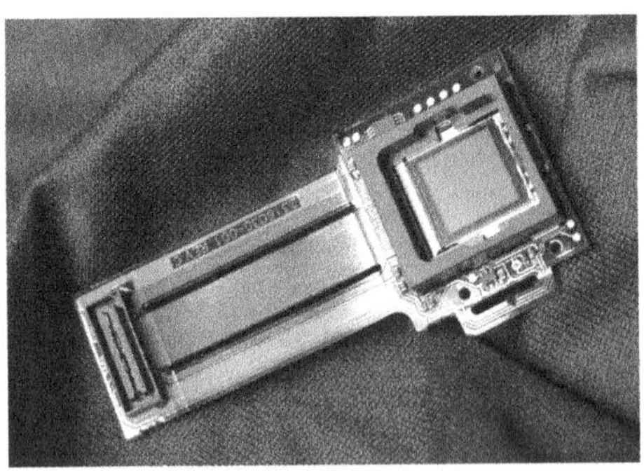

Figure 4.10 Displaytech SLM used in first InPhase drive. Reproduced with permission from Displaytech/Micron

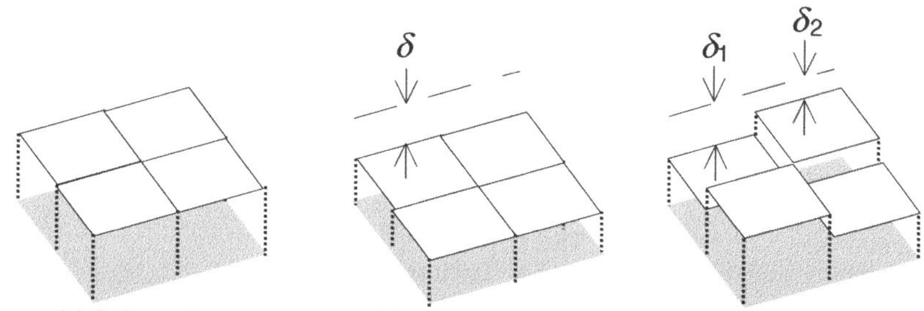

Figure 4.11 *Schematic of amplitude, phase, and amplitude phase modulation with MEMs structure. Courtesy of Silicon Light Machines*

modulation can be achieved by moving the grating or pixel structure along the optical axis of the optical system (usually perpendicular to the silicon surface). Combinations of amplitude and phase are achieved by combining these motions. For example, to incorporate the ability to modulate both amplitude and phase, a single pixel might comprise of four piston modulators moving in diametrically opposed pairs. Figure 4.11 shows various deflection states of the piston light value (PLV): (a) an undeflected state, mathematically '1'; (b) a uniform deflection for pure phase state '1 $e^{i\phi}$'; and (c) a differential deflection for amplitude and phase modulations '$A\ e^{i\phi}$'. Thus, pixel phase is controlled by the mean piston displacement $(\delta_1 + \delta_2)$ and pixel amplitude is controlled by differential piston displacement $(\delta_1 - \delta_2)$.

Overall, MEMs devices allow for very high speed, and they tend to operate well over large temperature ranges. Hermetically packaging them without inducing stress-related birefringence can be challenging however. In addition, some MEMs devices require illumination at awkward angles which can make drive packaging difficult.

4.3.3 Tapestry Drive SLM Specifications

Table 4.2 lists the high-level specifications of the current SLM. Both the SLM and the camera were developed under very tight timeline constraints. The SLM performs binary amplitude modulation with a high nominal CR. This allows for some degradation in contrast over the temperature range and aging of the device. The pixel size was designed to be 4/3 linearly larger than the smallest camera pixel that could be designed quickly and with little development risk. The frame rate supports several times the drive's initial transfer rate specification. The digital interface is double buffered, meaning that an image (data page) can be loading into the SLM while the previous data page is still being displayed. This is a very important requirement for both the camera and the SLM since it allows digital image transfers to be performed in parallel with the optical exposures.

As stated above, the major change for next-generation SLMs is to incorporate phase modulation directly into the SLM. This might be a single phase pattern that is kept on the chip and then shifted line by line across the modulator every frame. The pattern must be shifted in the correct direction as discussed in Chapter 13. Other possible future improvements include faster transfer rates, smaller pixel sizes, and larger pixel counts. In these latter

Table 4.2 First generation Tapestry drive SLM specifications

Parameter	Specification
SLM technology	Ferroelectric liquid crystal (FLC) on reflective CMOS
SLM mode	Binary
SLM format	1216 (H) × 1216 (V)
SLM active area	13.0 × 13.0 mm
SLM diagonal	18.4 mm
SLM pixel pitch	10.7 × 10.7 μm
Frame rate	≥1.1 kHz
Contrast ratio	≥100:1
Dimensions (L × W × H)	31.0 × 27.0 × 4.9 mm
Interface	High speed digital with double buffering on chip, parity on interface, test pattern, on-chip temperature sensor

cases, care must be taken to maintain backwards read compatibility in the Tapestry drive product roadmap, possibly by using the technique of asymmetric phase conjugation with an isoplanatic lens as described later in this chapter.

4.3.4 Consumer SLM Specification

The monocular HDS design described in Chapter 3 is focused on the consumer market. This means that the cost must eventually be reduced to approximately US$20 per drive, and the size should be reducible to a slim drive height (12.7 mm). These cost and size requirements result in a focal length for the storage lens of about 2.5 mm, with an NA of 0.85. Cost also determines the amount of silicon you can afford and using the experience gained by developing the HROM reader, SLM is targeted to eventually cost US$1. This requires very small pixels – a pixel pitch of 4 μm. Because one dimension of the object field has quite a bit of the angle space reserved for the reference beam, the SLM format is asymmetrical to maximize the number of pixels per page. To reach 100 MB s^{-1} with some margin, an SLM frame rate of approximately 7000 frames s^{-1} is required. A summary of these requirements is given in Table 4.3.

Table 4.3 Preliminary monocular SLM specification

Parameter	Specification
Format	1200 × 600: Binary amplitude with plus π and 0 phase per pixel (±1,0)
Pixel pitch	4 μm
Fill factor	>85%
Contrast	>60:1
Frame rate	7000 frames s^{-1} for 100 Mbytes s^{-1} user
Image load	Double buffered, load time less than 70 μs
Interface	High speed serial digital data interface to keep pin count down, serial control interface, on-chip temperature sensor
Global update	Switch from stored to display with single command line
Interface parity	For checking and syncing the interface, test pattern as well

4.4 Image Sensor

4.4.1 Introduction

An electronic imager is required to digitize the holographic image during data recovery. As discussed in Chapter 11, the imager must have finer resolution than the SLM in order to allow for oversampled detection and mechanical tolerances for interchange. While there are several detector technologies, CMOS sensor arrays are the primary technology for HDS systems. They are cheaper to produce, use less energy, generate less heat, can read-out subarrays of the image, and can be packaged cheaply a wafer at a time. These reasons are also why CMOS sensors are the camera of choice in cell phones. However, the fabrication processes are just now improving the blue sensitivity due to protection layer absorption, and current CMOS sensors have more noise than CCD technology.

The basic CMOS cell that is of use for HDS is a six transistor cell (6T) or similar, as shown in Figure 4.12. This structure has two nodes to accumulate charge. This first is used to capture the light and the other is a storage node. This pixel architecture [21,22] allows for double buffering the camera – transferring out one data page while simultaneously capturing another data page. Like the SLM, double buffering is required for reasonable system operation both for electronics speed as well as for optical integration times. The detector sensitivity is largely determined by the quantum efficiency, and for the current drive camera, the quantum efficiency was impacted by a fabrication process including an absorbing protective layer. The quantum efficiency was slightly less than 12%. The HROM camera employed a different fabrication process and achieved quantum efficiency closer to 25%. Thus both design and fabrication are critical for the performance of the detector. Microlenses were fabricated onto the silicon to improve the pixel fill factor of the device.

The next generation of CMOS image sensors should use a pixel structure called a buried photodiode [23] with a double-buffered pixel architecture. This architecture allows correlated double sampling (CDS) without loss in speed [24–26]. CDS results in very low noise (less than 10 noise electrons per pixel), no pixel fixed pattern noise (FPN, i.e. differences in the 'dark' level among the pixels), higher pixel fill factor, and much higher (~3–4x) sensitivity than the current 6T pixel architecture. In addition, by taking advantage of the decrease in noise with the buried photodiode, a much smaller voltage swing out of the pixel can be used. A small signal swing will allow the analog to digital converters (ADCs) to use a lower input range, which will dramatically lower power dissipation.

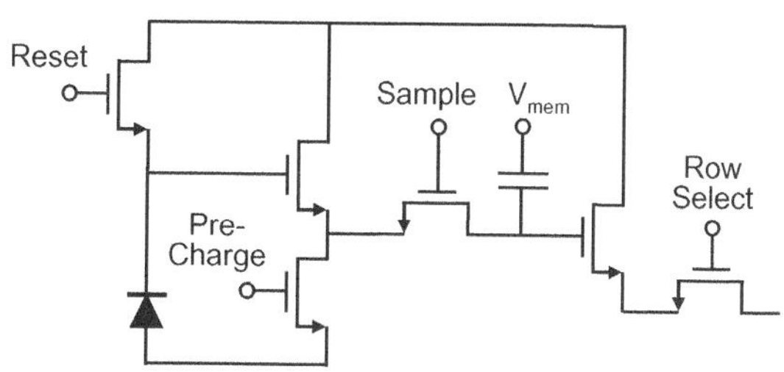

Figure 4.12 6T transistor pixel structure

Figure 4.13 *Photograph of the camera (LUPA-3000) and board*

4.4.2 Tapestry Drive CMOS Sensor

InPhase had a custom CMOS sensor designed to meet the system requirements. It is commercially available from Cypress Semiconductor as the LUPA-3000 CMOS image sensor. The detector is significantly larger than the SLM so that it can accommodate manufacturing tolerances as well as interchange tolerances across temperature and misalignments. Figure 4.13 is a photograph of the camera chip and the custom electronics board. The sensor is approximately the size of the US quarter (20 mm × 20 mm). The specifications listed in Table 4.4 include FPN and photo response nonuniformity (PRNU, i.e. differences in the sensitivity among the pixels). Currently the drive does FPN correction but does not benefit significantly from PRNU correction. This is likely due to the low signal level limiting the effects of PRNU. Keeping the storage node capacity small also helps to limit the thermal noise.

Table 4.4 *First generation Tapestry drive camera specifications*

Parameter	Specification
Format	1696 × 1710 pixels
Pixel pitch	8 μm
Shutter/pixel type	Global shutter, double buffered pixels, six transistor structure
Frame rate	485 fps at full field
Windowing	Blocks of 32 pixels for high frame rates
ADC resolution	8 bits
Interfaces	412 Mbps, 32 LVDS outputs
Package	BGA, 399 pin
Fill factor	95%
Gain per pixel	1×, 2×, 4× gain available
FPN	1.7% of V_{sat}
PRNU	2.2% of V_{signal}
Full well charge	27 000 electrons
Dark noise	21 electrons
Max power dissipation	1.1 W, drive usage never come close to this limit

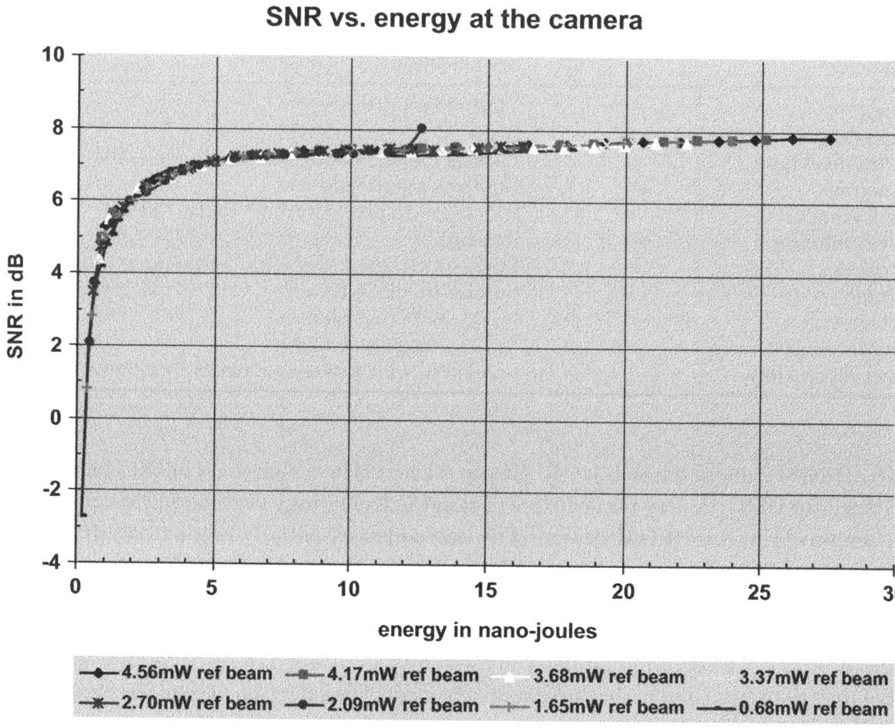

Figure 4.14 *SNR versus energy for different intensities*

The number of photogenerated electrons, or more importantly, the amount of energy needed to exceed the detector noise is a very important parameter. This is a function of the sensitivity of the sensor (quantum efficiency, fill factor, and pixel gain) and the noise in the detector array. ADC quantization noise also plays a role, especially at low signal levels. Figure 4.14 shows a plot of energy versus recovered page SNR for different intensities with the LUPA-3000 set at the lowest gain (1x), and it shows that regardless of intensity, >4 nJ is required for good detection. With a 2 ms integration time, this means that 2 μW of intensity is needed for good hologram detection. Exceeding this limit (for example, with longer exposures) results in only a modest improvement in SNR. This has been shown to be true at very high recording densities meaning that the detector noise is an important contributor to SNR if the signal to scatter ratio is large enough for good SNR. Curves like this must be generated for each camera gain setting in order to calibrate the optimal gain and energy requirements for each particular camera chip.

4.4.3 Image Sensors for Consumer HDS

As with the SLM, the size and cost of camera components are major determining factors for designing a consumer product. Silicon area and packaging are the driving factors for cost. The HROM camera specifications are summarized below, and then the specification for the monocular drive is presented.

Table 4.5 Preliminary specification for monocular image sensor

Parameter	Specification
Format	1696 × 864 pixels
Pixel pitch	3 μm
Shutter/pixel type	Global shutter, double buffered pixels, BD structure
Frame rate	6000 fps at full field
Windowing	Blocks of 32 pixels for high frame rates
ADC resolution	4–6 bits
Interfaces	High speed serial (>2 Gb s^{-1}) to reduce pin count
Package	Similar to HROM
Fill factor	~70% with microlenses
Gain per pixel	1x, 2x, 4x gain available
Power dissipation	<50 mW with real usage model (not part max)

The HROM camera has 640 × 640 4.6 μm 6T pixels, a design that could eventually be produced for US$1. A very inexpensive packaging technology developed for cell phones was employed in the HROM design. This technique essentially amounts to gluing the wafer between two glass substrates and then dicing it up. Another key to lowering cost and simplifying the package is reducing the number of IO pins required. One of the major differences between the Tapestry and HROM cameras was the number of bits of resolution of the ADCs. It was determined that only 4–6 bits are needed to fully capture the SNR, rather than the 8 bits used in the current drive sensor. The selection of 6 bits helped reduce the off-chip data rate. The aforementioned improvements in the fabrication process also yielded a much improved quantum efficiency. The resulting total sensitivity of the HROM camera at 405 nm is 0.036 A W^{-1}. In addition, a higher speed digital interface was used to reduce the number of pins. More details about the HROM camera and packaging are given in Chapter 15.

For the monocular drive camera, the desired pixel structure to use is a buried diode array with very small (3 μm) pixel pitch. The long term steady state cost goal is the same as HROM (US$1). Table 4.5 lists the preliminary camera specifications. The increase in sensitivity and decrease in noise that is expected comes from the different pixel structure. The decrease in noise should significantly help improve read transfer rate by lowering the required energy needed on read-out. The rest of the changes are necessary to achieve the 100 Mbytes s^{-1} user transfer rate, and to meet cost objectives. The chip will likely use the same packaging technology as the HROM camera.

4.5 Beam Scanners

4.5.1 Introduction

Quickly changing the angle of the reference beam is a requirement of any angle-multiplexed holographic system. Mechanical or optical devices that change the angle of a beam are called beam scanners. The range requirement for scanning in the angle multiplexing plane is about 30–35° (beam angle, as opposed to scanning mirror angle) with a high resolution. The current needed beam resolution is approximately 0.001° and is primarily a function of the media thickness. The speed of angle changes is also a very important parameter. Ideally, with current media packages, the scanner would step the beam angle about 0.06° and settle within less than 100 μs for sequential page access. Use of an internal sensor for the scanner position is required for calibration and accuracy. The current Tapestry drive uses a galvanometer ('galvo') mirror with a capacitive sensor to achieve this function. In addition, it is likely that such a system must implement pitch correction, i.e. angle adjustment perpendicular to the multiplexing plane. In the Tapestry drive this is done by using another galvanometer mirror rotating in this orthogonal direction. The requirements for pitch correction are a beam scan range of a few degrees (1–4°), with a resolution of 0.005°, ideally having the same speed as the multiplexing scanner. Algorithms for controlling the scanners during recovery are described in Chapter 14.

In addition to the scanner performance parameters discussed above, there are two other potential aspects of the scanner that should be considered. The first consideration is whether the scanner requires the beam to be imaged into the medium. For example the axis of rotation of a galvanometer mirror is centered on the axis of the motor. Thus, the downstream position of the beam changes as the mirror angle scans. For this reason, a 4F lens system (Keplerian telescope) is typically employed to create an image of the mirror in the medium, which is stationary even as the angle changes. InPhase has also designed a number of systems where the center of motion is centered in the media that employ two galvanometer mirrors rather than imaging optics.

The second aspect to consider is obliquity correction. As the angle of incidence into the media is changed, the projection of the beam on the media naturally expands or contracts geometrically. This change in size is not desired as it wastes media *M/#* and changes the intensity of the reference beam. The first generation Tapestry drive does not address this issue. However, there are two possible solutions for this problem. One solution is to mount a prism rather than a flat mirror on the galvanometer motor [27]. The second solution is to design the scanner such that it corrects the size of the beam versus the angle [28].

Generally, there are several technologies that can be used to make beam scanners. These include not only galvanometer mirrors, but also mechanical scanners, MEMs based scanners, liquid crystal based scanners, and acoustic optical beam scanners. Below are a few comments on these technology approaches.

4.5.2 Galvanometer

The Tapestry drive uses galvo mirrors for changing the angle to multiplex as well as to perform pitch correction. These galvanometers are commercially available and are produced by Cambridge Technology. They can be made reasonably small, and are controlled using an internal capacitive sensor with 16 bits of resolution. They can be configured for 40° of angle change (20° of mechanical rotation), with a resolution of 0.001°, and a sequential page move and settle time of 800 μs. Custom versions can reduce the movement time but require more current. To maximize speed, it is important to minimize the weight of the payload (mirror) by making it smaller and/or thinner. However, very high wavefront quality is required over the whole operating temperature range, which limits how thin the mirror can be made. Figure 4.15 shows a Cambridge galvanometer with a mirror attached.

Figure 4.15 Cambridge galvanometer with mirror. Reprinted with permission from Cambridge Technologies.

4.5.3 Mechanical Based Scanners

The HROM program drove the development of several mechanical actuators. Alps Electric designed and built the first custom voice coil mirror actuator. InPhase designed the feedback sensor and the control and calibration electronics. The feedback sensor consists of a Hall sensor and a small magnet placed on the yoke of the mirror actuator. The mirror actuator can rotate in two directions to allow for both angle multiplexing and pitch compensation. The angular range of motion was 11 and 8° for the two directions, respectively. The angular resolution was approximately 0.001° with use of the feedback sensor. The size of the actuator and mirror is 8 mm × 8 mm × 6 mm. Figure 4.16 shows the back and the front of the HROM mirror actuator.

Several other actuators have been designed with the center of beam rotation at the medium [29,30]. These include flexure based designs that implement a complex motion of the mirror yielding a stationary beam with only one actuator. A very promising design is shown in Figure 4.17. This actuator uses two voice coils to pull down a prism mounted on flexures. When both voice coils are equally activated, it rotates the prism in the multiplexing direction. Different amounts of force on the voice coils allow for orthogonal direction

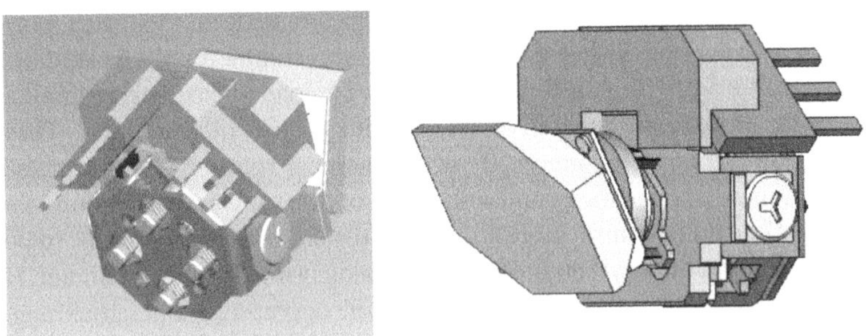

Figure 4.16 Back and front of mirror actuator

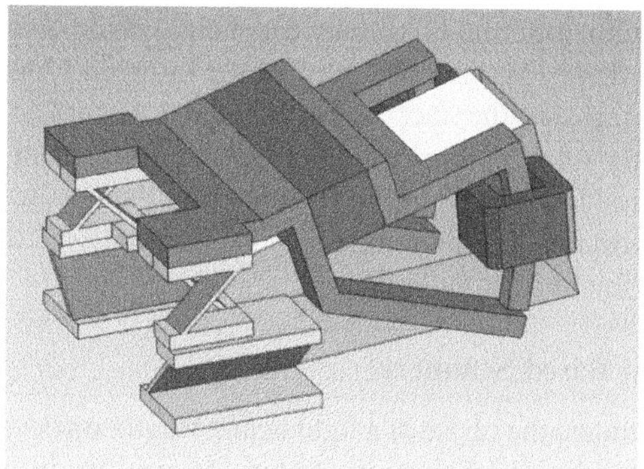

Figure 4.17 Next generation mirror actuator

control to be achieved. This actuator was designed for HROM, and it also circularized the beam by employing reflections off two different surfaces of a prism or mirrors. This eliminated the need for using anamorphic prisms or special collimation optics that are typically required to circularize a LD beam. The device has an extremely low profile, being only a couple of millimeters in height, yet still allowing for 16° of motion in the multiplexing plane and about 3° in the orthogonal direction. Since the payload is very small, the multiplexing speed is expected to be on the order of 100 μs.

4.5.4 MEMs Scanners

While there are undoubtedly many MEMs based approaches for building scanners, a particularly promising approach uses micromirror arrays [31]. A beam scanner based on a *single* MEMS mirror must comply with the severe requirements for HDS applications. The MEMS mirror must have a diameter of about 1.5 mm, and it must have a fast response rate with a large angular deflection range (> 10 kHz random access, and ±7.5°, respectively), high resolution (≤2 mdeg) and very good stability and optical quality ($\leq\lambda/10$ peak to valley wavefront error, <5% modulation, >85% optical throughput). One of the main limitations to the performance of a single mirror with these specifications is the presence of dynamic deformations that will occur when operating at high speed and large angles. In addition, mechanical clearance between the mirror and the silicon it is fabricated on will be an issue: a mechanical tilt range of 7.5° requires ~100 μm clearance. This is a problem as most surface micromachining processes can not accommodate more than 10 μm unless multi-wafer approaches are used. Due to the great difficulty in engineering relatively large, fast, and simultaneously dynamically flat mirrors, it is desirable to replace the *single* MEMS mirror scanner by a high fill factor *array* of electrostatically actuated piston-tilt mirrors. This technology is also mentioned in Section 4.3.

There are two main advantages to using MEMs arrays of small mirrors for a scanner. One advantage is that each individual micromirror now has a much smaller moment of inertia, and can therefore be positioned much more quickly to the desired angle. In order to obtain diffraction limited performance, the individual micromirrors must all be tilted at the same angle using the same driving circuitry. Additionally, the piston positions must be chosen such that the path length difference Δd between neighboring micromirrors fulfills the condition $\Delta d = N\,\lambda/2$, where λ is the wavelength of the laser, and N is an integer. The other advantage to mirror arrays is the fact that the angular accuracy requirements on the micromirrors are greatly relaxed. For example, the requirement of 2 mdeg accuracy for a single mirror is reduced to approximately 100 mdeg for the smaller elements of a mirror array. Thus, high performance beam scanners could potentially be designed using MEMs technology developed for SLMs.

4.5.5 Liquid Crystal Based Scanners

Liquid crystals can modulate the phase of a light beam. Raytheon developed a custom beam scanner using this principle called an *optical phased array for beam steering* [32]. This device had a few hundred microsecond response time and a fairly large angle scan range. It was architected as a one-dimensional liquid crystal cell array.

4.5.6 Acousto-Optic Beam Scanner

Acousto-optic deflectors (AODs) operate on the principle of diffraction by acoustically propagating waves. They are capable of high-speed deflection of optical beams without the use of mechanically moving mirrors such as those used in galvanometer and MEMs scanners. This not only makes them more robust to vibration, but also increases reliability, lifetime and scanning speeds. For these reasons, a custom slow-shear wave TeO_2 (tellurium dioxide) Bragg cell in conjunction with a magnification system might achieve the required beam scanning functionality for HDS. There are several commercial vendors for these cells.

Figure 4.18 shows a schematic of the proposed acousto-optic scanner at two different scan angles. Magnification is required to increase the scan range of the deflector.

For a 25° scan range, at least 2500 resolvable angles are required. Such a number is achievable with a well-designed high-performance slow-shear wave TeO_2 cell with a realistic but difficult to achieve bandwidth of 40 MHz, and a moderate aperture of 40 mm. Furthermore, the slow-shear mode of TeO_2 has one of the highest efficiencies of practical acousto-optic materials, resulting in nearly perfect diffraction efficiency, with 85% or more being readily achievable.

The output angle of the AOD is selected by applying a radio-frequency (RF) tone to a piezoelectric transducer. This tone is provided by a voltage controlled oscillator (VCO),

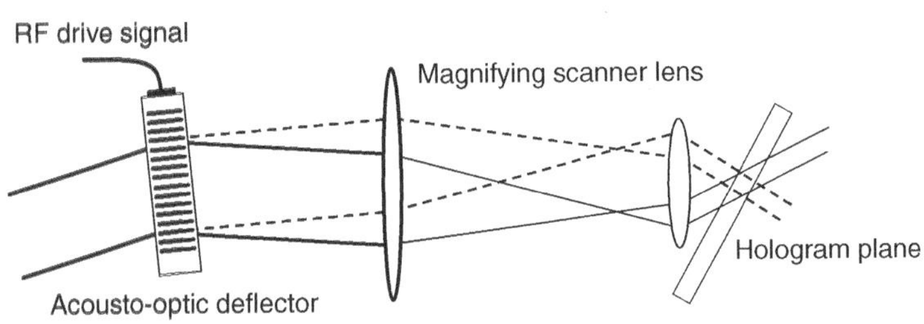

Figure 4.18 *Beam scan system incorporating an acousto-optical deflector*

and typically varies between 40 MHz and 80 MHz (40 MHz bandwidth). The most significant drawback of this scanner approach is the resulting Doppler frequency shift in the diffracted beam. Since the optical frequencies of the signal and reference beams will then be different, the resulting interference patterns move, causing the holographic gratings to wash out without adequate compensation. Such compensation can be performed by using another acousto-optic device in tandem with the scanner AOD to induce an equal, opposing Doppler shift. Fortunately, this second device can be driven in a resonant mode, ensuring efficiencies of over 90%.

The switching time for the entire acousto-optic scanner is limited by the transit time of the RF wave through the AOD cell, which would be 64.5 μs for the described TeO_2 cell. The only remaining drawback of such an AOD is the typical 4° scan range instead of the required 25°. However this range can be extended by the use of a scanner relay lens with a magnification of 0.16, as indicated above. This scanner lens can furthermore be very shallow (referring to the out of page dimension in Figure 4.18), since the scanning only occurs in one plane. This would allow the scanner to have a very low profile, which can decrease the scanner volume. However, these beam scanners can only be consider for applications where performance is critical due to their cost and size.

4.6 Isoplanatic Lenses

4.6.1 Introduction

Angle-polytopic phase conjugate HDS systems are useful for consumer applications. Normal lens designs that do not incorporate phase conjugation have also been studied [33]. Phase conjugation allows for simpler optics to be used since aberration correction occurs during conjugate hologram recovery. This aberration cancellation is typically only exact for the case where the recovered signal precisely retraces the path it took during recording, and is limited in cases of medium misalignment and drive-to-drive interchange where the retracing is inexact. Presented below is an isoplanatic lens design that largely cancels aberrations even in the presence of medium misalignment and interchange. This design concept is ideal for HROMs and consumer HDS systems where tolerances lead directly to cost. The design form is especially well-suited for HROM systems as it allows for asymmetric phase conjugation [34] – holograms recorded with a complex, expensive mastering lens can be recovered and almost perfectly phase conjugated using a simple, cheap reader lens.

The design presented here exploits the principle of isoplanatism [35], or spatial invariance. This means that aberrations in the point spread function do not vary significantly across the field [36]. Almost all lenses have limited isoplanatism due to the tiny 'isoplanatic patches' required to linearize the response of the lens and allow it to generate a Fourier transform [37]. The sizes of these patches are typically slightly larger than the lens point spread function (PSF) [38], meaning a few micrometers square. The lenses presented in the referenced paper may have isoplanatic patches several millimeters square, and are thus said to be extremely isoplanatic. The larger the isoplanatic patch, the more a shift or tilt of the storage lens with respect to the recorded hologram can be tolerated while still phase-conjugating the data perfectly.

4.6.2 Characteristics of Isoplanatic Lenses

A definition of extreme isoplanatism is readily obtained by extending existing definitions of 'infinitesimal' isoplanatism as defined in the literature. Systems with infinitesimal isoplanatism have the following characteristics:

- Infinitesimal translations in object space produce infinitesimal translations in image space without change in quality of the corresponding in image [39].
- Infinitesimal rotations in object space produce infinitesimal rotations in image space without changing the quality of the corresponding image [40].
- Wavefront aberration for a given point in the pupil is constant [41].
- The wavefront aberration corresponding to a given PSF in image space is constant.

To extend these definitions of infinitesimal isoplanatism to cover extreme isoplanatism, we simply change all infinitesimal rotations and translations to finite and change all instances of the word point to patch. As mentioned earlier, extremely isoplanatic patches can be orders of magnitude larger in area than the infinitesimal patches associated with a lens PSF. These definitions, once modified to cover extreme isoplanatism, can be used as design time constraints when optimizing a lens with a modern lens design program [42].

In HDS systems, the most convenient metric to measure the performance of the system is the SNR. Because SNR is largely a function of the PSF in the recording and recovering systems, the last characteristic of extreme isoplanatism is stated as: The SNR in a phase conjugate system is constant in the presence of finite shift or tilt of the system phase conjugation optics.

InPhase Technologies has developed an optical model of holographic storage systems that predicts the recovered page SNR by simulating the PSF using Huygens' method and the *k*-sphere formulation of volume holography. The model shows good correlation with experimental data for media shifts and rotation, and has been adapted as a Zemax® software plug-in that can be used to simulate SNR during the design of HDS optics.

4.6.3 Extremely Isoplanatic Holographic Storage Lens

Figure 4.19 shows an extremely isoplanatic FT storage lens and the arrow shows the direction of the reference beam on recording. The lens was optimized for isoplanatism by constraining the lens performance using the definitions above. Using the third characteristic of isoplanatic lenses listed above, the size of the isoplanatic patches can be directly examined by using the changes in Zernike polynomial coefficients or root mean square (RMS) wavefront error as a function of field.

Figure 4.20 shows the first nine Zernike coefficients as a function of field while Figure 4.21 shows the RMS wavefront error. The Zernike terms describe the first and third order wavefront properties for propagating a single SLM pixel through the storage system. The slope of these curves can give insight into how the wavefront of the different SLM pixels changes across the lens field. Figure 4.21 shows how the RMS magnitude of the wavefront error changes across the lens field. An isoplanatic patch is the area where the wavefront shape and magnitude do not change significantly.

To evaluate the size of the isoplanatic patch for this lens, an empirical criteria of 1/50th wave RMS is used. While this value is much more stringent than the 1/14th wave Marechal

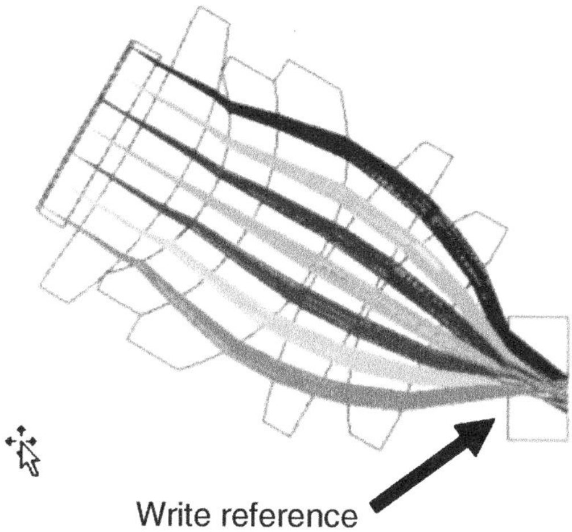

Figure 4.19 *Extremely isoplanatic storage lens with an effective focal length of 2.4 mm and 1.7 mm field*

criterion [43] for diffraction-limited performance, it has been proven to better predict SNR constancy during phase conjugation. Note that this value gives SNR constancy for a SLM with 4.6 μm pixels, and may be relaxed when larger pixels are used. Figure 4.21 shows a large isoplanatic patch between 0 and 1.4 mm of field where there is less than 1/50th wave

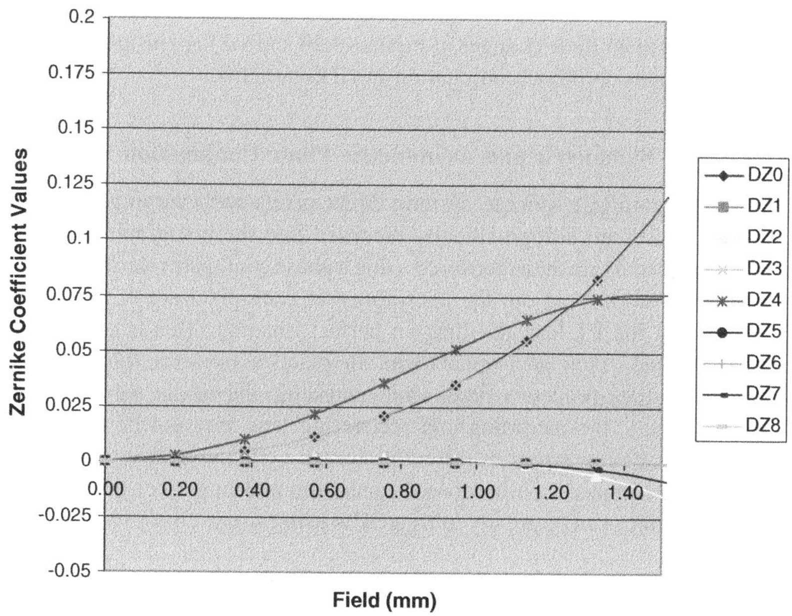

Figure 4.20 *Zernike polynomial coefficients as a function field for the storage lens*

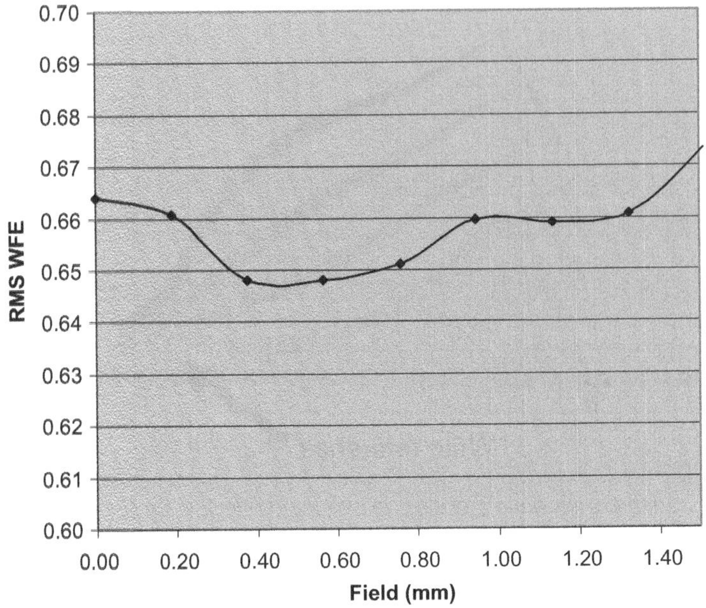

Figure 4.21 *RMS wavefront error versus field for the isoplanatic storage lens*

RMS variation. At the edge of the field the variation corresponds to an isoplanatic patch of 400 μm. It can therefore be concluded that over neighborhoods of order 400 μm wide, the wavefront shape and magnitude changes of Figures 4.20 and 4.21 are insignificant. This has profound consequences on phase conjugation in HDS systems.

4.6.4 Examples – Symmetric and Asymmetric Phase Conjugation

In an ideal phase conjugate storage system, SLM pixels with varying wavefront (see Figures 4.20 and 4.21) are holographically recorded into the holographic medium. The recorded pixel wavefronts are then recovered using a phase conjugate reference beam and an identical storage lens. On read-out the aberrations in each pixel are negated by reverse propagation through the FT lens resulting in perfect imaging. This is true for any lens regardless of attributes such as isoplanatism. In practice however, phase conjugation involves an intermediate process where, after recording, the media may shift, tilt, and/or shrink. Additionally, the recording and recovering lens may not be identical due to manufacturing or assembly errors in different storage systems, or due to environmental conditions. In these instances, errors do not cancel out and imperfect phase conjugation results. These conditions can be greatly mitigated by using an extremely isoplanatic storage lens.

In the first example, tilt errors are considered that introduce field shifts of about 400 μm (see Figure 4.22) in a symmetric phase conjugate system where the recording and recovery lens are identical. Because the tilt-induced shift is less than the size of our isoplanatic patch,

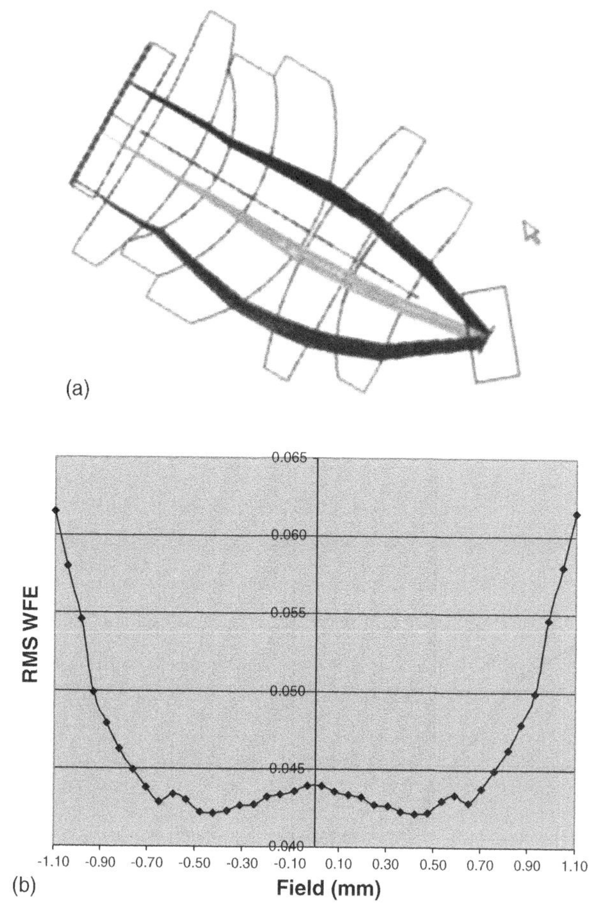

Figure 4.22 *(a) Symmetric read-out using the lens in Figure 4.19 with a 9.5° media tilt. (b) RMS wavefront in a symmetric system with 9.5° tilt error in media position*

the performance of the system is still diffraction limited as predicted [see Figure 4.22(b)]. Tilt insensitivity is important when performing wavelength compensation for thermal effects, though tilts encountered in a conventional HDS system are much smaller than 9.5° [44].

In the second example, asymmetric phase conjugation is investigated (see Figure 4.23), where a simple three-element lens is used to recover holograms written with the five-element lens of Figure 4.19. Without media shifts, tilts, or rotations, the three-element lens was designed to perfectly phase conjugate the pixel wavefronts recorded by the lens in Figure 4.19. With a medium axial shift of 80 μm, the lens is still diffraction-limited over the lens field (Figure 4.24). This three-element spherical lens assembly could be further simplified and replaced by a two-element aspheric lens assembly without loss of performance. See Chapter 15 for more details on the HROM reader system.

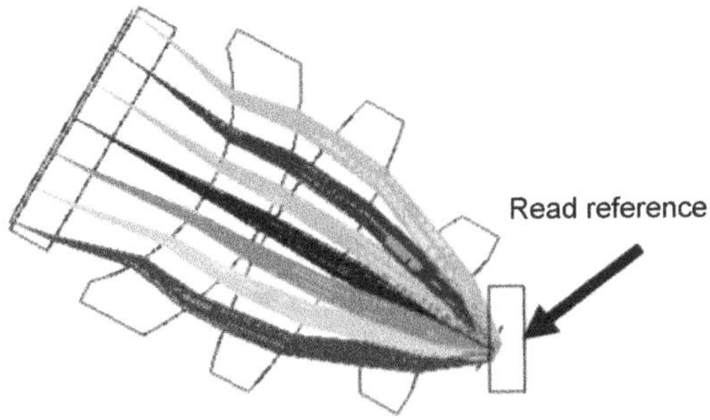

Figure 4.23 Asymmetric read-out using a three-element lens to recover holograms written with the five-element lens of Figure 4.19. Medium is shifted 80 μm to illustrate shift insensitivity

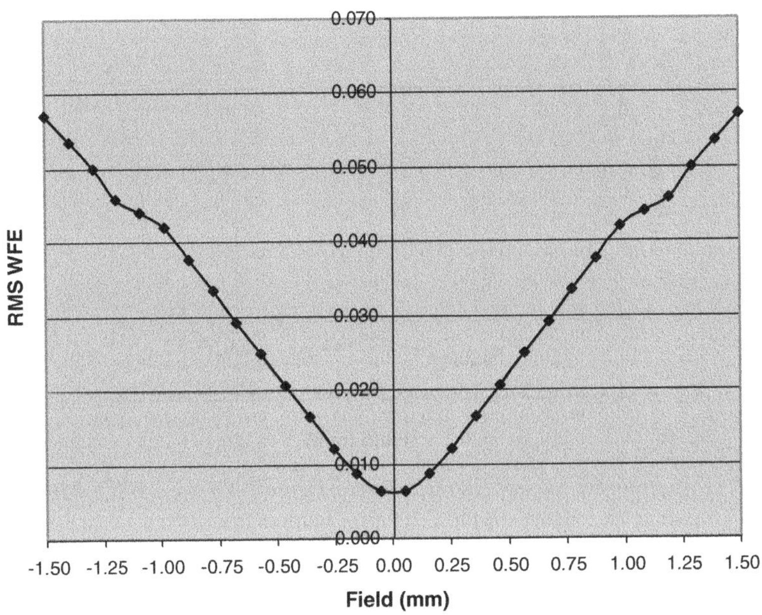

Figure 4.24 RMS wavefront error versus field in asymmetric system with 80 μm axial shift in media

4.6.5 Lens Design Notes: Phase Conjugation and Extreme Isoplanatism

Phase conjugation of FT lenses may be modeled in a modern lens design program [42] using a back propagation technique as follows:

(1) Trace rays from object space (SLM) to system stop (hologram) using a candidate recording lens.

(2) Consider the system stop to be a representation of the recorded hologram. The wavefront error of a given field point corresponds to the pixel wavefront holographically recorded at the FT plane.
(3) Reverse direction of propagation at the stop by making the next airspace negative and equal in magnitude to the airspace prior to the stop. This reversal represents reconstruction of the recorded hologram, propagating all pixel wavefronts back towards the FT lens. Do not reverse direction using a mirror as this does not model holographic reconstruction. Also note that pixel wavefronts are not required to be plane waves in a phase conjugate system.
(4) Next, enter a read lens. In the case of symmetric phase conjugation, simply negate all airspaces and element thicknesses but not curvatures as you incrementally assemble a read lens identical to the recording lens. Stop when the image plane (camera) is reached.
(5) At this point you should notice diffraction-limited performance regardless of the candidate lens design. This is the beauty of phase conjugation: any wavefront error recorded by the hologram is canceled by the back propagation of holographic reconstruction.
(6) Before attempting optimization, constrain all parameters describing the recording lens geometry to equal those of the read lens so that recording and read lenses will evolve in unison and symmetry will be preserved.

A design layout of an FT lens modeled by this process is shown in a Figure 4.25(a). The read lens has been shifted 200 μm so it can be differentiated from the recording lens.

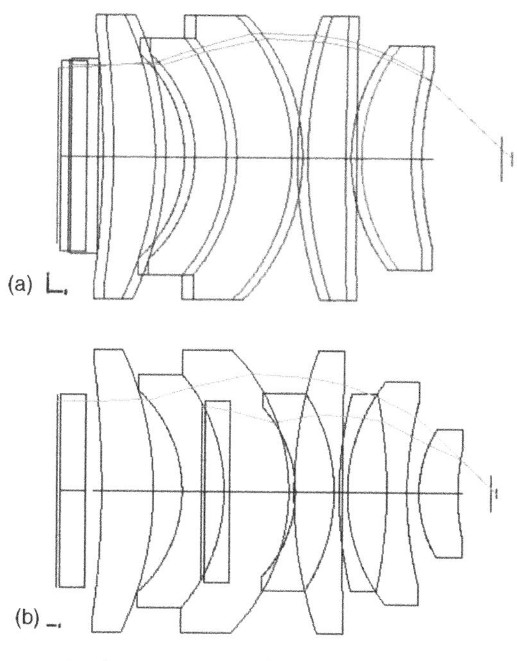

Figure 4.25 *(a) Symmetric phase conjugate design layout for the lens of Figure 4.19 with a 200 μm media shift. (b) Asymmetric phase conjugate design layout used to simultaneously design the five-element lens of Figure 4.19 and the three-element lens of Figure 4.23*

Figure 4.25(b) illustrates an asymmetric design layout. The modeling technique is substantially similar to that used for the symmetric system except that the read lens is entirely different from the recording lens. Using such a layout, the recording lens and the read lens are optimized simultaneously. Because these two lenses give diffraction-limited performance when used as a pair, they are said to phase conjugate perfectly.

Extreme isoplanatism can be accomplished by two techniques. The first technique is to shift and tilt the read lens with respect to the system stop (hologram) during optimization. Figure 4.25(a) illustrates such a shift. The read lens has been shifted with respect to the system stop by 200 μm. Two rays are depicted: one is a forward propagating ray: the other is its corresponding back propagating ray. Observe that back propagation follows a different path than forward propagation, effectively 'stretching' an isoplanatic point into an isoplanatic patch. Optimizing in the presence of such shifts (or tilts) automatically creates isoplanatic patches. The sizes of the isoplanatic patches may be increased by optimizing over a number of tilted and shifted systems of varying magnitude and directions, simultaneously. This technique effectively mimics assembly errors or errors in hologram placement. It therefore creates lenses that perform well in the presence of such errors.

The degree of isoplanatism achieved may be determined by plotting Zernike coefficients versus field as in Figure 4.20. Areas of the field where these coefficients are zero or slowly varying are isoplanatic. Similarly, isoplanaticism is implied when RMS wavefront error versus field is small in the presence of shifts and tilts (see Figure 4.21).

The second technique for achieving extreme isoplanatism is implemented by constraining Zernike coefficients versus field to be zero or slowly varying. This is done directly in the merit function used for optimization. Since first- and third-order wavefront properties need to be constrained, nine Zernike terms per field point must be constrained to equal the corresponding terms of neighboring field points. In this technique, the size of isoplanatic patches maybe increased by increasing the number of field points across the field. RMS wavefront error versus field may be used to check the outcome of such an optimization. This technique gives similar results to the first, and also produces lenses that perform well in the presence of assembly errors or errors in hologram placement.

4.7 Polytopic Filter

4.7.1 Introduction

The polytopic filter is an angular low-pass filter applied in the signal beam path in order to implement polytopic multiplexing [45]. During recording, the polytopic filter limits the spatial frequency band that will be recorded, thus minimizing the size of the holograms. During recovery, the polytopic filter blocks diffracted light from neighboring books of holograms, permitting substantial volumetric overlap among the books (i.e. polytopic multiplexing).

In the Tapestry drive implementation of the polytopic filter, the SLM and camera planes are imaged to the back focal plane of the storage lens with a 4F optical system. This provides access to an intermediate FT plane where the polytopic filter is placed. The filter is a small square aperture in a metal plate aligned on the optical axis. The filter is only slightly larger than the minimum (Nyquist) size required to pass the fundamental frequency components of the data pattern. Because the 4F relay lenses must work over a wide temperature and wavelength range, they are multi-element lens assemblies. An optical configuration of this type would be too big and expensive for a low-cost consumer product. In addition, as changes in the height of the medium vary the focal planes, a mechanical polytopic filter will vignette spatial frequency components and reduce SNR. Implementing a system without the relay lenses but with a filter in a Fourier plane requires that the filter be placed outside of the medium, which has significant disadvantages in terms of hologram size, medium consumption, etc. (see Chapter 15 for more details). For a consumer product, the filtering must clearly be done without the extra 4F relay. This effectively means that the filtering function must be accomplished on the image side of the storage lens rather than in FT space. There are two general approaches to filtering in image space that will be presented – mechanical filtering and interference filtering. Interference filters on curved surfaces are a particularly promising approach.

4.7.2 Current Polytopic Filter

The polytopic filter using the relay lens approach described above is shown in Figure 4.26(a).

Tolerances for the size and position of the polytopic filter were investigated by recording holograms and/or imaging though a 4F optical system while varying the size and position of the filter. Figure 4.27(a) shows a plot of SNR versus aperture size for a pixel matched system. SNR drops as the aperture size decreases below the whole zeroth order of the SLM (0.5 on scale). By the time the aperture is at Nyquist (0.25 in plot) the SNR is down 1.5 dB. However, this loss is entirely due to blurring of the image. Various equalization schemes were applied with the best getting back to the same SNR as the entire zeroth order (solid curve). This

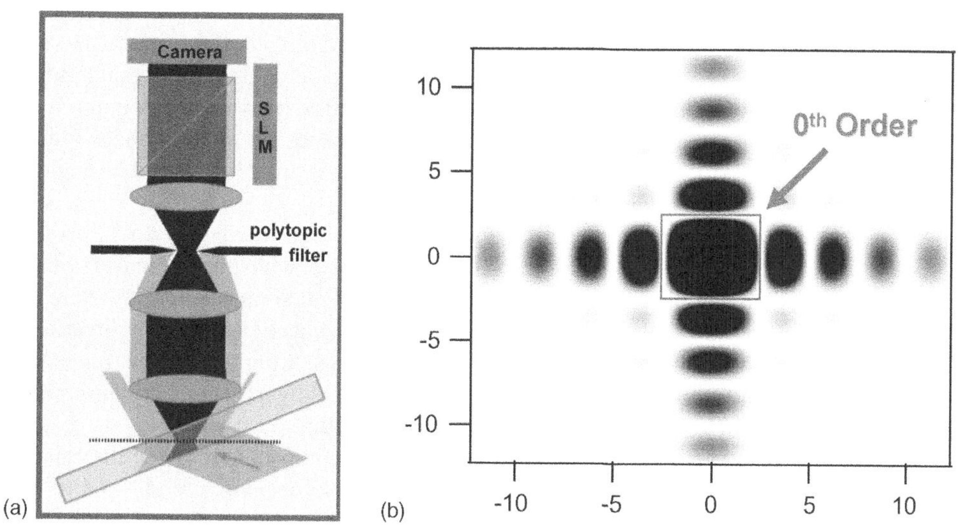

Figure 4.26 (a) Relay lenses and polytopic filter shown in the data path during recovery. (b) Diffractive orders in the Fourier plane of the SLM

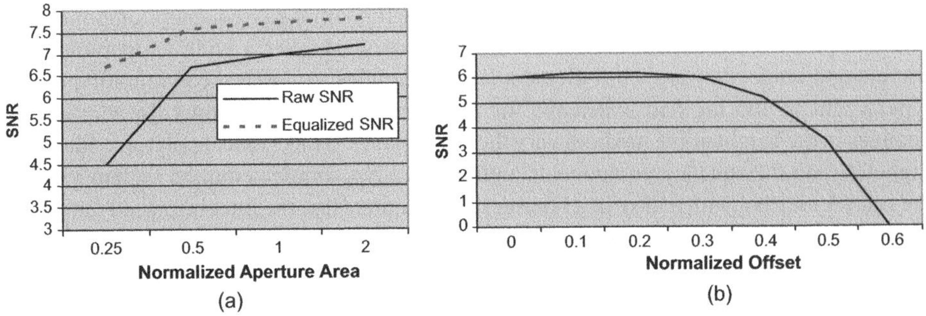

Figure 4.27 *(a) Plot of aperture size with raw data and various filtering schemes. (b) Misalignment of aperture along diagonal (same result along x or y) in the plane of the filter*

means that for a well designed oversampled detection scheme there is little SNR loss until the Nyquist limited area is reached. Figure 4.27(b) shows a plot of misalignment in the plane of the filter. The units of misalignment are normalized to the aperture size. The particular aperture used for this is twice the Nyquist area (1.4 linear). The tolerances to misalignment of the filter are quite large with ~30% relative misalignments being where the SNR starts to drop off.

Typical sizes for the polytopic filter might range from 1.1 to 1.3 Nyquist (linear dimension) in order to increase the tolerance to drive-to-drive misalignment during interchange.

4.7.3 Mechanical Filtering

The FT polytopic filter blocks light from unwanted neighboring holograms or higher order light that comes from the SLM. In the FT plane the larger angle associated with neighboring holograms and higher orders results in the light being located further off the optical axis and therefore easily blocked due to this light being spatially separated from the signal light. In image space the unwanted light (from higher orders of the SLM, unwanted holograms from adjactent books, or some scatter) spatially overlaps. However, the angular bandwidth of these components is still separate from the signal. Thus, by filtering light based on angle of propagation, a polytopic filter can be implemented.

The *acceptance angle geometry* may be understood by considering a camera pixel at the bottom of a hole. Figure 4.28 shows the geometry for a pixel with infinitesimal area at the bottom of an elliptical hole with minor axis length dx and major axis length dy. Here the acceptance angle in the x direction is atan(dx/r) and the acceptance angle in the y direction is atan(dy/r). The acceptance angle is therefore the angular spectrum of rays that propagate through the hole and reach the pixel. The acceptance angle of light that reaches the camera may be controlled by selecting the hole diameter and depth.

Figure 4.29 shows acceptance angle geometry for the complex case. Here the camera pixel has finite extent A_D and the pixel, as well as the hole opening, are allowed to tilt. Acceptance angle for this case is larger than for the case in Figure 4.28 because of the finite extent of the camera pixel. Acceptance angle, for this case, may be controlled by selecting hole diameter, hole depth, pixel size, and tilts of hole opening and pixel. Therefore, on read-

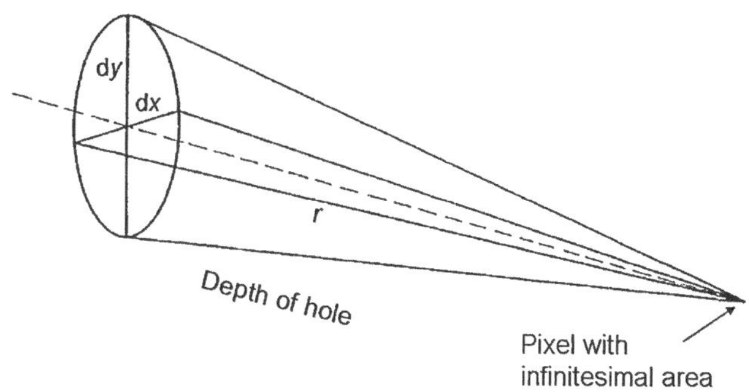

Figure 4.28 *Acceptance angle geometry – simple case*

out an angle filter can be implemented with macrotubes, lenslet arrays, and other such structures placed right on the camera. More details on mechanical filters for non FT filtering are available in Bell *et al.* [41]. For example, 3 μm wide square shaped tubes can be etched in silicon through a 41 μm depth and with a 4.6 μm pitch. Tubes can be matched to the CMOS pixels with a lithographically patterned etch. These angle filters are for use on the camera and therefore can function as polytopic filters on read-out.

4.7.4 Interference Filters

An interference filter consists of multiple thin layers of dielectric material having different refractive indices. There also may be metallic layers. In its broadest meaning, interference filters comprise also etalons that could be implemented as tunable interference filters. Interference filters are wavelength-selective by virtue of the interference effects that take place between the incident and reflected waves at the thin-film boundaries. InPhase designed a multilayer thin film that rejects high angles while passing low ones. The thin-film stack has about 100 layers, and was designed for a green wavelength. This works

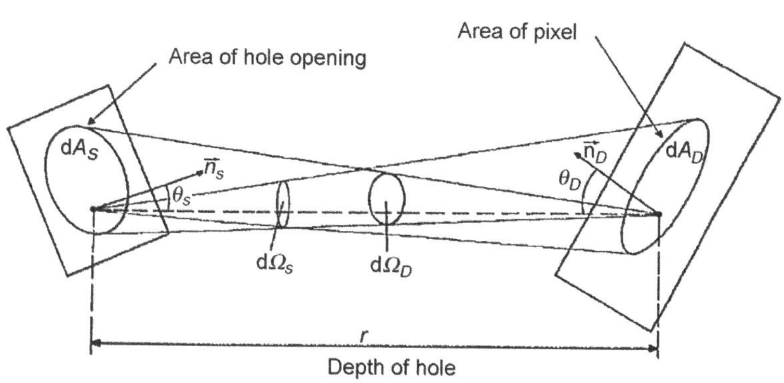

Figure 4.29 *Acceptance angle geometry – complex case*

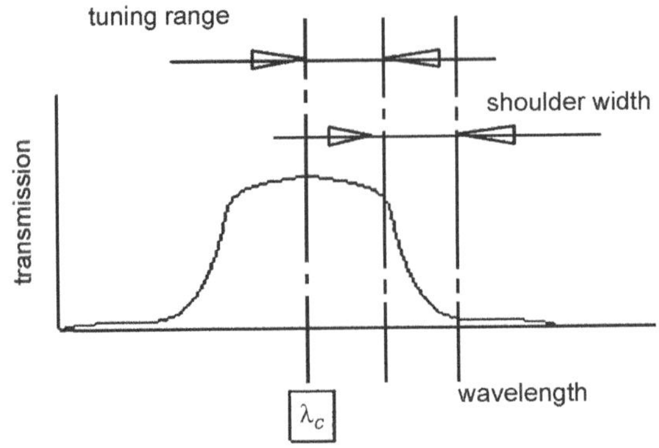

Figure 4.30 *Typical narrowband filter (transmission versus wavelength)*

for read-out, using a camera, but higher angles generated by higher orders of the SLM would not be blocked when recording as it has a limited angular stop band. Therefore, this filter is considered a read-out filter only. It would be possible to use such an element for recording if it were combined with a holographic element that filtered out higher diffracted orders. However, this looks to be a very difficult approach. The next embodiment using a curved surface appears to be a much more promising approach due to the use of mechanical amplification of the angles.

4.7.5 Thin Film Coating on Curved Surface

A simple Fabry–Perot narrowband filter placed on a curved surface may be used for polytopic filtering. The principle of operation for this configuration exploits the filter tuning effect in conjunction with the mechanical effects of the curved surface.

Typical transmission of such a filter is shown in Figure 4.30.

Narrowband filters are usually designed for normal incidence. When the angle of incidence of the incoming light deviates from normal incidence, the central wavelength of the filter (λ_c) decreases. This results in partial tunability of the filter. If λ_c is the central wavelength at an angle of incidence of $\tau < 20°$, λ_0 is the central wavelength at normal incidence, and n_e is the filter equivalent index of refraction, then:

$$\lambda_c = \lambda_0\left(1-\frac{\tau^2}{2n_e^2}\right) \tag{4.1}$$

Pidgeon and Smith [47] introduced the concept of equivalent index and have shown this idealization to accurately predict the shift in peak wavelength of practical Fabry–Perot narrowband filters. In practice, the value of n_e is intermediate between the high and low indices of the filter layers.

Figure 4.31 [48] shows the equivalent index versus order number for Fabry–Perot filters incorporating zinc sulfide ($n_H = 2.35$) and cryolite layers ($n_L = 1.35$). Using these conventional materials, tuning performance of a narrowband filter may be customized to give an equivalent index between $n_e = 1.4$ and $n_e = 2.2$.

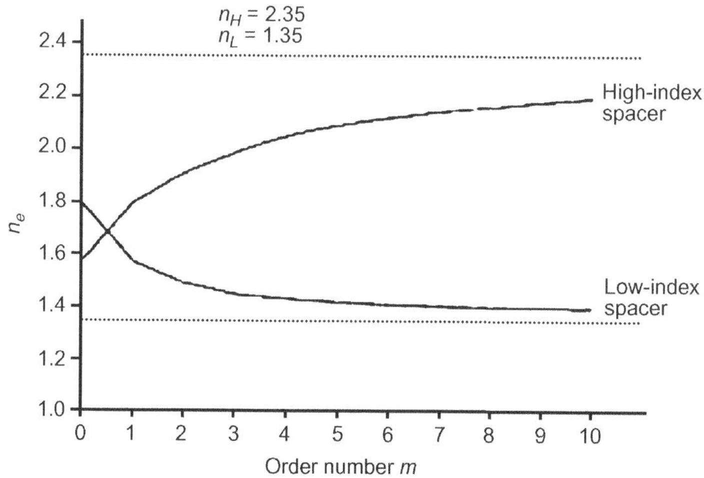

Figure 4.31 *Equivalent index versus order number for Fabry–Perot filters incorporating zinc sulfide (n = 2.35) and cryolite (n = 1.35) layers*

Figure 4.32 shows how a narrowband filter may be placed on a spherical surface to eliminate crosstalk and serve as a polytopic filter. Here, two adjacent books with Nyquist spacing value *nq* are considered. The addressed book is at the center of the coated spherical surface of radius ρ. In a practical system, this could be a glass surface belonging to the FT lens. Neighbor books, with NA = sin(θ), must have all rays extinguished by the spherical filter surface to eliminate crosstalk with the addressed book. The minimum angular separation between the rays of the neighbor book and the addressed book is shown by α.

Noticing that α is the angle between the neighbor book's extreme ray and the normal to the spherical surface, one can equate α to τ, the tuning angle. Crosstalk will be eliminated

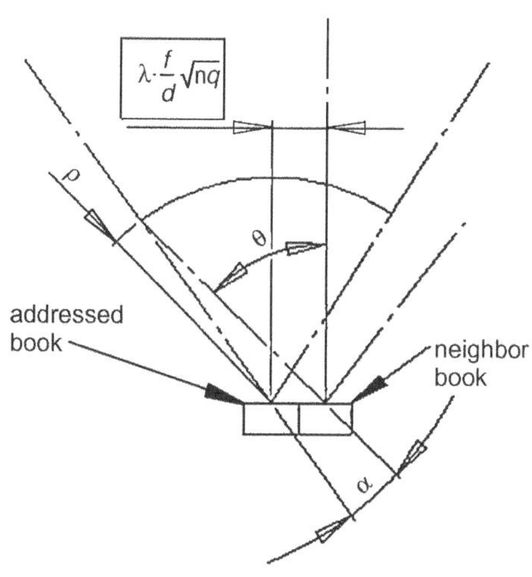

Figure 4.32 *Crosstalk geometry for thin film polytopic filter*

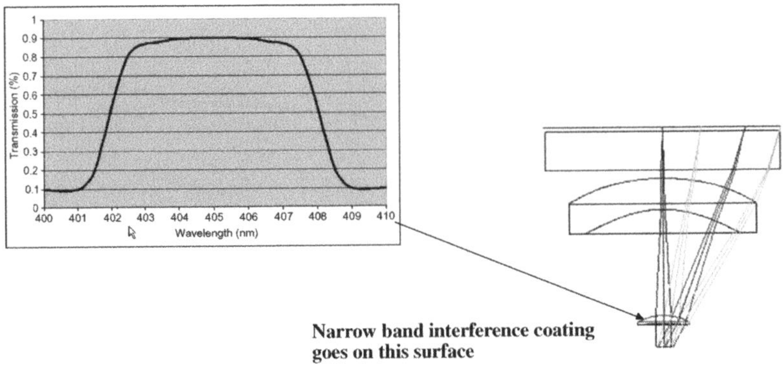

Figure 4.33 Diagram of coating and optical design of curved surface and lens in drive

when tilt α shifts the filter's passband to exclude the wavelength in use, that is, passband shift must be greater than tuning range plus shoulder width, see Figure 4.30. Since hologram tilt with respect to FT lens can diminish α, greater passband shift may be required in some cases.

Figure 4.33 is an FT lens that can be used as a quantitative example. This lens has focal length $f = 2.4$ mm and a concentric spherical surface with $\rho = 0.7$ mm. It is designed for 405 nm using 6.4 µm pixels. NA of this lens is 0.6. Book spacing, $nq = 1.5$ gives 186 µm between books. In this case $\alpha = 12.27°$. The tuning range for this system is 3 nm and shoulder width is expected to be 1.5 nm. An equivalent index, $n_e = 1.43$ makes $\lambda_c = 400$ nm. The 5 nm shift ensures the elimination of crosstalk because the system's tunable laser cannot produce 400 nm light. Reducing equivalent index to $n_e = 1.4$ would also allow 0.25° of hologram tilt for improved mechanical robustness. Referring to Figure 4.32, an $n_e = 1.4$ is readily obtained using conventional materials.

As the previous example shows, a thin film polytopic filter can be implemented simply using a Fabry–Perot narrowband filter on a curved surface with conventional materials. Other advantages include:

(1) Relay lenses are eliminated.
(2) Since relay lenses are eliminated, telecentricity of FT lens is unimportant.
(3) Critical alignment of hard polytopic filter is eliminated.
(4) Filtering is z-shift tolerant – no polytopic servo required or improved focus and media height tolerances with fixed polytopic filter.
(5) Filtering is insensitive to media or camera tilts.
(6) FT lens can provide the curved surface for the interference filter.
(7) Stray light is filtered prior to camera.

Acknowledgements

At InPhase we gratefully acknowledge contributions from the following people: Jason Ensher, Paul Smith, Kirk Cook, Susan Hunter, Mark Ayres, Chris Bergman, Bernard Bell, Alan Hoskins, Michael Derks, David Dodds, Mike Hartman, Vlad Krneta, Paul Morrison, Brock Myers, Dean Peters, Timothy Reed, David Samuels and Aaron Wegner.

The laser work was jointly done with Nichia. From Nichia we gratefully acknowledge: Masaki Omori, Shigeki Okauchi, Hideki Kondo, Tadaaki Miyata, and Naoki Mori. Thanks to Tomonori Morizumi for designing and providing the AR coated LD, Takashi Sasamuro, Hidenori Matsuo, and Keiji Fujita for evaluating the AR coated LDs and ECLDs, and Shingo Omura and Toshio Matsushita.

We also wish to thank Hal Zarem and Clint Carlisle at Silicon Light Machines, Mark Handschy and Michael O'Callaghan at Displaytech, Roel Aerts, Joost Seijnaeve, Cliff Drowley, and Donato Montanari at Cypress, and Flavio Pardo at Bell Laboratories for their help in explaining their respective technologies.

We gratefully acknowledge the contributions of Tatsuro Ide, Toshiki Ishii, Ken-ichi Shimada, Takeshi Shimano, Yukinobu Tada, and Masahiko Takahashi from Hitachi.

We sincerely thank Alps Electric Co., in particular, Yoshihiro Someno, Naoyuki Tokuchi, Shinji Mitsuya, Takuya Nagai, Yuichi Umeda, Itaru Hatano, Yasuyuki Hirose, and Motohiro Shimaoka.

References

1. N. Nakayama, *et al.*, Room-temperature continuous operation of blue-green laser-diodes, *Electronics Lett.*, Vol. 29, No. 16, pp. 1488–1489 (1993).
2. T. Laulia, *et al.*, Tunable external-cavity diode laser at 650 nm based on a transmission diffraction grating, *Appl. Opt.*, Vol. 41, No. 27, pp. 5632–5637 (2002).
3. M. Merimaa, *et al.*, Compact external-cavity diode laser with a novel transmission geometry, *Opt. Commun.*, Vol. 174, pp. 175–180 (2000).
4. M. de Labachelerie and G. Passedat, Mode-hop suppression of Littrow grating-tuned lasers, *Appl. Opt.*, Vol. 32, No. 3, pp. 269–274 (1993) with correction in *Appl. Opt.*, Vol. 33, pp. 3817–3819 (1994).
5. L. Nilse, H.J. Davies and C.S. Adams, Synchronous tuning of extended cavity diode lasers: the case for an optimum pivot point, *Appl. Opt.*, Vol. 38, No. 3, pp. 548–553 (1999).
6. G. Zhou, F. Mok, and D. Psaltis, Beam deflectors and spatial light modulators for holographic storage application, in *Holographic Data Storage*, H.J. Coufal, D. Psaltis and G. Sincerbox,eds. Springer-Verlag, New York, pp. 241–258, 2000.
7. J.T. LaMacchia and D.L. White, Coded multiple exposure holograms, *Appl. Opt.*, Vol. 7, pp. 91–94 (1968).
8. C.B. Burckhardt, Use of a random phase mask for the recording of Fourier transform holograms of data masks, *Appl. Opt.*, Vol. 9, pp. 695–700 (1970).
9. M.R. Ayres and R.R. McLeod, Intra-signal modulation in holographic memories, *Joint Int. Symp. on Opt. Memories and Opt. Data Storage*, Waikoloa, Hawaii, July (2008), paper ThC3.
10. M.J. O'Callaghan, *et al.*, Spatial light modulators with integrated phase masks for holographic data storage, *Opt. Data Storage Conf.*, Montreal, Canada (2006), pp. 23–25.
11. D. Armitage and D.K. Kinell, Liquid-crystal integrated silicon spatial light modulator, *Appl. Opt.*, Vol. 31, pp. 3945–3949 (1992).
12. M. Handschy and M.J. O'Callaghan,Spatial light modulators with changeable phase masks for use in holographic data storage, US Patent Application US 2007/0247687.
13. M. Inoue, Magneto-optic spatial light modulators for collinear holography, *Int. Symp. on Opt. Memories*, Singapore, October (2007), paper Tu-F-1.
14. M. Inoue, Magneto-optic spatial light modulators for ultra-high density storage with collinear holography," *ISO/IEC JTC 1/SC 23, Holographic Storage 12th SC 23 Plenary Meeting*, Kyoto, Japan (2005).
15. B.A. Kast, *et al.*, Implementation of ternary phase amplitude filters using a magnetooptic spatial light modulator, *Appl. Opt.*, Vol. 28, pp. 1044–1046 (1989).

16. S. Campbell, *et al.*, Digital holographic memory using a digital micromirror device SLM and a CMOS active pixel camera, *Opt. Data Storage Conf.*, Aspen, Colorado (1998), paper WC5.
17. D.R. Collins, *et al.*, Deformable mirror device spatial light modulators and their applicability to optical neural networks, *Appl. Opt.*, Vol. 28, No. 22, pp. 4900–4907 (1989).
18. J.I. Trisnadi, *et al.*, Overview and applications of Grating Light Valve based optical write engines for high-speed digital imaging, Photonics West 2004 - Micromachining and Microfabrication Symp., San Jose, California (2004), paper 5348-05.
19. F. Pardo,*et al.*, Mirror array characteristics, *Solid State Sensor, Actuator and Microsystems Workshop*, Hilton Head Island, South Carolina (2006).
20. S.D. Senturia, *et al.*, Programmable diffraction gratings and their uses in displays, spectroscopy, and communications, *J. Microlithogr. Microfabricat. Microsystems*, Vol. 4, pp. 041401–041406 (2005).
21. P. Willems, High Speed CMOS sensors, Planet Analog (www.planetanalog.com) and Cypress publications (www.cypress.com).
22. R. Aerts,1.4Gpix/s, 2.9Mpix CMOS image sensor for readout of holographic data memory, *IEEE Image Sensor Workshop,* Ogunquit, Maine (2007), paper P10.
23. B. Diericks,Buried, fully depletable, high fillfactor photodiodes, US Patent 7,253,019, August 7, 2007.
24. E.R. Fossum, *et al.*, Active pixel sensor array with multiresolution readout, US Patent 5,949,483, September 7, 1999.
25. S.A. Buhler, *et al.*, Highly uniform five volt CMOS image photodiode sensor array with improved contrast ratio and dynamic range, US Patent 5,742,047, April 21, 1998.
26. B. Diericks, *et al.*, Method and pixel for performing correlated double sampling, US Patent Application US 2005/0270394.
27. C. Shuman, *et al.*, Obliquity corrected scanning, using a prism mirror, US Patent 7,149,015, December 12, 2006.
28. C. Shuman,Obliquity correction system, US Patent 6,914,704, July 5, 2005.
29. B. Sissom, Miniature flexure based scanners for angle multiplexing, US Patent 7,336,409, February 26, 2008.
30. B. Sissom and R. Harris, Miniature single actuator scanner for angle multiplexing with circularizing and orthogonal scanning capability, US Patent 7,405,853, July 29, 2008 and US Patents 7,453,618, November 18, 2008.
31. F. Pardo, *et al.*, Advanced MEMs spatial light modulators for communications, imaging, and targeting, *NASA Tech Days* (2006); see optics.nasa.gov/tech_days/tech_days 2006/docs/53%20Lucent%20Advanced%20MEMS%20SLM%20for%20Communications%20Imaging%20and%20.pdf.
32. D.P. Resler, *et al.*, High efficiency liquid crystal optical phase array beam steering, *Opt. Lett.*, Vol. 21, No. 9, pp. 689–691 (1996).
33. Y. Takashima and L. Hesselink, Designs and tolerances of 2-element N.A. 0.8 objective lens for page based holographic storage, *Joint Int. Symp. on Opt. Memories and Opt. Data Storage*, Waikoloa, Hawaii, July (2008), paper TuB04.
34. E. Chuang, *et al.*, Demonstration of holographic ROM mastering, replication, and playback with a compact reader, *Int. Symp. on Opt. Memories*, Singapore, October (2007), paper Mo-D-6.
35. B. Sissom, A. Hoskins, T. Deslis and K. Curtis, Isoplanatic lens design for phase conjugate storage systems, *Optical Data Storage Conf.*, Waikolua, Hawaii, July (2008), paper TuP12.
36. W.T. Welford, *Aberrations of Optical System,* Adam Hilger Ltd, Bristol, 1986.
37. J.W. Goodman, *Introduction to Fourier Optics*, 3rd edn. Roberts & Company, Greenwood Village, Colorado, 2005.
38. R.S. Longhurst, *Geometrical and Physical Optics*. Longman, London, 1957.
39. T. Smith, A large aperture aplanatic lens not corrected for colour, *Trans. Opt. Soc. London* Vol. 24, p. 22 (1922).
40. W.T. Welford, The most general isoplanatism theorem, *Opt. Commun.*, Vol. 3 No. 1, pp. 1–6 (1971).

41. H.H. Hopkins, Canonical coordinates in geometrical and diffraction image theory, *Jpn. J. Appl. Phys.*, Vol. 4, Suppl. 1, p. 31 (1965).
42. W. Smith, *Modern Optical Engineering*, 2nd edn. McGraw-Hill Inc., New York, 1990.
43. J. Wyant and K Creath, Basic wavefront aberration theory for optical metrology, in *Applied Optics and Optical Engineering*, R. Shannon and J.C. Wyant,eds. Academic Press, Inc., New York, Vol. XI, Ch. 1, 1992.
44. A. Hoskins, *et al.*, Tolerances of a page-based holographic data storage system, *Opt. Data Storage Conf.*, Portland, Oregon (2007), paper WB1.
45. K. Anderson and K. Curtis, Polytopic multiplexing, *Opt. Lett.*, Vol. 29, pp. 1402–1404 (2004).
46. B. Bell, *et al.*, Non FT plane angular filters, US Patent Application 11/872,051.
47. C.R Pidgeon and S.D. Smith, Resolving power of multi-layer filters in non-parallel light, *J. Opt. Soc. Am.*, Vol. 54, p. 1459 (1964).
48. H.A. Macleod, *Thin Film Optical Filters*. Institute of Physics, London, p. 287, 2001.

5

Materials for Holography

Kevin Curtis, Lisa Dhar and William Wilson

5.1 Introduction

The primary challenge throughout the history of holographic data storage (HDS) has been the development of a recording material that could enable the promise of this technology. Holographic storage requires a material that can rapidly image the complex optical interference patterns generated during the writing step such that the imaged patterns are (i) generated with adequately high contrast, (ii) preserved (both dimensionally and in their contrast) during subsequent, spatially co-located writing, (iii) are unchanged by read-out of the holograms, (iv) are robust to environmental effects, and (v) survive over long periods (many years) of time.

Early efforts [1–3] in HDS sought to demonstrate the basic capabilities of the approach, i.e. the ability to record and recover data pages with high fidelity and to multiplex to achieve storage density. Because of this initial focus on the feasibility, efforts used photorefractive crystals which were well known, through the electro-optic effect, to translate optical patterns into refractive index gratings within the material. While valuable demonstration vehicles, photorefractive crystals suffer from limited refractive index contrast, slow writing times, and volatility (the read-out of holograms leads to erasure of recorded information). Attention soon turned to more commercially realistic alternatives such as photochromic, photoaddressable, and photopolymer materials where, similarly, optical interference patterns could be imaged as refractive index modulations. Photopolymers [4–9] quickly became the leading candidates as, in addition to their imaging properties, they exhibited high contrast and recording speeds and were nonvolatile. The first photopolymer systems to be used were borrowed from the holographic display market where single (or small numbers

Holographic Data Storage: From Theory to Practical Systems Kevin Curtis, Lisa Dhar, Adrian Hill, William Wilson and Mark Ayres

Table 5.1 Key system parameters and the related material characteristics

High data density	• Dynamic range • Millimeter thickness
High speed record and read performance	• Dynamic range • Photosensitivity (record)
High quality record and read performance	• Dimensional stability – no shrinkage • Optical flatness • Low scatter – low levels of noise in data recovery • Nonvolatile read-out • Linear spatial response
Long shelf and archive life	• Environmental/thermal stability
Low cost and high quality manufacturability	• Heat and solvent free manufacturing process • Short manufacturing cycle time

of) holograms are recorded in thin (typically $\leq 10\,\mu m$) materials. However, because these materials were optimized for display applications, i.e. recording a small number of high diffraction efficiency holograms, they cannot be used without modification. Display applications are relatively insensitive to dimensional changes, optical quality, and scatter; further development of this materials platform was required to meet the more stringent requirements for HDS.

The preliminary work in the different types of materials, especially in photopolymers, established a basic understanding of materials requirements. The dependence between the performance, (such as storage density, transfer rate, read-out fidelity, environmental robustness, and lifetime), and material characteristics, (such as refractive index contrast, photosensitivity, dimensional stability, and optical quality), could be outlined. Table 5.1 maps the critical performance parameters against material properties.

The storage density is a function of the number of holograms multiplexed into a volume of the recording media. Determining this number are several material parameters – principally its refractive index contrast and thickness. When multiplexing, the total available refractive index contrast of a material is partitioned among the multiple holograms recorded. As detailed in Chapter 2, the fractional index contrast of each hologram and material thickness determine its diffraction efficiency or signal strength. Given a minimum required signal strength defined by the system hardware, the combination of the material's index contrast and thickness place an upper limit on the number of holograms that can be multiplexed in a particular volume, i.e. the storage density. In addition, this number is also geometrically determined by the thickness. For example, as discussed in Chapter 2, in angle multiplexing, through the Bragg effect, the angular spacing between holograms is primarily determined by the thickness of the material which in turn yields the number of holograms that can be stored throughout a defined angular range.

The recording and read-out transfer rates depend on the energy efficiency of the material – how well it uses light to image the optical interference patterns (the photosensitivity) and to convert incoming light into signal (the diffraction efficiency). High photosensitivity results in rapid recording rates in that short (fast) exposures from a low power diode based laser can generate holorams. The diffraction efficiency or signal strength of each determines the read-out rate again highlighting the importance of the material's refractive index contrast and thickness.

Material properties such as dimensional stability, optical quality, scatter level, linearity and volatility affect the fidelity of the recording and read-out process. The dimensional stability of the material is reflected in the resistance of the material to shrinkage or expansion during the recording process and thermal changes.

For commercial systems, the manufacturing process and usage must also fit the needs of the application. For example, archival storage demands long (>50 years) lifetimes (i.e. data retention after recording). Adequate shelf life is also required. (The media has to last for several years after being manufactured but before being used to record data. It also has to survive shipping transients before recording.) Thermal or solvent-based post-recording processing (common for display materials) should be avoided. Finally, the manufacturing must be inexpensive. This requires the use of inexpensive plastic substrates, and fast fabrication times for high volume manufacturing.

Table 5.2 gives a comparison of the characteristics of the various candidate material systems. Later sections of this chapter describe the most commonly used material systems in detail.

5.2 Requirements for Materials for HDS

This section reviews the main characteristics of the recording material in more detail.

5.2.1 Index Change (*M/#*)

The diffraction efficiency of multiplexed holograms scales with the *M/#* of the material [10]. The *M/#* depends on the material index change and the thickness of the media. *M/#* was precisely derived for photorefractive materials and now is used for all materials defined by the equation:

$$\eta = \frac{M/\#^2}{N^2} \tag{5.1}$$

where η is the diffraction efficiency of the holograms, N is the number of holograms multiplexed in the same volume and *M/#* is the constant that relates the two quantities. *M/#* is critical because as you multiplex holograms (increase storage density) the diffraction efficiency goes down as $1/N^2$, (the number of superimposed holograms.) As shown in Figure 5.1, the diffraction efficiency of the holograms must be higher than the scatter noise generated from reading out the holograms. For a commercial drive, the typical diffraction efficiencies required are of order 1.4×10^{-3} which sets the required *M/#* at 30 for a 1.5 mm thick recording layer to achieve the capacity desired.

5.2.2 Dimensional Stability

Most recording processes such as a photo-initiated polymerization result in some physical shrinkage and/or bulk index change of the material system used. These recording based effects evolve as more holograms are recorded. These changes can be nonlinear across the multiplexed holograms (data pages) and are not desirable. If the shrinkage and bulk index

Table 5.2 *Characteristics of different material classes*

	Dynamic range	Photo sensitivity	Dimensional stability	Thickness	Optical quality	Nonvolatile/ post-processing	Media cost
InPhase Tapestry™ media	✓	✓	✓	✓	✓	✓	✓
Photorefractive crystals (e.g. $LiNbO_3$)	×	×	✓	✓	✓	×	×
Photorefractive polymers	×	×	×	×	–	×	×
Inorganic glasses	×	×	✓	✓	–	–	×
Biological/photochromic	×	×	–	–	–	×	×
Conventional photopolymers	–	✓	×	×	×	✓	–

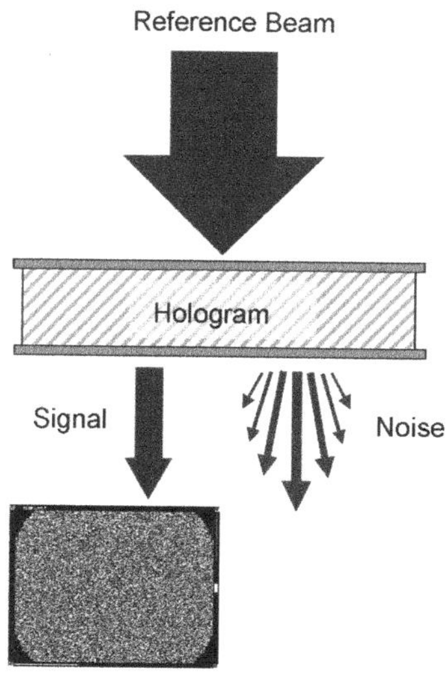

Figure 5.1 Diagram of signal and scatter

changes are small enough, they can be compensated for by changing the wavelength and angle of incidence of the reference beam, (if they are multiplexed with a plane wave reference). The shrinkage causes local deformation of the media impacting optical quality of the recorded images.

Other dimensional changes are also very important. Temperature changes cause the media to expand and contract. This distortion affects our ability to reconstruct the recorded images with high fidelity using the original reference beam parameters. Depending on the shape and construction of the media (i.e. disk shaped) the expansion and contraction can also be anisotropic. This expansion and contraction can be compensated for by reference beam wavelength and angle changes (see Chapters 14 and 15 for plane wave references). Unfortunately, lasers have a limited tuning range and increasing the tuning requirement can make the laser and the system more complex and expensive. Thus it is desirable to design the media platform to have low thermal expansion coefficients to minimize the required wavelength tuning. In addition, if the substrates and recording material are different materials the anisotropic thermal expansion can be decreased by making the substrate and media thermal expansion coefficients more similar. The anisotropic expansion and contraction depends on the media geometry. Thus a large disk will be more anisotropic (require more tuning) than a small card even if both are made with the same substrate and recording layers.

Humidity is another mechanism that can affect the media volume. Again, these changes can be compensated for by angle and wavelength changes (Chapters 14 and 15), if the optical quality of the media is not distorted. This distortion can happen if moisture is taken up in a nonuniform way. Thus the edges of a polymer disk should be sealed to prevent moisture uptake from the sides. This does not prevent moisture from entering the disk structure through the plastic substrates; it just makes sure that the uptake is uniform across the media.

5.2.3 Photosensitivity

High photosensitivity is required to enable fast transfer rates even when using diode lasers and highly inefficient optical systems. Photosensitivity can be defined as the slope at the bottom of the recording curve or the energy required to consume 80% of the *M*/#. (See Chapter 8 for more details.) However, for transfer rate and system reasons it is better to have the exposure curve to 80% be linear to within a factor of ~1.6 or less. If you have very nonuniform recording sensitivities then the average transfer rate (actual rate seen by the user) will be lower even if the sensitivity as measured by the first part of the curve shows the media to be very sensitive. In addition, the transfer rate and exposures will vary significantly as more holograms are recorded and more *M*/# is used. A change in exposures is needed to keep the diffraction efficiency of all the holograms approximately equal even in the presence of media nonlinear behavior with exposure energy (see Chapter 13). Large variations in exposure times are hard for the system to manage. In general at InPhase, we have found that keeping the ratio of longest exposure to shortest exposure to within a factor of three makes for reasonable record reproducibility and system management. Examples of materials that typically have a very large ratio of longest to shortest exposures are photorefractive materials where the exposure schedule is exponential.

5.2.4 Scatter

Media scatter and scatter in general are very important parameters in HDS systems. The natural Rayleigh scatter of the media typically will be a large component of scattered light in the system. This baseline scatter is due to the recording material and substrates. The substrates need to be of high optical quality material for the InPhase Tapestry drive since the light passes though the substrates on writing and reading. In addition the physics of the recording processes also affect the extent of the Rayleigh background given random index fluctuations in the volume determine the Rayleigh contribution. Ideally the recording layer material should be low scattering and very homogenous. If polymers are used, the starting materials should be filtered to clean the materials before manufacturing. The drive itself has to be designed to have clean and uniform illumination of the media as holographic media will record the spatial intensity of the beams. Intensity driven random spatial variations will, through the Rayleigh mechanism, cause unintended scattering centers to be recorded into the media.

5.2.5 Absorption Properties

Absorption of the light is required for recording but can result in signal loss during read-out. During recording, ideally the absorption would be uniform throughout the media volume. Typically different wavelengths are used to record at different temperatures to pre-compensate for any temperature variability of the media extinction coefficient. Low to modest absorption is desired to maximize the record uniformity throughout the volume resulting in optimal Bragg selectivity or hologram selectivity for multiplexing holograms. On read-out, ideally the media would not absorb the light so that the maximum amount of light can be diffracted by the holograms, and with no absorption any, additional light exposure would have no chance to cause any media changes.

5.2.6 Birefringence

Hologram formation does depend on polarization, so keeping the two recording beams polarized correctly can improve the measured sensitivity of the material. While most photopolymers are homogeneous and therefore birefringence free, substrate materials such as polycarbonate can have significant birefringence. This effect can also decrease the read-out light efficiency and thereby hurt read transfer rate. In the InPhase professional drive, the light is steered to the detector by phase conjugate reconstruction and the use of a polarizing beam splitter. With our current implementation of phase conjugate reconstruction the beams pass through the media structure twice allowing for even small amounts of birefringence to change the polarization and decrease the light that makes it back to the camera.

5.2.7 Optical Quality

Using phase conjugate geometry the optics in the drive can be made simpler and more robust with the image signal to noise ratios (SNRs) improved. However, to have good phase conjugate read-out using a plane wave, the plane wave must have reasonably good wavefront quality – see Chapter 14 for tolerances. In addition, for large data pages the optical quality (flatness) of the media over the hologram area can also affect the SNR of the recovered hologram. The Zerowave® manufacturing process described in Chapter 7 results in very high optical quality media even when using plastic substrates of DVD quality.

5.2.8 Manufacturability

Finally, the media must be able to be manufactured in high volume inexpensively. We detail the key issues driving the material and material manufacturing costs in Chapter 8. For example, a CD costs (without licensing), approximately 10 US cents to make and can be made in 2–3 seconds.

5.3 Candidate Material Systems

5.3.1 Photorefractive Materials

In broad terms, the photorefractive effect is defined as the generation of a modulation in the refractive index of a material by light. Most commonly, photorefractive materials refer to inorganic ferroelectric crystals such as doped lithium niobate ($LiNbO_3$) or organic photoconductors where, through the electro-optic effect, light can induce a change in index driven by charge redistribution [11–13].

The physics of this phenomenon is very simple; illumination excites electrons of a donor impurity into the bulk crystal conduction band. This leads to diffusion of mobile carriers into the nulls of the interference grating creating a space-charge electric field with the k-space characteristics of the grating. Using the electro-optic properties of the crystal, the electric field in turn leads to a density redistribution or refractive index grating that is 90° out of phase with the original interference pattern. Figure 5.2 illustrates this overall process while Figure 5.3 shows the excitation mechanism.

In inorganic crystals, linearly polarized light is typically used to generate the optical interference pattern with the polarization with respect to the crystal axes carefully

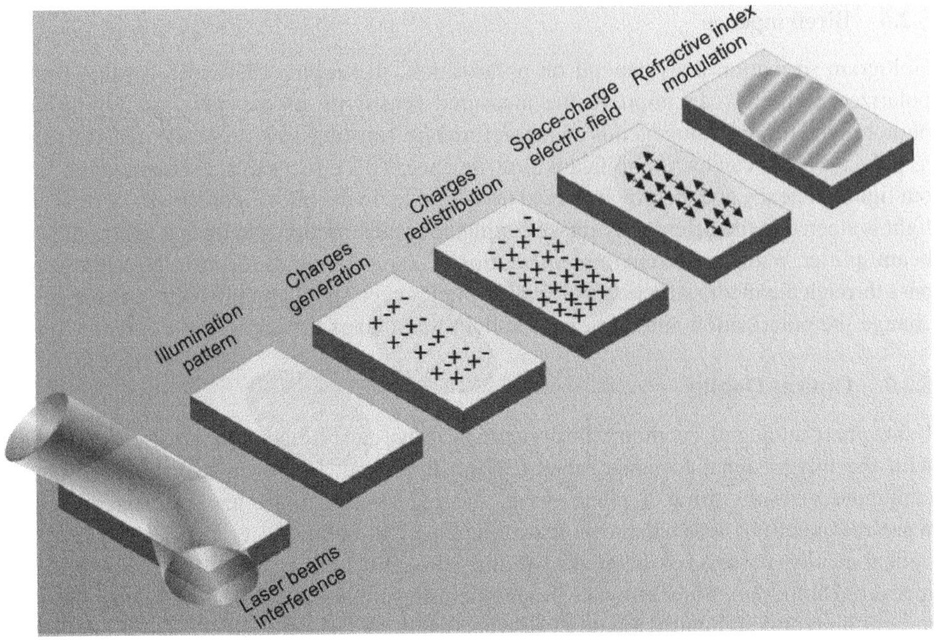

Figure 5.2 Overall photorefractive process

controlled (to maximize coupling to the electro-optic axis of the media). Organic materials designed as photorefractive polymers typically contain dipolar chromophores with an optical anisotropy which aligns with an external electric field. Here the second-order nonlinearity of the aligned chromophores gives us the electro-optic coupling needed.

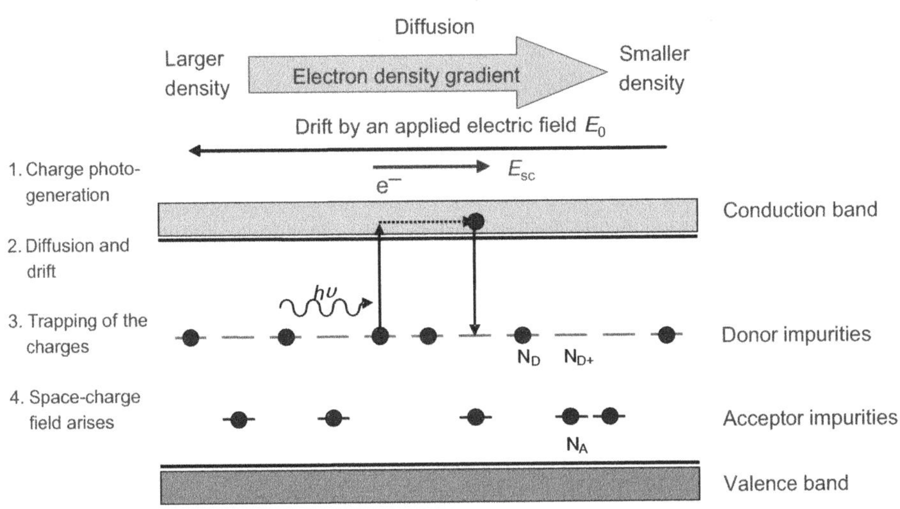

Figure 5.3 Details of the excitation process

The photorefractive effect in crystals was discovered in 1966 by Ashkin *et al.* [14] and first exploited for HDS in 1968 by Chen *et al.* [15]. Less than a decade later, Staebler *et al.* recorded 500 angle-multiplexed volume phase holograms in iron-doped $LiNbO_3$ [16] demonstrating the density capabilities of HDS. The material, while far from meeting all of the requirements for holographic storage, supported much of the research and development efforts of this time due to its instantaneous response, adequate (for demonstration purposes), refractive index contrast, absence of recording-induced dimensional changes, and erasability (which allowed reuse and the potential for rewriteable applications).

However, several issues – the dominant ones being the volatility of recorded holograms and the material's low photosensitivity – prevented these types of materials from fully enabling holographic storage. Thermal and electrical (for ferroelectric crystals) fixing can impart long-time and thermal stability [17–20] but involve extra steps not practical for a commercial system. Approaches such as two-photon excitation [21,22] and two-wavelength recording [23] have been proposed to overcome volatility but can introduce complications into the recording process. The most recent work on inorganic photorefractive materials has focused on tuning the dopants of $LiNbO_3$ to optimize the *M/#* and sensitivity of the material [24–27].

The photorefractive effect in organic materials was first reported in 1990 [28] and within 4 years their refractive index contrast matched that of inorganic materials. These materials consist of dipolar chromophores with an optical anisotropy dispersed within a matrix. Here again the optical interference pattern generates holes in the peaks of the grating. An externally applied electric field induces small-scale diffusion of the holes setting up an internal space-charge field that mimics the grating pattern. The nonlinear chromophores respond to the internal field and yield a refractive index pattern. Much work has focused on the optimal design of the chromophore and host material. The photorefractive effect has been demonstrated in materials such as polymers [29–31], liquid crystals [32] and organic glasses [33,34] where photoconductivity is provided by the matrix and the optical nonlinearity by the guest molecules. Some specialized systems include a photorefractive polymer composite [35], fully functionalized polymer systems [36], and organic glasses in which the chromophore acts as the charge transporting species, the optically birefringent species and additionally as the amorphous host [30].

Organic photorefractives have been of interest not only for their applications in data storage but also in optical processing where phenomena such as beam amplification and switching are employed and for displays [37]. Although the materials can exhibit high refractive index contrasts ($\Delta n \sim 0.01$), their use for storage is limited due to the high poling voltages ($\sim 50\ \mathrm{V\,\mu m^{-1}}$) required to induce charge diffusion. Because of the magnitude of the required field, a usable thickness is $\sim 200\ \mu\mathrm{m}$ thereby limiting the achievable storage density. In addition, volatility remains an issue with recorded materials typically stored in the dark to preserve the holograms. Erasable digital data storage at 0.5 Mbit $\mathrm{cm^{-2}}$ density was demonstrated in an organic photorefractive glass based on dihydropyridines with read-out stability over 6 h at room temperature [30].

5.3.2 Photoaddressable Systems

Photoaddressable systems were introduced in the 1990s by Bayer [38,39]. In these materials, holographic writing occurs through photoinduced birefringence. The materials

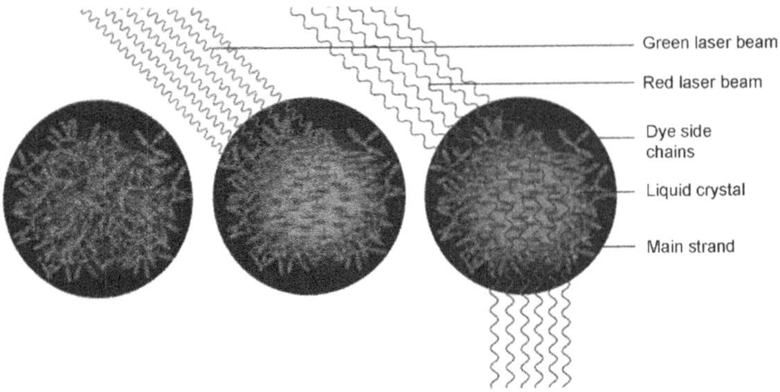

Figure 5.4 *Diagram of photoaddressable process*

typically consist of copolymer systems with alternating side groups of an isomerizing chromophore and a stabilizing mesogenic chain. The photoinduced isomerization of the chromophore leads to an orientation of the mesogenic side group which in turn leads to orientation of the polymer backbone resulting in a birefringence. The mechanism is shown in Figure 5.4 and a representative polymer main chain is shown in Figure 5.5.

These elegant materials can be designed to exhibit high refractive index contrast and long time and high temperature stability but are limited for data storage applications by their low photosensitivity and high optical absorption. In addition, because photoaddressable systems do not 'saturate', illumination with light of the same wavelength as used during recording can lead to continued reorientation and therefore volatility of recorded holograms.

These materials are currently primarily used for security applications or for smart cards. In collaboration with Optostor in the early 2000s, Bayer introduced a desktop reader and card using these photoaddressable polymers [40].

5.3.3 Photochromic Systems

Holographic recording in photochromic systems relies on direct absorption by chromophores in the material. These materials typically image the optical interference pattern of holographic recording as absorption changes within the material or phenomena such as

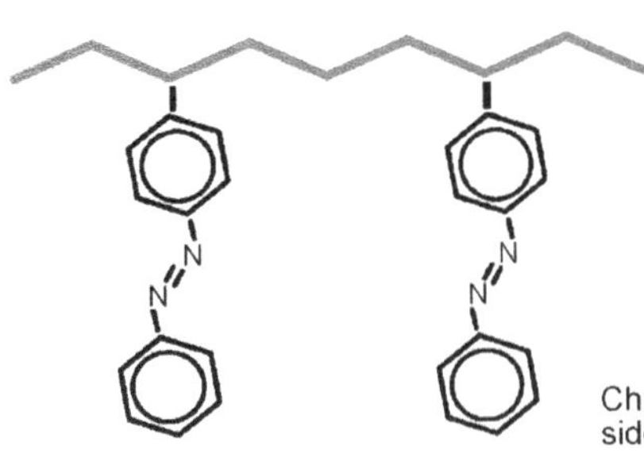

Figure 5.5 *Example of a photoaddressable polymer*

modulations in birefringence resulting from absorption-induced isomerization of the chromophore.

The 1990s witnessed much work in holographic recording in azo-benzene containing systems where the cis–trans isomerization of the material allowed holograms to be imaged into the material [41]. The European company Optolink was formed to further develop the materials. However, as with any absorption-based material, issues such as low photosensitivity, volatility (at the recording wavelength), and high optical absorption limit their utility for holographic data storage.

More recently, GE has developed photochromic materials based on chromophores within polycarbonate hosts [42–44]. Because of the issues of volatility, GE has targeted these materials for bitwise applications where a second wavelength can be used for read-out. Bitwise storage is discussed in Chapter 1.

5.3.4 Photopolymer Systems

These materials are essentially photodefinable polymer solid solutions. Photopolymers consist of a photoreactive, polymerizable system dispersed within a polymer host. During holographic recording, the optical interference pattern initiates a pattern of polymerization in the photoreactive system – polymerization is induced in the light intensity maxima of the interference pattern while no polymerization occurs in the nulls. This patterned polymerization sets up a concentration gradient in the unreacted species. Unpolymerized species diffuse from the nulls to the maxima of the interference pattern to equalize their concentration in the recording area, creating a refractive index modulation set up by the difference between the refractive indices of the photosensitive component and the host material [45–47]. Figure 5.6 shows the steps in the formation of the refractive index modulation.

The polymerization that occurs during recording changes the dimensions and the bulk refractive index of the material [48]. Changes such as these, if not controlled, can distort the

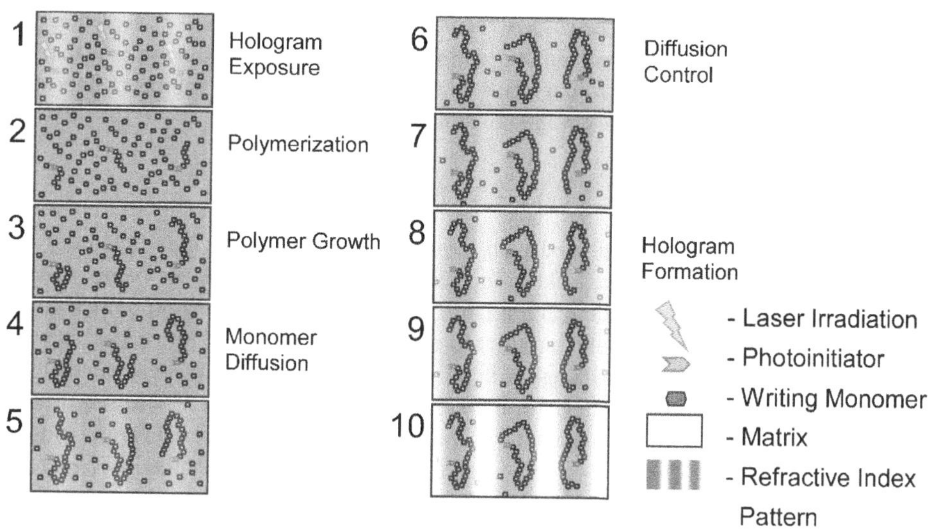

Figure 5.6 Photopolymer recording mechanism

imaging process, degrade the fidelity with which holographic data can be recovered, and ultimately limit the density of data the material can support. The design of photopolymer media for holographic storage applications must therefore balance the advantages of photopolymers (photosensitivity and large refractive index modulations) against the impact of the changes that accompany the polymerization.

There have been two main approaches to designing photopolymer systems for HDS systems. One approach, pioneered by Polaroid and then later Aprilis, Inc., uses a low-shrinkage chemistry known as a cationic ring-opening polymerization (CROP) to control the dimensional changes that occur during holographic recording [49]. These materials consist of a photoacid generator, sensitizing dye, CROP monomers and binder [50]. The CROP monomers typically have a functionality of greater than two and are based on cyclohexene oxides. In this strategy a pre-imaging exposure is used to increase the viscosity of the material to prepare for holographic recording. The CROP imaging chemistry is also tuned for high photosensitivity.

The other approach, introduced by Bell Laboratories, Lucent Technologies, and InPhase Technologies, is known as a two-chemistry approach [51]. The material is composed of two independently polymerizable systems – one system reacts to form a three-dimensional cross-linked polymer matrix in the presence of the second photopolymerizable monomer system. The photopolymerizable monomer is the imaging component, as it reacts during holographic recording. This strategy produces high-performance recording media as a result of several important attributes. The matrix is formed *in situ*, which allows thick and optically flat formats to be formed. The three-dimensional cross-linked nature of the polymer matrix creates a mechanically robust and stable medium. The matrix and photopolymerizable monomer system are chosen to be compatible in order to yield media with low levels of light scatter. The independence of the matrix and monomer systems avoids cross-reactions between the two that can dilute the refractive index contrast due to premature polymerization of the imaging component.

Both types of materials have been used to demonstrate digital HDS [52,53]. Recent work on the CROP materials has focused on fingerprint imaging and correlators rather than data storage applications [54]. The two-chemistry materials are the basis of InPhase's Tapestry media and are described in detail in the next chapter.

5.3.5 Other Materials

Many other materials have been tried with limited success. These include liquid crystal doped polymers and polymer glass-like composites [55–57]. The composites were used as an attempt to address the shrinkage and relatively larger thermal expansion properties of photopolymers. By using a glass or glass-like matrix, the thermal properties can be dramatically reduced and the shrinkage with polymerization limited. However, challenges with diffusion in these materials have resulted in limited sensitivity and *M/#* demonstrated so far. In addition, scatter between the matrix and the photopolymer can be an issue. Liquid crystal doped materials [58–60] can have large index changes and potentially could be reversible or used as rewritable holographic material. So far very limited performance has been demonstrated and issues with scatter and with the volatility of the holograms have been identified.

5.4 Summary

While many materials have been tried over the last 40 years, so far only photopolymers have been shown to be commercially viable materials for HDS. This is due to both performance and the costs involved. Chapter 6 will go into more detail of the photopolymer work and the design trade-offs involved in photopolymers used for HDS.

References

1. F.S. Chen, J.T. LaMacchia and D.B. Fraser, Holographic storage in lithium niobate, *Appl. Opt.*, Vol. 13, No. 7, pp. 223–225 (1968).
2. L. d'Auria, *et al.*, Experimental holographic read-write memory using 3D storage, *Appl. Opt.*, Vol. 13, No. 4, pp. 808–818 (1974).
3. D.L. Staebler, *et al.*, Multiple storage and erasure of fixed holograms in Fe-doped $LiNbO_3$, *Appl. Phys. Lett.*, Vol. 26, No. 4, pp. 182–184 (1975).
4. R.A. Bartolini, A. Bloom and J.S. Escher, Multiple storage of holograms in an organic medium, *Appl. Phys. Lett.*, Vol. 28, No. 9, pp. 506–507 (1976).
5. B.L. Booth, Photopolymer material for holography, *Appl. Opt.*, Vol. 14, No. 3, pp. 593–601 (1975).
6. D. H. Close, A.D. Jacobson, J.D. Margerum, R.G. Brault and F.J. McClung, Hologram recording on photopolymer materials, *Appl. Phys. Lett.*, Vol. 14, p. 159 (1969); E.F. Haugh, Hologram recording in photopolymerizable layers, US Patent 3,658,526, April 25, 1972; W.K. Smothers, B. M. Monroe, A.M. Weber and D.E. Keys, Photopolymers for holography, *Proc. SPIE*, Vol. 1212, pp. 20–29 (1990).
7. K. Curtis and D. Psaltis, Recording of multiple holograms in photopolymer films, *Appl. Opt.*, Vol. 31, No. 35, pp. 7425–7428 (1992).
8. K. Curtis and D. Psaltis, Characteristics of DuPont photopolymer films for 3-D holographic storage, *Appl. Opt.*, Vol. 33, No. 26, pp. 5396–5399 (1994).
9. U.S. Rhee, *et al.*, Characteristics of the DuPont photopolymer for angularly multiplexed page oriented holographic memories, *Opt. Eng.*, Vol. 32, No. 8, pp. 1839–1847 (1993).
10. L.P. Yu, *et al.*, Synthesis and physical measurements of a photorefractive polymer, *J. Chem. Soc., Chem. Commun.*, Vol. 23, pp. 1735–1737 (1992).
11. J. Frejlich, *Photorefractive Materials*. Wiley-Interscience, Hoboken, New Jersey, 2007.
12. J. Solymar, D. J. Webb and A. Grunnet-Jepsen, *The Physics and Applications of Photorefractive Materials*, Oxford Series in Optical and Imaging Sciences. Oxford University Press, Oxford, 1996.
13. P. Gunter and J.P. Huignard, *Photorefractive Materials and their Applications*, Vols 2 and 3, Springer Series in Optical Sciences, Springer, Berlin, 2006.
14. A. Ashkin, *et al.*, Optically induced refractive index inhomogeneities in $LiNbO_3$ and LiTaO, *Appl. Phys. Lett.*, Vol. 9, pp. 72–74 (1966).
15. F.S. Chen, J.T. LaMacchia and D.B. Fraser, Holographic Storage in $LiNbO_3$, *Appl. Phys. Lett.*, Vol. 13, pp. 223–225 (1968).
16. D.L. Staebler, W.J. Burke, W. Philips and J.J. Amodei, Multiple storage and erasure of fixed holograms in Fe-doped $LiNbO_3$, *Appl. Phys. Lett.*, Vol. 26, pp. 182–185 (1975).
17. K. Buse, A. Adibi and D. Psaltis, Non volatile holographic storage in doubly doped lithium niobate crystals, *Nature*, Vol. 393, No. 6686, pp. 665–668 (1998).
18. H. Vormann, G. Weber, S. Kapphan and E. Krätzig, Hydrogen as origins of thermal fixing of $LiNbO_3$:Fe, *Solid State Commun.*, Vol. 40, No. 5, pp. 543–545 (1981).
19. F. Micheron and G. Bismuth, Electrical control of fixation and erasure of holographic patterns in ferroelectric materials, *Appl. Phys. Lett.*, Vol. 20, No. 2, pp. 79–81 (1972).
20. J. Ma, *et al.*, Electrical fixing of 1000 angle-multiplexed holograms in SBN:75, *Opt. Lett.*, Vol. 22, pp. 1116–1118 (1997).

21. D. von der Linde, A.M. Glass and K.F. Rogers, Multiphoton photorefractive processes for optical storage in $LiNbO_3$, *Appl. Phys. Lett.*, Vol. 25, No. 3, pp. 155–157 (1974).
22. H. Vormann and E. Krätzig, Two step excitation in $LiTaO_3$: Fe for optical data storage, *Solid State Commun.*, Vol. 49, No. 9, pp. 843–847 (1984).
23. R. McRuer, J. Wilde, L. Hesselink and J. Goodman, Two-wavelength photorefractive dynamic optical interconnect, *Opt. Lett.*, Vol. 14, pp. 1174–1176 (1989).
24. Y. Guo, *et al.*, Improvement of photorefractive properties and holographic applications of lithium niobate crystal, *Opt. Express*, Vol. 12, pp. 5556–5561 (2004).
25. X. Xu, *et al.*, Holography for rewritable data storage and 3D display, Int. Workshop on Holographic Memory, Tahara, Japan (2008), paper 21B1.
26. X. Liang, X. Xu, S. Solanki, M. Li, R. B. A. Tanjung and T.-C. Chong, Holographic recording with Ag-doped $LiNbO_3$ crystal, *Int. Workshop on Holographic Memory*, Tahara, Japan (2008), paper 21P8.
27. H. Shyuan, J.-P. Liu and W.-C. Su, Two color recording in a photorefractive lithium niobate crystal, *Int. Workshop on Holographic Memory*, Tahara, Japan (2008), paper 22P5.
28. K. Sutter, J. Hullinger and P. Günter, Photorefractive effects observed in the organic crystal 2-cyclooctylamino-5-nitropyridine doped with 7,7,8,8-tetracyanoquinodimethane, *Solid State Commun.*, Vol. 74, No. 8, pp. 867–870 (1990).
29. W.E. Moerner and S.M. Silence, Polymeric photorefractive materials, *Chem. Rev.*, Vol. 94, No. 1, pp. 127–155 (1994).
30. L. P. Yu, W. K. Chan, Z. H. Peng and A. Gharavi, Multifunctional polymers exhibiting photorefractive effects, *Acc. Chem. Res.*, Vol. 29, No. 1, pp. 13–21 (1996).
31. W. Li, A. Gharavi, Q. Wang and L.P. Yu, A multifunctional photorefractive material showing high optical gain and diffraction efficiency, *Adv. Mater.*, Vol. 10, No. 12, pp. 927–931 (1998).
32. G.P. Wiederrecht and M.R. Wasielewski, Photorefractivity in polymer stabilized nematic liquid crystals, *J. Am. Chem. Soc.*, Vol. 120, pp. 3231–3236 (1998).
33. P.M. Lundquist, R. Wortmann, C. Geletneky, R.J. Twieg, M. Jurich, V.Y. Lee, C.R. Moylan and D.M. Burland, Organic glasses: a new class of photorefractive materials, *Science*, Vol. 274, No. 5290, pp. 1182–1185 (1996).
34. Q. Wang, L. Wang, H. Saadeh and L. Yu, A new family of amorphous molecular materials showing large photorefractive effect, *Chem. Commun.*, pp. 1689–1690 (1999).
35. S. Ducharme, J.C. Scott, R.J. Twieg and W.E. Moerner, Observation of the photorefractive effect in a polymer, *Phys. Rev. Lett.*, Vol. 66, pp. 1846–1848 (1991).
36. Y. Zhang, R. Burzynski, S. Ghosal and M. Casstevens, Photorefractive polymers and composites, *Adv. Mater*, Vol. 8, No. 2, pp. 111–125 (1996).
37. S. Tay, *et al.*, An updatable holographic three-dimensional display, *Nature*, Vol. 451, pp. 694–698 (2008).
38. J. Eickmans and T. Bieringer, Photoaddressable polymers for optical data storage, Materials Research Society Fall Meeting, Boston, Massachusetts (1998), paper T1.9.
39. J. Eickmans, *et al.*, Photoaddressable polymers: a new class of materials for optical data storage and holographic memories, *Jpn. J. Appl. Phys.*, Vol. 38, pp. 1835–1836 (1999).
40. Keeping a keen eye on security, www.research.bayer.com/edition_17/17_encryption.pdfx.
41. E. Lorincz, *et al.*, Rewritable holographic memory card system, *Opt. Data Storage Conf.*, Whistler, British Columbia, Canada (2000).
42. M. Dubois, *et al.*, Characterization of micro-holograms recorded in a thermoplastic medium for three-dimension optical data storage, *Opt. Lett.*, Vol. 30, pp. 1947–1949 (2005).
43. X. Shi, *et al.*, Improved sensitivity of dye-doped thermoplastic disks for holographic data storage, *J. Appl. Phys.*, Vol. 102, pp. 14907–14911 (2007).
44. C. Erben, *et al.*, Ortho-nitrostilbenes in polycarbonate for holographic data storage, *Adv. Funct. Mater.*, Vol. 17, No. 15, pp. 2659–2666 (2007).
45. G. Zhao and P. Mouroulis, Diffusion model of hologram formation in dry photopolymer materials, *J. Mod. Opt.*, Vol. 41, pp. 1929–1939 (1994).
46. A. Bloom, R. A. Bartolini and D. L. Ross, Organic recording medium for volume-phase holography, *Appl. Phys. Lett.*, Vol. 24, No. 12, pp. 612–614 (1974).

47. W.S. Colburn and K.A. Haines, Volume hologram formation in photopolymer materials, *Appl. Opt.*, Vol. 10, No. 7, pp. 1636–1641 (1971).
48. L. Dhar, *et al.*, Temperature-induced changes in photopolymer volume holograms, *Appl. Phys. Lett.*, Vol. 73, No. 10, pp. 1337–1339 (1996).
49. D.A. Waldman, H.-Y.S. Li and M.G. Horner, Volume shrinkage in slant fringe gratings of a cationic ring-opening holographic recording material, *J. Imag. Sci. Tech.*, Vol. 41, pp. 497–514 (1997).
50. D.A. Waldmann, C.J. Butler and D.H. Raguin, CROP holographic storage media for optical data storage at greater than 100 bits/μm^2, *Proc. SPIE*, Vol. 5216, p. 10 (2003).
51. L. Dhar, *et al.*, Recording media that exhibit high dynamic range for digital holographic data storage, *Opt. Lett.*, Vol. 24, pp. 487–489 (1999).
52. R. Shelby, D. Waldman and R. Ingwall, Distortions in pixel-matched holographic data storage due to lateral dimensional change of photopolymer storage media, *Opt. Lett.*, Vol. 25, No. 10, pp. 713–715 (2000).
53. L. Dhar, *et al.*, Holographic storage of multiple high-capacity digital data pages in thick photopolymer systems, *Opt. Lett.*, Vol. 23, pp. 1710–1712 (1998).
54. Aprilis Co. imager products home page, http://www.aprilisinc.com/biometrics_imager.htm.
55. A. Bloom, *et al.*, Non-polymeric organic host for recording volume phase holograms, *Appl. Phys. Lett.*, Vol. 29, No. 8, pp. 483–484 (1976).
56. M. Schnoes, *et al.*, Photopolymer-filled nanoporous glass as a dimensionally stable holographic recording medium, *Opt. Lett.*, Vol. 24, No. 10, pp. 658–660 (1999).
57. N. Hayashida, *et al.*, A new organic/inorganic-hybrid photoactive material for holographic data storage media, *Int. Symp. on Opt. Memories*, Singapore (2007), paper Mo-D-6.
58. G.S. Chen and D.J. Brady, Liquid crystal doped polymers as volume holographic elements, *SPIE Dev. Opt. Process.*, Vol. 1562, pp. 128–135 (1991).
59. H. Yoshizawa, *et al.*, Azoplymer/liquid crystal complex for polarization holograms, *Int. Symp. on Opt. Memories*, Takamatsu, Japan (2006), paper Th-I-05.
60. G.S. Chen and D.J. Brady, Surface stabilized holography in azo-dye-doped liquid crystal, *Opt. Lett.*, Vol. 17, No. 17, pp. 1231–1233 (1992).

6

Photopolymer Recording Materials

Fred Askham and Lisa Dhar

6.1 Introduction to Photopolymers

6.1.1 The Holographic Recording Process

Holographic photopolymers are in essence polymeric solid solutions. They consist of a photoreactive polymer system dispersed within a distinct polymer host. During holographic recording, the information bearing optical interference pattern initiates a pattern of polymerization in the photoreactive system, i.e. polymerization is induced in the light intensity maxima of the interference pattern while no polymerization occurs in the nulls. This patterned polymerization drives a concentration gradient in the spatial volume of the unpolymerized species. These species diffuse in the volume from the nulls to the maxima of the interference pattern in an attempt to equalize their concentration throughout the recording area, creating a refractive index modulation proportional to the difference between the refractive indices of the photoreactive component and the host material. Figure 6.1 shows the steps in the formation of a hologram in a photopolymer system of this type.

6.1.2 General Characteristics of Photopolymers

The central mechanism of 'permanent' hologram formation in photopolymers, *mass transport*, imparts one of the most important advantages over many other types of recording material, namely *nonvolatility*. Instead of relying on 'spatial-local' events such as absorption changes or orientational rearrangements upon recording to provide a refractive index

Holographic Data Storage: From Theory to Practical Systems Kevin Curtis, Lisa Dhar, Adrian Hill, William Wilson and Mark Ayres

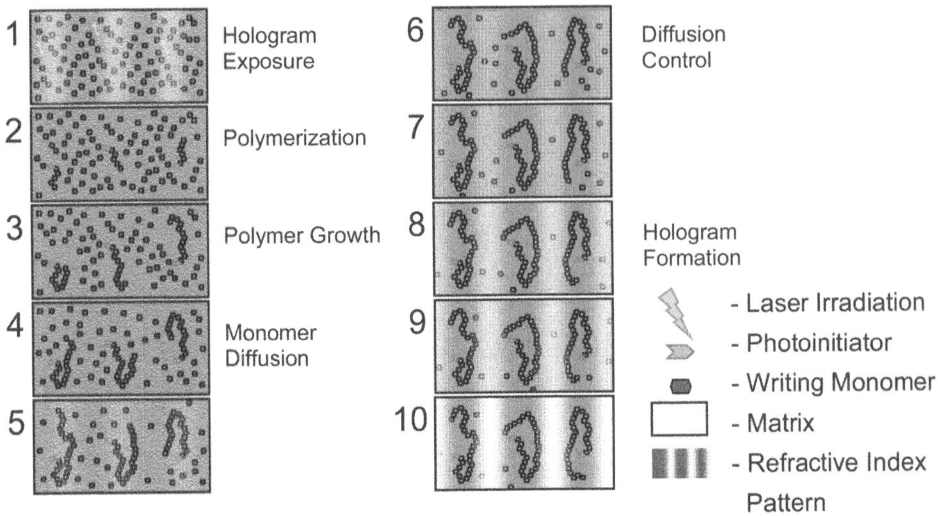

Figure 6.1 *Diagram of hologram formation*

modulation, photopolymers exploit the diffusion of species with contrasting refractive index to produce a material that is compositionally different than that before recording. This phenomenon is fundamentally different than those occurring in other materials where subsequent recording or read-out can reverse the absorption or orientational changes and erase the recorded holograms. The mass transport of photopolymers insures the permanence of the holograms.

Another distinguishing feature of photopolymers is the *amplification* of the recording process. The recording light initiates a polymerization reaction in the material as the first step towards establishing a refractive index modulation. Because polymerization reactions are chain reactions, i.e. each photon creates a reactive system that propagates through the material system until a termination step, photopolymers can exhibit significantly higher sensitivities than other materials which do not benefit from this 'amplification' of the recording light. The reactions are not single index changing events, as more monomers add to the reacting chain the index change per unit volume continues to increase.

The great flexibility offered potentially in the design of photopolymer systems also contributes to its superior performance. Photopolymers are multi-component systems. If properly utilized, their performance can be tuned through judicious choices of the photoreactive system and host matrix components. For example, choosing a photoreactive system with a significantly different refractive index than that of the matrix optimizes the dynamic range of the material. Here the index contrast that results from the photosystem polymerization maximizes the overall Δn possible. The polymerization kinetics is diffusion controlled. Choosing a host matrix with the appropriate glass transition temperature (T_g) and a high reactivity photoreactive system can further increase the sensitivity of the medium. The three-dimensional network of the host matrix can be adjusted to preserve the archival life of the material. Addition of inhibitors to the photoreactive system can enhance the shelf (unrecorded) life of the material. Photopolymers are an extremely powerful class of holographic recording materials. Because of the great latitude in their design.

6.1.3 Tapestry Two-Chemistry Photopolymer Materials

InPhase's Tapestry® media have been developed using a very flexible 'two-chemistry' platform. The 'first chemistry' is comprised of a thermoset polymer. Its polymerization forms the support matrix, or polymeric solvent in which the 'second chemistry' (a photopolymerizable system) can be induced to reaction at a later time. The components of the thermal polymerization make up a large majority ($\sim$85–90%) of the mass of the recording media layer and thus determine many of the physical properties of the media, while serving as an unusual solvent in which the subsequent photochemical reactions can take place.

The first chemistry is chosen such that matrix formation does not consume any of the components necessary for the second chemistry to proceed. The orthogonality of these two chemistries allows properties dominated by the matrix [e.g. T_g, cure rate, optical clarity, coefficient of thermal expansion (CTE), permeability, hydrophilicity, etc.] to be optimized largely independently of properties dominated by the photosystem (e.g. dynamic range, photosensitivity, recording induced shrinkage and scatter, etc.). It also ensures that potential index contrast is maximized since there are no opportunities for high and low index components to participate in the 'wrong' chemistry.

A further consequence of great practical importance for commercially viable media is that the concentrations of all photosystem components are precisely determined independently of the details of the first chemistry set-up conditions. Since these concentrations remain stable for long periods (years) after matrix formation is complete, media with consistent performance are readily produced.

From the perspective of new materials development, the potential range of formulation choices is very broad. Concentrations and identities of photoinitiator, monomer and terminator can be varied over a wide range, typically limited only by their solubilities. Systematic variation of formulation parameters and correlation to performance trends is readily accomplished. Competing choices for active components can be compared directly while other parameters are held constant. The great flexibility of this two-chemistry platform has allowed development of different media formulations with sensitivities from red wavelengths to blue and appropriate for thick data storage formats or thin display films.

6.2 Photopolymer Design

6.2.1 Host Matrix Systems of Photopolymers

The Tapestry 300R recording medium consists of a three-dimensional, cross-linked matrix and the writing chemistry dissolved in that matrix.

The diffusion rate of the monomer within the matrix is mainly governed by the properties of that matrix. Its polymerization kinetics is controlled by interaction dynamics. The photosensitivity, i.e. the index change per unit photon flux, shows a strong dependence on the T_g of the matrix. A lower T_g increases the accessible free volume for the monomer to diffuse in the matrix, so lower T_gs yield higher photosensitivities (Figure 6.2).

It is generally advantageous to form the optical article near room temperature since this is within the temperature range anticipated for the media's storage and use. This helps to

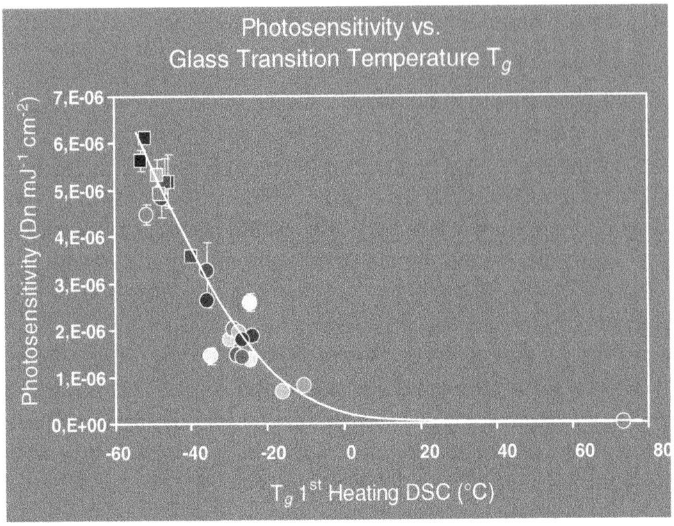

Figure 6.2 *Photosensitivity versus glass transition temperature of the matrix*

ensure that the material's physical characteristics (e.g. thickness, flatness) are maintained after manufacture. (For this reason, thermosets are obvious candidates.) For mass production, rapid matrix set-up times are essential in achieving low cost. While the structure of precursor components affect the rate at which thermal polymerization forms the matrix, use of an appropriate catalyst is typically required for optimization of the cure speed.

When considering the choice of catalyst it is obvious that it must possess sufficient activity and stability that low concentrations mixed into precursor solutions remain effective until use. This is especially true if the catalyst is considered to be toxic. It also should not absorb light at the wavelengths of interest either initially or upon aging.

Perhaps less obvious, it is equally important that the compound chosen has low activity toward catalyzing undesirable reactions. Finally, it must be realized that fast cure rates are only feasible if excellent mixing rates can be achieved. If substantial cure precedes full mixing, media homogeneity will be compromised.

6.2.2 Photoreactive System of Photopolymers

6.2.2.1 Monomer Systems

The photoreactive component of photopolymers typically contains three components – a photoinitiating system, the monomer system that undergoes polymerization, and a polymerization retarder system that prevents premature initiation of the polymerization and can control the extent of polymerization, i.e. the molecular weight distribution. The monomer systems of conventional holographic materials polymerize through either a free-radical or cationic mechanism. In both, recording light interacts directly with the photoinitiating system generating the reactive species – either a free radical (a fragment containing unpaired electrons) or a cation (positively charged species). Activated free-radical initiators attack unsaturated monomers such as acrylates (shown in Figure 6.3) while cationic initiators primarily target epoxide-type ring monomers (shown in Figure 6.4) [1].

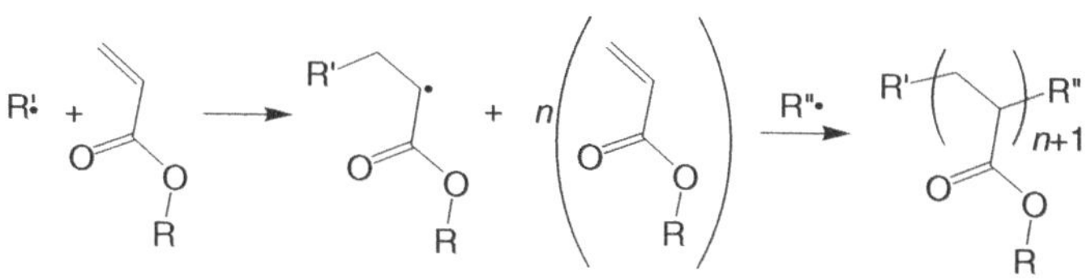

Figure 6.3 Acrylate polymerization

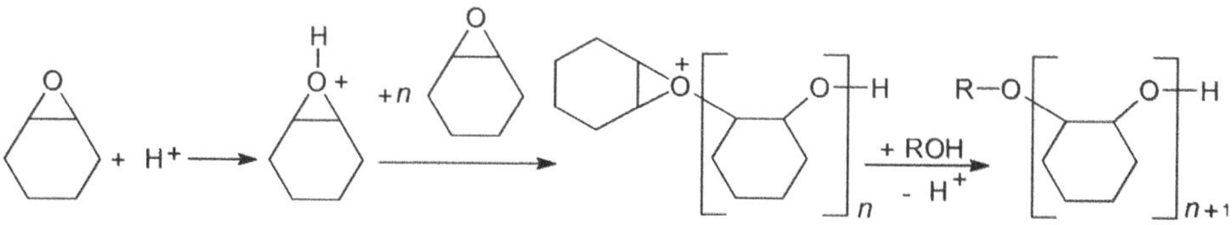

Figure 6.4 Cationic epoxide-type polymerization

The choice between free-radical and cationic systems for a holographic recording material depends on various factors. Cationic reactions can be of higher reactivity and exhibit lower recording-induced shrinkage than free-radical processes but are marked by difficulties in termination (ending of the growth of the polymer chain). In addition, the cationic systems are not often compatible with host matrices that can provide the mechanical properties needed for environmental and archival robustness. Because of the issues with termination and matrix compatability, recent Tapestry two-chemistry materials development has focused on free-radical photoreactive systems.

The most basic function of a writing monomer is to provide refractive index contrast relative to the supporting matrix. Maximizing this contrast is essential for achieving high dynamic range media. Typically, index contrast is developed using monomers of high refractive index and matrices of low index. While the opposite arrangement is possible, this choice is largely dictated by the desire to have support matrices with very low T_gs. The polymer structures most readily satisfying this low T_g requirement tend to produce materials with relatively low refractive index. Therefore, high performance monomers typically incorporate polarizable groups meant to boost the refractive index. Hence functionalities that have large molar refractivities were chosen. Heavy atoms (e.g. sulfur, phosphorus, halides), aromatics (e.g. phenyl, naphthyl) and heteroaromatics are commonly employed.

The most common limitation in the use of these and other more exotic high index groups is that of monomer solubility in the matrix. (The 'goal' is to increase local polarizibility without inducing macroscopic phase separation.) While an anthracene derivative may seem a better choice than a phenyl, or iodide better than bromide, often times these substructures reduce solubility to impractical levels. A monomer of lower intrinsic refractive index may actually be capable of providing media formulations of higher dynamic range simply because it can be incorporated at higher weight percents.

Unfortunately, simply increasing the weight percent of a given monomer in a formulation in order to increase dynamic range often suffers from the drawback of a corresponding increase in the physical shrinkage and/or bulk index change which occurs upon exposure to light. Constraints on the maximum allowable writing induced shrinkage for holographic data storage (HDS) media tend to be severe, typically less than 0.1%. The shrinkage that occurs depends upon both the identity of the polymerizing functionality and its concentration. For example, if polymerization induces a ring system to open or a small molecule to be released, the shrinkage per mole of polymerized functionality is reduced relative to a simple acrylate polymerization. However, these groups can change the bulk index of the material and use of such groups in free radical based polymerizations leads to difficulty in producing media with sufficient photosensitivity, presumably due to reduced propagation rates. Because of the requirement for high photosensitivity, acrylate based monomers are usually preferred.

To keep the shrinkage at acceptable levels, the concentration of acrylate functionality must be minimized. To provide sufficient dynamic range, the number and function of high index groups must be maximized. These facts taken together have led to the development of high molecular weight monomers incorporating many index groups which balance the desire for high refractive index with the need for good solubility. There is, of course, a limit to the extent that monomer size can be increased. This derives from the necessity that the monomer's diffusion rate be sufficient to support the polymerization kinetics. If, on the timescale of the polymerization, monomer concentration cannot be sufficiently replenished in the bright zones by diffusion from dark areas, then the fidelity of the recording suffers.

Recent work [2,3] has indicated the benefits of using nanoparticle dispersed monomer (acrylate) systems. In this approach materials such as TiO_2 and ZrO_2 nanoparticles are incorporated into the photopolymer formulation providing both increases in dynamic range and decreases in recording-induced shrinkage. The nanoparticles counter-diffuse with respect to the monomer system providing enhancements in performance. One challenge in using these materials is the level of scatter caused by the particles. Efforts to reduce the scatter have focused on the use of ZrO_2 rather than TiO_2 materials.

6.2.2.2 Photoinitiators

The photoinitiators are chosen primarily based on their absorption properties and quantum efficiency, although poor solubility or chemical stability in the matrix or precursor solutions can eliminate seemingly good choices. A holographic recording medium should exhibit uniform absorption throughout its thickness – too high an optical density can lead to nonuniform gratings and degrade the fidelity of the recording. However, it is also necessary to have a photoinitiator concentration that is sufficiently high to allow generation of enough radicals that the writing monomer can be completely utilized or dynamic range will be wasted. In practice it is better to include considerably more than this minimum amount of photoinitiator in order to lessen the severity of the fall off in photosensitivity typically experienced as the dynamic range is consumed during recording. For these reasons, an initiator of high efficiency but with the maximum of its absorption offset from the recording wavelength is desired. If it is anticipated that recording will occur over a range of wavelengths, it is desirable that the absorbance spectrum be as flat as possible through this region. Finally, once the recording is complete the photoinitiator must bleach effectively so that diffracted power is not lost to absorption during read-out.

The degradation in holographic performance when the absorbance of the media is too high is shown in Figure 6.5. In this example, the dynamic ranges of media with increasing

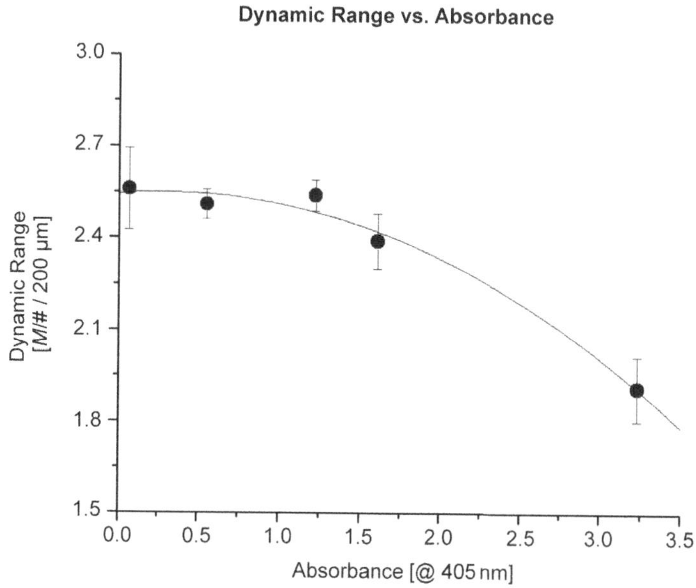

Figure 6.5 *Graph of M/# versus absorbance showing that too much absorption limits achievable M/#*

amounts of photoinitiator were measured demonstrating the sharp decline in *M*/# when the optical density exceeds 1.

6.3 Holographic Recording in Photopolymers

6.3.1 Hologram Formation Through Diffusion in Photopolymers

Several classes of theoretical models have been constructed to elucidate the formation of holograms in photopolymers, each with an increasing level of complexity. The basic model of monomer diffusion was proposed in 1971 by Colburn and Haines [4] and further treated by Zhao and Mouroulis [5]. More recent treatments have included the dependence of diffusion upon spatial position within the material [6], the effects of dark reactions [7] (polymerization that occurs after the recording light is turned off), the changes in the density of the photoreactive system upon polymerization [8],the impact of oxygen inhibition, dye bleaching and the interaction of inhibitors with monomer radicals [9].

6.3.2 General Use in a HDS System

The basic use model for a photopolymer is given in Figure 6.6. As explained in Chapter 13, typically media is pre-cured (illuminated with light) before hologram recording can commence. This pre-cure exposure is usually achieved using an incoherent light source such as a light emitting diode. This energy uses up oxygen in the material that prevents the initiation of polymerization. After recording the data holograms, an incoherent exposure is used to react any remaining monomers that were not polymerized during the hologram

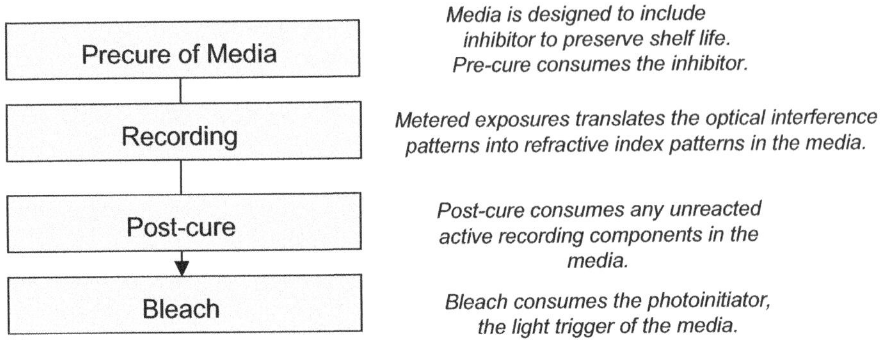

Figure 6.6 *Media usage model*

recording stage. This finishing exposure is called the post-cure. If the media is exposed with more energy, the remaining photoinitiator can be reacted. This exposure step is called 'bleaching' the media. The goal of this step is to minimize the residual adsorption of the fully recorded media. (The drive typically does not fully bleach the media but rather the final exposure represents a compromise between a post-cure and bleach exposure energy, for transfer rate reasons.) During the recording phase, the monomers diffuse in response to polymerization. This diffusion time is much slower (seconds) than the actual exposure times (milliseconds) used to record holograms. The holograms develop after the book has been recorded for several seconds. Other books can be recorded while the first book is still experiencing monomer diffusion. However, for the final book recorded, a few seconds must be allowed before the final post-cure exposure is applied.

A basic model of photopolymerization is given in the equations below. The first step in hologram recording is the absorption of a photon by the photoinitiator (PI) to form radicals:

$$PI \xrightarrow{h\nu} P\cdot + I\cdot \tag{6.1}$$

The second step is the termination of these radicals by oxygen [Equation (6.2)]. This termination results in the need for pre-cure before actual recording takes place. In addition to oxygen, other species [T as in Equation (6.3)] can be added to effect termination before, during and after recording of the holograms.

$$P\cdot + O_2 \rightarrow PO_2\cdot \quad \text{and} \quad I\cdot + O_2 \rightarrow IO_2\cdot \tag{6.2}$$

$$P\cdot + HT \rightarrow PH + T\cdot \quad \text{and} \quad I\cdot + HT \rightarrow IH + T\cdot \tag{6.3}$$

The next step in the process is the actual polymerization of monomer by primary radicals:

$$P\cdot + M \rightarrow M_1\cdot \quad \text{and} \quad I\cdot + M \rightarrow M_1\cdot \tag{6.4}$$

Polymerization occurs when a radical oligomer reacts with a monomer resulting in chain growth:

$$M_n\cdot + M \rightarrow M_{n+1}\cdot \tag{6.5}$$

Eventually the polymerization must be terminated. Termination is required for high quality data holograms with low scattering noise and good signal to noise ratio (SNR). With oxygen removed prior to chain growth the polymerization can be terminated by primary radicals [Equation (6.6)], other species [Equation (6.7)], and combination/disproportionation [Equation (6.8)].

$$M_n \cdot + P \cdot \rightarrow M_n P \quad \text{and} \quad M_n \cdot + I \cdot \rightarrow M_n I \tag{6.6}$$

$$M_n \cdot + HT \rightarrow M_n H + T \cdot \tag{6.7}$$

$$M_m \cdot + M_n \cdot \rightarrow M_{m+n} / M_m \cdot + M_n \cdot \rightarrow M_m + M_n \tag{6.8}$$

An oligomer chain can also be terminated by scavenging H from a chain transfer agent (backbone matrix, oligomer, added species, etc.) or a chain transfer agent can act as a free-radical initiator and start polymerization. Ignoring chain transfer agents, assuming only P- is an active radical, and assuming the O_2 reactions with the photoinitiator radicals are *very* fast and will dominate the chemistry until the O_2 is completely used up, nothing interesting will happen in the media until after the O_2 is gone. With these simplifications and assuming the diffusion and rate coefficients are constants, the diffusion and reaction equations can be written as shown below where $k_{x,y}$ is the rate constant for species x and y and D is the diffusion coefficient. For the photoinitiators the equations are given in Equations (6.9) and (6.10) where I_o is the intensity of the light with frequency ν:

$$\frac{\partial PI}{\partial t} = -\vec{\nabla} \cdot \left[D_{PI}(x,T) \vec{\nabla} PI \right] - \Phi \frac{I_o}{h\nu} PI \tag{6.9}$$

$$\frac{\partial P \cdot}{\partial t} = -\vec{\nabla} \cdot \left[D_{P \cdot}(x,T) \vec{\nabla} P \cdot \right] + \eta I_o PI - k_{P \cdot M} P \cdot M - \sum_n k_{M_n \cdot P \cdot} M_n \cdot P \cdot \tag{6.10}$$

The equation for radical monomer chains of length n is:

$$\frac{\partial M}{\partial t} = -\vec{\nabla} \cdot \left[D_M(x,T) \vec{\nabla} M \right] - k_{P \cdot M} P \cdot M - \sum_n k_{M_n \cdot M} M_n \cdot M \tag{6.11}$$

Equation (6.12) is for radical oligomer chains of length $n = 1$ and Equation (6.13) is for radical oligomer chains of length $n > 1$:

$$\begin{aligned} \frac{\partial M_n \cdot}{\partial t} = {} & -\vec{\nabla} \cdot \left[D_{M_n \cdot}(x,T) \vec{\nabla} M_n \cdot \right] - k_{M_n \cdot M} M_n \cdot M + k_{P \cdot M} P \cdot M \\ & - k_{M_n \cdot HT} M_n \cdot HT - \sum_m k_{M_n \cdot M_m \cdot} M_n \cdot M_m \cdot - k_{P \cdot M} P \cdot M_n \cdot \end{aligned} \tag{6.12}$$

$$\begin{aligned} \frac{\partial M_n \cdot}{\partial t} = {} & -\vec{\nabla} \cdot \left[D_{M_n \cdot}(x,T) \vec{\nabla} M_n \cdot \right] - k_{M_n \cdot M} M_n \cdot M + k_{M_{n-1} \cdot M} M_{n-1} \cdot M \\ & - k_{M_n \cdot HT} M_n \cdot HT - \sum_m k_{M_n \cdot M_m \cdot} M_n \cdot M_m \cdot - k_{P \cdot M} P \cdot M_n \cdot \end{aligned} \tag{6.13}$$

Polymer chains of length n are described by:

$$\frac{\partial M_n}{\partial t} = -\vec{\nabla} \cdot \left[D_{M_n}(x,T)\vec{\nabla} M_n \right] + \sum_{m=1}^{floor(n/2)} k_{M_m \cdot M_{n-m}} \cdot M_m \cdot M_{n-m} \cdot$$
$$+ k_{M_n \cdot HT} M_n \cdot HT + k_{P \cdot M} \cdot P \cdot M_n \cdot \tag{6.14}$$

Terminating species are governed by:

$$\frac{\partial HT}{\partial t} = -\vec{\nabla} \cdot \left[D_{HT}(x,T)\vec{\nabla} HT \right] - \sum_n k_{M_n \cdot HT} M_n \cdot HT \tag{6.15}$$

Measuring the various coefficients in these equations independently is very difficult. The citations above apply similar equations and assumptions and attempt to fit data and extrapolate polymerization performance. It is not clear how accurate these assumptions are throughout the entire recording conditions. Certainly during post-curing the assumptions must be different as the monomer concentration is significantly depleted and the polymerization and diffusion conditions change.

6.4 Rewritable

InPhase developed rewritable holographic media based on the architecture of the 'two-chemistry' materials used for its write-once media. This 'two-chemistry'" architecture is characterized by a photopolymer, that undergoes reaction during holographic recording, distributed throughout an inert, thermally cured matrix. This strategy is attractive for media with large dynamic range, dimensional stability, high optical quality, and mechanical and environmental robustness. Making the material rewritable (erasable and then able to record again) allows for a rewritable version of the Tapestry drive to be developed.

In these preliminary systems, over 300 plane wave holographic cycles have been demonstrated. The results of the cycling (record, read, erase, etc.) are shown in Figure 6.7.

Digital pages were multiplexed and cycling of entire books of holograms was demonstrated with extremely high fidelity digital recovery. Figure 6.8 shows an example of a recovered data page from the fourth write-erase cycle.

Environmental robustness was demonstrated as the system was cycled from atmospheric conditions to 80 °C, 95% relative humidity, maintaining its holographic performance. Figure 6.9 shows the maintenance of the material's dynamic range after several write-erase and environmental cycles.

The largest research and development problem that remains is to improve the fatigue resistance of the reversible systems. Currently, attention is focused on optimizing both the imaging component and the matrix system to extend the cyclability of the systems. In parallel, we are investigating reductions in matrix components that contribute to the material's fatigue.

In addition to pursuing routes to improve the *M/#* and photosensitivity of the rewritable materials, the largest challenge is ensuring the highest recording performance is coupled with fatigue resistance.

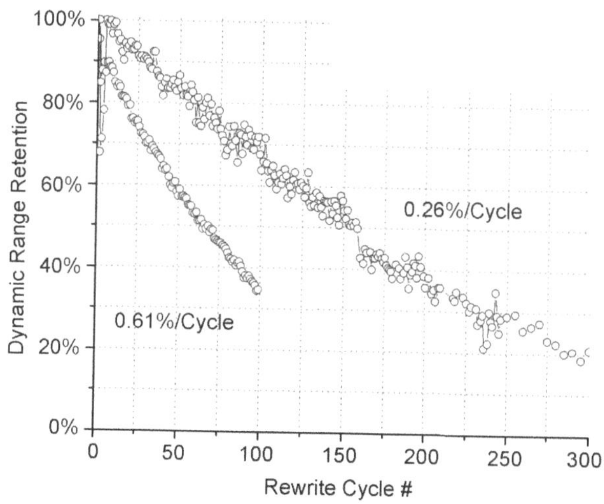

Figure 6.7 Results from the plane wave holographic cycling on the system shown in Figure 6.5. The results showing a 0.26% decrease per cycle are from an optimized matrix system while the decrease of 0.61% per cycle represents an unoptimized matrix system

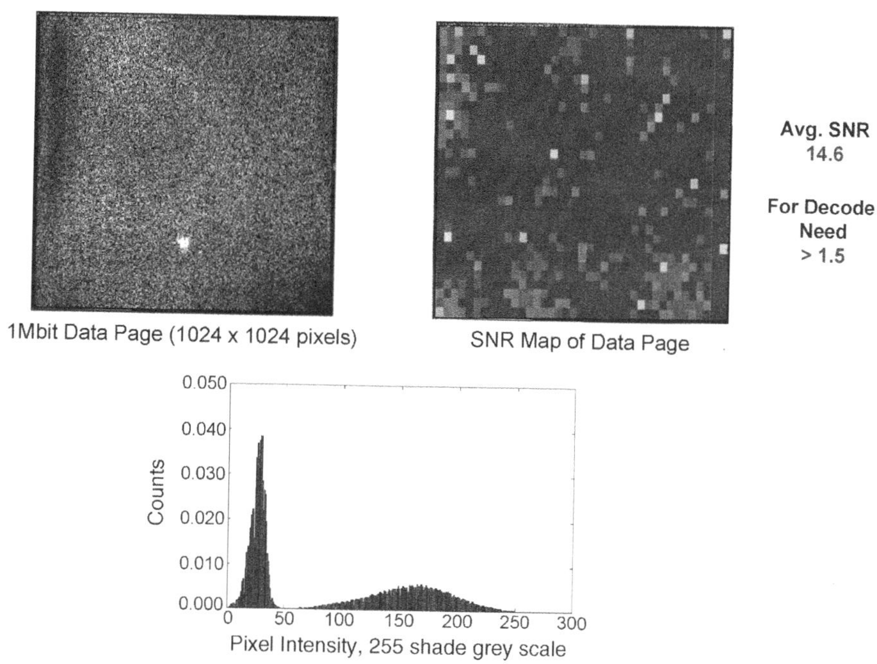

Figure 6.8 A recovered digital data page, its SNR map, and intensity histogram from the fourth write-erase cycle of a cycling set

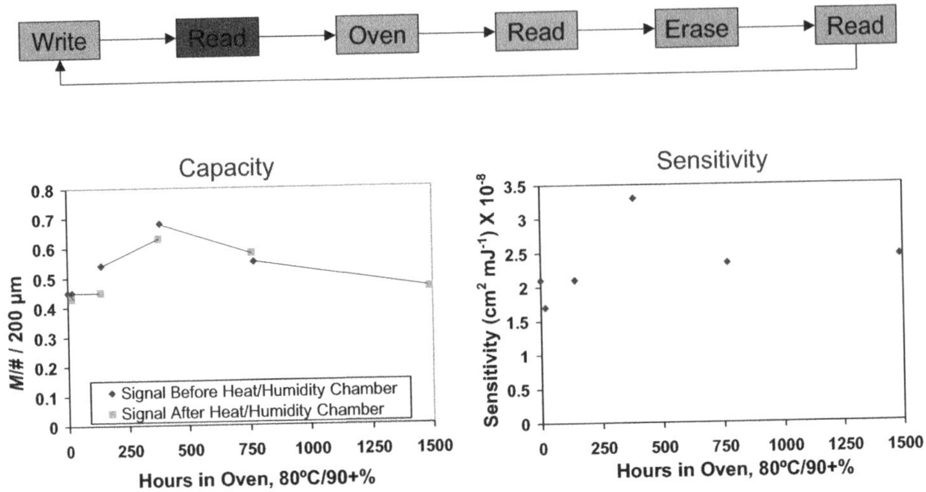

Figure 6.9 *Results from environmental cycling of an acenaphthylene system. Finally M/#s up to 2.8 per 200 μm were demonstrated in the acenaphthylene system shown*

References

1. P. Bahadur and N.V. Sastry, *Principles of Polymer Science*. Alpha Science International Ltd, Harrow, 2005.
2. N. Suzuki and Y. Tomita, Holographic recording in TiO_2 nanoparticle-dispersed methacrylate photopolymer films, *Appl. Phys. Lett.*, Vol. 81, pp. 4121–4123 (2002).
3. N. Suzuki and Y. Tomita, Highly transparent ZrO_2 nanoparticle-dispersed acrylate photopolymers for volume holographic recording, *Opt. Express*, Vol. 14, pp. 12 712–12 719 (2006).
4. W. S. Colburn and K. A. Haines, Volume hologram formation in photopolymer materials, *Appl. Opt.*, Vol. 10, No. 7, pp. 1635–1641 (1971).
5. G. Zhao and P. Mouroulis, Diffusion model of hologram formation in dry photopolymer materials, *J. Mod. Opt.*, Vol. 41, pp. 1929–1939 (1994).
6. J.T. Sheridan, M. Downey and F.T. O'Neill, Diffusion-based model of holographic grating formation in photopolymers: generalized non-local material responses, *J. Opt. A: Pure Appl. Opt.*, Vol. 3, pp. 477–488 (2001).
7. M. Toishi, *et al.*, Analysis of photopolymer media of holographic data storage using non-local polymerization driven diffusion model, *Jpn. J. Appl. Phys.*, Vol. 46, No. 6A, pp. 3438–3447 (2007).
8. J.H. Kwon, H.C. Hwang and K.C. Woo, Analysis of temporal behavior of beams diffracted by volume gratings formed in photopolymers, *J. Opt. Soc. Am. B*, Vol. 16, pp. 1651–1657 (1999).
9. M.R. Gleeson, *et al.*, Modeling the photochemical effects present during holographic grating formation in photopolymer materials, *J. Appl. Phys.*, Vol. 102, pp. 023108-1–023108-9 (2007).

7
Media Manufacturing

David Michaels and Lisa Dhar

7.1 Introduction

The processes for manufacturing holographic recording media represent a balance between the requirements for the desired performance of the drive and the cost and capabilities of the fabrication equipment. Issues associated with fabrication capabilities such as the existing infrastructure and media yield also factor into the strategy for manufacturing.

In this chapter, we will discuss the overall flow of the manufacturing process, the specifications for the media, the methods and equipment used for manufacturing, and the trade-offs or compromises that sometimes are required to achieve a manufacturing process for a commercial system. InPhase Technologies collaborated with Hitachi-Maxell during the development phase to construct the system described here.

7.2 Tapestry Media Overview

7.2.1 Overview of Disk Structure

A schematic of the Tapestry media for the professional archival drive is shown in Figure 7.1. The active recording layer (the photopolymer material) is sandwiched between two plastic substrates. Both substrates are coated on their outer surfaces with anti-reflection (AR) thin films to minimize reflection losses of the incident blue laser light during both recording and read-out. (The coating is a nonmetallic, transparent, dielectric stack.) A tracking pattern is

Holographic Data Storage: From Theory to Practical Systems Kevin Curtis, Lisa Dhar, Adrian Hill, William Wilson and Mark Ayres

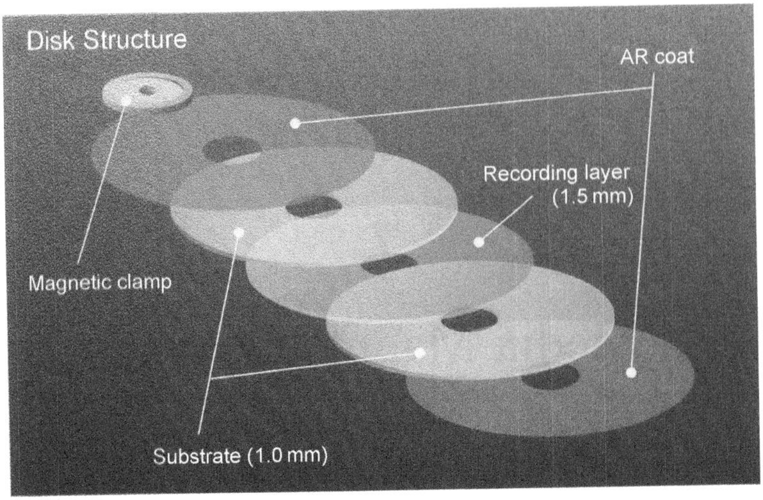

Figure 7.1 *Schematic of media construction*

molded onto the outer surface of the bottom substrate and read out through reflection of a servo laser at 850 nm. A metal hub is attached for magnetically clamping to the spindle of the drive and the outer edge of the disk and the center hole area are sealed against moisture.

The Tapestry media and drive are designed for archival storage of digital assets for use in libraries and jukeboxes. Because of this, the media is housed within a magneto-optical-style cartridge [1] for compatibility with the robotic handling of existing library systems. (In addition, the cartridge protects the unrecorded disk against exposure to ambient light.)

Figure 7.2 is a photograph of the media and the cartridge.

Figure 7.2 *Photograph of media and cartridge*

7.3 Media Manufacturing Process

7.3.1 Flow of the Manufacturing Process

Figure 7.3 illustrates the general flow of the manufacturing process [2] from the incoming raw materials for the plastic substrates and photopolymer recording layer, to the molding (with stampers), AR coating, bonding, sealing, cartridging, to the post-production testing and processing. The stamper for the bottom substrate has a tracking pattern that enables accurate determination of the disk theta (rotational) position using a 850 nm laser diode. The thermoplastic raw materials for the substrates are injection molded into disk substrates. Once molded, the substrates are coated with a multilayer stack of inorganic materials designed to provide the specified reflectivities at the wavelengths of the recording and servo lasers. Inner center seals are attached to the substrates and the substrates are bonded with the photopolymer recording layer using the ZeroWave [3] process into their sandwiched configuration. After bonding, the media is scanned to check for defects and the extent of the fill of the recording layer to the outer edge. The center plug seal and edge seal are attached to the disk and the media is cartridged. The final inspection processes are then performed. Other inspections are performed throughout this manufacturing process flow. For more information on the various tests performed on the media and the test equipment, see Chapter 8.

The following subsections describe the details of a process that has been used to manufacture Tapestry media.

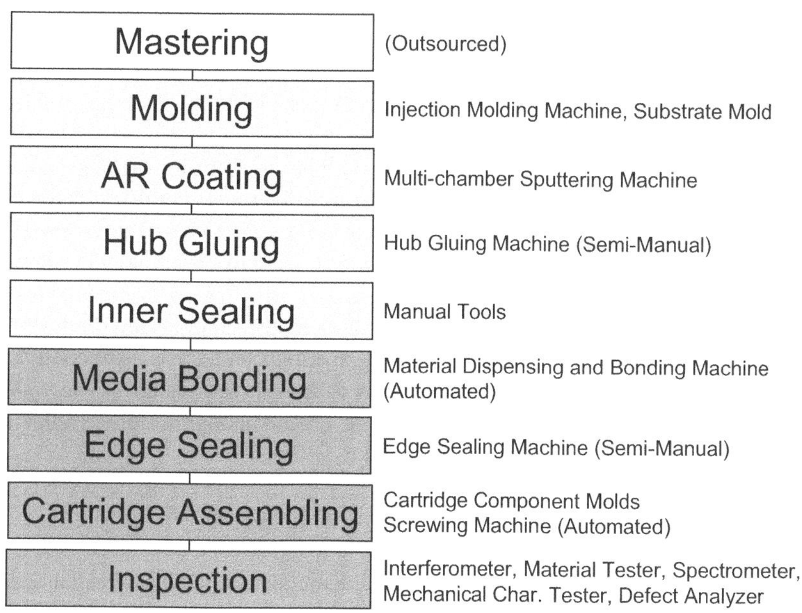

Figure 7.3 Manufacturing process flow diagram

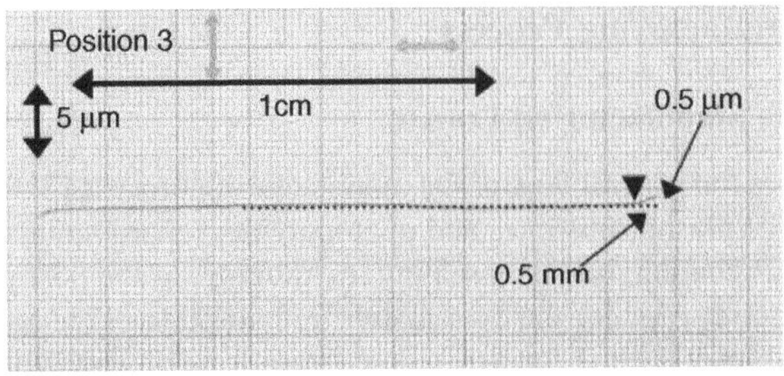

Figure 7.4 *Profilometer scans of the outer 15 mm diameter of a patterned substrate. The data illustrate the control of planarity of the substrates*

7.3.2 Molding of Substrates

Patterned and patternless stampers are used to mold the bottom and top substrates, respectively. The tracking pattern [4] is generated using a standard laser beam recorder (LBR) system. The pattern for the bottom substrate is located near the inner diameter and has a characteristic depth of 100 nm.

In an early production development line, a Sumitomo DISK 5 system (similar systems are also appropriate) with a 130 mm diameter die and the patternless and patterned stamper was used. Substrates of 1.0 mm thickness are produced. The molding conditions (temperature, pressure and other parameters) are adjusted to yield the required optical flatness, quality and edge wedge specifications. A surface profile of the outer 15 mm diameter of a typical molded substrate is shown in Figure 7.4.

7.3.3 Anti-Reflection Coating

A multi-layer stack coating is applied to the outer surfaces of both substrates to minimize reflection losses from both of the recording and read-out laser and to provide an adequately high reflectivity at the wavelength of the servo laser. (The specifications for the coating are listed in Section 7.4.) In order to fulfill the required specifications, a multiple layer stack is used. Typically a stack consists of an even number of layers, e.g. two or four layers. Here an additional bottom coating layer (next to the substrate) is used to strengthen the adhesion of the coating to the substrate to avoid generation of cracks during the environmental testing at extreme temperatures and humidities. These type of dielectric coatings are produced using standard multi-chamber sputtering systems. (The performance of a five-layer configuration is more robust against thickness variations than a three-layer system.) The reflection properties of a five-layer coating are shown in Figure 7.5 and histograms of the reflectivities for a run of 625 substrates are shown in Figure 7.6.

7.3.4 Hub and Inner Sealing

For the Tapestry media, a hub similar to that used with magneto-optic disks is attached to the bottom substrate. The hub, consisting of a metal plate in polycarbonate, is attached with an ultraviolet cured adhesive material. In addition, inner and center plug seals (as show in

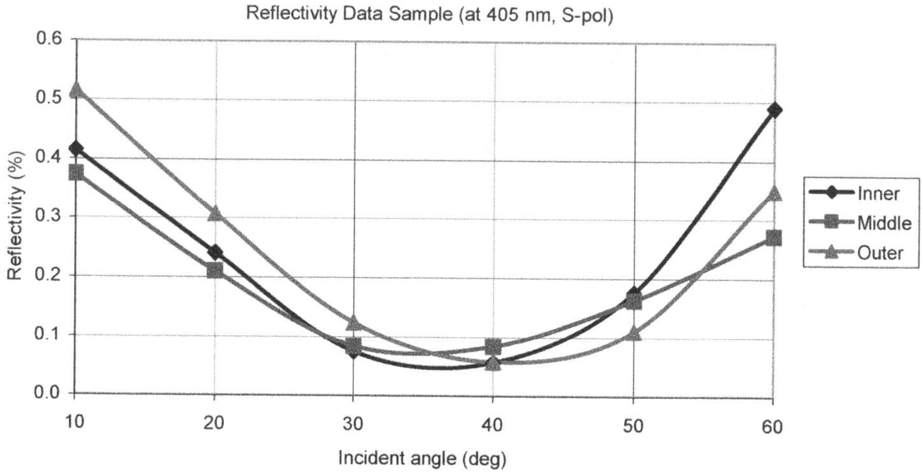

Figure 7.5 *Measured reflectivities for S-polarized, 405 nm light at the inner, middle and outer diameter locations of a substrate*

Figure 7.7) are heat welded onto the substrates. The inner seal and center plug seal material can be a laminate material such as EL-100 made by the Honeywell Corporation. The inner seals will protect the recording material from radial incursions of moisture while the center plug seal allows the recording material to be smoothly dispensed between the two substrates. Figure 7.8 shows a photograph of the heat welding equipment used to apply the seals.

7.3.5 Bonding

The recording material is bonded between the two substrates using the ZeroWave process [3]. In this method, the recording material is dispensed between the two substrates while the substrates are held in a flat configuration. This strategy allows fabrication of media with the required optical flatness, optical quality, and low mechanical tilt (see Section 7.4.3). The substrates are held in compliance, typically by vacuum, with a reference surface that is optically flat. The force of the vacuum flattens the substrate against the reference surface to an optically flat configuration largely independent of the quality of the original substrate. Figure 7.9 shows an example of the tooling that holds the substrate. Two such holders are placed at the appropriate spacing for the required media thickness as the recording material is dispensed and hardened between the substrates.

The recording material is stored, prepared (filtered, degassed and held at the correct temperature), mixed and dispensed using a commercially available dynamic mixing system. With this system, the material is dispensed in precise volumes minimizing any bubbles that may be introduced. The dispensing is also managed to control the extent of the fill of the material to the outer edge (see Section 7.4.2). The fill is achieved with a nozzle through the center hole of the top substrate.

For the Tapestry media, an automated, integrated system that connects the dispensing system to a robotically controlled bonding system, as shown in Figure 7.10, is utilized. The

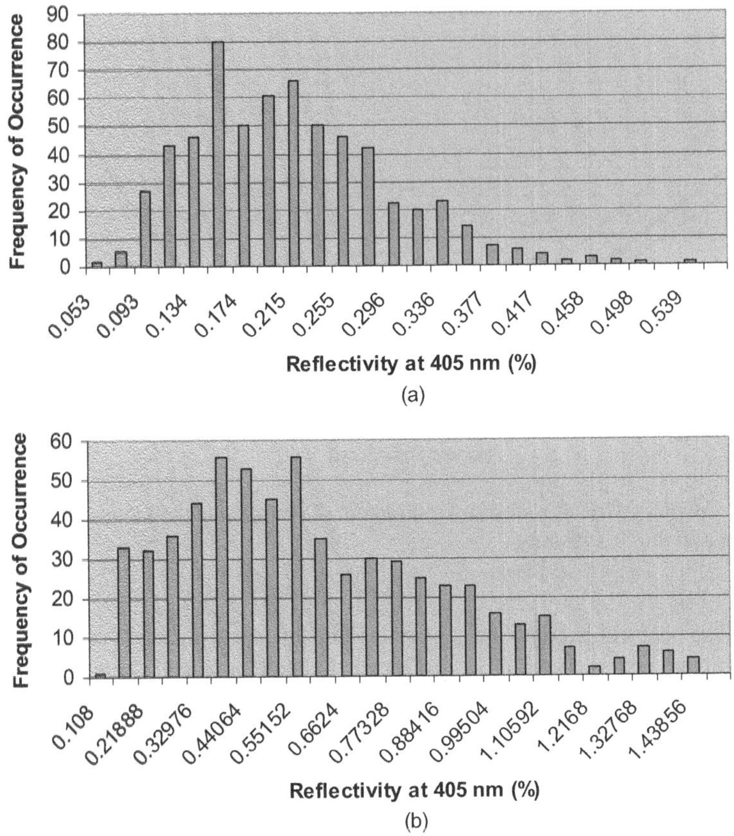

Figure 7.6 *Histograms of the measured reflectivities of 405 nm, S-polarized light at the inner diameter of a run of 625 substrates for angles of incidence of 10° (a) and 60° (b)*

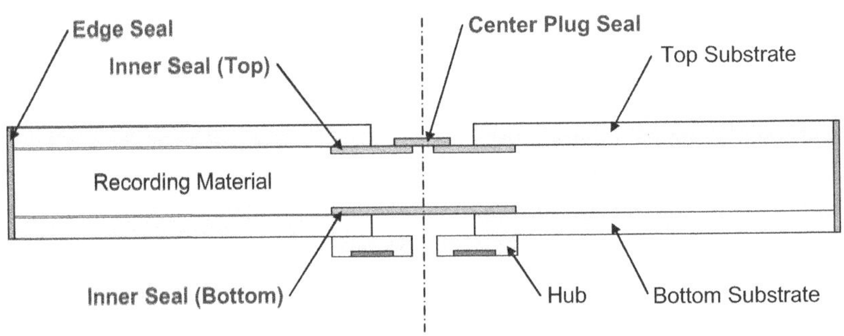

Figure 7.7 *Schematic of the placement of the hub and inner and center plug seals on the substrates*

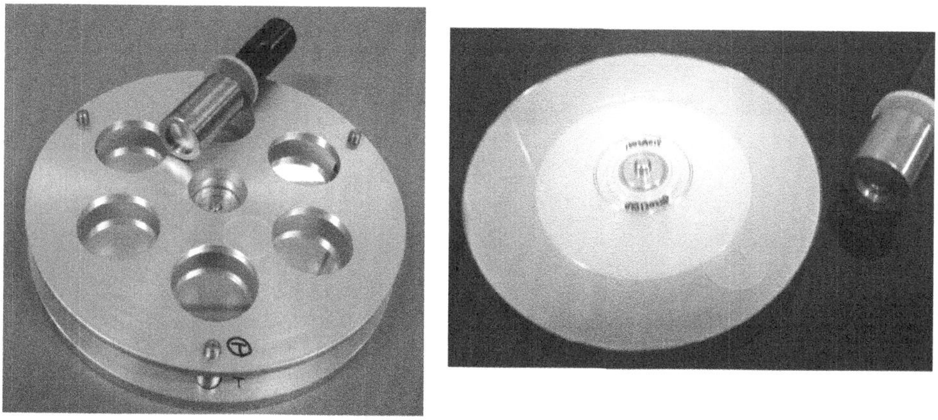

Figure 7.8 Equipment used for heat welding of the inner and center seals

dispensing of the single disk can be done in a few seconds; however, the disk filled with the recording material should be held in the correct configuration while it is curing. In order to achieve rapid overall cycle times in production of the bonded disk, several disks are fabricated in turn so that a disk can be held in the correct configuration while another is being filled. In this way, the disk once completely cured is released from the equipment every 10 s or less yielding manufacturing rates of 10 s or less per disk.

Figure 7.9 Photograph of the equipment used to bond the disks

Figure 7.10 Photograph of the automated bonding system

7.3.6 Edge and Center Plug Sealing

After the media is bonded a center plug seal is heat welded at the center hole of the top substrate to complete the protection against moisture at the inner diameter of the disk. To protect the exposed recording material at the outer edge of the disk, an edge seal is also applied. Several types of laminate materials are appropriate for the edge seal such as the EL100 used for the inner seal or a PAKVF4 material. Both of these sealing materials are available from Honeywell. A photograph of the system used to apply the edge seal is shown in Figure 7.11.

Figure 7.11 Photograph of the edge seal application system

7.3.7 Cartridging

The final step of the fabrication process is the assembly of the disk into the cartridge. The current design of the cartridge is a double-wall configuration to protect the unrecorded media against ambient light exposure. A schematic of the cartridge construction is shown in Figure 7.12. The inner and outer shells of the double wall are molded from polycarbonate. A shutter which moves between the inner and outer shell is released only when an interlock release tab built into the cartridge is depressed. A radio-frequency identification tag which can store tracking information and meta-data about the disk is inserted into the cartridge. Testing of the light-tightness of the cartridge has shown the robustness of the unrecorded media to exposure of at least 5 years of ambient light. The test results are summarized in Figure 7.13.

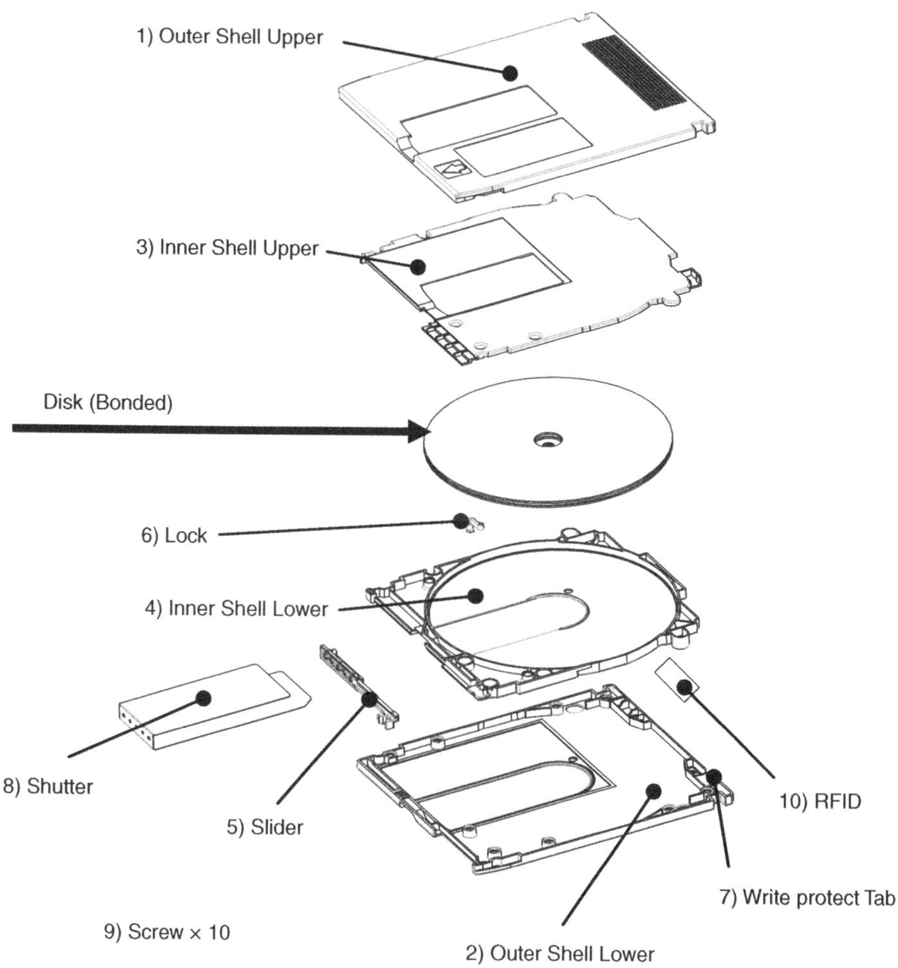

Figure 7.12 Exploded diagram of the parts of the cartridge. RFID, radio-frequency identification

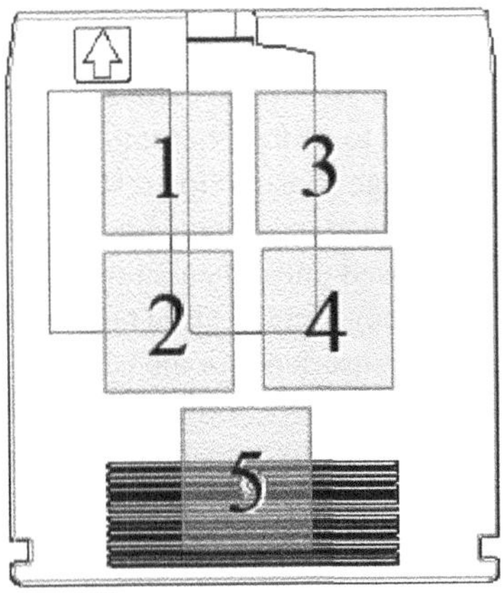

Conditions	*M*#/200 µm	% Retain
Reference	6.30	100
1	6.27	99.5
2	6.24	98.4
3	6.29	99.8
4	6.25	99.2
5	6.15	97.6

Figure 7.13 *Unrecorded media in the double wall cartridges were tested to exposure of 4 Mlx-h of a xenon lamp (corresponding to 5 years of ambient office lighting). The measured M/# (storage capacity) at several locations (marked as 1–5) remained within 95% of the unexposed media*

7.4 Specifications for the Tapestry Media

The primary specifications for the media for use in the professional archival drive are given in the tables of this section. The specifications are derived from various factors:

- the physical dimensions of the drive;
- the targeted performance of the drive storage system (a user storage capacity of 300 GB and write and read transfer rates of 20 MB s^{-1} for the first generation product);
- the performance of the individual optical components of the drive such as the power and wavelength of the recording laser, the efficiency of the camera, the light throughput of the optical components, and the angles used by both the reference and data beams;
- and the tolerances of the holograms as well as the servo and feedback capabilities of the drive system (Chapter 14).

7.4.1 Substrates

Table 7.1 gives the various specifications for substrates used in the media.

The choice of thermoplastic material for the substrates is governed by both bulk properties and the ease of processing such as the molding characteristics and the coating properties of the produced substrates. Currently, polycarbonate, poly(methyl methacrylate) and amorphous polyolefin are the most commonly used thermoplastic materials in the optical industry. In the case of the Tapestry media, the amorphous polyolefin (an example of this material is Zeonex 480R from Zeon Chemicals) is chosen for the substrates. Table 7.2 shows the approximate properties of these materials.

Of the three candidate materials, the amorphous polyolefin provides the best combination of properties for the Tapestry media – the match of the refractive indices of the substrate and recording material, transparency at 405 nm, low water absorption properties, birefringence after molding, and processing or glass transition temperature.

Table 7.1 Specification of the substrates

	Description	Units	Aim	USL	LSL
1.0 Substrates					
1.01	Material	n.a.	Zeonex 480R		
1.02	Diameter	mm	129.8	129.9	129.7
1.03	Thickness	mm	1.000	1.040	0.96
1.04	Center hole diameter	mm	15.0	15.1	15
1.05	AR Coat (% at 405 nm) ($0 \leq$ AOI ≤ 30) (s-pol) (in data area)	%	<3	3	0
1.05b	AR Coat (% at 405 nm) ($30 <$ AOI $\leq$ 60) (s-pol) (in data area)	%	<1.5	1.5	0
1.06	AR Coat (% at 405 nm) ($0 \leq$ AOI ≤ 55) (p-pol) (in data area)	%	<5.0	5	0
1.06a	AR Coat (% at 405 nm) ($55 <$ AOI $<$ 60) (p-pol) (in data area)	%	<11.0	5	0
1.07	AR Coat (% at 850 nm) (AOI 0 to 10) for bottom substrate	%	>35	45	35
1.08	AR Coat ID start	mm	18	19	17
1.09	AR Coat OD end	mm	128.0	129	127
1.10	Refractive index ($d = 587.56$ nm, at 25 °C)	n_D	1.534	1.536	1.532
1.11	Light transmission (at 405 nm × 1.0 mm thick)	%	>88.9	100	88.9
1.12	Birefringence at 0 deg AOI (633 nm, double pass)	nm	<80	100	–50
1.13	Birefringence at +/–30 deg AOI tangential (633 nm)	nm	<190	150	0
1.14	Birefringence at +/–30 deg AOI radial (633 nm)	nm	<100	50	–100
1.15	Surface roughness (Ra average line)	nm	<0.8	0.8	0
1.16	Concentricity of center hole to OD	mm	<0.050	0.05	0
1.17	Radial tilt	deg	0	0.3	–0.3
1.18	Tangential tilt	deg	0	0.1	–0.1
1.19	Allowable OD ski jump start diameter on stamper side	mm	>126	130	126
1.20	Allowable OD ski jump height deviation from data plane on stamper side	μm	0	5	–5

n.a., Not applicable; n_D, refractive index at the sodium line.

The performance of the AR coating affects both the writing and reading transfer rates of the system. In order to achieve the desired transfer rates of the first generation Tapestry product of 20 MB s^{-1}, the losses due to reflection upon writing and reading should be kept below 3% and 11%, respectively, over a wide angle of incidence (0–60°) of the incoming 405 nm light. (Details of the calculations of the dependence of both write and read transfer rates upon optical power incident to the media, and other drive and media parameters are given in Chapter 9.) The specifications for the AR coating of the substrates represent the interplay between minimizing reflective losses at the media of the laser light, minimizing

Table 7.2 Properties of substrate materials

	Amorphous polyolefin	Polycarbonate	Poly(methyl methacrylate)
Refractive index (632 nm)	1.524/1.53	1.58	1.49
Transmission at 405 nm (%)	>90	88	91
Water absorption (%)	<0.01	0.3	0.3
Glass transition temperature (°C)	138/136	149	110
Birefringence @ 30° incident angle, 405 nm	~55 nm	~−175 nm	~25 nm

the complexity of the multilayer stack of the coating, and ensuring the robustness of the performance of the coating to process variations. In order to achieve the required level of reflection losses for both S- and P-polarizations of the light, a five-layer coating is used. The specifications were adjusted allowing higher reflectivities at the lower angles (<30°) of incidence for S- and higher angles (>55°) of incidence for P-polarization in order to achieve a relatively balanced light throughput performance for both polarizations and include any thickness variations that may occur in the deposition of the layers. Figure 7.14 shows the dependence of the reflectivity of S-polarization as a function of the angle of incidence of the incoming light for a five-layer coating design and for configurations where there are small variations in the thickness of one of the layers.

The specification for the birefringence of the substrate contributes to the overall birefringence of the bonded media which is covered in Section 7.4.4. The remaining specifications for the substrates, the surface roughness, the radial and tangential tilts, and the allowable 'ski jump' (deviation from planarity) at the edge of the substrates were determined by balancing what is routinely achieved in optical media manufacturing, compatible with the ZeroWave process, and produces the required physical profile for the bonded disk.

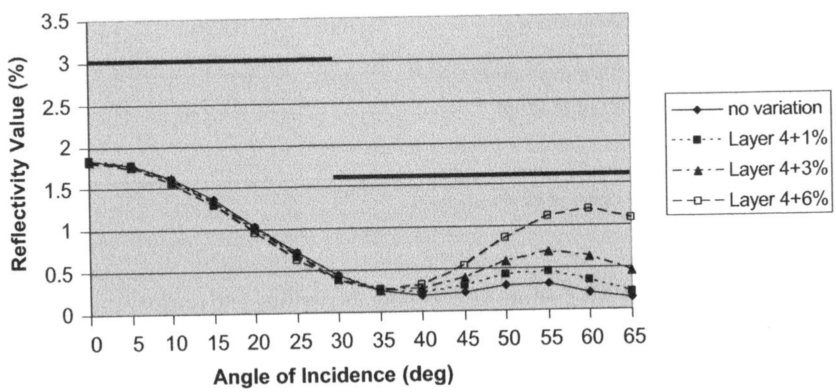

Figure 7.14 *Simulated reflectivities of S-polarized 405 nm light from the five-layer AR coating as a function of angle of incidence. The plots illustrate the dependence of the reflectivity upon variations (1, 3 and 6%) in the thickness of the fourth layer of the coating. The AR coating specification is set to account for such process variations*

Table 7.3 Specifications for the recording layer

	Description	Units	Aim	USL	LSL
3.0 Recording Layer					
3.01	Recording layer thickness	mm	1.5		
3.02	Allowed under fill in from OD edge	mm	Fill to OD	Fill to OD	0.9
3.03	Minimum fill to ID	mm	Complete	0	0

7.4.2 Recording Layer

Table 7.3 gives the specifications for the recording layer.

The recording layer thickness is determined by the target capacity of the system, the angular spacing required to achieve the capacity, and the geometries of the optical layout. (Details of the calculations of the dependence of the density on the recording layer thickness, the angular addressing, and other parameters of the media and drive are given in Chapters 2 and 3 and the resulting capacity formats using these densities are described in Chapter 13.)

The extent of fill of the recording layer at the outer diameter of the disk affects the final ski jump of the bonded disk with an edge seal applied. Too large an underfill leads to a mechanical strain when the seal is applied leading to a non-negligible ski jump. Figure 7.15 illustrates the dependence of the ski jump (measured by the metric, pitch error) upon the extent of underfill supporting a specification of the outer diameter underfill of 0.9 mm or less.

7.4.3 Assembled Media

Table 7.4 gives the specifications for the assembled media.

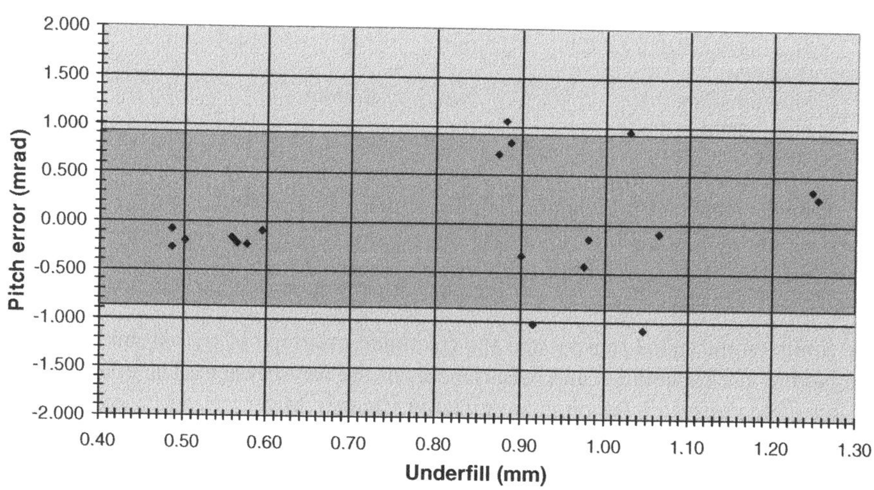

Figure 7.15 Dependence of the measured pitch error upon the measured underfill of a disk

Table 7.4 Specifications for the assembled media

4.0 Assembled media					
4.01	Overall media external thickness (in data zone) measured at 25 °C	mm	3.5	3.505	3.495
4.02	Maximum rotation run-out of outer edge (pk to pk)	μm	0	200	0
4.03	Maximum outer radius rotational envelope	mm	na	65.28	na
4.04	Maximum radial pitch error in data zone	mrad	0.00	0.50	–0.50
4.05	Maximum tangential pitch error in data zone	mrad	0.00	0.50	–0.50
4.06	Allowable OD ski jump start diameter	mm	>126	29.9 (or NA)	126
4.07	Allowable OD ski jump height deviation from data plane	μm	0	5	–5
4.09	Maximum edge seal over hang (hub side)	mm	0	0.05	0
4.09a	Maximum edge seal over hang (non-hub side)	mm	0	0.1	0
4.10	Maximum edge seal distance inward from outer edge	mm	<0.2	0.2	0
4.11	Edge seal over lap length	mm	50	70	50
4.12	Edge seal peel strength	g	>50	>50	50
4.13	Edge seal peel strength at overlap	g	>20	>20	20
4.14	Bottom inside center hole seal diameter	mm	22	22.5	21.5
4.16	Top inner nozzle seal OD diameter	mm	22	22.5	21.5
4.17	Top inner nozzle seal ID hole diameter	mm	6.5	7	6
4.18	Top outer center seal OD diameter	mm	14	14.5	13.5
4.19	Hub location	None	Bottom		
4.20	Hub height from disc surface	mm	1.35	1.35	1.15
4.21	Media radial tilt margin	deg.	0	0.3	–0.3
4.22	Media tangential tilt margin	deg.	0	0.15	–0.15

The limits on the radial run-out and the rotational envelope of the assembled disk are determined by the capabilities and requirements of the servo system. The effects of pitch error and servo correction for pitch are given in Chapter 14.

Ski jumps or deviations from planarity of the disk require changes in the angle of the incidence of the reference beam (through use of a galvo mirror system). The extent of the ski jump can be measured optically through the metric, the pitch error. (See Section 8.8 for a detailed discussion of the measurement of pitch error.) The allowable magnitudes of the

Table 7.5 *Specifications of media performance and characteristics*

5.0 Performance and characteristics					
5.01	*M*/# (per 200 μm)	None	4.00	4.20	3.80
5.02	Volume shrinkage (%)	%	0	0.14	0
5.03	Sensitivity energy to 80%	$cm^2\,mJ^{-1}$	6×10^{-6}	7×10^{-6}	5×10^{-6}
5.04	Scatter (at 405 nm, 70 deg deviation) post cure	BSDF	$<6 \times 10^{-4}$	6×10^{-4}	0
5.05	Scatter (at 633 nm, 70 deg deviation) pre-cure	BSDF	$<5 \times 10^{-5}$	5×10^{-5}	0
5.06	Birefringence (at 405 nm) (0–60 AOI)	nm	<28	28	0
5.07a	Mapped maximum bubble diameter allowed	μm	<100	100	25
5.07b	Mapped number of maximum diameter bubbles allowed	Count	D1	—	—
5.07c	Mapped maximum bubble diameter allowed	μm	<400	400	100
5.07d	Mapped number of maximum diameter bubbles allowed	Count	D2	—	0
5.07e	Mapped maximum bubble diameter allowed	μm	<800	800	400
5.07f	Mapped number of maximum diameter bubbles allowed	Count	D3	—	0
5.07g	Mapped maximum bubble diameter allowed	μm	>800	—	800
5.07h	Mapped number of maximum diameter bubbles allowed	Count	0	1	0
5.08a	Maximum particle dimension	μm	<100	100	25
5.08b	Total number of particles allowed	Count	D4	—	0
5.08c	Maximum particle dimension	μm	<400	400	100
5.08d	Total number of particles allowed	Count	D5	—	0
5.08e	Maximum particle dimension	μm	<800	800	400
5.08f	Total number of particles allowed	Count	D6	—	0
5.08g	Maximum particle dimension	μm	>800	—	800
5.08h	Total number of particles allowed	Count	0	1	0
5.09a	Surface scratch	μm	0	<200	0
5.09b	Total number of scratches allowed	Count	D7		

ski jump and pitch error are governed by the speed and frequency with which the servo-galvo feedback system must be engaged to correct for deviations. As discussed in Chapter 14, if the pitch is within this specification the wobble servo can follow the changes in pitch. Since the wobble servo uses the captured data to close loop servo on the pitch, angle, and wavelength error, this servo can be closed without any penalty to data transfer rate.

The specifications for the edge, inner, and center plug seals, and hub are determined by the geometry of the disk-cartridge system, ensuring there is no interference between the various physical systems.

The allowed radial and tangential tilts of the media are determined by the servo feedback system and discussed in Chapter 14.

7.4.4 Media Performance and Characteristics

Table 7.5 gives the specifications of media performance and characteristics.

Media performance specifications such as the *M/#*, shrinkage, sensitivity and birefringence are dictated by the budgets for the capacity and the transfer rate which are summarized in Chapters 4 and 14.

The presence of defects affects the capacity of the disk yielding areas where data cannot be recorded. The specifications for the defects are given in terms of not only the allowable sizes of bubbles, particles and scratches but also their total count such that:

$$D1 + 3D2 + 4D3 + 2D4 + 3D5 + 4D6 + 2D7 < 106$$

The derivation of this defect function is outlined in Section 9.6. The value of 106 means approximately 0.5% of the books in a 300 GB format will be rewritten due to media defects.

7.5 Manufacturing of Higher Performance Tapestry Media

As the Tapestry media moves from 300 GB to 1.6 TB capacities, the manufacturing process remains unchanged. Unlike traditional optical media manufacturing where substrate thicknesses change, multiple layers are added, or other complications arise, the Tapestry media configuration, basis of manufacturing process and manufacturing equipment remain the same throughout the performance roadmap. The improvements are derived by improved drive components and integration, and improved recording materials. Section 3.3.2 covers the technology roadmap to support the product performance goals.

Acknowledgements

InPhase acknowledges the following people for contributing to the developments described in this chapter: Peter Whiteside, Andrew Lucas, Samuel Miller, Alex Harris and Rod Harris.

We would like to express a special thank you to Hitachi-Maxell for being a wonderful development partner. In particular we wish to thank Takashi Hanyu, Tatsuya Imazu, Tetsuo Morimoto, Shinichi Nakahara, Hiroshi Sakamoto and Eiji Takahashi.

References

1. Magneto-optical cartridge ISO 11560.
2. T. Imazu, *et al.*, Development of 300 GB holographic media, *Int. Workshop on Holographic Memories*, Penang Malaysia (2007), paper 27o3.
3. S. Campbell, A. Harris and N. Levinos, Method for fabricating a multilayer optical article, US Patent 5,932,045, July 6, 1999; L. Dhar, *et al.*, Method and apparatus for multilayer optical articles, US Patent 7,112,359, September 26, 2006.
4. I. Redmond, The InPhase professional drive OMA: design and function, *Opt. Data Storage Conf.*, Montreal, Canada (2006), paper MA1.

8
Media Testing

Kevin Curtis, Lisa Dhar, Alan Hoskins, Mark Ayres and Edeline Fotheringham

8.1 Introduction

The testing of holographic media is of critical concern for both developing new media as well as manufacturing media in volume. Basic material and holographic properties can be measured with the drive and plane wave recordings and are critical for actual performance. In addition, media edge wedge (radial pitch) requirements and other media defects are very important in holography and must be accurately characterized. The defect detection for volume production leverages current CD/DVD inspection equipment with some modifications. Holographic media shelf and archival lifetime needs to be well characterized for a holographic storage product to be realized. This chapter covers the media test equipment and processes for both the development and the manufacturing processes.

8.2 Plane Wave Material Testing

8.2.1 Introduction

Plane wave testing of materials is a fundamental tool for basic holographic material development. This simple tester can measure most fundamental parameters of holographic media, including:

- dynamic range ($M/\#$, Δn)
- photosensitivity
- initiation energy

Holographic Data Storage: From Theory to Practical Systems Kevin Curtis, Lisa Dhar, Adrian Hill, William Wilson and Mark Ayres

- shrinkage
- read-after write characteristics
- spatial frequency response
- optical thickness
- media absorption
- coefficient of thermal expansion (CTE) of material
- shelf and archival lifetime.

InPhase Technologies has developed and sold plane wave testers to other companies over the years, and most people in the optical storage industry now measure these parameters using this commercial tester. Since the tester has become the de-facto standard, a section explaining the plane wave tester and how it calculates the key parameters is necessary.

8.2.2 Plane Wave Tester Set-up

The plane wave tester (Figure 8.1) consists of two continuous-wave lasers, a recording laser operating around 405 nm and a 633 nm HeNe monitoring laser. The output of the recording laser is conditioned by a set of optics to produce a plane wave source. The light is then split between two arms of an interferometer that intersect in the recording medium. The recording medium is mounted on an angular rotation stage with 0.002° of resolution. Mechanical shutters under computer control provide precise (10 ms resolution) exposure times. The powers of the transmitted and diffracted light from the recording medium are monitored by photodetectors. Light from the HeNe laser is used to align the sample medium and to monitor dark reactions and grating dynamics during hologram recording. The power of the transmitted and diffracted light from the HeNe laser is measured by photodetectors.

The optical system is mounted on a 36″ × 48″ optical table and is enclosed in a light-protecting case. Due to the long exposures used to record holograms (tens of seconds) the tester is extremely sensitive to vibration. Because of this, the optical table is vibration

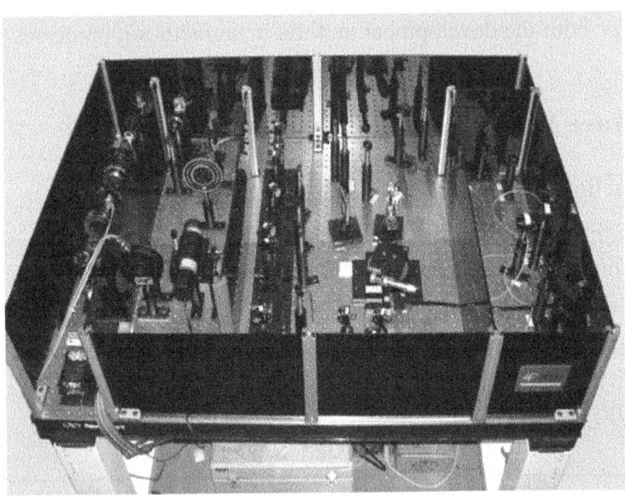

Figure 8.1 Material tester

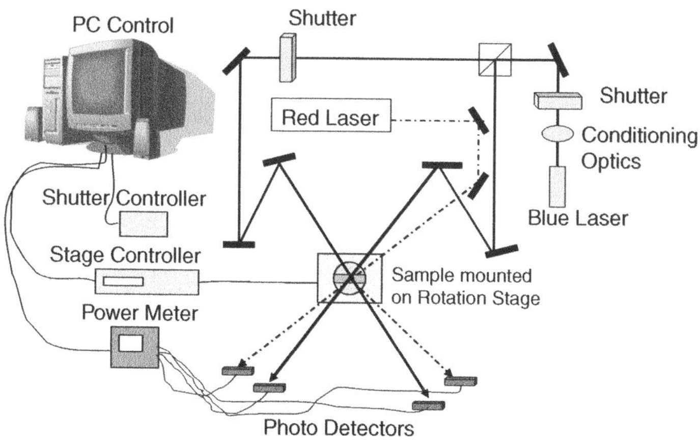

Figure 8.2 *Diagram of a plane wave holographic material tester (HMT)*

isolated and the shutters are also isolated from the table. In addition to protecting the system from stray light, the protecting cover also prevents medium vibration from acoustic noise. Figure 8.2 shows a basic schematic of the optical system and the computer control. The computer controls the mechanical stage used to angle multiple holograms, the shutter for turning on and off the two light beams, and monitors the various power meters that measure the transmitted and diffracted beams. Beam overlap issues are eliminated by irises (adjustable apertures not shown in Figure 8.2) that allow the read-out beam to be smaller than the recording beam. The media is held in a kinematic mount so that the holographic medium can be easily removed from the set-up and then accurately replaced for measurement.

The plane wave tester measures the medium parameters by multiplexing holograms and then measuring the hologram properties. The software and the set-up allow the user to automatically control:

Angular multiplexing experiments

- The number of holograms to be multiplexed.
- The angular separation between holograms.
- The angular range over which the holograms are multiplexed.
- The exposure schedule of the holograms.
- The time delay(s) or settling time(s) between recording events.

In-situ grating dynamics experiments

- Manual or computer-controlled recording exposures.
- Timing sampling intervals for monitoring dynamics.
- Overall time range for monitoring dynamics.

The data recovery is also automatic. The data collected are scans of the angle multiplexed holograms to determine diffraction efficiencies and peak shapes and locations. To get accurate measurements, the individual hologram diffraction efficiency must be kept low ($<15\%$) and the medium should not have a high absorption coefficient ($<40\%$ in

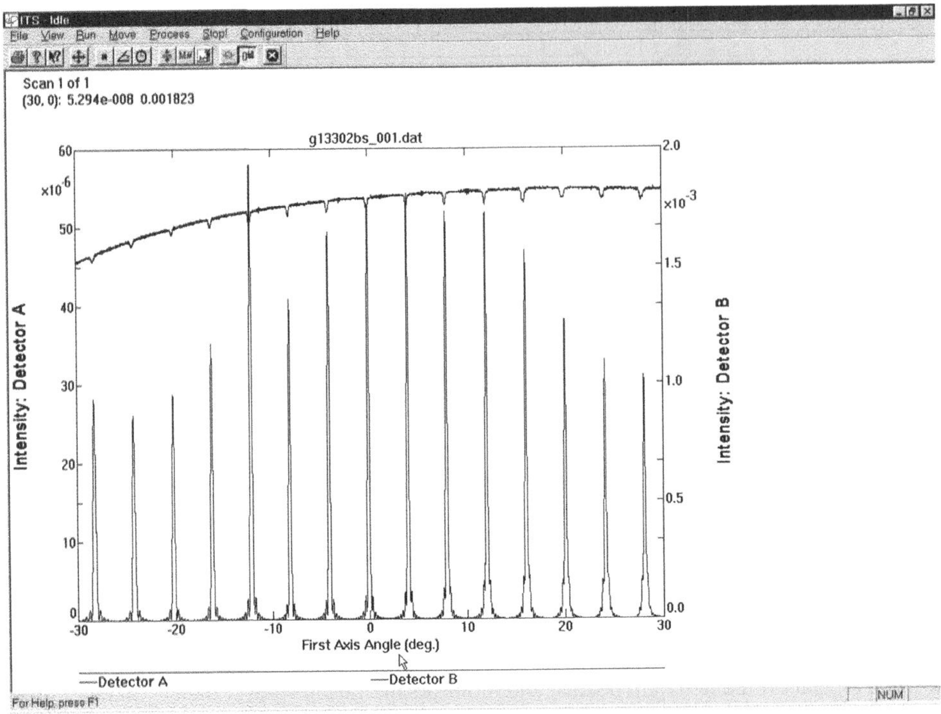

Figure 8.3 Typical scan of angle multiplexed holograms

unrecorded media). After the recording, the holographic medium is post-cured by an incoherent diode. For preliminary shelf-life tests, the medium can be put into an oven before recording to measure any changes from unheated media as explained in Section 8.11. For preliminary archival life experiments the recorded medium can be placed into an oven, after first measuring the basic parameters to see if the diffraction efficiency or peak shapes change after accelerated aging. Typical scans of the transmitted (top curve in Figure 8.3) and diffracted (lower curve in Figure 8.3 with diffraction peaks) signals are monitored.

8.2.3 Measurements and Analysis

After recording angle multiplexed holograms and curing the media, the data are collected. The basic data are angle scans of the holograms (Figure 8.3). This yields individual peak shapes, diffraction efficiency, and peak deviation from recorded angles due to shrinkage and index change. It is very important to have a separate measurement of the index of refraction of the media before recording and after cure. All photopolymers have some bulk index change and this has to be determined by standard methods covered later. Diffraction efficiency of the holograms is defined as the fraction of diffracted intensity divided by the sum of the transmitted and diffracted intensities. This definition removes absorption and beam overlap issues. For system use, this definition has to be modified or at least the losses due to overlap and absorption have to be accounted for as they can be significant. This tester is designed to isolate and measure the material performance. Examples of outputs yielding

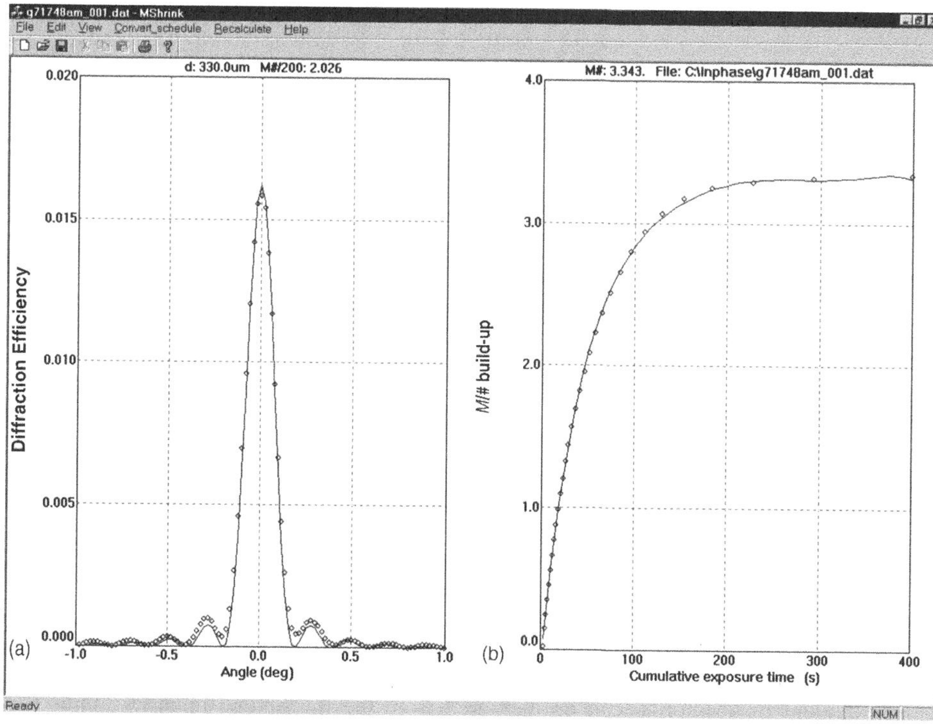

Figure 8.4 *(a) Peak shape. (b) M/# build-up with energy*

the *M/#* and optical thickness are shown in Figure 8.4. The *M/#* is calculated by summing the square roots of the diffraction efficiencies of the multiplexed holograms. The optical thickness is calculated by fitting the Bragg peaks of one of the multiplexed holograms to the curve shape predicted by Kogelnik's [1] coupled wave theory. *M/#* scales linearly with material thickness. In order to compare performance between formulations, the *M/#* must be scaled to a particular thickness. InPhase uses a standard thickness of 200 μm to compare *M/#*. The total equivalent refractive index change (Δn) of the holograms that the material recorded can be calculated from the *M/#* by the equation (also from Kogelnik):

$$\Delta n = \frac{M/\#}{\pi d} * \lambda * \sqrt{\cos(\Theta_R)\cos(\Theta_S)} \tag{8.1}$$

where d is the media thickness, λ is the wavelength and Θ_S and Θ_R are the respective signal and reference beam angles of incidence measured from normal. Figure 8.4(b) can be given as *M/#* versus exposure time or plotted as *M/#* versus the energy used. The media sensitivity is given by the energy needed to get to 80% of the *M/#* in Figure 8.4(b) when the initiation energy is subtracted. The initiation energy is the amount of energy required to start recording. Typically photopolymers have oxygen and other terminators that have to be used up before polymerization can begin. The inhibition is helpful for maintaining good media shelf life, but has a weak trade-off with transfer rate in the drive. The initiation energy is determined by measuring the slope at the beginning of the *M/#* curve and finding where that line crosses the energy axis.

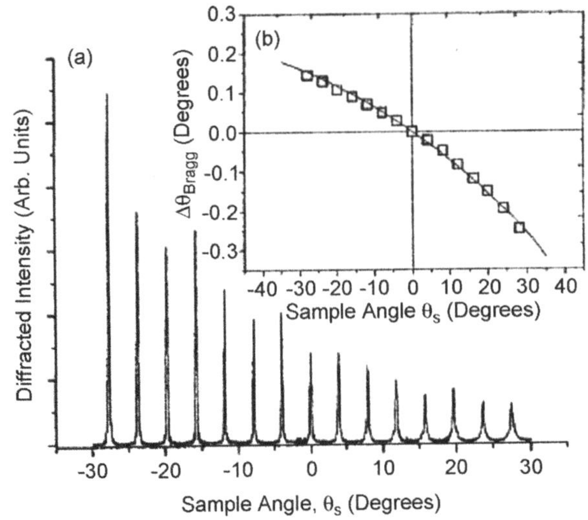

Figure 8.5 *(a) Plot of hologram angle versus diffracted intensity. (b) Plot of change in angle of holograms versus sample angle. Reprinted with permission from Applied Physics Letters, Temperature-induced changes in photopolymer volume holograms by Lisa Dhar et al., 73, 10, 1337–1339 Copyright (1998) American Institute of Physics*

The amount of absorption can be measured from the transmitted intensity minus the incident intensity. The spatial frequency response of the material can be measured by changing the angle between the beams to change the grating period. After changing this angle, the material performance can be measured to map out performance versus spatial frequency.

The dimensional stability (i.e. recording-induced dimensional change) is characterized by measuring the Bragg detuning (i.e. the extent to which the recorded holograms 'rotate' off of their original recording position) induced by recording in the material. A typical scan of holograms is shown in Figure 8.5(a) [2]. The hologram angular change (Bragg detuning) versus sample angle is plotted in Figure 8.5(b). Since temperature changes can affect this measure (as well as media sensitivity) the temperature should be controlled and measured to make sure the comparisons are fair.

Finding the effective shrinkage involves a calculation of the original grating (hologram), the grating after shrinkage, and the bulk index change. This calculation is given in detail in Dhar *et al.* [2] and is summarized here. The hologram is recorded with a grating spacing of Λ_o and ϕ_o grating angle, and then under goes an expansion/shrinkage $(1+\sigma)$ that is constrained to just be in the thickness direction. The geometry to write holograms is shown in Figure 8.6. The original grating parameters are given by:

$$\Lambda_o = \frac{\lambda}{2n_o \sin\left[\dfrac{\arcsin(\sin\theta_1/n_o) + \arcsin(\sin\theta_2/n_o)}{2}\right]} \tag{8.2}$$

$$\phi_o = \frac{\pi - \arcsin(\sin\theta_2/n_o) + \arcsin(\sin\theta_1/n_o)}{2} \tag{8.3}$$

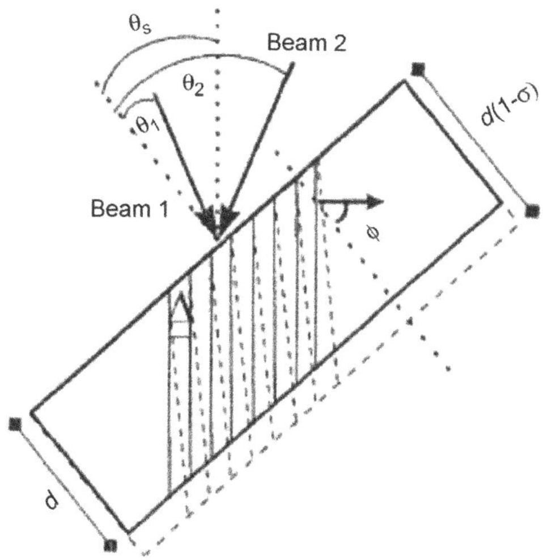

Figure 8.6 Recording geometry and parameters. Reprinted with permission from Applied Physics Letters, Temperature-induced changes in photopolymer volume holograms by Lisa Dhar et al., 73, 10, 1337–1339 Copyright (1998) American Institute of Physics

After expansion/shrinkage, the grating is rotated as seen in Figure 8.6 (dotted lines). The new grating spacing of Λ_1 and ϕ_1 grating angle is given by:

$$\phi_1 = \frac{\pi}{2} - \arctan\left[\frac{\tan(\pi/2 - \phi_1)}{1+\sigma}\right] \tag{8.4}$$

$$\Lambda_1 = \Lambda_o \frac{\sin\phi_1}{\sin\phi_o} \tag{8.5}$$

In these equations, λ is the wavelength, n_o is the original index of refraction, and the θ values are the angles shown in Figure 8.6. The expected shift in the angle from Bragg is given by:

$$\begin{aligned}\Delta\theta_{\text{Bragg}} = {} & \arcsin\left\{n_1 \sin\left[\phi_1 - \frac{\pi}{2} + \arcsin\left(\frac{\lambda}{2\Lambda_1 n_1}\right)\right]\right\} \\ & -\arcsin\left\{n_o \sin\left[\phi_o - \frac{\pi}{2} + \arcsin\left(\frac{\lambda}{2\Lambda_o n_o}\right)\right]\right\}\end{aligned} \tag{8.6}$$

where n_1 is the new index of refraction. The best fit curve to these data is found using Equation (8.6). From this fit, the equivalent shrinkage of the material can be determined.

In summary, the standard plane wave tester can measure many fundamental material parameters. Table 8.1 summarizes the most importance parameters and the range and resolution of the measurements.

8.2.4 Two Plane Wave Material Testing

While the HMT does a good job of measuring the dynamic range and photosensitivity of the holographic media, the resolution of index of refraction changes and media shrinkage

Table 8.1 *Typical resolution of plane wave measurements in transmission*

Holographic material tester capabilities			
Holographic media metric (units)	HMT automated capabilities	HMT range	HMT resolution
Media shrinkage (%)	Automated	>0.1	±0.02
Dynamic range (*M*/#)	Automated	>0.0	±0.1
Photosensitivity [$\Delta n/(\mathrm{mJ\,cm^{-2}})$]	Automatically determined through *M*/# analysis	$>4 \times 10^{-9}$	$\pm 1 \times 10^{-9}$
Media thickness (mm)	Automatically determined through *M*/# analysis	0.01–5.0	±5%

measurements can be improved by using a two plane wave material tester (TPMT). The TPMT, like the HMT, uses the hologram Bragg peak shift to calculate the bulk index change and the media shrinkage but it does it with a precision on the order of 10^{-5}. This resolution is required to accurately measure the index change and media shrinkage with photo-induced polymerization, temperature-induced expansion or contraction, and the swelling of the media due to water absorption.

To understand how the TPMT works, first note that the shift in the Bragg peak of a plane wave hologram is due to both shrinkage and the bulk change in the media index of refraction. Because this shift is a function of two variables, we should be able to solve for the shrinkage and index change by measuring the angular shift of the Bragg peaks of two holograms, *A* and *B*. Explicitly:

$$\begin{bmatrix} \Delta n \\ \Delta z \end{bmatrix} = \begin{bmatrix} \left.\dfrac{\partial \Delta\theta}{\partial \Delta n}\right|_A & \left.\dfrac{\partial \Delta\theta}{\partial \Delta z}\right|_A \\ \left.\dfrac{\partial \Delta\theta}{\partial \Delta n}\right|_B & \left.\dfrac{\partial \Delta\theta}{\partial \Delta z}\right|_B \end{bmatrix}^{-1} \begin{bmatrix} \Delta\theta_A \\ \Delta\theta_B \end{bmatrix} \tag{8.7}$$

where Δn is the bulk index change, Δz is the media shrinkage and $\Delta\theta$ is the Bragg peak shift for a plane wave hologram from its writing angle. To understand the effectiveness of this solution, it is necessary to define $\Delta\theta$ in terms of Δn and Δz. From the standard k-space formulation of holography, the diffraction efficiency of a plane wave hologram as a function of probe beam orientation is defined by:

$$DE \propto \mathrm{sinc}^2\left(\frac{\Delta\vec{k}_N L}{2\pi}\right) \tag{8.8}$$

where

$$\mathrm{sinc}(x) = \frac{\sin(\pi x)}{\pi x} \tag{8.9}$$

and L is the thickness of the holographic media. $\Delta\vec{k}_N$ is a function of the geometry of the system and is defined for each hologram by:

$$\Delta\vec{k}_N = \left|\left(\vec{k} \pm \vec{K}_G\right) \cdot \hat{N}\right| - \sqrt{\left|\vec{k}\right|^2 - \left|\left(\vec{k} \pm \vec{K}_G\right) - \left(\left(\vec{k} \pm \vec{K}_G\right) \cdot \hat{N}\right)\hat{N}\right|^2} \tag{8.10}$$

where $\hat{N}$ is the media normal unit vector, $\vec{k}$ is the probe beam propagation vector in the media, and $\pm\vec{K}_G$ is the grating vector for the different primary diffraction orders. For a single plane wave hologram, $\pm\vec{K}_G$ is defined by:

$$\vec{K}_G = \pm \begin{bmatrix} \dfrac{1}{1+\Delta x} & 0 & 0 \\ 0 & \dfrac{1}{1+\Delta y} & 0 \\ 0 & 0 & \dfrac{1}{1+\Delta z} \end{bmatrix} \left(\vec{k}_R - \vec{k}_{obj}\right) \tag{8.11}$$

where Δx, Δy and Δz are the media shrinkage (typically Δx and $\Delta y$are zero as they are constrained by the medium substrates), $\vec{k}_R$ is the propagation vector of the write reference beam in the media and $\vec{k}_{obj}$ is the propagation vector of the object beam in the media. If we assume that the media normal is collinear with $\hat{z}$ and that the object and reference/probe beams are in the $\hat{x}$–$\hat{z}$ plane, then the best Bragg matching angle can be solved for explicitly:

$$\begin{aligned} \theta &= \sin^{-1}\left(\frac{\lambda}{2\pi}\vec{k} \cdot \hat{x}\right) \\ &= \sin^{-1}\left\{\frac{\lambda}{2\pi} \frac{-(\vec{K}_G \cdot \hat{x})\left|\vec{K}_G\right|^2 \pm \left|\vec{K}_G \cdot \hat{z}\right| \sqrt{4\left[\left(\vec{K}_G \cdot \hat{x}\right)^2 + \left(\vec{K}_G \cdot \hat{z}\right)^2\right]\left|\vec{k}\right|^2 - \left|\vec{K}_G\right|^4}}{2\left|\vec{K}_G\right|^2}\right\} \end{aligned} \tag{8.12}$$

With this solution the partial differentials used in our matrix equation above can easily be found as the grating vector, $\pm\vec{K}_G$, is defined in terms of media shrinkage and the read probe beam propagation vector magnitude is defined in terms of the index change:

$$\left|\vec{k}\right| = \frac{2\pi(n + \Delta n)}{\lambda} \tag{8.13}$$

The linear systems of equations produced by the partial differentials that are needed to decouple media shrinkage and index change are ill-conditioned to one extent or another (i.e. the matrices are nearly singular). This can be evaluated by examining the condition number of the matrices, defined as the product of the norm (using any basis) of the matrix and its inverse:

$$\kappa = \mathrm{cond}(A) = \|A\|\left\|A^{-1}\right\| \tag{8.14}$$

Generally, the log of a matrix condition number, $\log_{10}(\kappa)$, is an estimate of the number of digits of precision lost during the solution of the linear system of equations. When $\log_{10}(\kappa)$ is equal or greater than the smallest number of digits of precision in the matrix or vectors of the system of equations, then the system is essentially not solvable. Because of this, it is important to have the condition number be as small as possible. By carefully choosing the object beam and write reference beam angles, it is possible to get matrices with condition numbers between 20 and 25. The consequence of this is that with the accuracy of the stages, power meters, and numerical fitting algorithms (used to locate the Bragg peak), the TPMT is capable of resolving Δn to ± 0.00016 and Δz to ± 0.00013 with a single pair of holograms. These limits can be further reduced by recording multiple pairs of holograms; with only five pairs of holograms the measurement uncertainty is reduced to 0.00007 and 0.00006 for index change and shrinkage, respectively.

A general schematic of the TPMT is shown in Figure 8.7. The tester interferes beams in a media sample. Two of the beams are fixed (object beams) while the other has variable angle and fixed position through the use of a mirror gavanometer and relay lenses (reference/probe beam). The angles of the two object beams and reference/probe beam with regard to the media normal are well characterized and are accurate to within 25 mdeg.

To measure the index change due to polymerization, one or more pairs of weakly diffracting holograms (to minimize the *M*/# usage) are simultaneously written and then media is flood cured with an external LED array. The angle of the probe beam is then scanned while the diffraction efficiency (DE) of the two holograms is monitored. The DE curve versus angle is fit to a sinc^2 profile and the peak Bragg angle is determined for both holograms to within 250 μdeg. The shift in Bragg matching angle for both holograms from the writing conditions is used to determine the index change and media shrinking during the flood cure.

The tester is also equipped with an oven allowing it to measure the CTE of the media as well as the dependence of the index of refraction on temperature. In this mode the CTE of the substrates is assumed to dominate the expansion of the media in the transverse directions. Both CTE of the media and the dependence of its index on temperature can be measured to a resolution of $\sim 15 \times 10^{-6}$ using five pairs of holograms. Since this mode of operation does not depend on minimizing the amount of monomer usage, more pairs of holograms can be written increasing the accuracy of the measurement.

The system can also be used to measure the expansion and index change when the media absorbs moisture from the atmosphere. In this set-up, the system writes pairs of holograms and the media is flood cured to use up the remaining monomer. The probe beam is scanned to find the Bragg peaks of the hologram pair. The media is then put into an environmental chamber with a known temperature and relative humidity and allowed to absorb a certain weight percent of water into the media matrix. The media is then loaded onto the TPMT and the shift in Bragg angles is measured from the previous flood cured measurement. This shift in hologram angles is then used to calculate the index change and expansion.

Because the system also has a tunable laser, it can also be used to determine the wavelength tuning requirements of different media packages. In this case the laser wavelength is tuned on hologram recovery until both hologram Bragg peaks are coincident. Doing this reveals the wavelength tuning requirements as a function of temperature or weight percent of water absorbed. This allows the media package to be designed to minimize the laser tuning requirements of the system.

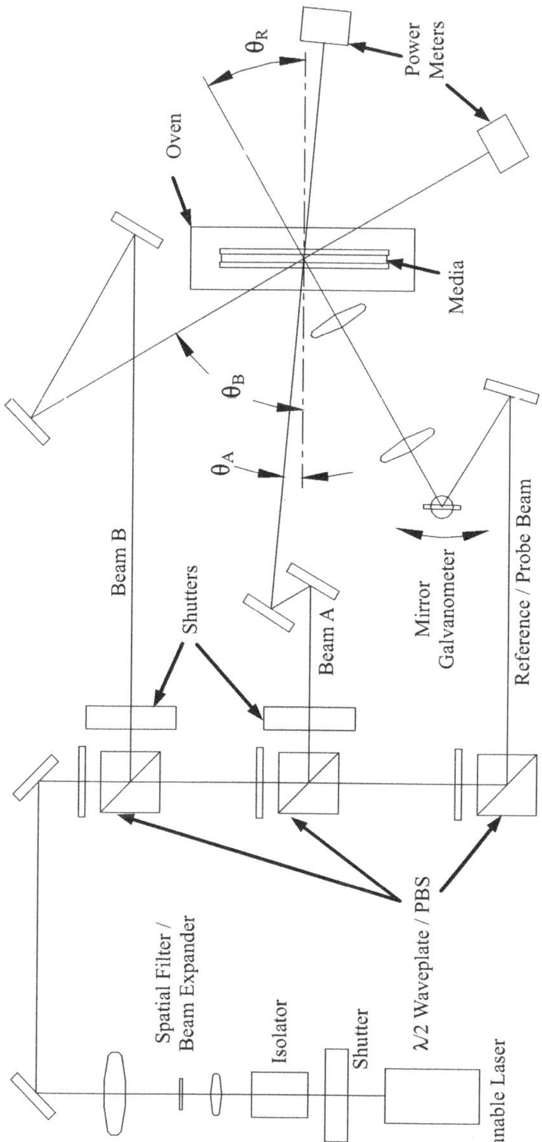

Figure 8.7 *Schematic of the InPhase Technologies TPMT*

8.3 Bulk Index Measurements

It is critical that the bulk index and bulk index change with post-cure is measured before using the plane wave tester. The standard way to measure bulk index is with a prism coupler. The prism couple inputs light at different angles and then measures the reflectance versus angle at that given wavelength. The film should be within 100 or 200 nm from the prism surface. Using this information, it is possible to calculate the bulk index of the film [3–5] to approximately the fourth decimal place. InPhase uses a commercial prism coupler measurement tool to measure the bulk index of the polymer films. It is possible to use this equipment to measure the refractive index change with temperature as well.

8.4 Scatter Tester

Scatter, both from the medium surface and volume, is one of the most important media parameters. The scatter tester developed by InPhase Technologies uses both blue (~405 nm) and red (~633 nm) lasers to measure the scatter of media and substrates. The blue light gives the appropriate scatter for the wavelength used in the drive but is destructive to the recording layer. The red light can be used for nondestructive testing of recording media. The media scatter tester measures the bidirectional scattering distribution function (BSDF) [6] with an angular resolution of 0.05°.

Each photosensitive formulation has a specific scatter profile as a function of measurement angle. The scatter measurement photodiode on the media scatter tester (MST) is located at the end of a swing arm, which provides the ability to measure scatter as a function of detector angle. The swing arm has 0.02° resolution and a 90° range, allowing for both very accurate and wide-ranging measurements. The sample holder is mounted on top of a computer-controlled X-Y stage, with 1 in. of travel in each axis. This allows the user to measure the scatter properties of a sample at a variety of positions, simply by moving the stages. This can be helpful in measuring the consistency of a sample. To get enough dynamic range, the system uses a chopper and a lock-in amplifier. The tester diagram is given in Figure 8.8.

The first set of calculations is the calibration data for the power meters and photodiode. The purpose of the calibration for power meters is to allow a measurement of the monitor power meter value to correlate with an input power incident on the sample.

To this end, the transmitted power meter is placed so as to be able to read the incident laser power, and a set of data is collected for the two values. This ratio changes as a function of micrometer position, because the reflective wedge has a reflectance that varies as a function of divergence/convergence angle. The data are collected for a series of micrometer positions and the ratio of monitor power to transmitted power is plotted versus micrometer position of the stage that holders the wedge (W).

The purpose of the calibration for the photodiode is to allow a measurement of the photodiode voltage by the lock-in amplifier to correlate to a particular power on the monitor power meter, and from there to a measurement of actual scattered power, by virtue of the power meter calibration done previously. This ratio is measured as a function of ND filter inserted into the laser beam path to calibrate the lower powers that will occur while scatter is being measured.

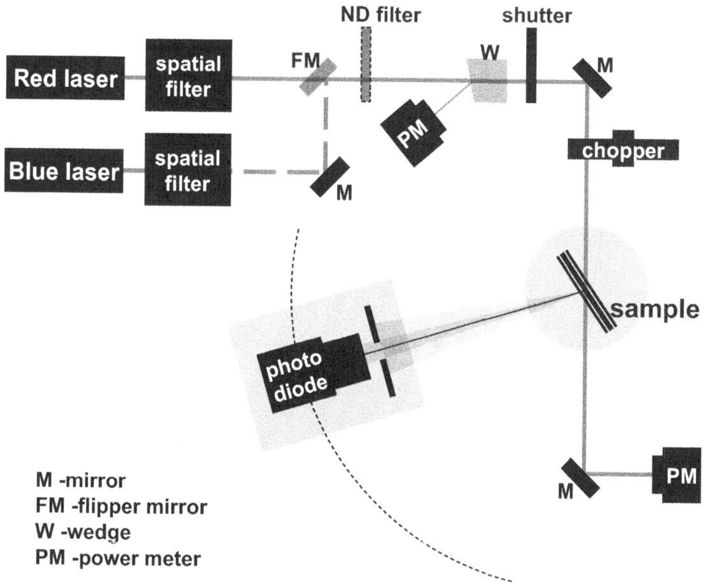

Figure 8.8 *Diagram of a media scatter tester*

Using these calibration constants, the measured off-axis photodiode voltage values can be correlated with a scatter power (in watts), and the value of the monitor power meter can be correlated to the incident power (in watts).

Scatter power must be calibrated versus the solid angle subsumed by the detector. The solid angle is calculated from the diameter of the iris (I7) and the distance from the sample to I7. The solid angle is defined as:

$$\Omega_s = \frac{\pi \left(\frac{d}{2}\right)^2}{r^2} \tag{8.15}$$

where d is the iris diameter and r is the distance from the sample to the iris.

The scatter function, BSDF, is defined as:

$$\text{BSDF} = \frac{P_{sc}}{P_{in}\Omega_s} \frac{1}{\cos\theta} = \frac{4r^2}{\pi d^2} \frac{P_{sc}}{P_{in}} \frac{1}{\cos\theta} \tag{8.16}$$

where P_{sc} is the scattered power and P_{in} is the incident power on the sample. Here θ is the angle between the monitor photodiode and the normal of the sample, reflecting a geometric factor of the laser beam shape 'viewed' by the photodiode.

Figure 8.9 shows the BSDF versus scatter angle for two media samples with different substrates. The scatter function is can only be measured from 10 to 90° as an angle of zero coincides with the straight through beam. Disks made using the first substrate material could not support high density recording due to the excessive scatter from that particular substrate material. Disks using the second substrate material could easily support high data densities with good SNR. Clearly scatter of the media package and substrates is important for high

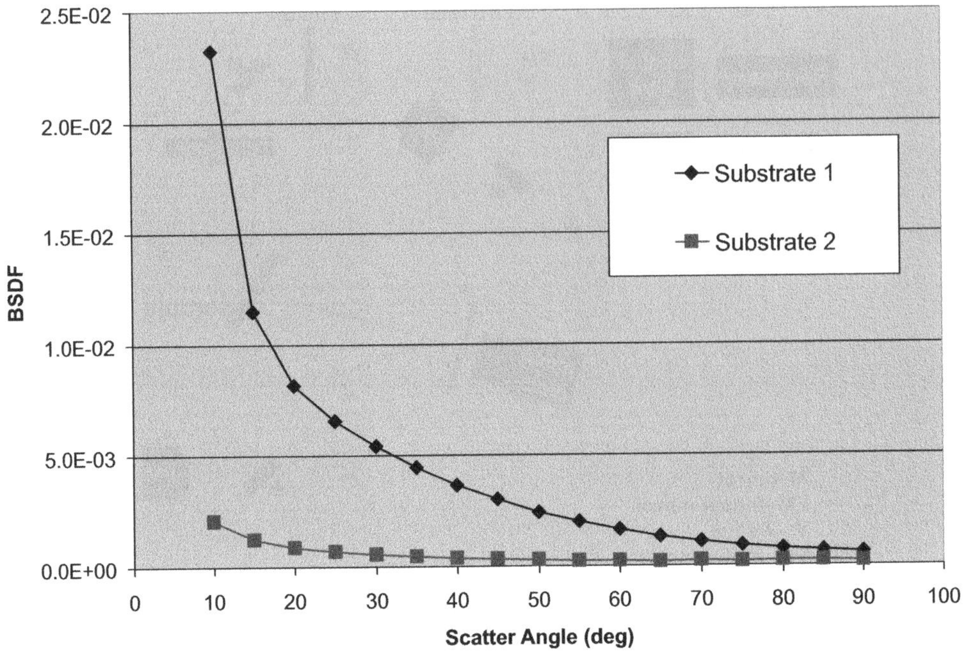

Figure 8.9 *Plot of BSDF versus scattering angle*

density storage. The MST allows for the various components of the media package and different formulations to be tested for scatter levels.

8.5 Spectrophotometers/Spectrometers

Spectrophotometers and spectrometers are very useful in media and material development. There are three main uses: absorption/fluorescence measurements at drive wavelengths, near IR measurements, and to measure AR coating performance versus angles. For absorption, there are three levels that are of interest in the drive. The first is the absorption of unrecorded media. This determines the materials sensitivity at that wavelength. The second is absorption after curing. Curing is defined as when all the monomer has been used up (all the *M*/# has been used up). The absorption after cure is the absorption that the drive will start with on read-out, and affects the diffraction efficiency on read. The third case is when the media has been bleached where bleached means that all the photo-initiator has been converted. This is the final state of the media after many reads and typically for the current photo-initiator has less absorption than after cure. Figure 8.10 shows the plots of relative absorption in arbitrary units versus wavelength from 390 to 450 nm. The virgin or unrecorded media starts out with approximately 35% total absorption at 405 nm. The final bleached state at 405 nm is when around 7–8% of the light from the media is absorbed. Fluorescence of the media can also be measured. Typically the fluorescence is red shifted by 50 to 100 nm.

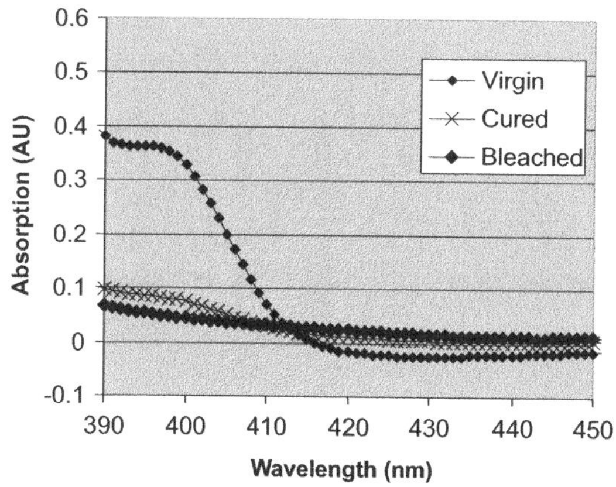

Figure 8.10 Absorption versus wavelength for holographic media in virgin, cured and bleached states

Near IR spectrometers can assist in measuring when the monomers have been polymerized. Since the concentration of monomers is low, it typically is difficult to quantitatively measure final cure. However, to reaffirm polymerization or determine other chemical reactions, near IR can be useful. Near IR spectrometers are also useful for measuring water uptake as explained in Section 8.11.4.

Spectrometers can also be used to measure the performance of AR coating. The reflectance versus wavelength or reflectance at certain incidence angles can be measured to confirm performance is at specification.

8.6 Scanning Index Microscope

8.6.1 Overview

In both holographic [7] and volumetric bitwise storage [8], the drive optics only read a portion of the three-dimensional (3D) dielectric perturbation. Index, absorption, or emission (e.g. fluorescence) outside the detection capability of the optics is therefore an unknown noise term that can perturb wavefronts or produce inter-symbol interference. In the case of holographic data storage (HDS), ambiguity and nonlinear medium response terms may be important and are not measured by the HDS system. The scatter from defects and other elements may record index structures that may demand depth-selective characterization. In addition, cure boundaries build up index profiles that are also of importance and can be characterized by this microscope. Devices such as the phase contrast microscope and the differential interference contrast (DIC) microscope produce images indicative of phase effects at an image plane within a volume but are typically unsuited for obtaining quantitative 3D measurements. Interferometers and wavefront sensors can be very sensitive to path-integrated index but cannot discriminate features in depth.

We describe a coherently illuminated transmissive scanning microscope [9,10] that employs a position-sensing detector (PSD), such as a lateral effect photodiode, rather than an ordinary photodetector. We show how the extra information provided by the PSD can be

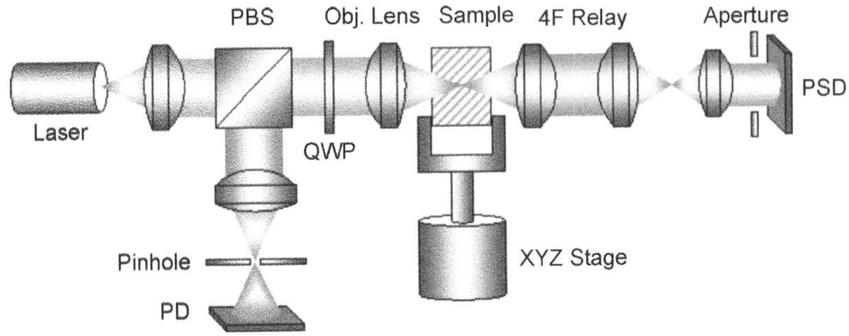

Figure 8.11 *Schematic diagram for ringdex scanner. PBS, polarizing beam splitter; QWP, quarter wave plate; PD, photodiode; 4F Relay, two lenses in Keplerian telescope; PSD, position-sensing detector. Reprinted with permission from Applied Optics, Image oversampling for page-oriented optical data storage by Mark Ayres, Alan Hoskins and Kevin Curtis,* ***45****, 2459–2464 Copyright (2006) Optical Society of America*

used to calculate the complex 3D dielectric distribution – i.e. the volumetric distribution of both refractive index and absorption – of a weakly diffracting optical sample. (A similar method of achieving phase sensitivity by differential detection has been proposed for data storage [11].) The instrument is noteworthy for its optical depth discriminating properties, and for the fact that its phase transfer function is relatively flat throughout the region of support.

8.6.2 System Layout

Figure 8.11 shows a schematic diagram of the system. A diode laser at 635 nm provides a coherent light source. The aspheric objective lens creates a 0.3 NA probe beam that is corrected for a focal depth of 1.2 mm. This configuration permits a total scanning depth range of about a millimeter without active spherical aberration correction.

Figure 8.11 also shows a confocal reflection path. Confocal microscopy is a well known method of quantitatively measuring the 3D distribution of reflection and thus provides complementary information. However, it does not separate the real and imaginary parts of the dielectric constant, and can not measure low spatial frequencies that would, for example, cause wavefront aberration. The light transmitted through the sample is collected by a 4F camera lens relay. The relay allows a large working distance on the exit side of the sample despite the small size of the PSD. The detector is apertured to collect only the angular extent of the original (undeviated) probe beam. The PSD detects the x and y centroids of the optical power falling upon the detector (i.e. spot position), as well as the total power.

Scanning is accomplished by mounting the sample on a high-precision x, y, z translation stage. A 3D image is collected by translating the probe focus through a 3D grid covering the volume of interest. At each point, the controller computer digitizes and records the x and y centroids, and the total optical power. Figure 8.12 shows a scanning transmission microscope. The disk is mounted vertically on the stages as shown in Figure 8.12.

8.6.3 System Response

The instrument's response may be understood by considering the coherent interaction of dielectric grating components with Bragg-matched, coupled probe components. The

Figure 8.12 Photograph of a scanning transmission microscope

complex dielectric modulation of a grating component dictates both the amplitude and the phase of the light diffracted between the two coupled modes. Weakly diffracted light mixes coherently with the undepleted probe producing an intensity modulation that is detected at a specific point on the detector in the far field, as illustrated in Figure 8.13.

Starting from the Bragg relationship for one-way diffraction:

$$E_d(\vec{k}_d) = j\kappa\varepsilon(\vec{k}_\varepsilon)E_p(\vec{k}_d - \vec{k}_\varepsilon) \tag{8.17}$$

where E_p and E_d are the complex amplitudes of the coupled probe components, ε is the complex dielectric component and κ is the approximately constant coupling factor. By coherently combine the undiffracted and once-diffracted components in *both* directions

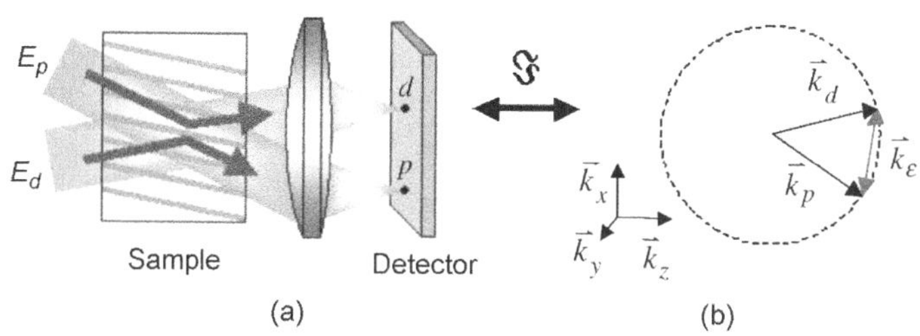

Figure 8.13 (a) Coupled plane wave probe components E_p and E_d interact with a Bragg-matched grating within the sample and are detected at points p and d on the detector in the far field. (b) k-space representation of the grating and wave vectors. Reprinted with permission from Applied Optics, Image oversampling for page-oriented optical data storage by Mark Ayres, Alan Hoskins and Kevin Curtis, **45**, 2459–2464 Copyright (2006) Optical Society of America

we eventually arrive at the 3D transfer functions:

$$H_{xPSD}(\vec{k}) \equiv \frac{C_x^*(\vec{k})}{\varepsilon_p(\vec{k})} = j\kappa 2k_x E_p(\vec{k})E_p(\vec{k}) \tag{8.18}$$

$$H_P(\vec{k}) \equiv \frac{P^*(\vec{k})}{\varepsilon_a(\vec{k})} = 4\kappa E_p(\vec{k})E_p(\vec{k}) \tag{8.19}$$

where C_x^* and P^* are the conjugates of the transforms of the PSD x centroid and power channels, respectively, ε_p and ε_a are the real (phase) and imaginary (absorption) components of the dielectric grating and $\otimes$ is the cross-correlation operator. [An expression analogous to Equation (8.18) also applies to the PSD y centroid channel.] The transfer functions H_{xPSD} and H_P are illustrated in Figure 8.14, along with their corresponding impulse responses. After the scan data are collected, the controller applies the inverse of these transfer functions to C_x and P to determine ε_p and ε_a (limited to the nonzero region of support).

The spatial resolution of the instrument is established by the manifold of the autocorrelation term, and is thus the same for both the phase and absorption transforms. For the 0.3 NA probe beam, the transverse resolution becomes 1.05 µm, and the maximum depth resolution becomes 13.8 µm. We estimate the path length differential sensitivity at approximately 1.5×10^{-4} waves. Generally, the instrument will be sensitive to the entire spatial band addressed by a linear transmissive optical data storage technology only if the resolution established by the probe beam NA and wave number exceeds that of the storage device. However, for advanced storage technologies, such a probe configuration will either require active spherical aberration correction, or have a severely limited depth scanning range.

Note that while the absorptive impulse response [Figure 8.14(c)] has the form of the probe beam (a typical result for microscopes), the phase response is instead a differentiated version of the probe [Figure 8.14(d)]. One consequence of this is that the strong DC bias of the absorptive response is removed, and the transfer function is in comparison flattened throughout the region of support. Another consequence is that, while the absorptive response integrated over any depth plane is constant (due to conservation of power in the probe beam), the phase response drops sharply with distance from the focal plane, thereby isolating the plane of interest. This desirable optical depth-sectioning property is typically effected with a spatial filter in confocal reflection microscopy (the pinhole in Figure 8.11) but is difficult to achieve in transmission [12].

8.6.4 Experimental Example

This instrument is being used to characterize volumetric index distributions in holographic recording media at InPhase Technologies. Figure 8.15 shows the refractive index at a focal plane in the middle of a recording medium exposed only with the data-bearing object beam. The index pattern has been recovered from the raw PSD signals by application of Equation (8.18). The fans of striations emanating from the central region are the low frequency gratings of the ambiguity ('intra-object modulation'). Note that

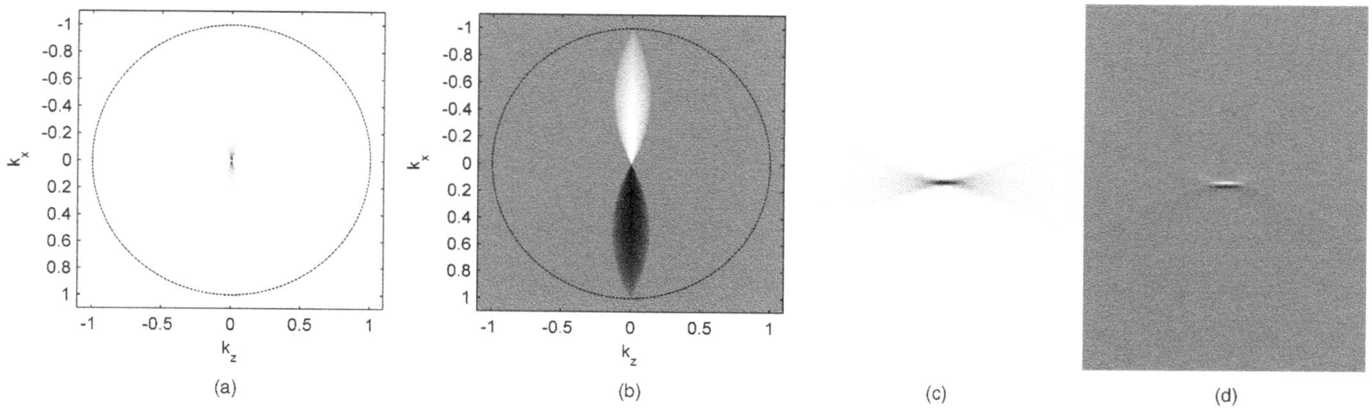

Figure 8.14 *Transfer function cross-sections: (a) H_P (absorptive); (b) H_{xPSD} (phase);. The corresponding impulse responses are shown in (c) and (d). Reprinted with permission from Applied Optics, Image oversampling for page-oriented optical data storage by Mark Ayres, Alan Hoskins and Kevin Curtis, **45**, 2459–2464 Copyright (2006) Optical Society of America*

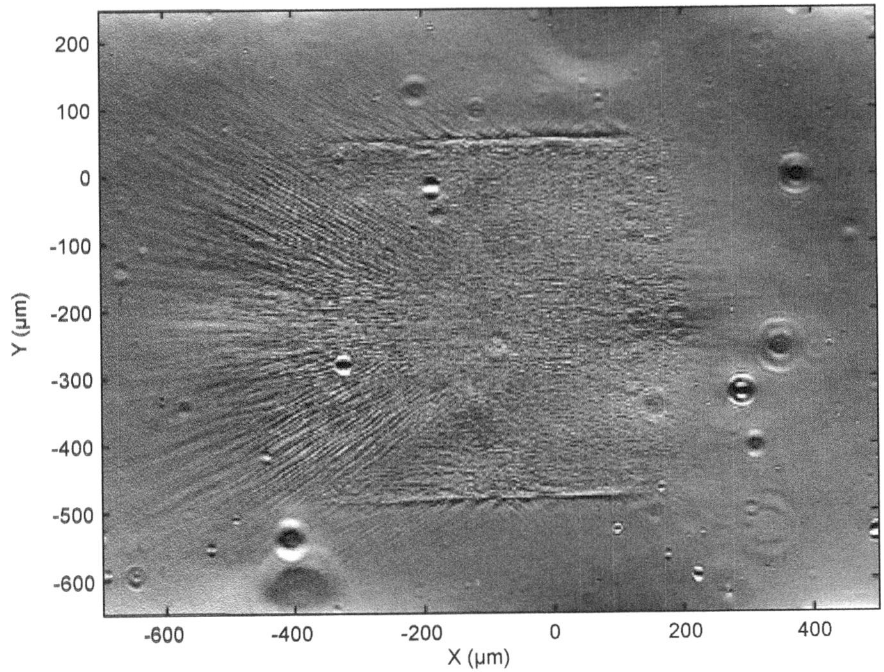

Figure 8.15 *Index modulation by the object beam in Fourier-plane holographic recording. Reprinted with permission from Applied Optics, Image oversampling for page-oriented optical data storage by Mark Ayres, Alan Hoskins and Kevin Curtis,* ***45****, 2459–2464 Copyright (2006) Optical Society of America*

these terms are never Bragg-matched directly by the reference beam during the normal operation of the storage device, hence motivating the use of 3D phase metrology to quantify their effects.

8.7 Interferometers

A large interferometer can be very useful for studying the large scale index changes and optical quality of the disk. In the past, InPhase and Bell Laboratries defined a Q parameter that was the Strehl ratio divided by the root mean square wedge. The goal was to keep the number close to one. Now the disks are made so well, that the Q factor is not used.

The interferometer is used in the transmission geometry at 632 nm which is nondestructive to the disk. The testing is done in transmission geometry with the interferometer then the disk and then a flat mirror. This allows the instrument to measure the entire disk for index changes and profiles of the optical path length changes. This is very good for noticing any problems with the mixing and bonding stage of disk manufacturing.

Figure 8.16(a) shows a typical disk. The minute variation shown is the highly reflective AR coating which is designed for blue wavelengths but the interferometer uses red light.

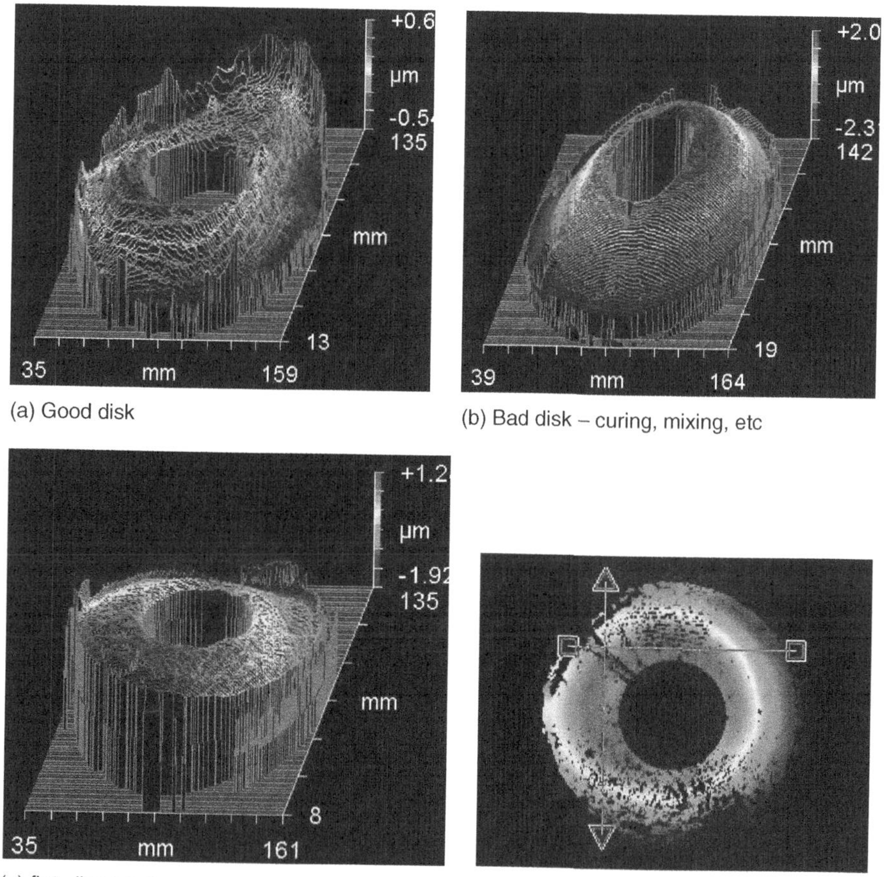

(a) Good disk

(b) Bad disk – curing, mixing, etc

(c) flat alignment

(d) splay pattern from injection misalignment

Figure 8.16 *Interferometer results*

This mismatch in wavelength causes some back-reflection that confuses the Zygo interferometer.

Figure 8.16(b) is a plot of a bad disk showing a large hump shape. This is either a disk that has cured too quickly on the flat (too much catalyst, too much heat in the mix head, or the bonding flat has residual heat), has been pulled off the flat early (the disk is not fully cured), or has a bad chemical ratio or bad chemicals or bad curing conditions for the backbone matrix. (Tests outside of the Zygo would give the exact cause.)

Figure 8.16(c) is of a disk where the flats have come out of alignment and need to be re-aligned (massive tilt on the disk). This is what a first trial disk would look like when bringing up a new bonding system.

Figure 8.16(d) is of a disk that has the signature of a 'splay' pattern that usually comes from misalignment of the injection head to the disk injection hole. The lines in the top right of the plot show the splay pattern. It can also be caused by a bad seal. If this effect is prevalent on numerous disks then it is due to injection head misalignment but if it only occurs occasionally then it is due to a bad seal.

8.8 Research Edge Wedge Tester

If the wedge of the disk exceeds a certain threshold, it can hurt the SNR of the holograms due to beam overlap or misalignment of the pitch across the hologram. In addition, if the induced pitch error is large enough that the wobble servo can not follow it, the drive would to have to recalibrate on that book which would reduce transfer rate significantly. For these reasons, radial pitch is a very important manufacturing parameter. The substrates themselves must be fabricated to within specification on radial pitch, the bonding process must yield the total disk pitch specification, and finally the edge seal must not affect the disk's edge wedge. In fact, the edge seal must allow the disk to maintain pitch conditions over the operating temperature range and survive the high temperatures needed for archival testing. Typically radial pitch is easy to achieve in the inside of the disk and harder to maintain at the edges of the disk. This is why it is called edge wedge or edge pitch.

In order to carefully measure edge wedge, a tester was developed to precisely measure the wedge. This tester uses the fact that a wedge causes a deviation in the angle of a beam:

$$\text{wedge angle} = 1/(n-1) \times (\text{beam deviation angle}) \tag{8.20}$$

With the index of refraction (n) known and a careful measurement of the deviation angle, the wedge angle of the disk can be calculated. Figure 8.17 shows the set-up used to measure edge wedge. The deviation is carefully measured by using a camera and measuring the centroid of the beam.

A measurement of edge wedge versus radius can be done by moving the disk in radius. The measurement can be repeated as the disk is rotated to measure along different radii. Figure 8.18(a) shows a disk that is out of specification (± 1 mrad). Figure 8.18(b) shows a disk that is within specification all the way out to 63 mm radius (edge of data area) even after a bake at 80 °C. The different plots are measurements along different radial spokes separated by 90° rotations around the disk. The dark box represents the initial specification range. This range was determined by not having to change pitch within a book and may be more stringent than required (see Chapter 14 for wobble servo and tolerances). If the wedge remains within the box then the rate of change of the edge wedge is also sufficiently good.

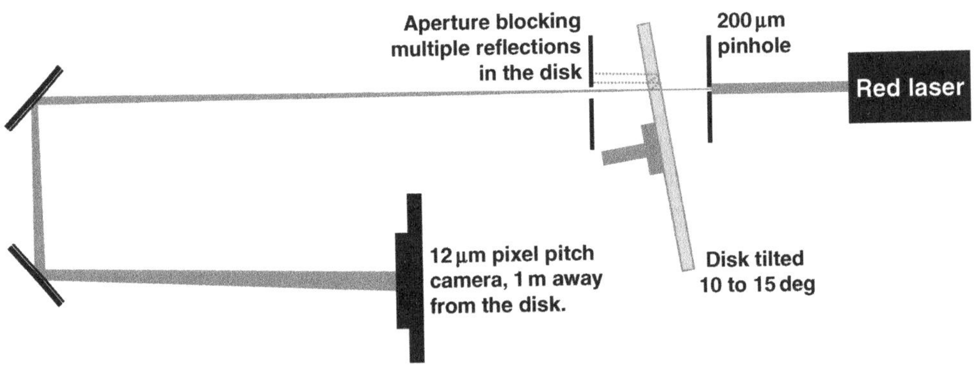

Figure 8.17 Diagram of a media edge wedge tester

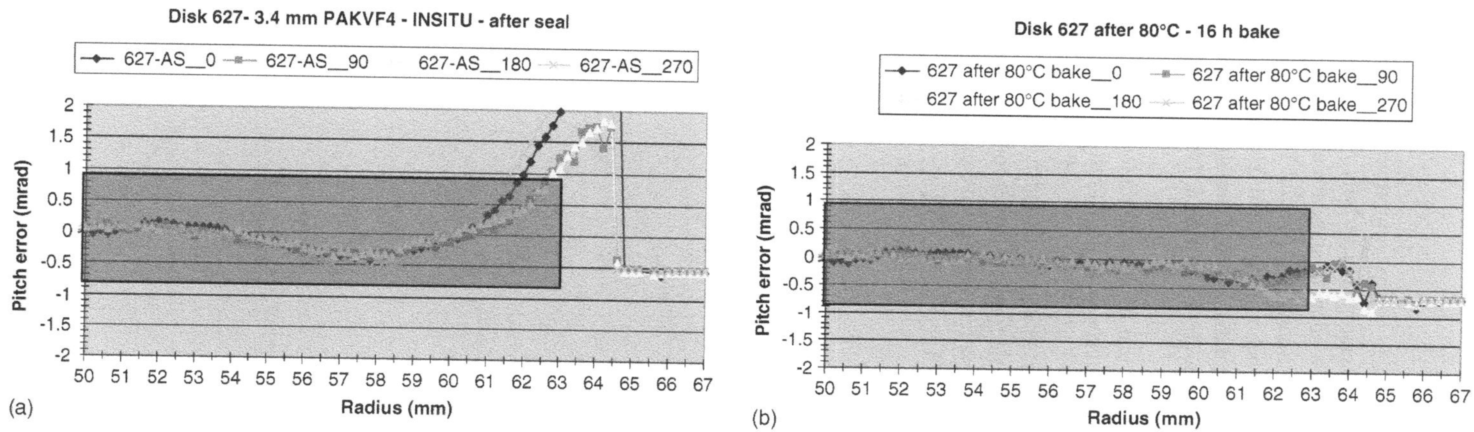

Figure 8.18 *Edge wedge versus disk radius for several different radial spokes: (a) disk that is out of specification; (b) disk that is in specification even after baking at 80 °C*

8.9 Defect Detection

Defect defection and classification is a very important volume manufacturing issue to determine if the media is in specification. Fortunately the traditional optical disk manufacturers face the same issues. Two vendors of standard inspection equipment can and have been used (Dr Schenk GmbH and Dr Schwab Inspection Technology GmbH). Other vendors could be used as well but have not been investigated. However, the inspection equipment must use red wavelengths to not affect the material, and the thresholds for determining what a defect is have to be optimized for InPhase media. Dr Schenk was used for the initial InPhase media inspection.

These machines can extensively test media parameters for sampling purposes or test a few critical parameters for in-line inspection. In addition to defects, these machines can inspect substrates and full media for birefringence, reflectivity, radial and tangential tilt, vertical and axial deviation, transmission, optical density, and substrate/full media/recording layer thicknesses. It is also possible for them to measure edge wedge of the disk as well as sync the defect data with the servo pattern so the drive can skip known defects. The defect map and size classification is shown in Figure 8.19.

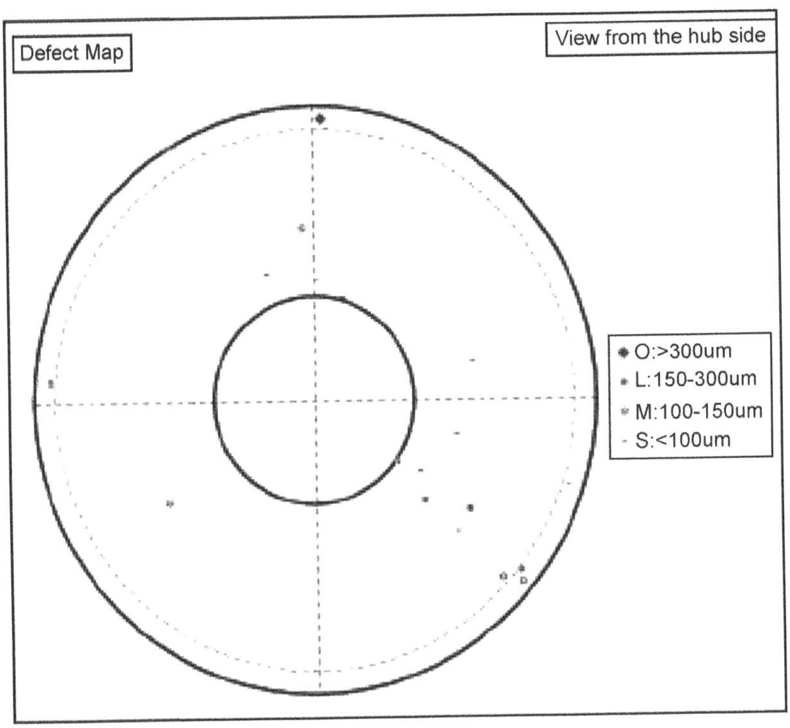

Figure 8.19 Defect data from an InPhase disk as measured by the Dr Schenk tester.

8.10 Digital Testing of Media Properties

InPhase Technologies can also use its professional drive to measure a variety of parameters while recording data at high densities. This is extremely useful as, obviously, the drive measures over exactly the correct angular bandwidth and has the fast exposures and timing that are used in a storage product. The speed of recording can cause important differences in the media response. With the current media, a number of important time limits on media usage can be determined and eventually optimized from a HDS system viewpoint.

8.10.1 Scatter

The InPhase professional drive has exactly the right angular bandwidth to measure scatter that will affect drive performance for various formulations. While the BSDF measurements made by the MST can be used to quantify the relative scatter properties of different media packages, it is difficult to use them to accurately predict the scatter level that will be seen by the drive. The procedure to measure scatter using a drive is:

1. Take several exposures of using the drive camera while keeping the shutter closed. This action will calibrate the camera and determine the amount of dark signal captured by the camera.
2. Repeat the exposures this time illuminating the media package with the reference/probe beam. Typically a grid of spots on the media is used so the scatter can be sampled using several reference/probe beam angles.
3. Calculate the scattering function over the solid angle subtended by the camera by subtracting the dark signal from the exposures captured in step 2, and dividing by the effective responsivity of the camera and the power in the reference/probe beam.

Typically the steps are performed for both virgin and cured media. It is very common that the post-cure media scatter is higher than the scatter from the unexposed media. Also the scatter from other material formulations can be compared in a very meaningful way. The scattered power for three different formulations is shown in Figure 8.20 for a single reference/probe

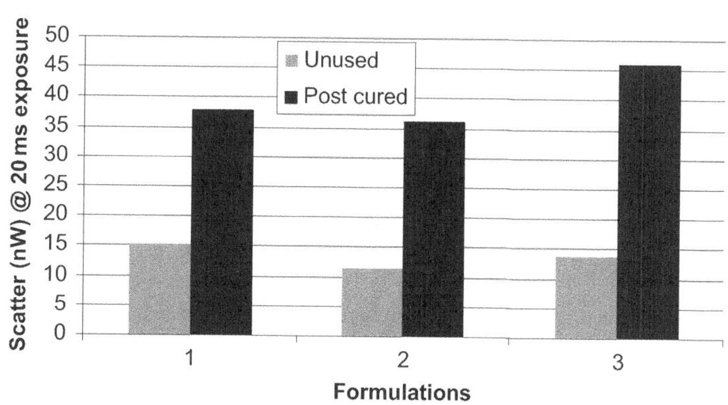

Figure 8.20 Scatter levels for three formulations

beam angle. The virgin media in all three formulations has roughly the same scatter but formulation number 3 has much more scatter after post-cure.

8.10.2 Media Sensitivities and *M/#* Usage

The drive can measure *M/#* usage and sensitivities by recording a grid of books at density. The *M/#* usage is a strong function of the recording speed. The faster the recording speed, the more efficiently the media is used. This is evident by the fact that the recording schedule flattens out with higher recording speeds which in turn lowers the overall average exposure time. Because of beam overlap, beam power ratios, and the complex nature of the data beam and the spatial frequency response of the media, the sensitivity typically seen in the drive is different than measured in the plane wave tester. The *M/#* and sensitivities are measured by the same procedure as the plane wave tester except that the *M/#* is normally taken to about 80% of saturation.

Finally, it is important to understand that the capabilities of a media to support density are not solely determined by the *M/#* and scatter. Shrinkage, which can cause media deformation, is also critical. In general, it is better to have a small margin for *M/#*. High excess of *M/#* is harmful in that it will have to be cured, causing a transfer rate hit, and it can induce media distortions making formatting more difficult. A higher *M/#* does not compensate for more shrinkage.

8.10.3 Media Timing Tests

The major media time effects that must be measured for proper drive operation are pre-cure and post-cure time-out. These parameters are needed to determine a format for this disk as covered in Chapter 14. In a phenomenon called pre-cure time-out, a piece of media sitting idle after pre-cure will see the inhibition gradually come back. This effect is most likely due to oxygen diffusing back into the area. To test the time constant of this phenomenon, books of holograms are written with different delays between pre-cure and the start of recording. As the inhibition returns, the first pages in the books will lose diffraction efficiency as they have to overcome the inhibition. A limit on the loss of diffraction efficiency can be set to define the time-out. With the current Tapestry formulations, pre-cure time-outs are typically around 10 min.

The post-cure time-out is a limit on the amount of time recorded data can sit before being cured. This time constant varies with the media formulation, recorded data density and temperature. It is measured by recording a number of areas at a given density and post-curing them after different delays. Typically, the average SNR falls with the amount of time between recording and post-cure. With the current InPhase Tapestry media formulation, the post-cure time-out is on the order of 3 h for a 0.25 dB SNR drop.

8.10.4 Media Termination Test

SNR often proves to be a more sensitive measure of grating deformation than diffraction efficiency. Most media formulations are designed to stop polymerization (i.e. terminate) within a reasonable duration to allow some diffusion. For some formulations this termination process is incomplete and continuous monitoring of the SNR will show a degradation of the SNR before any drop in the diffraction efficiency is evident.

8.11 Accelerated Lifetime Testing

8.11.1 Introduction

Media shelf life (the length of time the media can sit before recording without degradation) and archival lifetime (the length of time recorded data are recoverable without degradation) are critical parameters for the product. The use of accelerating aging techniques (higher temperatures and humidity levels) to estimate the lifetime is standard for storage media [13–15] and there are two international standards for optical media [16,17]. InPhase uses these procedures and standards to determine shelf and archival lifetimes of holographic media.

The general approach of using Arrhenius and Eyring models in accelerated aging studies in any technology is predicated on the idea that the principal mechanism determining the lifetime is known, that the metric is a valid measure of that mechanism, and that no other chemical or physical processes are significant. Since hologram formation in photopolymers is inherently driven by chemical (polymerization) and diffusion processes, these mechanisms can be characterized with activation energy as required by the Arrhenius model. InPhase and Maxell have done a lot of work to ensure that other physical effects, such as disk warping, hub adhesion, and so on, do not limit the lifetime of holographic media.

The general model is based on the Eyring equation:

$$t = aT^{\alpha} e^{\frac{b}{T} + (c + \frac{d}{T})RH} \tag{8.21}$$

where t is the archival lifetime, T is the temperature in degrees Kelvin, RH is relative humidity and a, α, b, c and d are constants. In the simplest case and as per the standards, a, α, d, RH and c are set equal to zero. In this case, the equation can be simplified by taking the natural logarithm to $\ln(t) = \ln(a) + b/t$. Given that the $T = 25\,°\mathrm{C}$ for storage, $t_{25°\mathrm{C}} =$ 50 years, $T =$ test temperature (i.e. 80 °C), and t_{test} is the measured failure time, the constants a and b can be found given the failure time. This procedure is repeated for several temperatures to improve accuracy. This whole procedure has to be repeated a few times with different relative humidity to decide on acceleration due to humidity.

InPhase initially focused on 80 and 90 °C as primary test temperatures with 100 °C as the third temperature that is used. The initial testing is being done at dry and 95% relative humidity (the two possible extremes). More relative humidity points will be added in time. Figure 8.21 shows test temperature and relative humidity versus expected lifetime. Since lifetime is a strong function of temperature, it should be possible to determine a 50 year lifetime in around 1000 hours at 80 °C.

8.11.2 Media Shelf Life Testing

As stated earlier, shelf life is the amount of time that recording media can still be used without degradation after manufacture. With holographic media, aging after manufacture can negatively affect sensitivity and *M/#*. The acceptable loss of *M/#* is arbitrarily set to 10% of the original *M/#* as the drive can still perform acceptably at this level. The measurements to find the shelf life time constant can be made with the plane wave tester and an oven to accelerate the aging. Figure 8.22 shows the effects of accelerated aging at 80 °C on sensitivity and *M/#*. The *M/#* is unaffected after 675 h; unfortunately the media sensitivity is down by 25%. Reductions in sensitivity up to about 10% do not affect recording speeds.

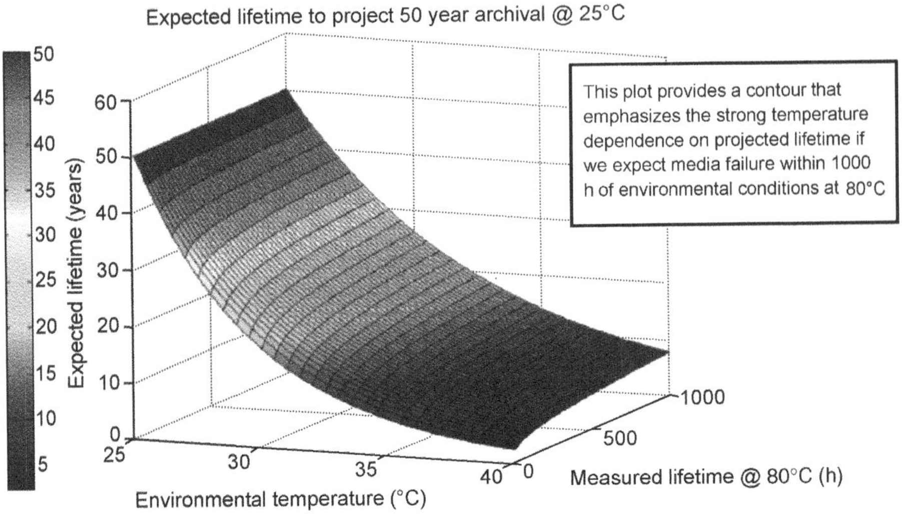

Figure 8.21 *Plot of test temperature and relative humidity versus expected lifetime*

(The InPhase drive can adjust for these changes by increasing power.) Sensitivity losses greater than 10% can, unfortunately, negatively impact the write transfer rate of the drive. The preliminary data presented suggest that shelf life is significantly longer (~10 years) than the original specification of 5 years.

8.11.3 Disk Archive Testing

The standard definition for archival failure is the device exhibiting unrecoverable errors after all error correction codes have been applied. This definition can be applied to holographic storage or any other optical storage technologies. While this definition is ultimately the most important criterion, it is informative to measure media degradation before this limit is reached. For both CDs and DVDs, it is possible to measure a raw error

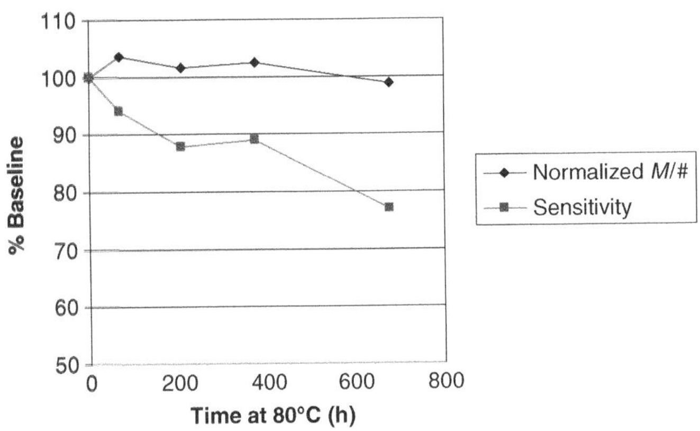

Figure 8.22 *Shelf life tests of sensitivity and M/#*

before error correction is applied. This is possible because both CDs and DVDs use a 'hard decision' error correction scheme: a Cross-Interleaved Reed–Solomon Code (CIRC) for CDs and a Reed–Solomon Product Code (RS-PC) for DVDs. In both these Reed–Solomon codes, a hard (firm) decision is made for each data bit to decide whether it is a 1 or 0. It is then the task of the Reed–Solomon error correction scheme to fix the decisions that were incorrect.

In recent years, so-called 'soft-decision', iterative, error correction codes have emerged. They are widely used in telecommunications, particularly in wireless applications, and have considerably more error correcting capability than hard-decision codes. Of the iterative codes available, Low Density Parity Check codes are most likely to be used in future data storage applications. Their performance improvement over Reed–Solomon codes is so great that it is probable that these codes will eventually become part of the data channel in future optical and magnetic drives, in spite of their considerably more complex implementation. The improved error correction capability of these iterative codes has led to their use by InPhase Technologies as the primary error correction code in the data channel.

Unlike hard-decision codes where a bit is determined to be a 0 or 1 prior to correction, the pertinent characteristic for determining archival lifetime for systems with soft-decision codes is the probability or likelihood of bit being a 0 or a 1. The decoder iterates over the tentative bit values, modifying them as it progresses, to finally arrive at a self-consistent set of bit values that have minimum probability of error. In the archive drive, the higher levels of error correction codes are hard-decision codes. However, the combination of soft and hard codes results in a very stong function of SNR versus corrected errors. Consequently, for archivability metrics, there is no intrinsic equivalent of the 'raw' error rate available with hard-decision codes. The primary tracking metrics that we use are SNR and diffraction efficiency of the pages and books as these give a better measure of degradation.

The basic material's archive properties can be tested with a plane wave test of *M*/# of multiplexed holograms and its degradation with time at elevated temperatures. Figure 8.23 shows a plot for data taken at 80, 90 and 100 °C with failure defined as 8% loss of *M*/# using all the holograms multiplexed. The natural log of the hours is plotted (1200 h at 80 °C) to allow the data to be fit. Using this, it is possible to estimate the archival lifetime of the material at 25 °C as 339 years.

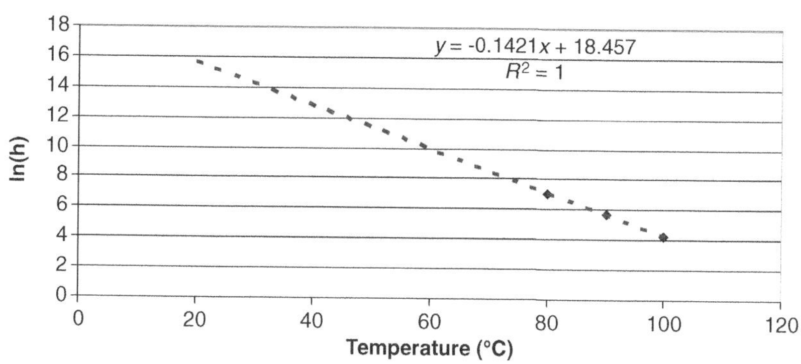

Figure 8.23 *Archive data on Tapestry material*

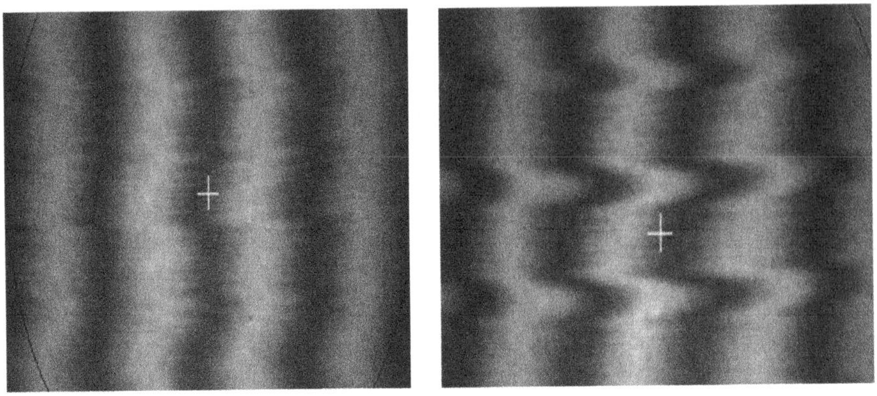

(a) 600 μm track spacing (b) 800 μm track spacing

Figure 8.24 *Interferometer photographs of media aged 265 h at 80 °C with different track spacing*

Thus it appears that the InPhase Tapestry material is very good at retaining its diffraction efficiency for a long time. Plane wave holograms are very uniform in the medium and thus cannot capture the SNR loss experienced by high angular bandwidth digital holograms caused by distortion of the media. These distortions are due to diffusion of components and cannot affect SNR before any diffraction efficiency losses are apparent (339 years later). This degradation is strongly dependent on the particular formulation of the holographic media and does not occur in some formulations. For media where this effect is evident, the key for good archive retention is to make the illumination of the media as uniform as possible. This uniformity includes track spacing, reference beam profile, and media post-cure profile. By spacing the tracks too far apart, steps in the index of refraction can build up between and across the tracks and can cause SNR loss (Figure 8.24). The 800 μm track spacing shown in Figure 8.24(b) illustrates more index change between the tracks than the 600 μm track spacing shown in Figure 8.24(a).

The InPhase professional drive uses an aperture with soft edges to define the reference beam size and to eliminate diffractive ringing of the beam in the media. If the reference beam is too small or the beam has not been properly apodized, then the intensity nonuniformity of the beam in the initial media can show slight degradation in the SNR as the media ages. Many of the writing techniques discussed in Chapter 14 are aimed at making the media usage more uniform to reduce these effects. Newer media formulations do not show any degradation due to these effects. Regardless, the SNR loss due to these factors can be estimated by examining interferometer pictures of the disk after accelerated aging. With media formulations where this is a factor, 90% of this effect is noticeable within the first 150 h at 80 °C.

An example of archive testing of the 300 GB recorded disk (Format Example 1; see Chapter 13) is shown in Figure 8.25. Here the entire disk was recorded and then bookcases were examined in detail immediately after recording and after accelerated aging. Collected data include diffraction efficiency as a function of probe beam angle (the diffraction peak shape can be used to estimate media distortion), page SNR, and raw camera images. Figure 8.25 compares a recovered bookcase after 165 h at 80 °C with the original recovery

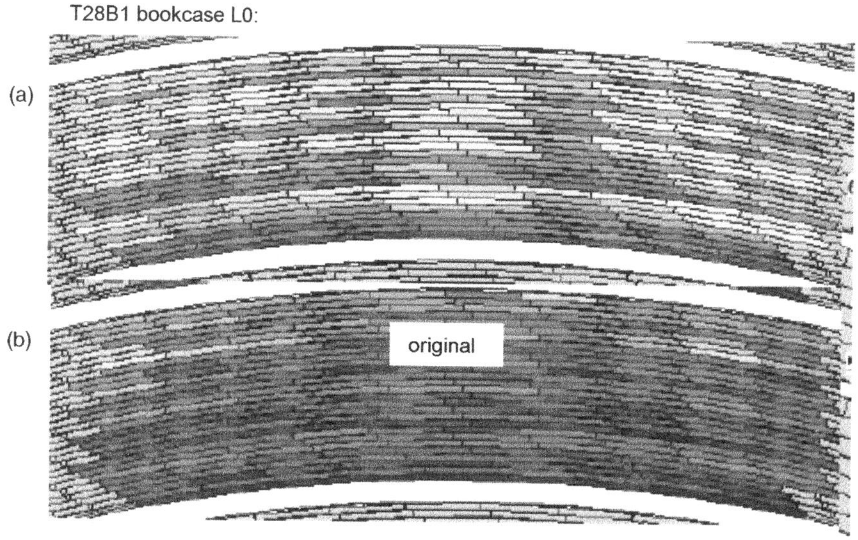

Figure 8.25 (a) Recovery after 165 h at 80 °C and (b) original recovery

after interchange. Each rectangle represents a book of 80 holograms. These books actually overlap in the media but have been separated radially in the figure for clarity. The shade of the rectangles is representative of the average SNR of the book. The SNR ranges from 4.5 (white)to 6 dB (black)in Figure 8.25. The average SNR loss for this bookcase and this media formulation is 0.4 dB from an original average SNR of 5.2 dB. This is close to the total loss expected over 50 years. For reference, the InPhase drive channel's lower recovery limit is 1.8 dB with three levels of error correction. While this is a very good result, newer media formulations show even better performance. More extensive thermal and relative humidity tests of full disk formats continue in final product qualification.

8.11.4 Edge Seal Testing

InPhase media packages have an edge seal to prevent water from coming through the media edges at a higher rate than through the polyolefin based substrates. Without an edge seal, the disk can develop a wedge around the outer circumference that distorts the media and destroys the recorded holograms. If the wedge occurs before hologram recording, the induced pitch of the reference and probe beams can make successful recording and recovery impossible. Thus, one of the most important aspects of edge seal testing is verifying the seal does not cause a wedge during application, during normal drive operation, and during the accelerated aging conditions used for lifetime testing. The edge wedge can be measured using the edge wedge tester shown in Figure 8.18. The wedge has to be identical for disks stored in both wet and dry conditions. Water absorption by the media is measured using a near IR spectrometer. The OH change is correlated with weight gain. Figure 8.26 shows the absorbance of water for an edge sealed disk at various different radii for a disk put into the oven for 80 h at 80 °C and 95% relative humidity. Figure 8.26 demonstrates a good edge seal as the water content is nearly uniform across the disk. Without an edge seal the outer parts of the disk would have several times the concentration of water as the inner tracks of the disk.

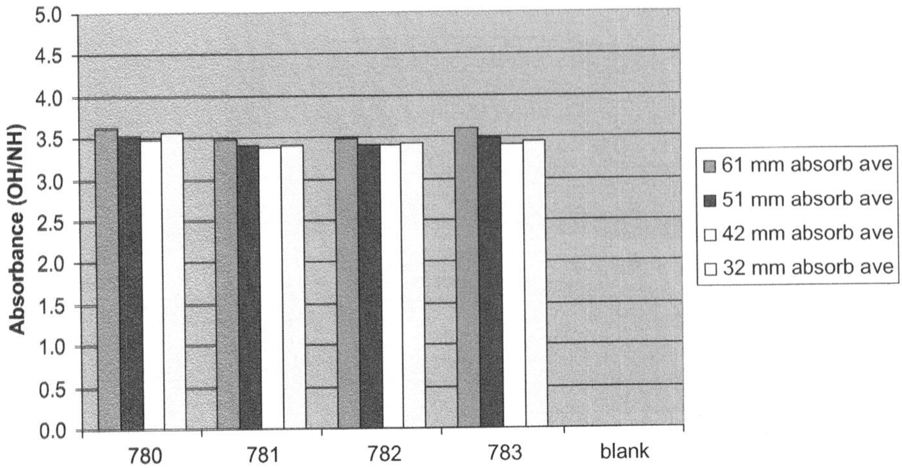

Figure 8.26 *Water absorption at various radii on the disk after 80 °C/95% relative humidity*

Acknowledgements

We would like to thank Paul Smith, Friso Schlottau, Peter Whiteside, Ben Ihas, David Michaels, Sam Miller, Ian Redmond, David Beal, Ernest Chuang, Sergei Orlov, Michael Tackitt, Cotton Anderson, Keith Farnsworth, Sean Quirin, Melinda Schnoes, Fred Askham, David Samuels, Charles Stanhope, Ken Anderson, William Grush, Adrian Hill and Martin Pane.

At Hitachi Maxell we sincerely thank Tatsuya Imazu, Takashi Hanyu, Tetsuo Morimoto and Tomizo Taniguchi.

References

1. H. Kogelnik, Coupled wave theory for thick hologram gratings, *The Bell System Tech. J.*, Vol. 48, No. 9, pp. 2909–2946 (1969).
2. L. Dhar, *et al.*, Temperature-induced changes in photopolymer volume holograms, *Appl. Phy. Lett.*, Vol. 73, No. 10 pp. 1337–1339 (1998).
3. P.K. Tien, *et al.*, 'Modes of propagating light waves in thin deposited semiconductor films, *Appl. Phy. Lett.*, Vol. 14, pp. 291–294 (1969).
4. P.K. Tien, Light waves in thin films and integrated optics, *Appl. Opt.*, Vol. 10, pp. 2395–2413 (1971).
5. R. Ulrich and R. Torge, Measurement of thin film parameters with a prism coupler, *Appl. Opt.*, Vol. 12, pp. 2901–2908 (1973).
6. F. B. Leloup, *et al.*, Design of an instrument for measuring the spectral bidirectional scatter distribution function, *Appl. Opt.*, Vol. 47, No. 29, pp. 5454–5467 (2008).
7. K. Anderson, E. Fotheringham, A. Hill, B. Sissom and K. Curtis, High speed holographic data storage at 100 Gbit/in^2, *Joint Int. Symp. on Opt. Memories and Opt. Data Storage*, Honolulu. Hawaii, July (2005), paper ThE1.
8. R. R. McLeod, *et al.*, Micro-holographic multi-layer optical disk data storage, *Appl. Opt.*, Vol. 44, pp. 3197–3207 (2005).

9. M. Ayres, Signal modulation for holographic memories, PhD Thesis, University of Colorado, Boulder, Colorado (2007).
10. M. R. Ayres and R. R. McLeod, Volumetric phase metrology for optical data storage, Opt. Data Storage Conf., Montreal, Canada (2006), pp. 132–134.
11. Y. Kawata, R. Juskaitis, T. Tanaka, T. Wilson and S. Kawata, Differential phase-contrast microscope with a split detector for the readout system of a multilayered optical memory, *Appl. Opt.*, Vol. 35, pp. 2466–2470 (1996).
12. J. W. O'Byrne, *et al.*, Adaptive optics in confocal microscopy, Proc. 2nd Int. Workshop on Adaptive Optics for Industry and Medicine, pp. 85–91 (1999).
13. O. Slattery, R. Lu, J. Zheng, F. Byers and X. Tang, Stability comparison of recordable optical discs – a study of error rates in harsh conditions, *J. Res. Natl Inst. Stand. Tech.*, Vol. 109, pp. 514–524 (2004).
14. C. Sodergard, J. Martovaara and J. Virtanen, Research on the life expectancy of the CD-R, *VTT Information Tech. (Finland)* (1995).
15. V. Navale, Measurements for predicting life expectancies of higher density magnetic and optical media, NARA Report (2004).
16. ECMA-379, Test method for the estimation of the archival lifetime of optical media, 1st edition (2007).
17. AES38-2000, AES standard for audio preservation and restoration – Life expectancy of information stored (2000).

9

Tapestry Drive Implementation

Kevin Curtis, Ken Anderson, Adrian Hill and Aaron Wegner

9.1 Introduction

InPhase Technologies has developed a 300 GB, 20 MB s^{-1} holographic recordable drive (Tapestry®) and associated removable disk media for professional archive applications. The InPhase Architecture, based on phase conjugate angle-polytopic multiplexing, is described in Chapter 3 and is not repeated here. This chapter describes the important components and subsystems that make up this drive and presents the associated write and read transfer rate models.

The Tapestry drive was developed following a standard product development life cycle, as shown in Figure 9.1. InPhase Technologies spun out of Bell Laboratories after more than 6 years of research focused on the data channel, materials, recording media, and basic digital systems. In 2001, InPhase built its first series of angle multiplexing test beds to provide a vehicle to develop the electronics architecture, design a compact storage lens, and define optomechanical tolerances for performance and interchange. These test beds were based on commercially available green (532 nm) lasers. In parallel, InPhase, in cooperation with Sony Corporation's Laser Division, started development of a single-mode, external cavity blue laser.

Once a sufficiently stable single-mode blue laser had been developed, test beds based around these lasers were constructed. These test beds incorporated phase conjugate reconstruction and angle-polytopic multiplexing, and were able to achieve localized high density data recordings. They also provided a platform for blue media testing, tolerance analysis, data detection and servo development.

Holographic Data Storage: From Theory to Practical Systems Kevin Curtis, Lisa Dhar, Adrian Hill, William Wilson and Mark Ayres

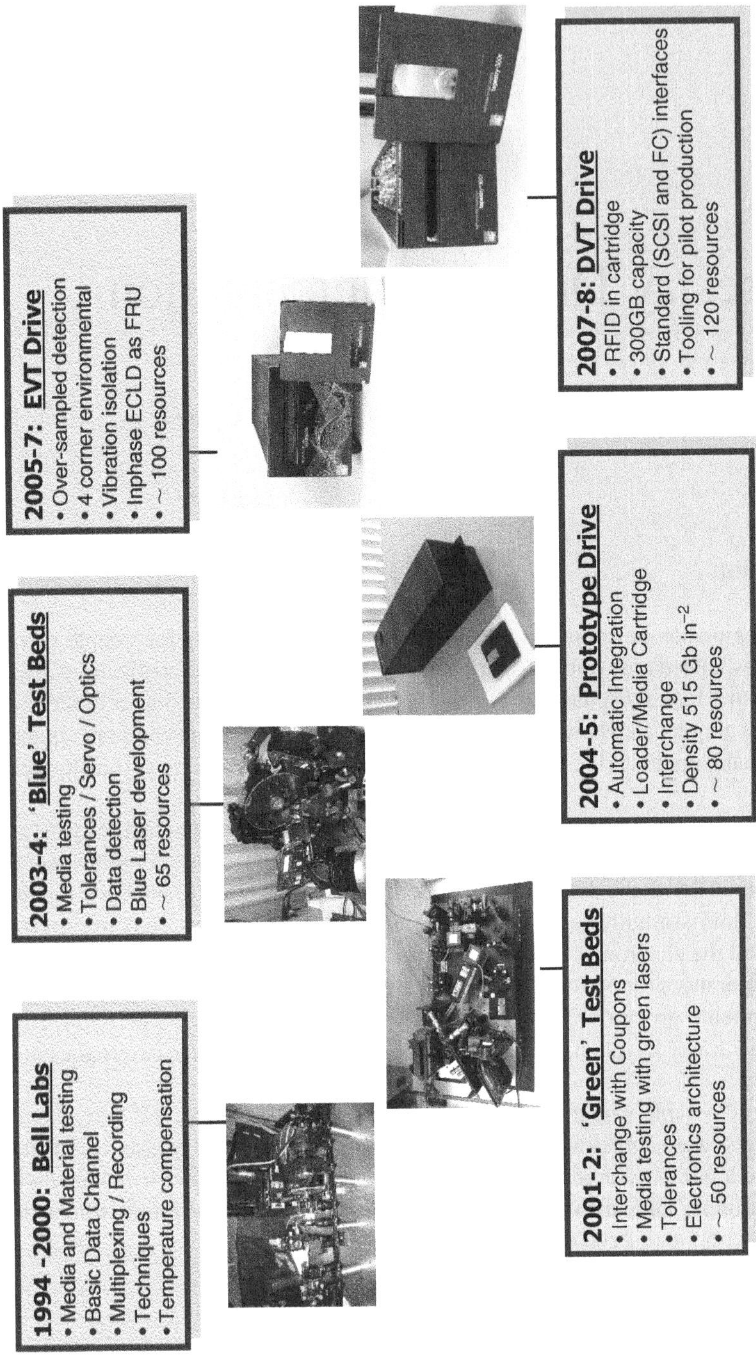

Figure 9.1 Development cycle

A semi-integrated prototype drive was developed at the same time as the test beds, and was used to demonstrate recording density and transfer rate. The prototype packaged all the optics and mechanics into a drive enclosure, with the electronics set contained in a separate frame. The majority of the optics and optical mounts were commercially available parts, while the electronics and loader were custom components. The prototype drive facilitated the development of the media cartridge and loader, and was used for investigating vibration isolation, the initial automated media interchange algorithms, and media formats.

With the knowledge gained from the high density test beds and the prototype drives, the Engineering Verification Test (EVT) phase of the program was started. The EVT drive was the first ever working holographic drive in form factor, with all the electronics, mechanics, and optics integrated inside the same enclosure. This drive integration step allowed for more comprehensive testing of the drive, including thermal tests in environmental chambers, and shock and vibration tests, in addition to the verification of the basic functionality.

As is typical in such product cycles, the results of these tests led to several design changes that were implemented in the Design Verification Test (DVT) phase of the product development. In addition, many design for manufacturing improvements were incorporated into the DVT design.

Other modifications included changes to the packaging, and electronics to improve robustness, vibration performance, cooling, manufacturability, and reliability. To further improve drive reliability and minimize drive down time, a replaceable laser diode system was incorporated. A pitch correction mechanism was added to reduce the drive assembly and thermal tolerance requirements. Vibration and shock performance was improved by changing the mounting system to coincide with the center of mass and by increasing the sway space. A new PCBA, the 'hood' board, was incorporated to improve cooling and make the electrical connections simpler and more reliable. Figure 9.2(a) shows the drive with the top of the enclosure removed. The hood board is visible, extending over most of the optical mechanical assembly (OMA). The metal frame at the rear of the drive houses the rest of the electronics, protects them from shock, and helps reduce electronic noise emissions from the drive.

Figure 9.2(b) shows the media cartridge with the shutter open to show the disk. The holographic recording media consists of 1.5 mm of InPhase Tapestry™ photopolymer sandwiched between two 130 mm diameter, 1.0 mm thick polyolefin substrates. One of the substrates contains a tracking pattern for tangential positioning. Both substrates have an anti-reflection (AR) coating that is optimized for 405 nm. The disk package has a magnetic hub for securing it to the drive spindle. The entire disk is encased in a light-tight cartridge.

The Tapestry drive is three standard half-heights tall (4.875 in.), one standard 5.25 in. form factor wide (5.75 in.), and 27.5 in. long.

By the end of 2008, the drive was performing to all product specifications except transfer rate. The transfer rate was about 5 MB s^{-1} for both read and write, limited by firmware and electronics. In 2009, the electronics speed was increased and the firmware improved for both speed as well as for error handling. By incorporating these improvements, InPhase has finally produced the first commercially viable holographic data storage product, and the first design is ready to go into low volume production. Future generations will increase performance and lower cost and complexity. This chapter describes in more detail the product design, components, and basic manufacturing of the first generation Tapestry drive.

(a)

(b)

Figure 9.2 (a) DVT with top of enclosure removed. (b) Disk inside cartridge

9.2 Optical Implementation

9.2.1 Architecture

CAD drawings of the top and bottom views of the Tapestry drive are shown in Figure 9.3. The foundation for almost all of the drive components is a solid metal frame, which also provides the reference for optical alignments. The media loader and the spindle for rotating the disk are clearly visible in the lower right of Figure 9.3(a). The optomechanical assembly

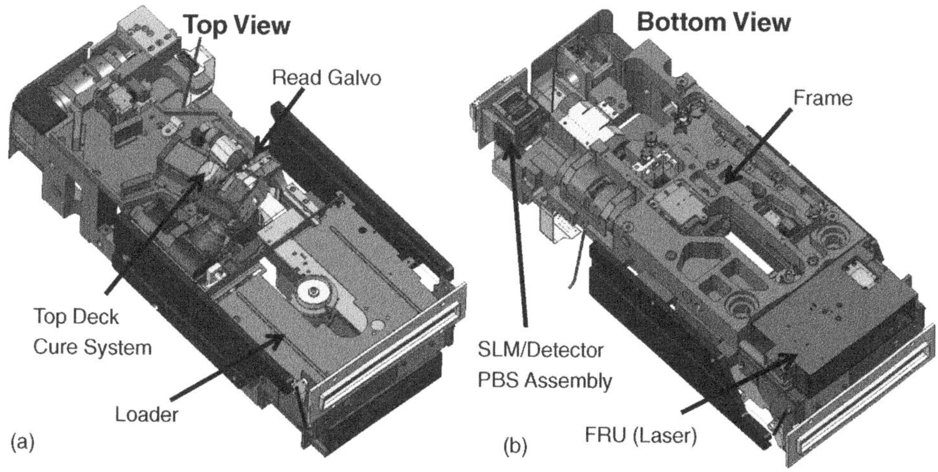

Figure 9.3 (a) Top and (b) bottom views of the OMA

(OMA) attaches to the vibration isolation system (on the long sides of the drive) which is also a part of the loader.

The cure/toaster subsystem is on the top deck of the drive, with the cure system, read galvo, and part of the toaster projecting over the loaded media position.

The part of the OMA which actually records and reads the data consists of the beam conditioning optics, the data path and the reference path. The SLM/camera and polarizing beam splitter assembly of the data path is visible in the top left portion of Figure 9.3(b). The illumination for the SLM enters from the top right of the figure.

The laser unit is the gray block in the front of the drive in Figure 9.3(b). To accommodate the alignment tolerances associated with removing and replacing the laser unit, several imaging systems that image the laser output into the drive are located under the loader. These imaging systems employ a series of mirrors under the frame to match the path lengths of the signal and reference beams.

Path length matching has proved essential for avoiding the negative effects of grating motions within the cavity of the external cavity laser diode (ECLD). Even though the laser has a large coherence length, it is very sensitive to tiny motions of the grating. (See Chapter 4 for an extensive description of the laser.)

In a precursor to the current drive, a path length mismatch of about 430 mm caused the laser to be susceptible to motions of as little as 10 nm. The allowable grating motion is given by:

$$\delta d = 2\Delta f_{tol} d^2 \Delta t / \Delta x q$$

where δd is the maximum grating movement during the exposure interval Δt, Δf_{tol} is the tolerable frequency shift (which is on the order of 100 Hz), Δx is the path length difference between the beams, d is the cavity length of the ECLD and q is the mode number.

The experimental validation of the improvement from path length matching is shown in Figure 9.4. Figure 9.4(a) shows the fringe visibility when the path length mismatch is 430 mm. The upper curve in Figure 9.4(a) is the fringe visibility and the lower curve

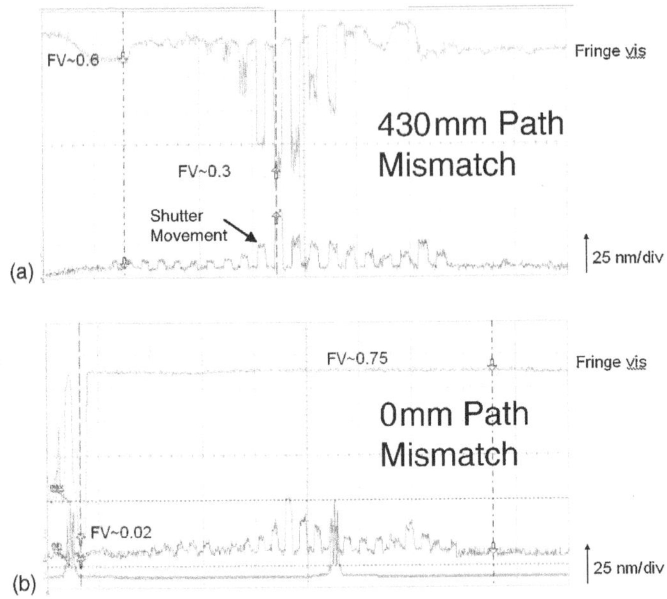

Figure 9.4 *Fringe visibility and grating perturbation for (a) 430 mm path length difference and (b) 0 mm path length difference*

in Figure 9.4(a) is the measured displacement of the grating. Ideally the fringe visibility would remain constant so that holograms can be recorded with high and stable visibility. Figure 9.4(b) shows that the visibility remains constant when the system is path length matched, even when the laser grating is mechanically displaced.

Maintaining high fringe visibility during recording is crucial to recording holograms with high SNR. As the fringe visibility decreases, the diffraction efficiency of the holograms also decreases, which results in a lower signal component of the signal to scatter ratio (SSR). The hologram SNR in turn is strongly dependent on the SSR. Figure 9.5 shows the dependence of SNR on fringe visibility for two values of SSR.

The laser subsystem continuously monitors the laser fringe visibility during recording. If a fringe visibility below a threshold is detected during a write operation, the current book is marked as 'bad' and is rewritten at the next location in the media.

Mitigating the effects of drive vibration was widely considered to be a major hurdle to producing a viable holographic storage drive. In practice, deleterious effects from vibration have proved to be controllable with good servo engineering, adequate isolation, and the use of short (approximately 1 ms) write exposure times.

The remainder of this section will highlight several key drive components whose designs are crucial to the drive performance.

9.2.2 Field Replaceable Unit (FRU)

The Tapestry drive is designed to have a lifetime of 100 000 h, which exceeds the expected lifetime of the existing blue laser. For this reason, the laser is part of a FRU. The

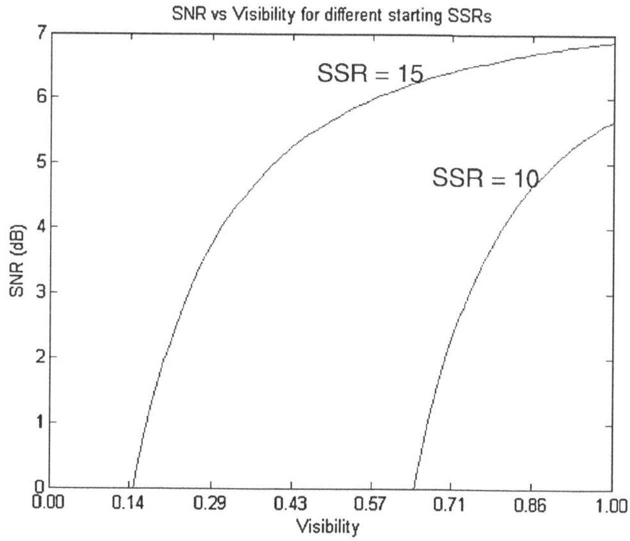

Figure 9.5 *SNR versus normalized fringe visibility for two SSR values*

FRU, shown in Figure 9.6(a), contains the laser, collimator, grating, isolator, and the electronics to monitor and control the laser. The FRU mounts kinematically to the OMA baseplate on the rails shown in Figure 9.6(b). The rails are at the front of the OMA so that the FRU can be replaced by removing the front panel, unlocking the mounting mechanism, and sliding the FRU off the rails. The electrical connection is a single flex cable that can be disconnected after the unit has been slid off the rails. The FRU can be replaced in the system by connecting to the flex cable, sliding the unit onto the rails until it touches the back, and locking in place using a half turn with a hexagonal wrench. No alignment of the system is needed. Each FRU is calibrated separately, and the calibration data are stored within the FRU electronics. The FRU can predict its failure by monitoring the current and power at nominal temperature, which significantly improves the reliability and availability of the drive. Since the FRU is a self-contained unit, it can be manufactured separately, and the performance can be enhanced without requiring changes in the rest of the drive.

9.2.3 Shutter

A galvanometer with a small blade attached is used as a shutter to control whether light from the laser is blocked or allowed to pass to the rest of the drive. The blade is placed near the focal plane of the first imaging relay lenses to minimize the required blade motion. These relay lenses image the light from the laser into the drive to improve the FRU alignment tolerances. Figure 9.7 shows a CAD drawing of the galvo shutter assembly located on one side of the frame. This shutter implementation dramatically reduces vibrations compared with more standard shutter designs using linear motion of one or more blades to block a beam.

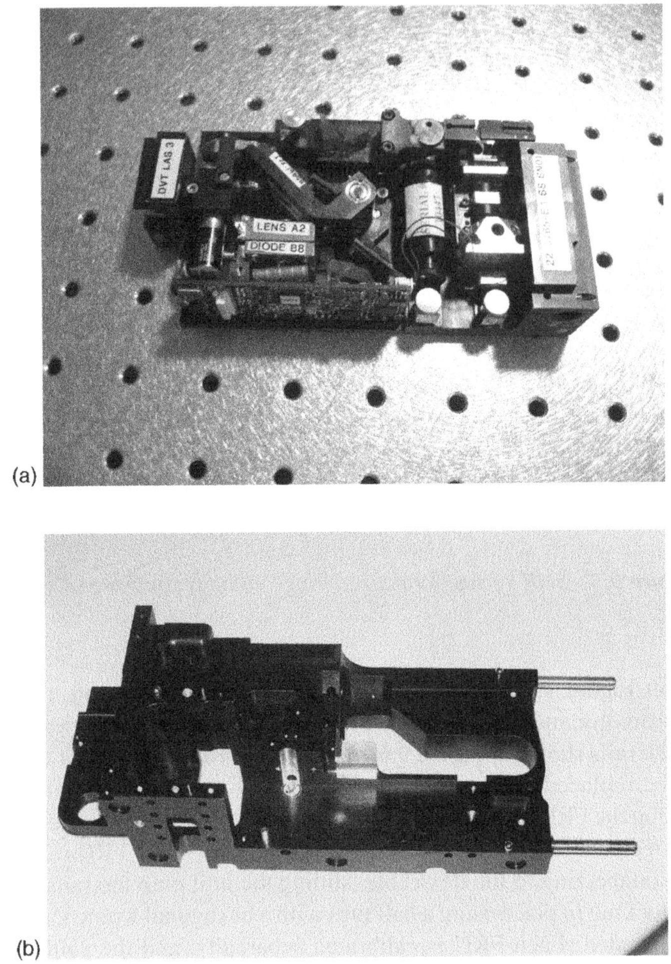

Figure 9.6 (a) Laser FRU without cover. (b) OMA baseplate with rails for mounting FRU

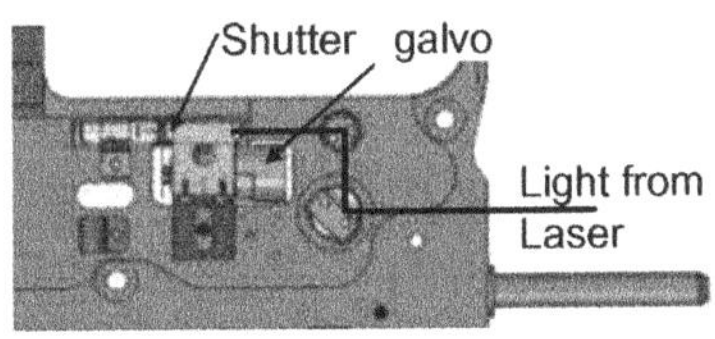

Figure 9.7 Galvanometer based shutter

9.2.4 Optical Divider

The optical divider splits the beam from the FRU into the data and reference beams in the OMA. Figure 9.8 shows the optical divider, which consists of a polarizing beam splitter (PBS), a half wave plate and a rotary motor [visible at the rear of Figure 9.8(a)]. The motor controls move the half wave plate into or out of the beam entering the PBS.

(a)

(b)
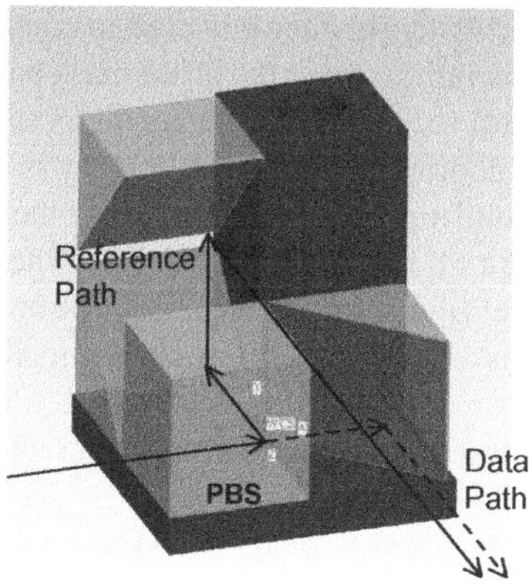

Figure 9.8 (a) Photograph of back side of optical divider. (b) CAD drawing of optics used in optical divider

For reading holograms, all of the energy should be directed into the reference path. The half wave plate is moved out of the incoming beam from the FRU, and the polarization of the beam is such that all the light reflects off the PBS and enters the reference path.

For writing, the half wave plate is positioned in the beam, and oriented to split the incoming beam so that the ratio of reference beam to signal beam power is approximately 4.5 : 1 at the media plane. Precise polarization control allows this ratio to be maintained over temperature and wavelength changes.

The power ratio of 4.5 : 1 is roughly equivalent to an intensity ratio of 1 : 1 in the media, which results in the optimal usage of the *M/#*. The optimum ratio is best determined experimentally (see Section 13.3.2) because the data beam in the media is not uniform intensity and the reference beam does not perfectly overlap the data beam throughout the volume.

The combination of the rotary motor, half wave plate and slide provides a high reliability solution. The equivalent optical performance could be achieved with a liquid crystal(LC) half wave-plate and a PBS, but it is difficult to achieve the required lifetime with a LC half wave plate at a wavelength of 405nm.

9.2.5 Data Path

The data path after the optical divider consists of an apodizer plus beam expander optics, phase mask, 4F imaging optics, SLM/camera PBS assembly, 4F relay lenses with polytopic filter, and the high numerical aperture (NA) storage lens.

The apodizer is a single element that converts a Gaussian beam into a beam with a flat top intensity profile so that the SLM is more uniformly illuminated, without wasting light [1,2]. The combination of the apodizer and the beam expansion optics transforms the Gaussian laser beam into a beam with a diameter slightly larger than the SLM, and with an intensity variation across the SLM of less than 30%. This conditioned and enlarged beam is incident on the phase mask which is imaged onto the SLM. The phase mask design and usage is covered in greater detail in Chapter 13.

The mechanism for moving the phase mask is shown in Figure 9.9. As explained in Chapter 13, the phase mask motion should be orthogonal to the reference beam sweep, as measured at the media plane. A stepper motor is used to continually move the phase mask across the data beam, using the rails to move the phase mask back and forth. The speed of the motion depends on the speed of recording and the geometry of the mask's phase features. If the motion is too fast, the hologram diffraction efficiency is reduced. If the motion is too slow, there is correlation build up between pages, and the formation of hot spots and distortion in the media, which all result in an SNR loss. The turn around when the phase mask reaches one of its travel limits is roughly timed to coincide with short stack moves or inter-book moves. The phase mask motion is automatically adjusted depending on the hologram exposure times.

The same relay imaging lenses are used before and after the SLM, except that there is no polytopic filter before the SLM.

The SLM/camera assembly consists of the SLM and camera glued onto a PBS (see Figure 9.10). The assembly is separately aligned as a subsystem in an alignment fixture that uses another camera and SLM with relay lenses to image the specially aligned SLM and camera onto the subassembly to be aligned. The SLM and cameras are used to display and

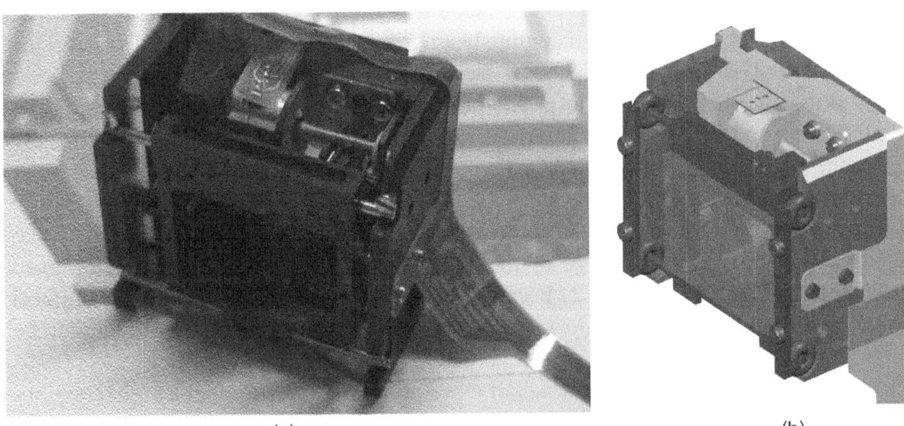

(a) (b)

Figure 9.9 (a) Photograph and (b) CAD drawing of phase mask and mechanism

Figure 9.10 SLM/camera PBS assembly

capture images in this set-up in order to align these components to within a pixel in x and y, with nearly zero relative rotation between them, and less than $10\,\mu m$ of focus error. The components are then glued to the PBS to fix their relative location.

After the SLM, a set of 4F relay lenses with a polytopic filter (Figure 9.11) are used to image the SLM and filter it before the storage lens. These relays are aligned as a 4F system in infinite conjugate configuration. Each lens consists of two singlets and a doublet, all with spherical surfaces. The effective focal length of a single lens is 40 mm. The effective NA of the relay system with the polytopic filter is less than 0.05. The focal length of these lenses was chosen so that the SLM/camera PBS could be moved out from under the media, to reduce the drive height.

The polytopic filter is made from a thin metal sheet, with a filter area of 1.3 times the Nyquist area to allow for misalignments on interchange. To align the polytopic filter, one side of the pair of lenses is illuminated with collimated light. The back reflection of the filter is used to center and place the filter in the correct focal position. This whole pre-aligned assembly is then placed into the drive during the overall drive build.

Figure 9.11 Relay lenses with polytopic filter

The storage lens is isoplanatic, with an NA of approximately 0.65, and a focal length of 12.395 mm. Using an isoplanatic lens increases the allowable assembly tolerances, and also increases the media alignment tolerances in the radial and focus directions. The lens consists of three singlet lenses and one doublet lens, all with spherical surfaces. If the storage lens elements were all centered, considerable aberration would be induced in the data beam because the beam's optical axis intersects the media at an angle of 25°. This in turn would enlarge the Fourier transform (FT) plane and the hologram. To minimize this effect, the lens element closest to the media is intentionally decentered to compensate for entering the media at an angle (see Section 13.2.3). This minimizes the media volume occupied by the holograms.

9.2.6 Reference Path

The principal components in the reference path after the optical divider are the soft aperture, the galvanometer group, the scanner lens assembly, and the read galvanometer. These components are all mounted to the frame, except for the read galvanometer, which is mounted on the top deck on the opposite side of the media.

The soft aperture defines the size of the reference beam. It is termed 'soft' because the transmission function has a gradual edge, with gray levels over a distance of approximately 110 μm, rather than an abrupt transition. The softer edge provides two benefits. First, the soft edge minimizes any diffractive ringing arising from the 40 mm of free space propagation before the write galvo, which is imaged into the media. (The soft aperture is not imaged into the media because this would require an additional set of optics. In addition, there is approximately 60 mm of propagation from the media to the read galvo and back to the media on read which would also contribute to diffractive ringing. The diffractive ringing would decrease the wavefront and amplitude quality of the reference beams.) Secondly, even if a hard aperture were imaged into the media, it would not be advantageous because of the media response. Imaging a hard edge would record bulk index distortions in the media. Currently the aperture size as defined by full width at half maximum (FWHM) is 1300 μm in the theta direction, and 580 μm in the radial direction. It is rectangular to cover the data beam volume in the media.

The galvanometer group, consisting of two galvanometers, is shown in Figure 9.12(a). The first galvanometer (the pitch corrector) has a glass wedge, with an AR coating, mounted on it. The rotation of this slight prism results in small changes in angle even with large galvanometer angle changes, resulting in high resolution. The total optical deviation is approximately $\pm 4°$. A close-up view of the pitch corrector is shown in Figure 9.12(b). This pitch corrector changes the angle of incidence of the reference beam in the radial direction of the disk (radial pitch). There are several reasons this is needed.

The radial media tilt tolerance of a page based system is relatively tight [3]. Holographic sensitivity to pitch (the holographic Bragg degenerate dimension) is on the order of 0.5 mrad for a 0.3 dB SNR loss. This means that if the accumulated radial tilt error is too large, the drive will start to lose SNR from this effect. There are a few contributors to a pitch variation or error: pointing changes of reference beam at the media due to thermal expansion or contraction of upstream components, reference beam mirror axis not parallel to rotational axis, radial tilt in disk manufacturing or disk wedge at the outer diameter of the disk. Pointing changes due to thermal expansion or contraction are typically caused by some

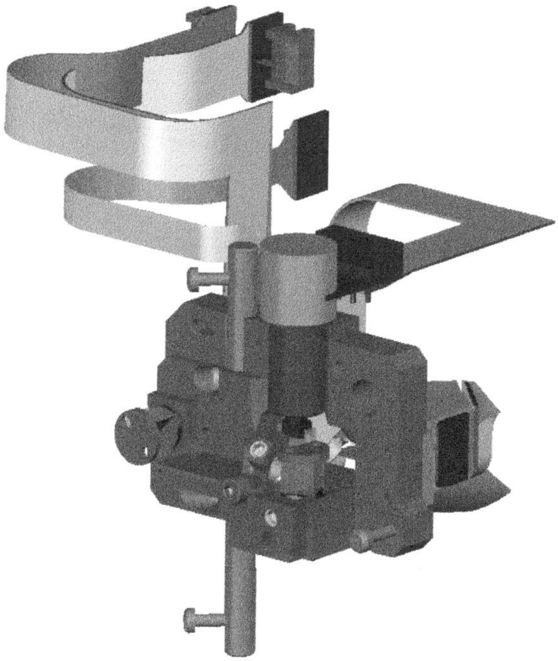

(a)

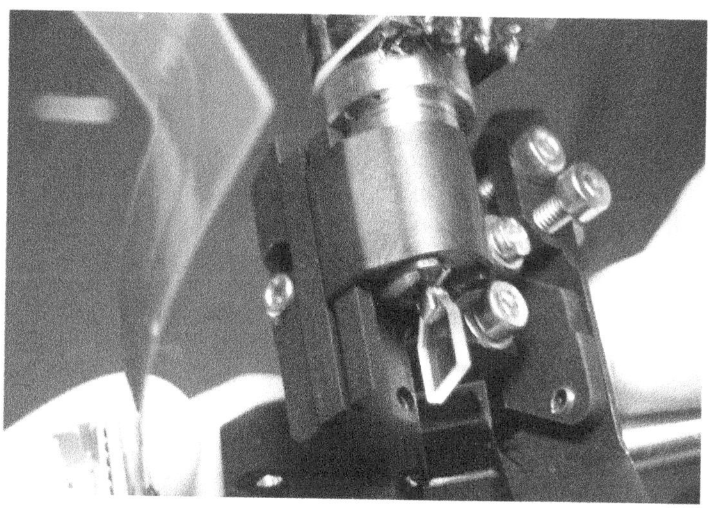

(b)

Figure 9.12 *(a) Galvanometer group. (b) Close-up pitch corrector galvanometer with slight prism*

asymmetric thermal expansion of glue joints in mirror mounts. Extreme care and analysis of mechanical designs and gluing techniques must be taken to ensure thermal sensitivity is minimized. The problem is exacerbated by the fact that there are tens of mirrors in the optical path; each of which contribute to the variation. For this reason, InPhase developed proprietary spherical mounts for mounting mirrors and glue bonding procedures to ensure

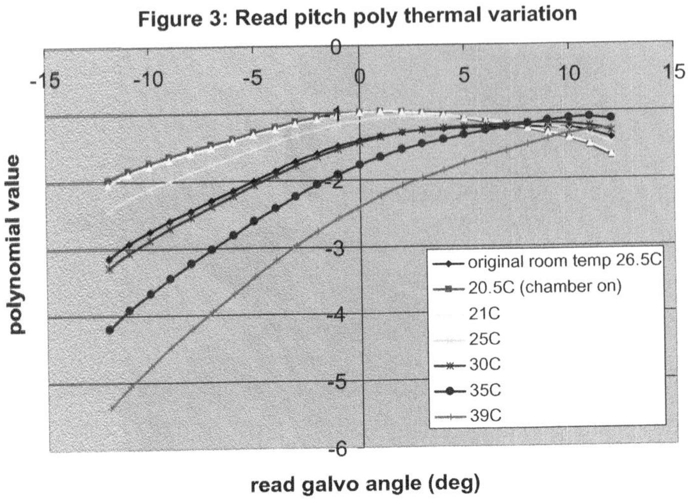

Figure 9.13 Residual pitch in a drive as temperature is changed from 20.5 to 39 °C

even expansion during thermal cycling. To limit radial tilt due to media manufacturing, a wheel that stabilized the disk and removed the radial tilt of the disk is incorporated into the loader. This stabilizer wheel removes nearly all of the disk tilt as well as setting the media height. If the media has a wedge toward the outside of the disk, this wedge causes a pitch change as well. Careful manufacturing and edge sealing can limit this to around 1 mrad. Even with all these measures to control pitch, the pitch corrector is necessary to calibrate out any residual pitch that is left over and to compensate for changes in pitch with temperature. Figure 9.13 illustrates the amount of pitch correction compensated for with the pitch corrector for drive ambient temperatures ranging from 20.5 to 39 °C. (The units are in galvo angle not optical angle.)

The light from the pitch corrector reaches the second ('write') galvanometer after reflecting off a stationary mirror. This mirror is designed to not be a perfect reflector. The light that is not reflected off the mirror impinges on a power sensor which provides an estimate of the amount of light from the reference beam that is incident on the media.

The mirror is in the image plane of the scan lenses and is imaged into the middle of the media. This configuration allows the reference beam angle to be scanned to multiplex holograms in a book, without changing the location where the beam is incident on the media. The scan lenses form a 4F imaging system, with a bend in it so that it fits into the drive (see Figure 9.14). The usable scan range for the lenses is $\pm 12.5^\circ$.

The read galvanometer on the other side of the media contains a quarter wave plate and a mirror. The combination of the mirror and quarter wave plate has three purposes. The primary purpose is to retro reflect the read-out beam to cause phase conjugate reconstruction. Secondly, it changes the polarization of the beam so that the reconstruction can be directed to the camera by the PBS in the SLM/camera/PBS assembly. Lastly, during recording it is used to deflect the reference beam away from the media. The read mirror incorporates a sub-wavelength structure and a reflective coating from Nano-Optics that

Figure 9.14 Scanner lens assembly

changes the polarization and reflects the beam. The advantage of this structure is that it maintains good optical flatness and polarization properties over large changes in temperature and wavelength.

9.2.7 Cure System and Toaster

The cure system is used to pre cure the media before writing holograms, and to post cure the media after recording. The cure beam spot size in the media (measured by FWHM) is 3.9 mm in the radial direction by 4.5 mm in the theta direction. The beam uniformity in the media should be better than 15%. The cure system consists of a high power blue laser diode (200 mW in the current drive, and expected to be 500 mW in the next generation drive), with a selected center wavelength range of 400–403 nm, where the media absorption is highest. The light from the laser is collimated and then passes through a moving diffuser to remove any spatial coherence. The beam is folded by three mirrors, one of which is slightly leaky so that the beam power can be measured. The light is then incident on a lenslet array (focal length of 1.74 mm, with a pitch of 200 μm) that determines the size of the beam. The lenslet is roughly imaged into the center of the media by a single lens. The efficiency of the light delivery is over 50%, with most of the losses attributable to apodization. The profile of the cure beam is sharpened spatially to reduce the cure track spacing, which allows for seven data tracks in a single cure beam width. A CAD drawing of the top deck with the cure system is shown in Figure 9.15.

For the first Tapestry version, the drive allows multiple write sessions as long as the disk remains in the drive. However, to ensure the archival life of the media, the disk will be cured entirely before the disk is ejected, even if the disk is only partially recorded. To speed up the full disk cure, which includes the center and edges of the disk, a 'toaster' cure system is used. The toaster comprises three separate groups of 390 nm LEDs arranged roughly in a pie shaped area so that in a single rotation the entire disk can be completely post cured, even if no data were written.

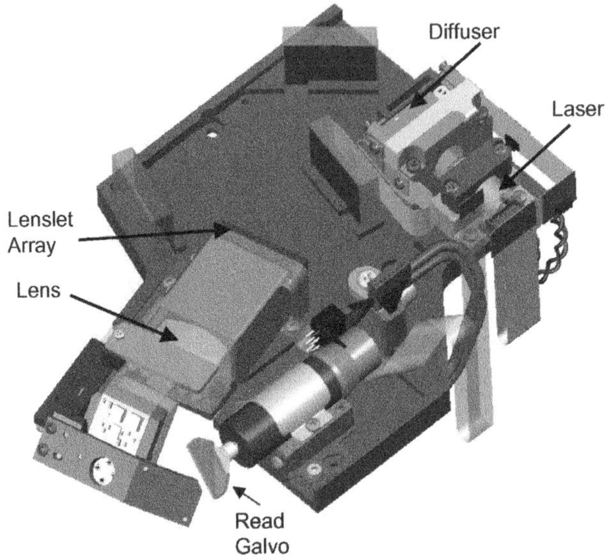

Figure 9.15 Close-up of top deck with cure system, part of toaster, and read galvanometer

9.3 Mechanical Implementation

9.3.1 Loader

The loader moves the media into the drive, opens the cartridge shutter, and sets the disk's radial and theta positions relative to the optical head. The large size of the optical head means that it is preferable to move the disk relative to the fixed head, rather than moving the head over a fixed disk hub. The loader also has a small wheel outside the data area at the edge of the disk that pre-tilts the media. The wheel is 200 µm higher than the spindle to consistently set the height of the media in the drive, which removes any excess radial tilt resulting from disk manufacturing variations. A CAD drawing of the loader is shown in Figure 9.16.

A sled is used to move the media in the radial direction during cartridge loading and unloading, and for disk radial positioning during reading writing or curing. The radial motion is effected by a geared DC brush motor. The sled has an optical encoder and the radial motion has a resolution of 1 µm and a repeatability of approximately 0.7 µm. The measured radial run-out of the media in the drive is 3–5 µm.

The disk is rotated by a DC brush spindle motor (MicroMo 2342S012CR) that is inverted so that the coils are closer to the hub.

The loader has a door over the opening that closes after the media have been loaded. This is for safety purposes, and also to keep light and contaminants out of the OMA.

9.3.2 Cooling

For a drive in a library, cooling air must flow from front to back. A long board, called the 'hood' board, was designed to fit across the top of the OMA. This hood board simplifies cabling and

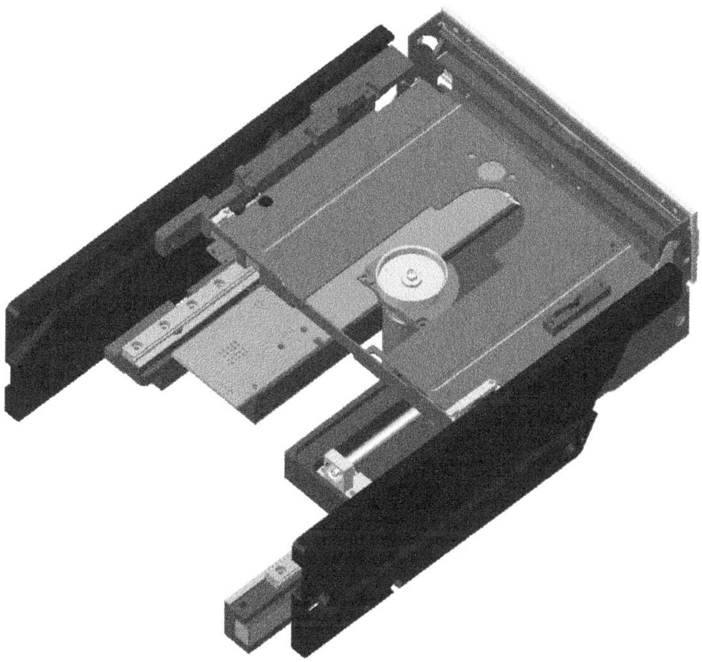

Figure 9.16 CAD drawing of the loader

creates an air channel near the top of the drive that allows for cooling air to flow front-to-back. The hood board seals the OMA from the air channel on top to prevent contamination. A recirculating system cools the OMA chamber and transfers heat into the air flow above the hood board with a small fan, to cool the SLM and camera. The air is pulled across the top of drive and into the electronics box by a larger fan located inside the electronics housing. The air flow channel can be seen in Figure 9.2(a) at the top of the front panel.

9.3.3 Integrated Vibration Isolation System and Sway Space

Initial vibration and shock testing were performed on the EVT units to determine the important structural resonances. A vibration system was designed and installed on the outside of the EVT frame for testing. For the DVT units, the system was integrated into the frame. The isolation system consists of nonlinear damped spring attachments that hold the OMA to the frame from the sides (see Figure 9.3). These attachments support the sides of the OMA at four points. The sway space around the loader is approximately 3 mm to provide shock and vibration protection.

As a rule of thumb, the maximum amount of disk motion that can occur while a hologram is being recorded is approximately one-tenth of the grating period. Larger motions lead to decreased hologram SNR. For a given level of vibration, shorter write exposures result in smaller disk displacements. Figure 9.17 shows the experimental dependence of recorded hologram SNR on the root mean squared media motion. This dependence roughly corresponds to the rule of thumb. For an exposure time of 1 ms, the vibration-induced

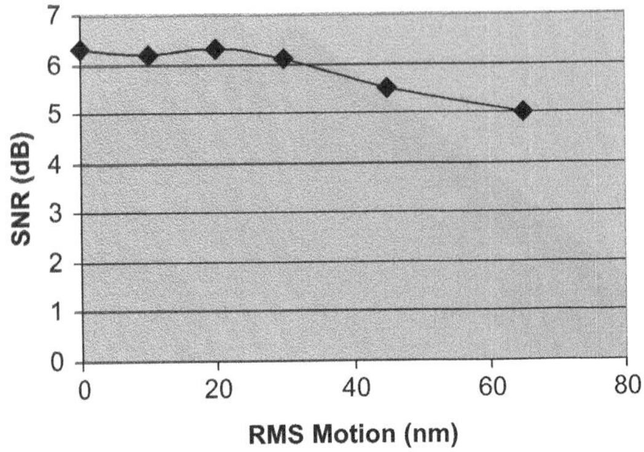

Figure 9.17 *Plot of SNR versus RMS disk motion*

disk velocity should be less than 40 nm ms^{-1}, or about 40 μm s^{-1}, to limit SNR degradation to less than 1 dB. In the first generation Tapestry drive, the disk servo system using a red laser can detect disk motions as small as 5 nm. If the velocity limit is exceeded during a write, the current book is marked as 'bad' and the book is rewritten at the next available disk location.

For next generation archive drives and consumer products, the write exposures will be approximately 0.1 ms, which will provide greater tolerance to vibration.

9.4 Electronics and Firmware

9.4.1 Electronics

The drive electronics consists of six circuit boards (printed circuit board assemblies, PCBAs), five of which are mounted in an enclosure at the rear of the drive. The sixth, the 'hood' board, is positioned over the top of the OMA and extends to the front of the drive (see Figure 9.2). Each board addresses a discrete aspect of the drive's functions (for example, the Write2 board controls only the write data channel), to simplify parallel development and debugging in the first generation Tapestry drive. In future generations, application specific integrated circuits (ASICs) will be used to integrate several functions and eliminate PCBAs from the drive. Figure 9.18 shows the major electronic components in the drive.

The drive hardware is controlled by the 'hood', laser, and 'skid' boards. The hood board connects to all the main electronic components except for the camera, SLM and loader. The hood board has the servo digital signal processor (DSP), and a field-programmable gate array (FPGA) that implements many hardware controls and assist functions, to reduce the computational load on the servo and main drive processors. The 'skid' board provides electrical access to the front panel, the loader's spindle and radial motors, and the optical encoders, including the optical encoder for reading the disk servo pattern. It also provides access to the media's radio-frequency identification (RFID) chip for recording and reading the library map. The laser board and its FPGA control the laser, and interface with both the main and servo processors. There are laser commands for setting laser power

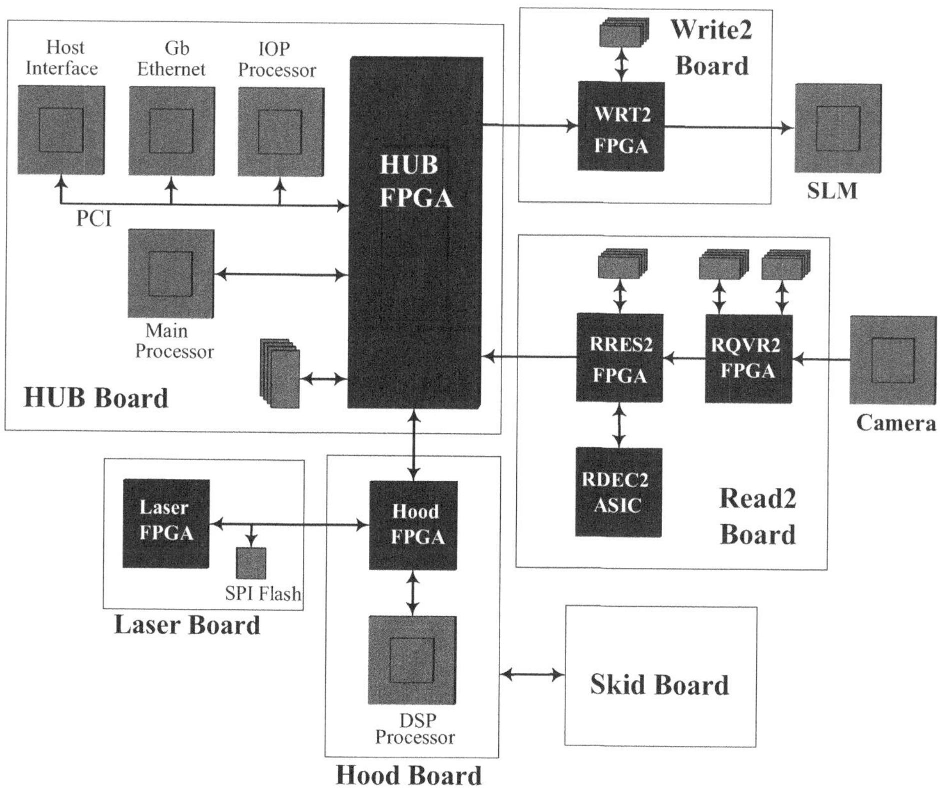

Figure 9.18 *Electronics, partitioned into six circuit boards*

and wavelength, automatic mode control, automatic single mode find functions, and laser error reporting.

The read and write channels each have their own PCBAs with FPGAs. Only the LDPC decode block (RDEC2) has been implemented in an ASIC (by our supplier, Trellisware). The write data channel receives user data from the host, adds cyclic redundancy check (CRC) protection, and places the user data into a large buffer memory (around 1 GB) on the HUB board. The buffer facilitates pipelining to speed up write sessions.

The HUB board contains the interface processor (IOP) and the main processor, as well as the big buffer. The host interface is provided on a separate card that plugs into the PCI bus, so that different drive interfaces (for example, Fibre Channel or parallel small computer system interface, SCSI) can be provided by simply changing this one card. The different host interfaces all use commercially available interface cards.

Figure 9.19 shows a flow diagram of the write channel through the hardware. The data are all processed in hardware under firmware control. On write, the user data come into the system over the host interface and a CRC is added to detect errors. Parity bytes for the two outer ECC levels are added before the data stream is sent to the big buffer. Data passes from the big buffer into the write FPGA, encodes the data, applies the page-wise LDPC parity information, and formats the page for presentation on the SLM. This formatted page is then

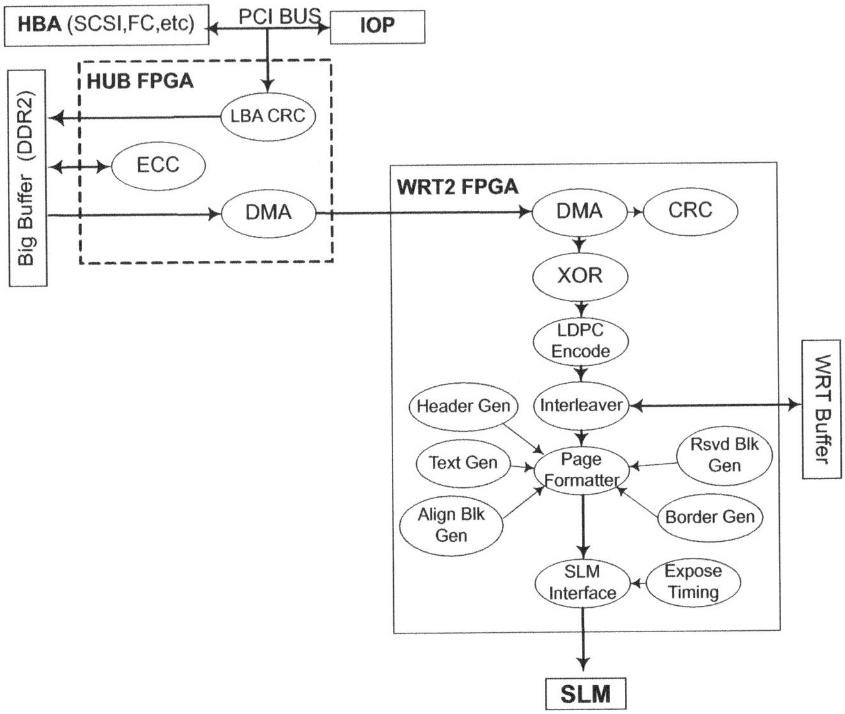

Figure 9.19 *Write channel flow diagram with FPGA labeled*

transferred to the SLM as needed. More information on the data channel is provided in Chapter 11.

Figure 9.20 shows the flow diagram for the read process. Again, all data are processed in hardware, with flow control provided by the firmware. The image detected on the camera is transferred to the read quiver (RQVR2) FPGA. This FPGA provides low level processing on the detected image. Fixed pattern noise (FPN) is removed from the raw image data by subtracting an image that is stored when the drive is first calibrated. The image is then linearized by taking the square root of each pixel's value. Next the headers (bar codes) are located in the image, and decoded. Using the header information, the quiver (misalignment) information is calculated for the reserved blocks in parallel, using 19 separate microprocessors in the FPGA. These low level steps are shown as unshaded ovals in the upper left of Figure 9.20.

The quiver results allow the image to be resampled. After resampling, the log likelihood ratios (LLRs; see Section 11.5.1) are calculated, deinterleaved and then fed into the LDPC decoder. The shaded ovals in Figure 9.20 represent these steps. The page based ECC, derandomization (XORing) and CRC check are shown as striped ovals in Figure 9.20. The outer levels of error correction are only used if the CRC detects that there are errors present after the inner LDPC decoding step. Finally, the CRC is removed, and the data are returned to the user.

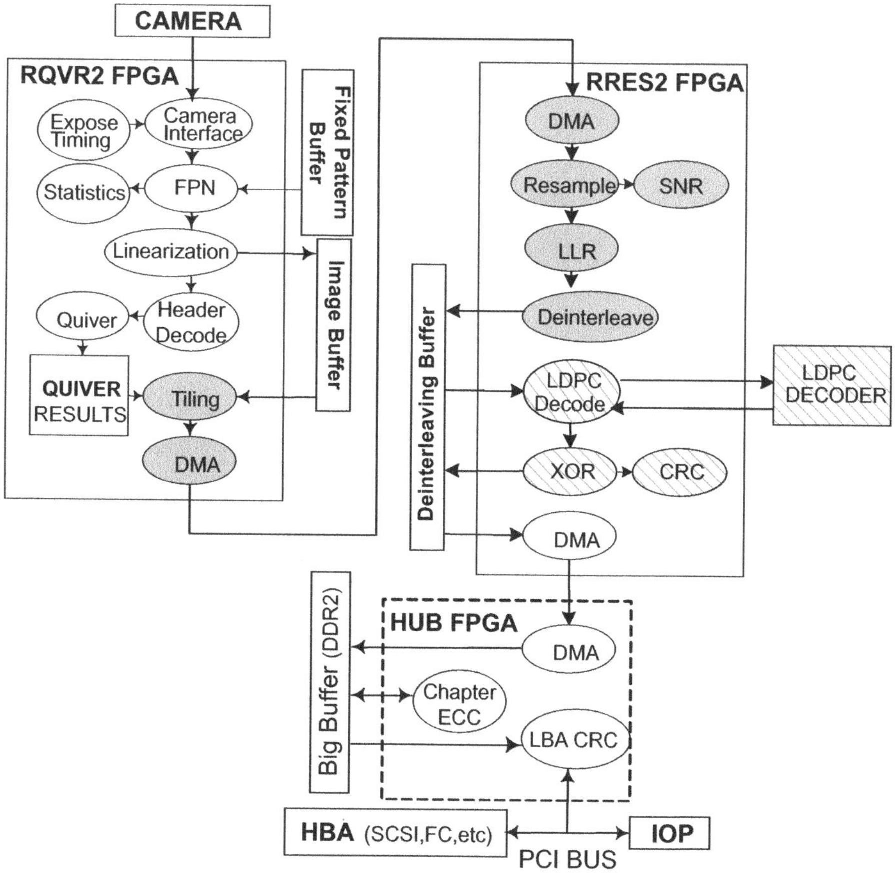

Figure 9.20 *Read channel flow diagram*

During the read process, a number of channel metrics, such as SNR and intensity centroids, are also calculated, and fed back to the firmware for use by the servo system.

9.4.2 Firmware

The drive electronics has three different processors. The main system control uses the MicroC/OS-II real time operating system (RTOS), running on a Motorola Coldfire processor. The second processor is a PowerPC, running embedded Linux, which provides interfaces (such as GbE, SCSI, FC and ftp) to the drive host. Servo control of the radial, theta and galvanometer axes, as well as shutter control is provided by a Texas Instruments DSP.

The custom firmware consists of more than 1.5 million lines of C++, while the DSP code and the operating systems are written in C. Large sections of the read and write data channels are implemented in FPGAs, so the processors and most of their code provide servo and drive control, interface functions and overall drive management.

The C++ firmware running on the main Coldfire processor is architected as three layers on top of the underlying hardware. Figure 9.21 shows the three layers (Hardware

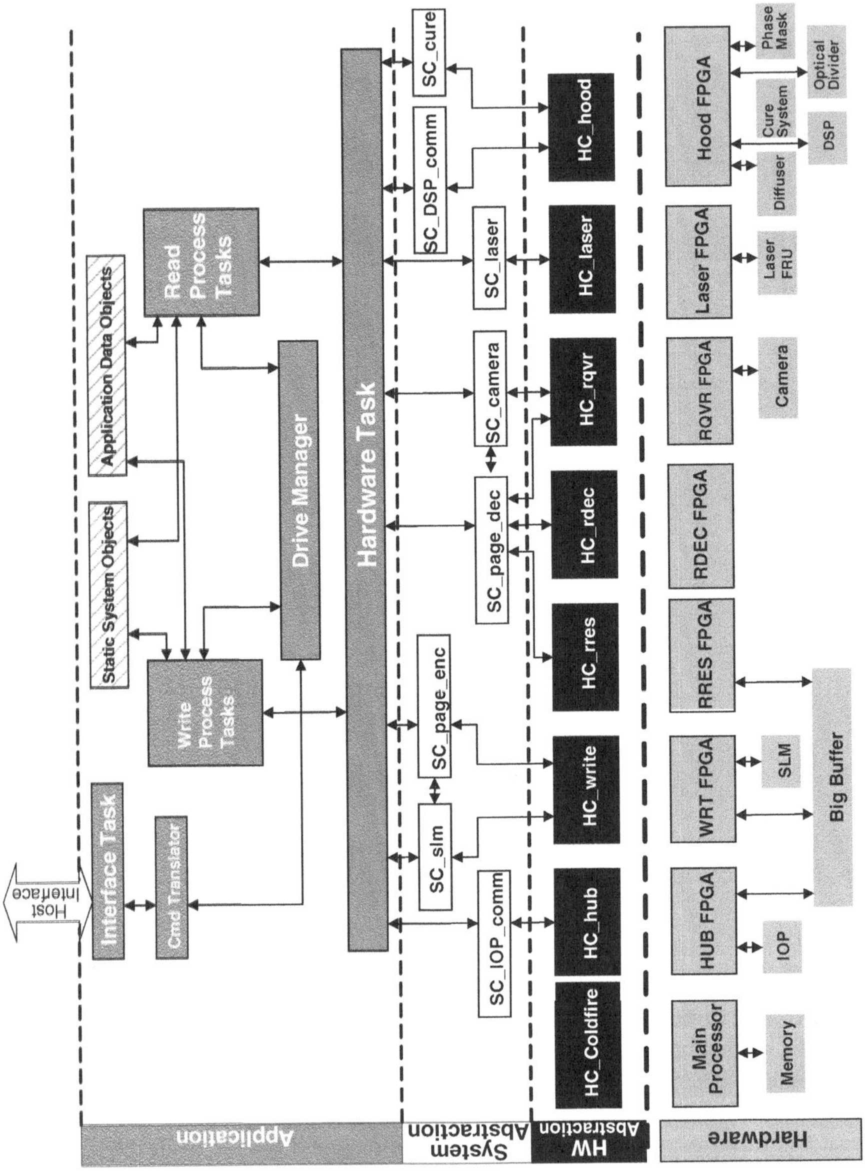

Figure 9.21 Main firmware architecture, showing the three layers above the hardware and the major components

Abstraction Layer, System Abstraction Layer and the Application Layer) above the hardware.

The Hardware Abstraction Layer provides an interface to the electronic hardware, such as the FPGAs, complex programmable logic devices (CPLDs), and PCBAs. It provides access to Command Service Requests (CSRs) for the system abstraction layer objects, and implements the common services that these hardware components require, such as FPGA loading and initialization, Interrupt Service Requests (ISRs) set-up, and board serial number access.

The System Abstraction Layer presents an Application Programming Interface (API) to the tasks for the system objects, such as:

- SLM
- camera
- laser
- cure diffuser
- cure LED
- phase mask
- optical divider
- DSP and IOP link layer
- temperature sensors
- beam power sensors
- hardware DMA engines
- hardware ECC encoder/decoder
- hardware page encoder/decoder.

The Application Layer consists of the command service tasks, the static system objects and the application data objects. Static system objects are helper objects for the main tasks, and they include:

- CFI manager. Controls the processor flash file system.
- Big buffer cache manager. Manages user data cache RAM.
- Fault manager. Tracks HW error/fault state.
- Media defect detection. Implements media defect detection algorithm used when writing.
- Feedback/feedforward. Implements the algorithms used for feedback/feed-forward during reads.
- Trace log. Event logging facility.
- Write pre-compensation. Implements the temperature/wavelength compensation algorithms used during writes.

The Librarian is the major task of the Application Layer. It is a collection of objects locating and tracking user data physical disk location and enforcing writing and reading rules. It accesses the format file, stored in the processor's flash memory, which contains the translation between physical addresses and axes coordinates. The Librarian also reads and writes the Library map, stored in the RFID and on the media, which contains the data on current disk usage.

The Librarian task is divided into these major objects:

- Locator. Interprets the format file and translates between physical address and physical disk location as well as providing physical address to cure address relationships.
- Format Manager. Enforces writing rules, such as sequencing and layer diffusion timing.
- Cure Manager. Enforces cure rules, such as pre- and post-cure time-outs and cure site neighbor rules.
- Card Catalog. Contains the information of the locations written so that they are read back correctly. A binary version of this is stored in the Library Map.
- Branch Manager. Controls the physical disk zones.
- Drive Emulation Table. Translates host logical block addresses (LBAs) to drive LBAs to enable random access writes.
- Partition Manager. Facilitates having the drive act as a multi-partition device (for future enhancements).

The Application Layer's application data objects contain the data and state for entities that need to be shared between multiple tasks. Three application data objects (Anthology, Chapter, and Page) are defined in the logical format.

The application tasks service the external and internal drive commands. These tasks are statically prioritized and run preemptively under the operating system's scheduler. The major application tasks are:

- interface tasks
- command translator
- drive manager
- hardware task
- write process tasks
- read process tasks.

The interface tasks handle commands to and from the host over various interfaces. The main drive firmware currently supports a serial interface, and an interface to the IOP. The IOP currently provides parallel SCSI and Fibre Channel external drive interfaces. External commands are received by the interface task and forwarded to the appropriate command translator. Interface tasks also handle block data transfers to and from the host.

The command translator translates external commands (such as SCSI commands) received from the interface tasks into internal InPhase Protocol commands, and forwards them to the drive manager. This allows the drive to be developed using InPhase Protocol commands, or commands that are native to the drive. Only the command translator has to be modified to support other external interfaces and protocols.

The drive manager provides the highest level drive control. It receives all external commands from the interface/command translator tasks and routes them to the appropriate task. It maintains the drive's operational state (Idle, Write, Read) and controls the switching between these states. In addition, it receives all internal commands that requiring state changes and queues these to ensure proper drive operation. For example, it handles read after write (RAW) cases by forcing a write session closure and write state exit before allowing another read.

The hardware task is the hardware API for the other tasks and handles all hardware commands. It also provides task context for the system abstraction layer objects.

The write process consists of four tasks: the Logical Write task, the Chapter Write task, the Book Write task and the Page Write task. Similarly, the read process consists of three tasks: the Logical Read task, the Chapter Read task and the Page Read task

The decision to write the main drive firmware in C++ rather than C has provided great flexibility and maintainability to the code development, without compromising speed. Drive firmware and FPGA binaries can be easily updated by transferring a single binary file to the drive by ftp, and executing an update script on the IOP system.

9.5 Basic Build Process

9.5.1 Overview

The DVT holographic drives and FRUs are built in an assembly line fashion using a pilot line that was developed at InPhase over a 2-year period. The pilot line uses both standard and unique alignment techniques, with an implementation that can be transferred to a volume manufacturer. Going from EVT to DVT, a large optical drive manufacturer was consulted and over 80 basic Design For Manufacturing (DFM) changes were made to help the build process. Many significant DFM improvements remain to be implemented. Figure 9.22 shows a flow chart of the Tapestry drive and FRU build process. The FRU has eight build stations including a final calibration and test station. The OMA (drive minus FRU and electronics) also has eight build stations, with an additional end of line (EOL) testing stage that is external to the build line. At each station, the OMA frame or FRU baseplate is kinematically mounted into the station so that particular station's tooling and equipment can be used with minimal alignment. The dotted box around the first three FRU stations represents the assembly of a complete ECLD. The remaining stations are used to finish the FRU build, and to calibrate it for integration into the OMA. Several key subsystems that are aligned elsewhere feed into the OMA/drive build. Examples include the phase mask, optical divider, galvo group, top deck, loader, SLM/camera PBS assembly (PBS SCL), and electronics. These are assembled, aligned, and tested at separate build stations.

For both FRU and OMA build lines, the maximum time of any station (takt time) is close to 2 h. Thus, a complete drive can be built every 2 h when all stations are operating. The current build method uses a number of manual alignment techniques that with automation could reduce these takt times significantly. For the FRU build, the total build time is approximately 12 h from start to completion. The OMA build requires a total time of about 16 h. The various subassemblies also require a total of about 20 h per drive to complete.

Figure 9.23 shows each OMA/drive station's critical output (what is assembled or aligned in that step), critical exit criteria that must be passed before the drive moves on to the next station, and the verification of the previous station that is done before starting the current station's work. Station 1 installs a number of components into the frame to prepare for the more complex alignments that take place in the following steps. These include installing the parts for the kinematic mounting used at each of the assembly stations. Station 2's primary purpose is to align the laser delivery optics. Station 3 installs the data path imaging optics.

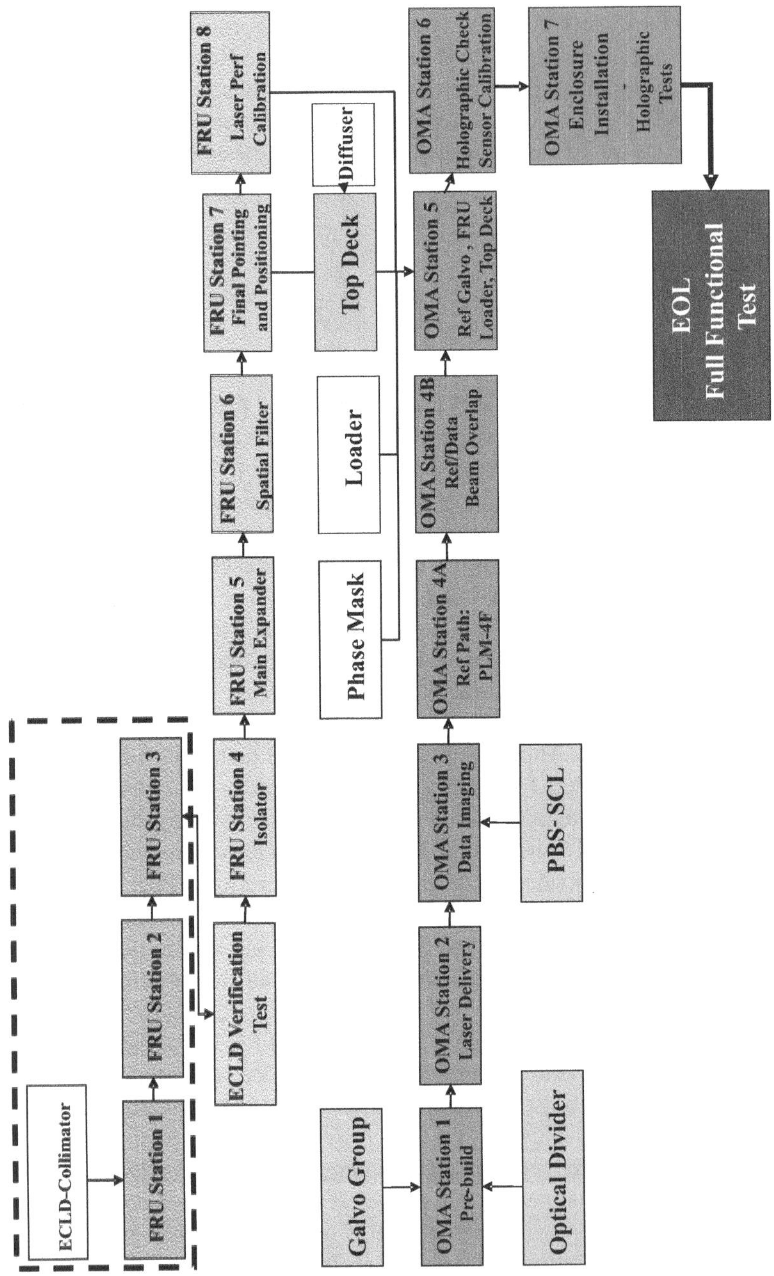

Figure 9.22 *Flow chart for drive and FRU build*

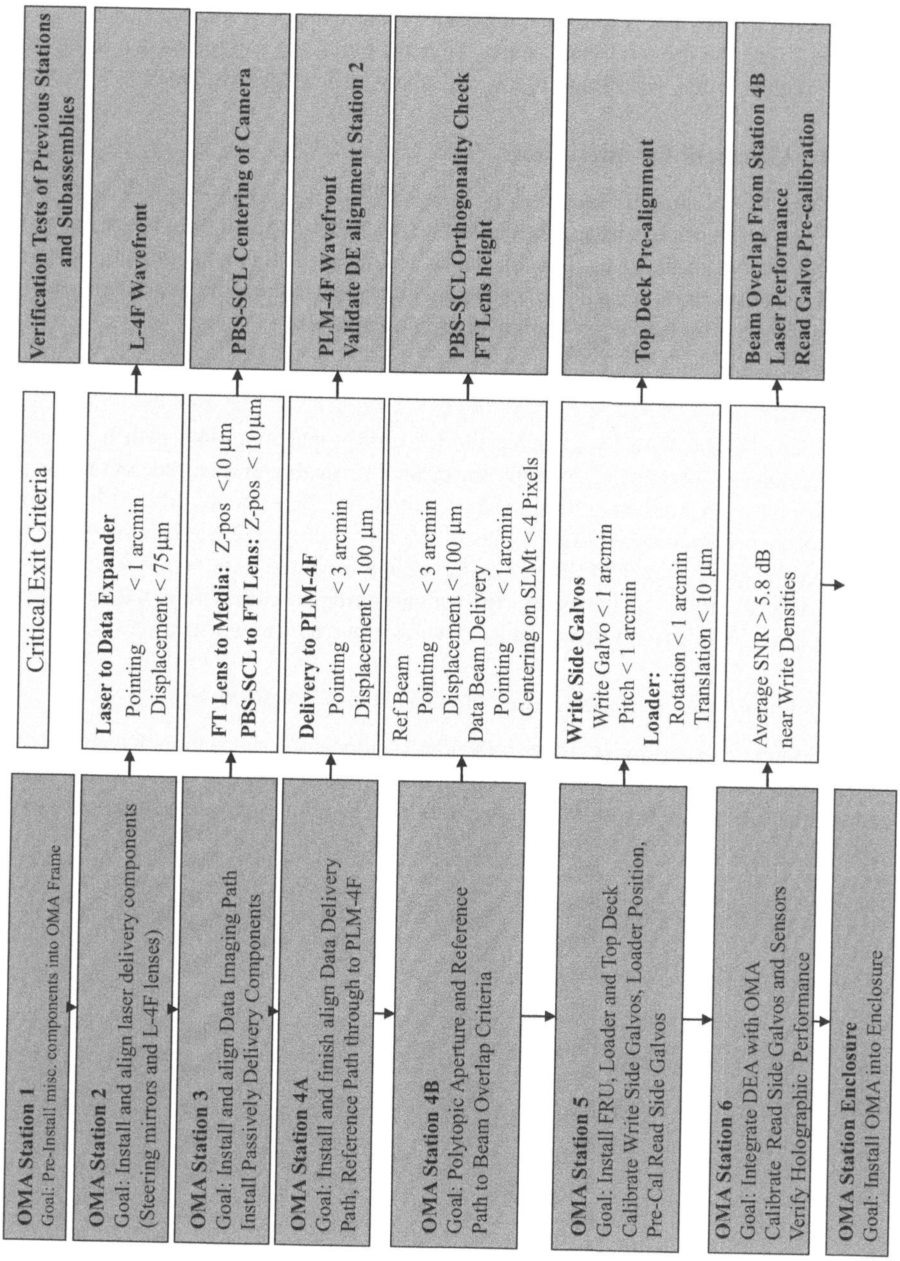

Figure 9.23 *Station output, critical criteria and verifications*

Station 4A installs the first part of the reference path delivery optics, while station 4B installs the high NA storage lens, the reference scan lens, and checks for beam overlap in the media. Station 5 installs the FRU, loader, galvo group, top deck, and calibrates the galvos. Station 6 integrates the electronics and verifies that the basic holographic performance exceeds a set level. Station 7 puts the drive into enclosure. After the build, the completed drive goes to EOL testing with full disk writes and reads, and many basic diagnostic tests.

9.5.2 Drive Alignment for Interchange

For interchange of holographic media to be successful in the InPhase architecture, the angular sweep of reference beams and the cone of angles that make up the data page must be closely matched between all drives. To achieve this, a number of alignment techniques were developed. The height, pitch and roll of the media with respect to the reference and data path are closely controlled by use of several novel alignment tools.

9.5.2.1 Height of Data-Spot Focus with Respect to Drive

The storage lens height is set by focusing the forward-propagating data path light onto a mirror that is set at a specific height above the critical vertical reference pads of the drive. As the storage lens is moved up and down in its mount, the focus moves across the mirror surface, resulting in a counter-propagating beam that is re-collimated when the storage lens is at the correct height. Figure 9.24 shows the storage lens mounted in the frame. The mirror-tool is omitted for clarity, as are the diagnostics for measuring the collimation of the counter-propagating beam. The five reference pads are shown as dark circular surfaces.

9.5.2.2 Reference Beam Fan with Respect to Drive and Data Beam

The fan of reference beams must intersect the disk at almost the same angle for all drives. This means that the pitch and galvo angles both have to be set and calibrated within each drive to angular tolerances of approximately $\pm 0.005°$. The tight tolerance is achieved using

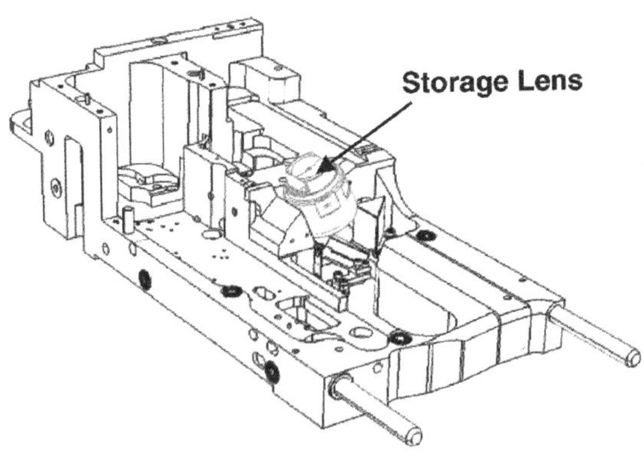

Figure 9.24 The storage lens is mounted in the frame and adjusted until its focus is at a preset height above the three vertical reference pads. These pads are the same surfaces that the loader will eventually sit on

a 'golden galvo' tool that consists of a galvanometer mounted on a precision tooling block. The galvanometer is precisely positioned and calibrated to the 'golden galvo' block on an additional station, so that its axis of rotation is in line with the precision reference pads machined into the block. The calibrated 'golden galvo' tool is located on the frame while referencing the five pads shown as dark circular surfaces in Figure 9.24. This guarantees that the galvanometer's axis of rotation lies parallel to the line defined by the intersection of the two planes given by the two reference pads on the side of the frame and the three pads on the top of the drive. Once the angles are set, the reference beam is centered on the data spot projected by the storage lens. The residual 'out of plane' misalignments of the reference and data beams are removed by calibrating the pitch galvo.

9.5.2.3 Loader Alignment on the Frame

The loader is constructed as a separate subassembly. In that process, the height, pitch and yaw of the media are all set with respect to three mounting posts in the loader that will eventually mate with the drive frame at the upper three reference surfaces noted in Figure 9.24. At this point, only the lateral placement and the rotation of the loader remain to be adjusted. By rotating the loader in the theta axis, the loader's motion parallel to the plane defined by the reference surface on the left side of the drive can be aligned. This sets the loader's motion back into the fan of reference beams, since the normal to the plane defined by the reference beams was also aligned to the reference pads. This process is shown in Figure 9.25.

The final step of the loader alignment sets the tangential position of the loader with respect to the drive. This step employs a calibrated, nonholographic disk with a precisely positioned transmissive radial feature, which is moved over the focused spot from the storage lens. By

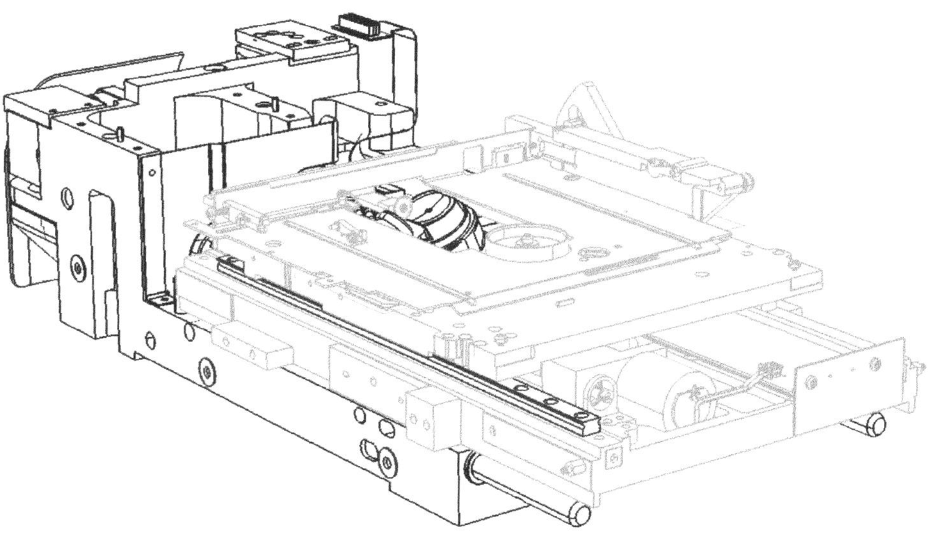

Figure 9.25 *The loader's rotation is set by aligning the main linear bearing rail of the loader parallel to the reference pads on the left side of the drive. The loader is shown in gray*

monitoring the light transmitted through the radial feature while moving the disk to different radii and scanning the disk in theta, the tangential offset of the loader is established, adjusted accordingly, and re-tested. Once the loader is in the correct tangential position, it is locked down, and the drive's servo can be calibrated with additional features on the loader alignment disk. This calibration removes residual offsets in radial and theta to within 5 μm. The calibration data are stored on each individual drive.

As described above, the rotation between camera and SLM is set during the build of that unit. Placing this unit into the drive requires no further alignment.

9.6 Defect Detection

Of all the media-related factors that must be taken into account when optimizing a holographic system, the most fundamental is scatter from the photopolymer itself. The relationships between media scatter, diffraction efficiency, SSR and SNR were described in Section 9.2.1, and are detailed further in Chapter 11.

Two media-related defects compromise hologram recording: inclusions of foreign particles, and bubbles. In the current media fabrication process, black flecks from the substrate molds are occasionally present in the photopolymer. These flecks are highly absorptive and hence do not result in appreciable scatter unless the fleck is large. Bubbles, which are caused by the bonding process, result in large local index changes in the media that act as scattering centers. Bubbles with diameters of 30–40 μm can degrade the SNR by a few tenths of a dB because of the local increase in SSR. Figure 9.26 shows examples of recovered data pages recorded at media locations with bubbles of various sizes, and their corresponding SNR loss. Generally, scattering from a bubble of a given size results in an SNR degradation much larger than for a black particle of similar size. Even though bubbles are deleterious to SNR, manufacturing processes have been developed that include less than 20 bubbles per disk.

Two strategies are used in the drive to mitigate the effects of bubbles. The first and preferred method is to produce a defect map during manufacture of the disks that is stored in the RFID chip contained in the media cartridge. The defect map defines 'keep out' zones that are not used for recording holograms. The second method is to detect media defects during recording. Defects are detected by monitoring scattered light from the incident data and reference beams as the books are being written. If the scatter level exceeds a threshold value, the entire book is marked as 'bad' and is rewritten at the next available media location. Figure 9.27 shows the scatter from a bubble and the sensor on the top deck that is used to measure this scatter.

Monitoring the media scatter level during the write process to determine media defects is more difficult than using the media defect map that is generated during media manufacturing.. Consequently, the drive fomat must have more capacity than the product specification to allow for media defects and other drive errors (such as excess vibration) that cause book rewrites.

Estimation of the extra data capacity required for a disk uses a worst case analysis that treats all books that contact a bubble as 'bad'. The media manufacturing specification requires that there are no bubbles larger than 800 μm. Bubbles with diameters less than 30 μm are assumed to have no deleterious effect. Bubbles with diameters between 30 μm

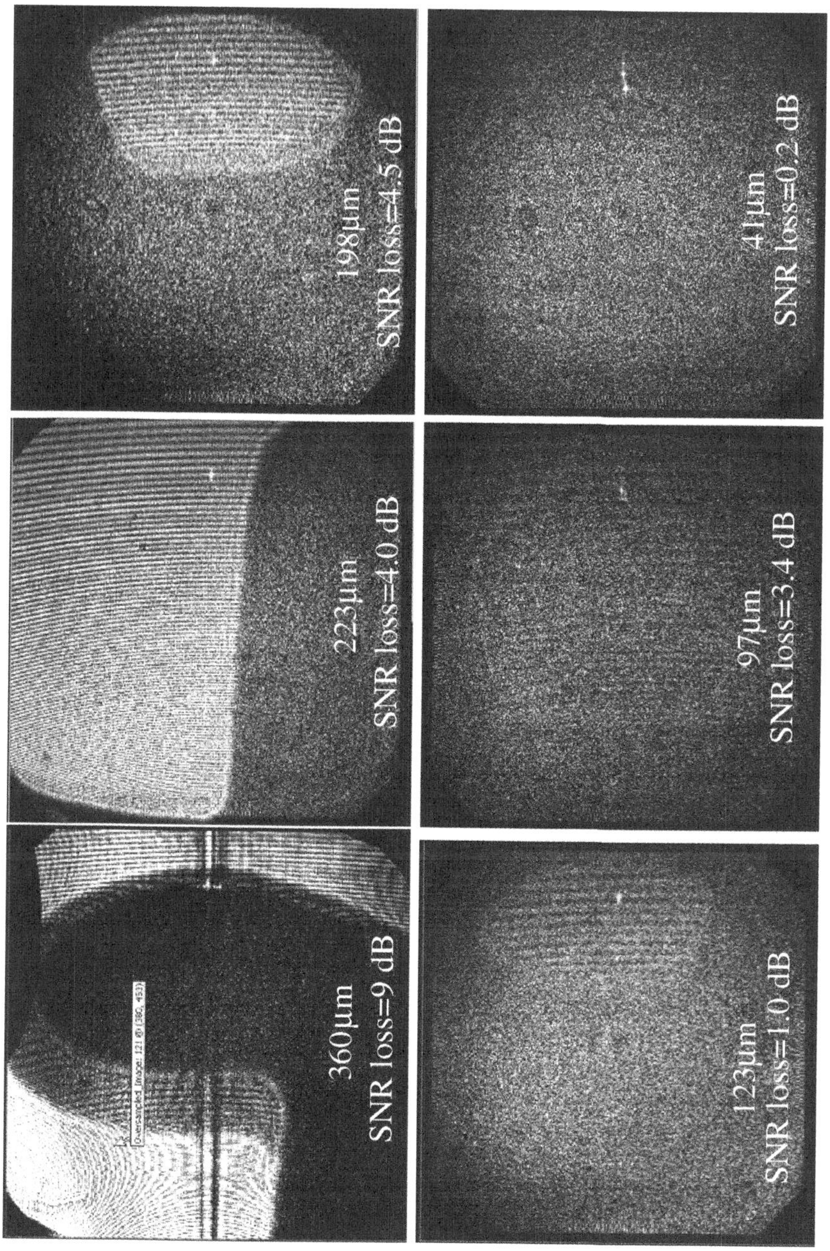

Figure 9.26 *Photograph of recovered data pages with different sized bubbles and the corresponding SNR loss*

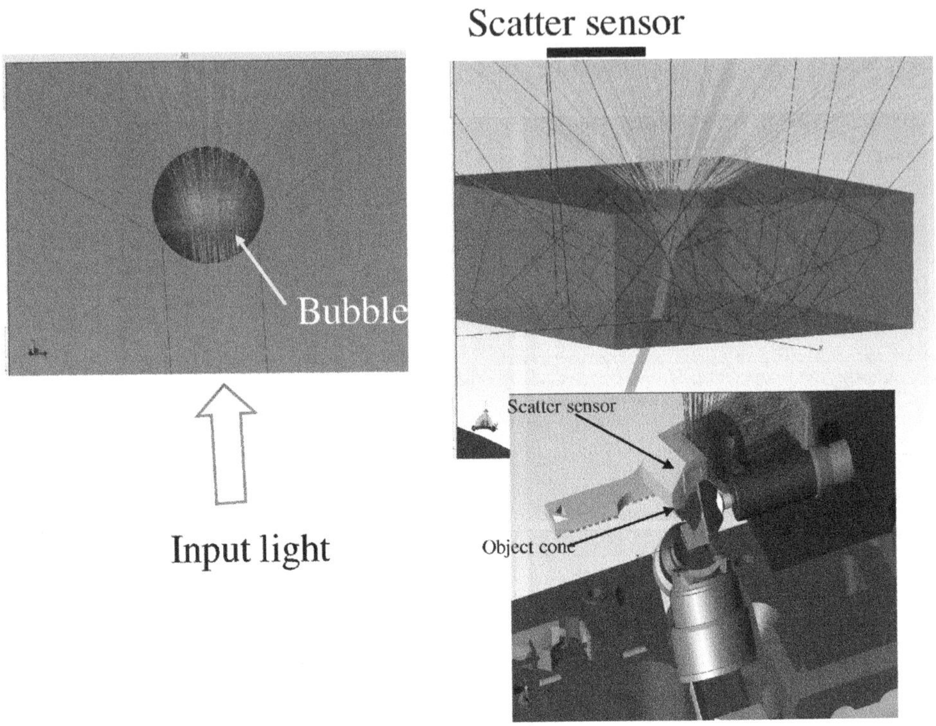

Figure 9.27 Scatter from bubbles is detected with a sensor on the top deck

and 100 μm are designated 'D1' defects, and are assumed to prevent recording at two book sites. Similarly, bubbles with diameters between 100 μm and 400 μm and those with diameters between 400 μm and 800 μm are labeled 'D2' and 'D3' defects, and are assumed to prevent recording at three and four book sites, respectively. For the Tapestry drive, the allocated excess data capacity requires that the numbers of D1, D2 and D3 defects is less than 106, that is $2D1 + 3D2 + 4D3 \leq 106$. The same number of book locations (106) may be unusable because of laser or servo errors while maintaining the 300 GB user capacity on a disk.

9.7 Read and Write Transfer Rate Models

For a given disk capacity, the disk format specifies the number of pages in each short stack, the number of short stacks per book, the number of books, the number of cure sites, and finally the number of dummy books that have to be written to complete a disk.

The overall read and write transfer rates can be calculated using these counts, together with the individual hologram exposure times, the number of user data bytes per page, delays for servo moves and the firmware overhead. The rest of this section provides sample calculations for an example disk format.

9.7.1 Simple Write Transfer Rate Model

The average number of user data bytes on a page depends on the overall error correction code (ECC) code rate. In addition, books that are rewritten because of media defects, and laser and servo errors, reduce the effective average user bytes per page.

A single data page contains 77 672 bytes of user data together with ECC parity bytes and a cyclic redundancy check. To estimate the average bytes per page, this byte count is reduced by 10% to allow for the overhead associated with the chapter level error correction. The byte count is reduced by a further 2% for the overhead of the anthology level error correction. Lastly, an additional 1% is allocated for book rewrites. Taken together, the effective average number of user bytes per page is 67 512 bytes.

Holograms are not all recorded with the same exposure time, as shown in the Tapestry drive's exposure schedule in Figure 9.28. The average hologram exposure is approximately 1 ms. Using the disk format that yields the lowest transfer rate (Format Example 1 in Chapter 13), Table 9.1 shows the transfer rate calculation. This format uses 320 pages per book grouped in four short stacks.

The critical page to page times of 1.6 and 2.4 ms are set by the exposure times, galvanometer move and settle time, and the firmware overhead. The book and stack move times are less critical because they occur only every 80 pages. The toaster time is an upper limit that would be used to toast an entire blank disk. The basic format parameters that strongly affect transfer rate are the number of pages per book and per short stack, and the average number of user data bytes per page. Making the single change from four to three short stacks in a book can improve transfer rates by over 2 MB s^{-1}. As the bookcases are made larger, the transfer rate and capacity increase because fewer dummy books are required. For the 'wave' format shown in Chapter 13, increases of 30–40% in capacity and 10–20% in transfer rate are possible.

9.7.2 Simple Read Transfer Rate Model

For reading, the average number of user data bytes per page is larger than for writing. The number of data bytes on any individual page is of course unchanged (77 672 bytes) but there are no rewrites for media defects and drive errors. Also, the anthology parity pages are

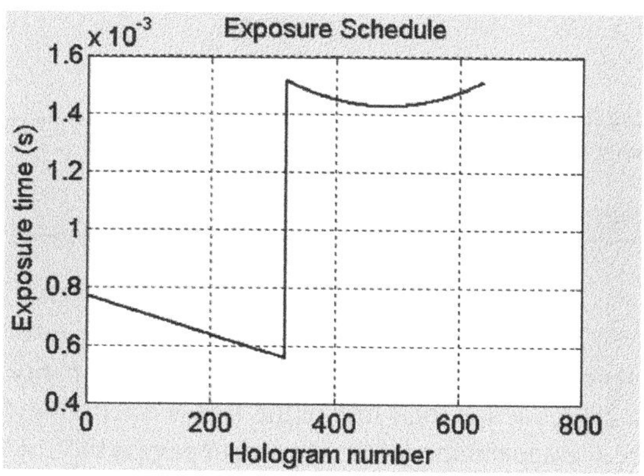

Figure 9.28 Example exposure schedule

Table 9.1 *Write transfer rate model*

Page to page	Layer 1	Layer 2
Exposure time (L1/L2)	600	1400 μs
Galvo move	800	800
Firmware overhead	200	200 μs
	1600	2400 μs
Book to book		
Book to book move time	50 000	50 000 μs
short stack move time	50 000	50 000 μs
Firmware overhead	2000	2000 μs
	714 000	970 000 μs
Write Times		
Page capacity with overhead	67 512 bytes/page	
# pages/book	320 pages/book	
# shortstacks/book	4	
# layers	2	
Total books	16 524	
Total dummy books	1414	
Total data Books	15 110	
Total write time	13 913.208 s	
Totals	231.887 min	
	3.86478 h	
Cure + cure move times		
Toast time (blank media)	900 s (15min)	
Total cure sites	624	
Pre-cure move time per site	20 mS	
Pre-cure time per site	75 ms	
Post-cure move time per site	75 ms	
Post-cure time per site	1000 ms	
	1630.08 s	
	0.4528 h	
Totals		
Total time (write, moves, cure)	4.31758 h	
Total data	326.43402 GB	
Write transfer rate (worst case)	21.001607 MB s^{-1}	

almost never read. Typically, about one-fifth of the chapter ECC parity pages are read out, which is effectively a 2% reduction (one-fifth of the 10% written) from the single data page byte count. Hence, the average number of user bytes per page is 98% of 77 672 bytes, which is 76 048 bytes.

Table 9.2 shows the read transfer rate calculation.

Table 9.2 Read transfer rate model

Page to page	Layer 1	Layer 2
Exposure time (L1/L2)	2000	2000 μs
Galvo move	800	800
Firmware overhead	200	200 μs
	3000	3000 μs
Book to book		
Book to book move time	50 000	50 000 μs
short stack move time	50 000	50 000 μs
Firmware overhead	5000	5000 μs
	11 65 000	11 65 000 μs
Read times		
Page capacity with overhead	76 048 bytes/page	
# pages/book	320 pages/book	
# shortstacks/book	4	
# layers	2	
Total books	16 524	
Total dummy books (not read)	1414	
Total read books	13 447.9	
Total read time	15 666.804 s	
Totals	261.113 min	
	4.35189 h	
Totals		
Total time	4.3518899 h	
Read transfer rate	25.276375 MB s^{-1}	

The calculations in Table 9.2 use read exposure times of 2 ms to ensure that the reference power incident on the camera exceeds 2 μW, which is required to achieve high SNR. The book to book firmware overhead is higher for reading than for writing because of the more complex calculations associated with the read feedback algorithms. The wobble servo (see Chapter 14) requires no additional overhead.

9.8 Summary

This chapter has outlined most of the principal components of the first generation Tapestry drive. As detailed in Chapter 3, there is a clear roadmap for future products. Along with increasing performance in future generations, there will be accompanying simplifications to the system. More than half of the current optics can be eliminated by further component integration and design evolution, while maintaining backward read compatibility. In addition, most of the electronics can be integrated into two or three ASICS. These changes will result in dramatic reductions in size, cost and complexity of future generations.

By proving the technology and implementing a fully functional drive, the InPhase team has also provided a springboard for the evolution of simpler consumer products such as the monocular architecture and holographic read only memory (HROM).

Acknowledgements

The Tapestry drive is the primary result of the work of all the employees and contractors at InPhase Technologies. We would like to give special thanks to Kirk Cook, Tolis Deslis, Tod Earhart, Jason Ensher, Michael Johnson, Demetri Lignos, Keith Malang, Ian Redmond, Brad Sissom, Paul Smith, Jimmy Wade and Tom Wilke. We would also like to thank Maxell Corporation for their work on media development, and Alps Corporation, Sanyo Corporation and Lite-on Corporation for their help and support with various drive components and aspects of drive manufacturing.

References

1. C.M. Jefferson, *et al.*, IBM Holographic digital data storage, in *Holographic Data Storage*, H. J. Coufal, D. Psaltis, and G. Sincerbox, eds. Springer-Verlag, New York, pp. 369–382, 2000.
2. C. Shuman,Single component aspherical apodizer, US Patent 6,825,992, November 30, 2005.
3. A. Hoskins, *et al.*, Tolerances of a page-based holographic data storage system, *Opt. Data Storage Conf.*, Portland, Oregon (2007), paper WB2.

10

Data Channel Modeling

Lakshmi Ramamoorthy, V.K. Vijaya Kumar, Alan Hoskins and Kevin Curtis

10.1 Introduction

Typical data storage devices record one bit at a time on to the surface of a disk or maybe tens of bits at a time on to the surface of a tape. In holographic data storage (HDS), typically a million or more bits are recorded and recovered at the same time from the volume of the media. This requires a very different system and therefore a very different data channel than what is used in other storage devices.

Figure 10.1 shows a diagram of the data channel at a very high level. First the data bit stream must have error correction codes (ECCs) applied to it and then be arranged into a block of data which we call a page. During the formatting of a page, modulation coding and other page features are added. It is this page that is presented to the optical system by a spatial light modular (SLM). The SLM typically modulates the light with a dark (zeros) or bright (ones) pixel pattern that corresponds to the data page. This optical pattern is low pass filtered and stored as a quasi-Fourier transform in the media. It is then read-out from the media and detected on a camera sensor. This pattern which has been low pass filtered now has coherent and incoherent noise added to it. The coherent scatter from the media and the lens is an important noise source for this channel. This pattern must be detected and converted back into the original data bits with bit error rate (BER) of less than 10^{-12} for consumer devices and 10^{-18} for professional or enterprise storage devices. As you can imagine this is a significant challenge.

This chapter will describe a physical channel model for HDS. The components of the model will be explained and important noise sources identified. The results of the model will be compared with experimental data. A good physical model can be helpful to better

Holographic Data Storage: From Theory to Practical Systems Kevin Curtis, Lisa Dhar, Adrian Hill, William Wilson and Mark Ayres

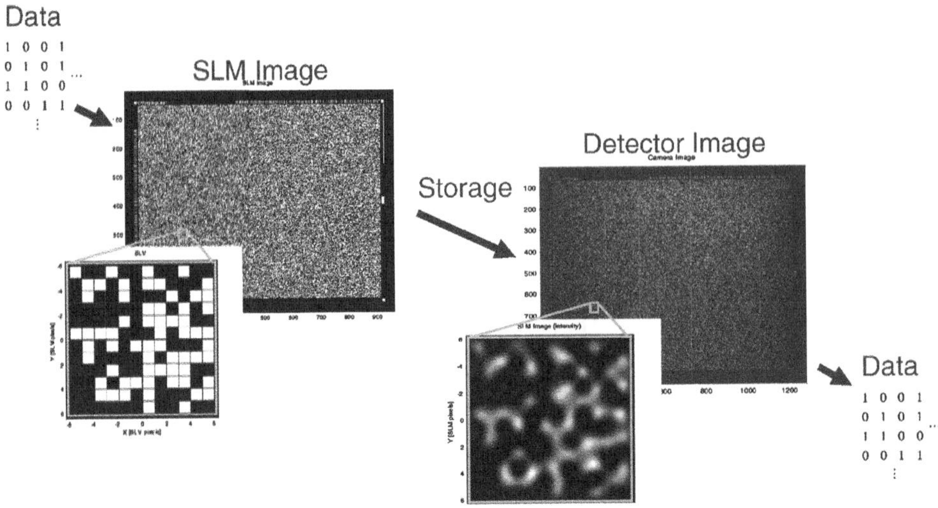

Figure 10.1 *Overview of HDS data channel*

understand where performance improvements can come from as well as understand the importance of the relative noise sources. Some metrics for comparing the model and the experiments are point spread functions (PSFs) and signal to noise ratio (SNR) of the recovered pages. An analysis of real data pages is also given to identify the channel. After this, a simpler channel model will be presented that is useful for generating a lot of data in order to allow the effective evaluation of the ECC. The physical channel model and channel estimation is taken from Lakshmi Ramamoorthy's PhD thesis and publications [1,2].

Chapters 11 and 12 go over the details of the data channel such as page formatting, modulation, detection, and logical format. In addition, Chapter 12 presents some future research directions for the HDS data channel.

10.2 Physical Model

10.2.1 Introduction

The conventional HDS optical system is a two lens arrangement (called a 4F or four focal length system) where a page of binary data is placed in the SLM plane and its Fourier transform (obtained with the first lens system) is recorded in the holographic medium. The schematic of the HDS system is shown in Figure 10.2. During retrieval, the medium is illuminated with the appropriate reference beam and a light wavefront containing the two-dimensional Fourier transform (2D FT) of the data page emerges from the medium and it passes through the second lens system. This produces the original recorded data page corrupted by noise and inter-symbol interference (ISI) as an amplitude wavefront in front of the detector (camera). The intensity of this wavefront is then detected by the camera. This model is still accurate for phase conjugate systems where the signal goes back through the first lens rather than the second.

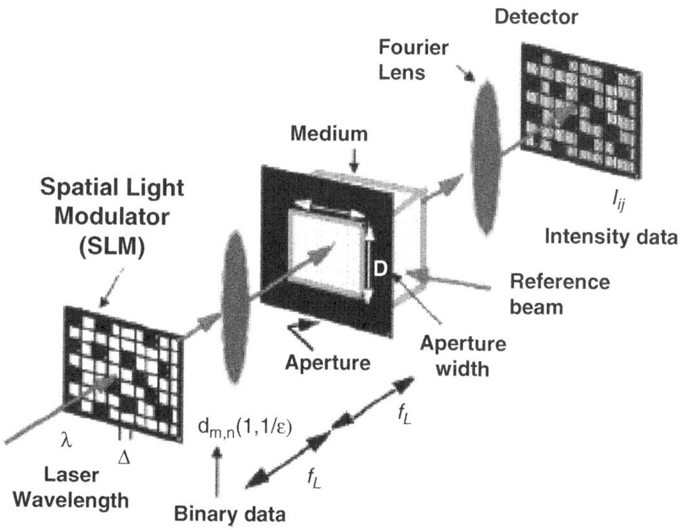

Figure 10.2 *Schematic of the HDS system*

In this chapter we introduce and discuss the various components and impairments present in the HDS system in order to model the channel and develop a channel simulator. A channel simulator is important for our analysis because we would have control over the levels of the channel impairments and hence will be able to investigate their effects on the performance of the system. To compare the results, channel quality metrics are introduced such as channel SNR, page and blockwise histogram SNR, normalized root mean squared error (NRMSE), raw and decoded BER, which are used for performance evaluation.

10.2.2 Details of Model

The light that illuminates the SLM has a wavefront that is not exactly flat because the light source is at a finite distance from the SLM. Even when using apodizers to make the illumination more constant, there still is an appreciable variation in intensity on the SLM, otherwise too much light is thrown away by over-expanding the beam and using only a small portion of it near the center. This non-flat illumination effect is included by modeling the incident light by a Gaussian wavefront. The degree of flatness can be varied by changing the standard deviation of the Gaussian function. Default value corresponds to 10% decrease in intensity from the center to the corners. For the SLM, the main impairments are the finite contrast ratio, nonuniformity, and nonfull fill-factor. Also of importance is that a random phase mask is usually placed next to the SLM to reduce the dynamic range requirements in the frequency plane. Details of each of these are described below.

The input to the SLM is a page of binary data, but in the SLM, pixels cannot achieve zero intensity. The amplitude contrast ratio (ACR) is defined as the ratio of the average amplitude of 'one' bits to the average amplitude of 'zero' bits. In the computer simulations, the zeros in the SLM data page are replaced by $1/\varepsilon$ with $\varepsilon = 10$ as the default choice. Typically the ACR specification for an SLM is greater than 80 : 1. Pixel intensities vary across the entire SLM. In this simulator, nonuniformity is modeled by a random number, drawn from a uniform distribution. Nonuniformity is a tunable parameter and we use the default value of standard deviation to be 1% of the mean of the input binary data page (which is 0.5). The field in the back focal plane of the first lens is the 2D FT of the input light leaving the SLM. This FT exhibits a large DC peak that can saturate the medium. One way for making intensity more uniform in the frequency domain is to use a phase mask [3] in the SLM plane. The transmittance function of the SLM, in the presence of a phase mask can be described as follows:

$$f(x,y) = \sum_{n=-N/2}^{n=N/2} \sum_{m=-M/2}^{m=M/2} f(m,n)\exp^{j\phi(m,n)}\mathrm{Rect}\left(\frac{x-m\Delta}{\alpha\Delta}\right)\mathrm{Rect}\left(\frac{y-n\Delta}{\alpha\Delta}\right) \tag{10.1}$$

where $\varphi(m, n)$ is the phase mask value in $(-\pi, \pi)$, M is the number of columns and N is the number of rows in the data page, Δ is the SLM pixel pitch, and $0 < \alpha \leq 1$ is the SLM linear fill factor (i.e. square-root of areal fill factor for a square pixel). In an SLM the active portion of the pixel is typically less than the entire pixel size. The ratio of active area to total pixel area is called the fill factor. The phase mask steps are drawn from a uniform distribution with K levels, with default value of $K = 16$. In the presence of nonfull fill factors, we can express the transmittance function of the SLM as follows:

$$f(x,y) = \left[\sum_{n=-N/2}^{n=N/2} \sum_{m=-M/2}^{m=M/2} f(m,n)\exp^{j\phi(m,n)}\delta(x-m\Delta, y-n\Delta)\right] * \left[\mathrm{Rect}\left(\frac{x}{\alpha\Delta}\right)\mathrm{Rect}\left(\frac{y}{\alpha\Delta}\right)\right] \tag{10.2}$$

where $*$ denotes 2D convolution and $\delta(.)$ denotes a delta function. The field in the back focal plane of the first lens is the FT of the transmittance function [in Equation (10.2)], i.e.:

$$F(u,v) = K\left\{\left[FT\left[f(m,n)\exp^{j\phi(m,n)}\right] * \sum_k \sum_l \delta(u-k/\Delta, v-l/\Delta)\right]\right\}\mathrm{sinc}(u\alpha\Delta)\mathrm{sinc}(v\alpha\Delta) \tag{10.3}$$

where K is a proportionality constant, $\mathrm{sinc}(x) = \sin(\pi x)/(\pi x)$, and u and v denote the spatial frequencies corresponding to x and y. Based on Equation (10.3), the SLM fill factor effect is simulated in the frequency domain as a sinc function point-wise multiplying the FT of the SLM output. The default value for the areal SLM fill factor is 0.95 which is similar to the Displaytech SLM (see Chapter 4).

For the storage medium, the frequency plane aperture (i.e. the polytopic filter) and the optical scatter noise are the major sources of impairment. By using a smaller aperture, the HDS system can offer more density but will suffer from more ISI since a frequency plane aperture acts as a spatial lowpass filter. An aperture of width $D = \lambda f_L/\Delta$ where Δ is the SLM pixel pitch, λ is wavelength and f_L is the focal length, is known as the Nyquist aperture. In our simulator, the aperture is square, centered at the origin. The aperture width is a tunable parameter with a default value of 1.21 times the Nyquist width which has been found experimentally to be robust to misalignments due to interchange, etc. The various sources of noise (e.g. light scatter, lens reflections, stray light, etc.) that affect the light amplitude are lumped together and called the optical noise. Optical noise is modeled as a circularly symmetric complex Gaussian noise random variable given by:

$$n_o = n_r + jn_i \tag{10.4}$$

where j is equal to $\sqrt{-1}$. Independent, additive white Gaussian noise (AWGN) components are added to the real (n_r) and imaginary (n_i) parts of the light amplitude incident on the camera. It is important to note that since these noise sources are added to the amplitude of the light this means they are a coherent noise source.

In the physical model, electronic noise, dark noise, nonfull fill factor and nonuniform quantization are the sources of impairment for the detector (camera). Camera electronic noise, modeled as AWGN, is added to the amplitude of the camera value which is related to the intensity of the light. The variance and mean are tunable. Another source of noise in the camera is the dark noise, produced in the absence of any incident light and is due to thermally generated electrons rather than optically generated electrons. In this simulation, dark noise is modeled by a uniform distribution independently for each page of data. To simulate the dark noise, these random numbers are added to the output intensity of the camera. The percentage of dark noise intensity is a tunable parameter. Camera fill factor is the ratio of sensitive area to the total pixel area. Camera fill factor is also simulated in the frequency domain. After adding optical noise to the light wave front impinging the camera, FT is applied to the noisy wavefront. A frequency-domain sinc function corresponding to the detector fill factor is multiplied point-wise by the frequency-domain wavefront. An inverse FT is applied to determine the effective light amplitude and thus the intensity of the wavefront incident on the camera. Areal camera fill factor is a tunable parameter but the default value is 0.40. The latest cameras have higher effective fill factors by using microlenses in front of each pixel fabricated on the silicon wafer. Another camera digitalization effect is the quantization error. In this simulation, we consider uniform quantization with fixed upper and lower limits of mean $\pm 6 \times \sigma$, where σ, where σ is the standard deviation of the unquantized page of data.

The channel simulator parameters used are: ACR = 10, light drop from center to corners = 10%, SLM nonuniformity = 1%, SLM fill factor = 95%, frequency plane aperture = Nyquist size, camera fill factor = 40%, optical noise variance = 0.012, optical noise mean = 0, electronic noise variance = 0.001, electronic noise mean = 0.43, dark noise standard deviation = 1%, dark noise mean = 0 and quantization = 10 bits.

10.2.3 Quality Metrics for the Model

In order to evaluate the results, we use a number of different quality metrics. One is the channel signal to noise ratio (cSNR) [4], which considers only the optical and electronic noise. Then there is hSNR which can be of two types; page histogram SNR and blockwise histogram SNR. Blockwise histogram SNR is a quality metric introduced by InPhase Technologies based on observation of real HDS data pages. We are interested in the histogram or output SNR which is determined from the histograms of the recorded ones and zeros because there are several channel impairments, the signal degradation due to these are captured in the histogram or output SNR. Then for pixel-wise comparison there is NRMSE. Finally, we need to compare the number of errors in the recorded page after detection, which is the raw BER; bit error rate and also the errors after decoder which is the decoded BER. We define cSNR (first introduced in [3]) as follows:

$$\mathrm{cSNR} = \frac{\mu_1}{4\sigma_o^4 + 4\mu_1\sigma_o^2 + \sigma_e^2} \tag{10.5}$$

where μ_1 is $a + 2\sigma_o^2$, a is the mean of the camera output pixel values, σ_o^2 is the optical noise variance and σ_e^2 is the electronic noise variance. This SNR captures the signal to the cumulative noise variance (which is the denominator of the cSNR expression) that includes the optical and electronic noise. Next we define the page histogram SNR:

$$\mathrm{hSNR} = 20\log_{10}\left(\frac{\mu_1 - \mu_o}{\sigma_1 + \sigma_o}\right) \tag{10.6}$$

where μ_1 is the mean of the 'ones' in the page, μ_0 is the mean of the 'zeros' in the page, σ_1 is the standard deviation of the 'ones' in the page and σ_0 is the standard deviation of the 'zeros' in the page. Further the larger the separation between the means of the ones and zeros, the larger the hSNR value gets. This makes the hSNR consistent with the resultant smaller probability of error as the overlap in histograms will be reduced. Also, the smaller the standard deviation of the ones and zeros, the higher the hSNR will be which is also consistent with the intuition that the tail probabilities are small resulting in lower probability of error.

The cSNR and hSNR serve different purposes. The cSNR of the channel simulator can be controlled directly, whereas the hSNR is the output SNR which is the combined effect of all the channel impairments. For blockwise page SNR, we determine the theoretical BER for the blocks of a page using the relationship:

$$\mathrm{BER} = \frac{1}{2}\mathrm{erfc}\left(\frac{Q}{\sqrt{2}}\right) \tag{10.7}$$

where erfc is the complementary error function defined as:

$$\mathrm{erfc}(x) = \frac{2}{\sqrt{\pi}}\int_x^\infty e^{-t^2}\,\mathrm{d}t \tag{10.8}$$

and Q is given by:

$$Q = \frac{\mu_1 - \mu_o}{\sigma_1 + \sigma_o} \tag{10.9}$$

where μ_1, μ_0, σ_1 and σ_0 are calculated for blocks in a page. The theoretical block BER is averaged for the entire page to get $\mathrm{BER}_{\mathrm{page}}$ and converted to SNR using the relationship:

$$\mathrm{SNR} = 20\log_{10}\left[\sqrt{2}\,\mathrm{erfc}^{-1}(2\mathrm{BER}_{\mathrm{page}})\right] \tag{10.10}$$

where erfc^{-1} is the complementary error function inverse.

For pixel-wise comparison we use the NRMSE metric, which is defined as follows:

$$\mathrm{NRMSE} = \sqrt{\frac{\sum_{m,n} |R(m,n) - \alpha S(m,n)|^2}{\sum_{m,n} |R(m,n)|^2}} \tag{10.11}$$

where (m, n) is the pixel location, $R(m, n)$ is the real recorded data page and $S(m, n)$ is the simulated data page and α is the scaling constant selected as follows to minimize the NRMSE in Equation (10.11):

$$\alpha = \frac{\sum_{m,n} R(m,n)S(m,n)}{\sum_{m,n} S(m,n)^2} \tag{10.12}$$

The raw BER is the number of bits in error divided by the total number of bits after the detection step. The decoded BER is the same, however it is after the decoding step.

10.2.4 Implementation Details and Effects of Parameter Variations

We use a basic set of impairment values and vary each of them to observe the impact of each parameter on the simulated data pages. The channel simulator parameter values used are the same as mentioned earlier.

For comparison of the simulated data pages, the page histogram SNR and the corresponding raw BER using the adaptive threshold detector are reported. The parameters are varied uniformly across the entire page, hence it is sufficient to observe the page histogram SNR and the blockwise histogram SNR would be about the same.

10.2.4.1 Amplitude Contrast Ratio

A random input page of zeros and ones is generated. The ACR is introduced by assigning the zeros of the data page a nonzero value $1/\varepsilon$, where ε typically takes a value like 5, 10, 15, etc. The impact of ACR on the simulated data pages is apparent by observing the page histogram SNR and raw BER of the pages. The histogram SNR is shown in Figure 10.3(a), and the corresponding raw BER is shown in Figure 10.3(b). As expected, SNR increases for increasing values of ACR, however, the SNR saturates to about 3.17dB after an ACR of about 15. This is because the other impairments limit the SNR. A similar trend can be observed in the raw BER as well. It decreases from 0.0746 for an ACR of 5 to about 0.0565 for ACR of 15. Beyond an ACR of 15 the BER does not decrease much.

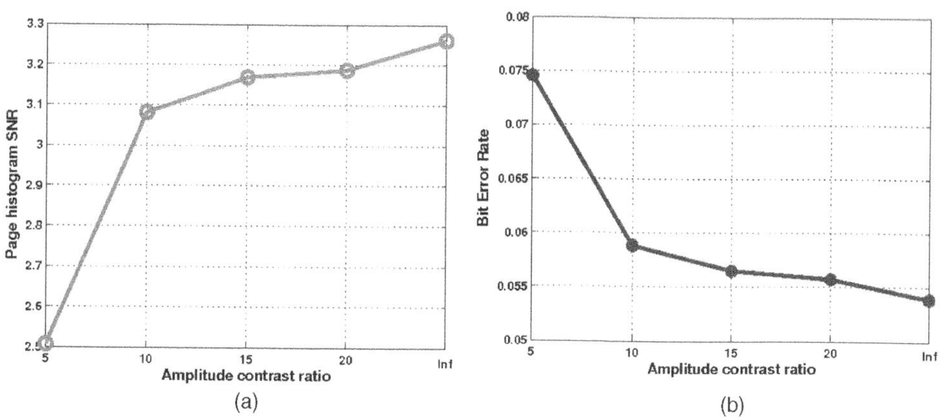

Figure 10.3 *Range of amplitude contrast ratio: (a) page histogram SNR; (b) bit error rate*

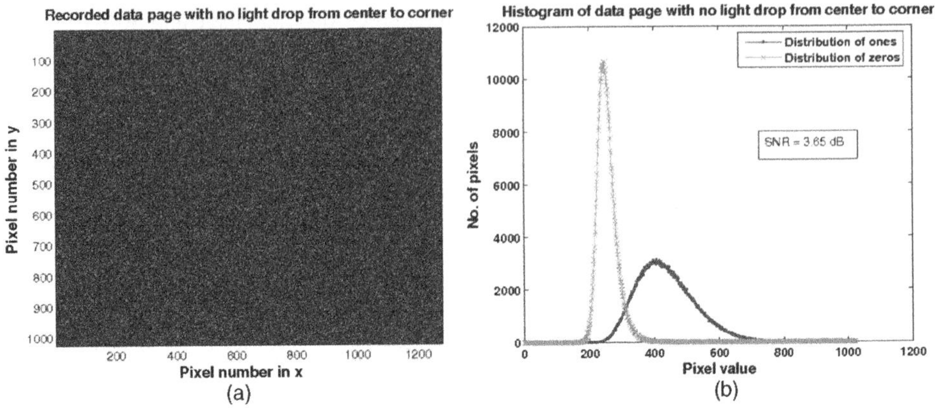

Figure 10.4 Data page without light level drop from page center to corners: (a) simulated data page; (b) histogram of simulated data page

10.2.4.2 Light Drop from Page Center to Corners

A Gaussian wavefront is point-wise multiplied to the input data page to simulate the effect of the light intensity decreasing from page center to corners. The Gaussian variance is determined based on how much the input illumination amplitude drops from the center to the corners. Results for a simulated data page, when all impairments except light level drop are present, are shown in Figure 10.4(a). The page has a histogram SNR of 3.65 dB and the histograms of the ones and zeros in the page are shown in Figure 10.4(b). The histogram on the left corresponds to that of zeros and the one on the right corresponds to the histogram of ones. However, when we introduce light level drop of 40% from page center to corners, the simulated data page has darker regions in the corners as shown in Figure 10.5(a). The corresponding histogram SNR is –0.13 dB and the histograms are shown in Figure 10.5(b). The histogram SNR and raw BER for various levels of light decrease from page center to corners are shown in Figure 10.6(a) and (b),

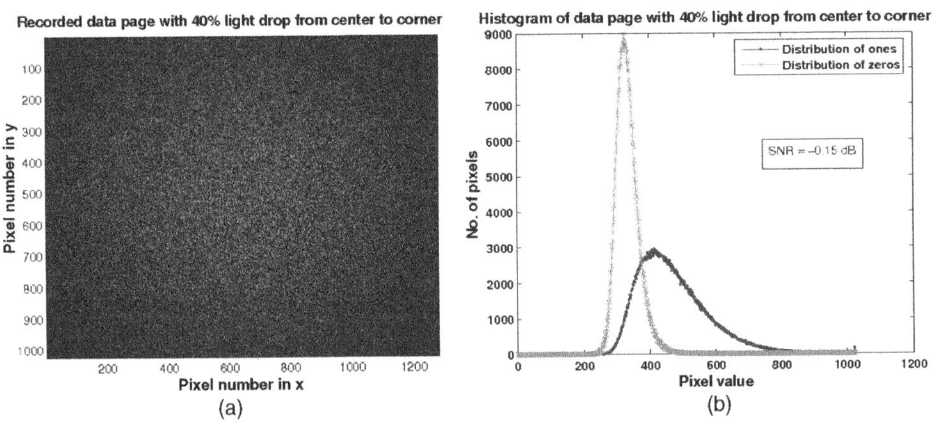

Figure 10.5 Data page with 40% light level drop from page center to corners: (a) simulated data page; (b) histogram of simulated data page

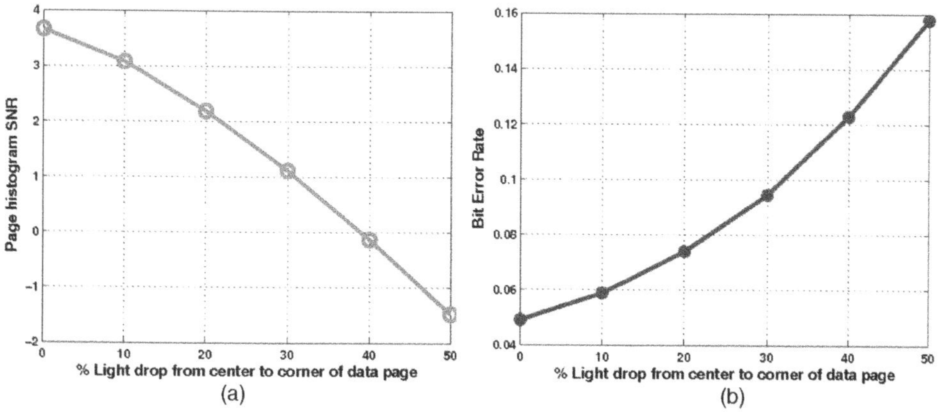

__Figure 10.6__ Range of light level drop percentage from page center to corners: (a) page histogram SNR; (b) bit error rate

respectively. The light level drop has a significant impact on the SNR and BER. The SNR goes from 3.66 dB for no light level drop to –0.13 dB for 40% light level drop from page center to corners and the BER increases from 0.0553 to 0.1297.

10.2.4.3 Spatial Light Modulator Nonuniformity

Pixel intensities vary across the entire SLM. Nonuniformity is modeled as a random number drawn from a uniform distribution ranging from –1 to 1. The input standard deviation of the random number generator corresponds to the percentage of the mean (which is 0.5) of the binary data page. Values smaller than zero are truncated to zero as intensity values cannot be negative.

The histogram SNR (in dB) decreases almost linearly with the increase in the nonuniformity percentage as shown in Figure 10.7(a). Similarly the raw BER also increases almost linearly with the nonuniformity percentage as shown in Figure 10.7(b). For 1% page

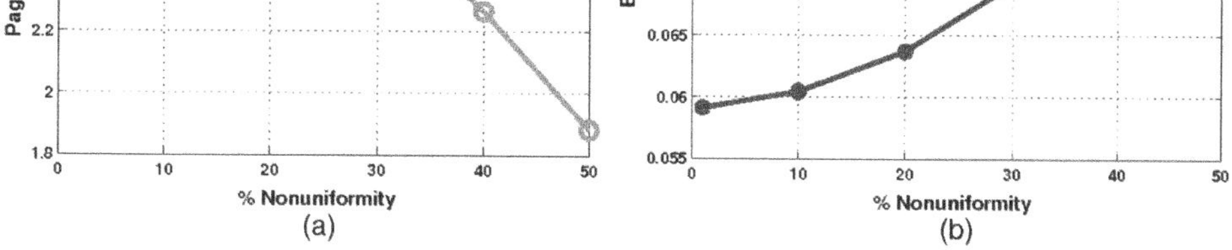

__Figure 10.7__ Range of SLM nonuniformity percentage: (a) page histogram SNR; (b) bit error rate

nonuniformity the page histogram SNR is 3.06 dB and the corresponding raw BER is 0.0591. However, if we increase the nonuniformity percentage to 20% of the mean of the binary page, the page histogram SNR goes down to 2.85 dB and the corresponding BER is 0.0637. For nonuniformity of 50% the SNR is 1.88 dB and the BER is 0.0856.

10.2.4.4 Spatial Light Modulator Fill Factor

The SLM fill factor is simulated in the frequency domain as mentioned in earlier. The convolution of the data page with a rect function in the space domain is effectively like multiplying by a sinc function in the frequency domain. The SLM fill factor simulator input is the areal fill-factor percentage, the linear fill factor α in Equations (10.2) and (10.3) is the square root of the fractional value of the areal fill factor. From Equation (10.3) we observe that the FT spectrum replicates every $1/\Delta$ distance spacing. With the Nyquist filter or in other words the frequency plane aperture, ideally only one replica of the spectrum is needed. As shown in Figure 10.8, the simulation of the SLM fill factor is effectively point-wise multiplication of the central FT spectrum (which is the FT of the data page) with the sinc function limited to the region between the dashed lines. This graphic shows the one-dimensional visualization of the fill-factor simulation. This is implemented in two dimensions. The amplitude of the 2D sinc function in will be α^2. Typically, we ignore the scaling constants involved in the simulator and later determine the scaling constant required to match with the real data closely by minimizing the NRMSE. However, when we study the fill factor effects on the SNR and raw BER it is important to consider the scaling constant.

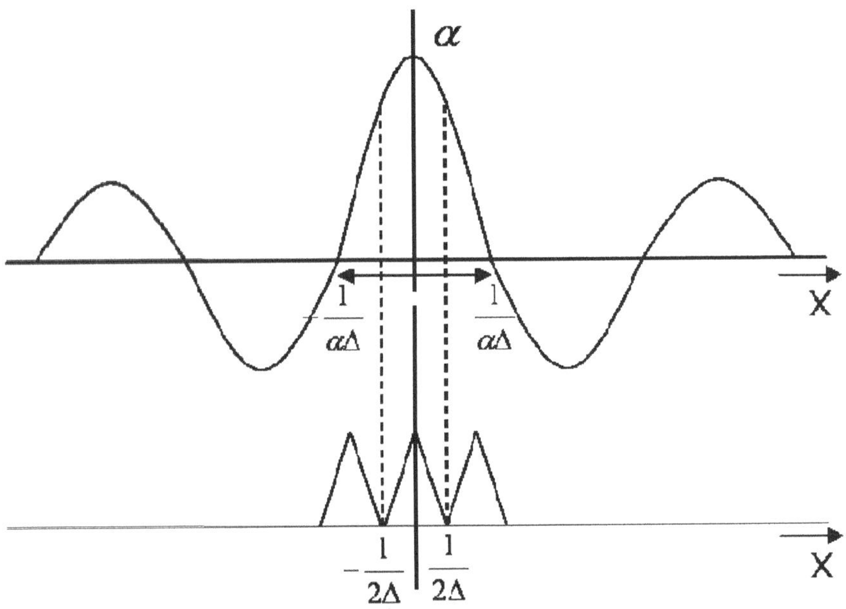

Figure 10.8 Simulation of the SLM fill-factor effects. Between the dotted lines is the sinc envelope that gets multiplied with the central FT spectrum of the data page which is the portion of the FT that is retained

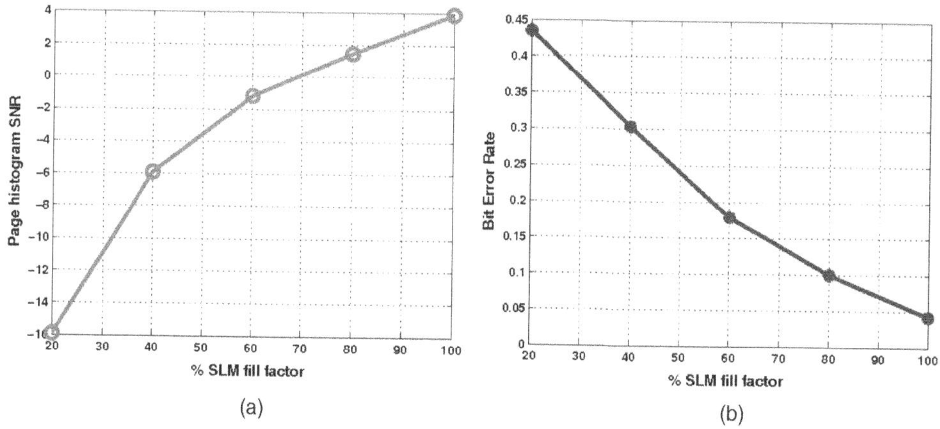

Figure 10.9 *Range of areal SLM fill factor percentage: (a) page histogram SNR; (b) bit error rate*

The ratio of the light levels to noise is an important factor and is captured through the amplification constant in the sinc function.

Now, we observe the resulting page histogram SNR [Figure 10.9(a)] and the raw BER [Figure 10.9(b)] for a range of SLM fill factors. The areal SLM fill factor increase from 20 to 60% results in a steeper increase in the page histogram SNR from –15.92 to –1.20 dB. After 60% SLM fill factor, the increase is slower and the SNR is 3.88 dB for 100% SLM fill factor. The corresponding raw BER is 0.0425.

10.2.4.5 Aperture

The frequency plane aperture is primarily used to filter out the unwanted replicas of the FT spectrum of Equation (10.3). Simulation of apertures up to the Nyquist size is possible with the Fourier transformed data pattern. However, it is sometimes desirable to see the performance at aperture sizes greater than the Nyquist size. To obtain such apertures, we over-sample the data page by a factor of 2. In this case, the frequency spectrum is compressed by a factor of 2 and hence using the FT of the over-sampled data page we can simulate apertures greater than the Nyquist size. Basically, a mask with ones in the central rectangular region which is required to be recorded and zeros outside that region acts as the aperture mask. This is point-wise multiplied with the Fourier transformed data pattern. After the aperture effects are applied and the inverse FT is obtained, the data page is down-sampled by averaging the 4 pixels that represented a single pixel during over-sampling.

As seen in Figure 10.10(a), the page histogram SNR increases rapidly with the increase in aperture size from 0.8 $\times$ Nyquist size up to the Nyquist size and beyond that it increases relatively slowly. This is because below the Nyquist size useful signal is cut-off by the aperture and that is reflected in the SNR and BER values. Above the Nyquist size there is no additional signal information, however more light is admitted by the aperture, hence the signal level compared with the noise level is more. The SNR at the Nyquist size is 3.06 dB and the corresponding raw BER is 0.0593 [Figure 10.10(b)]. SNR increases to 5.82 dB for 1.4 $\times$ Nyquist size with the raw BER being 0.015. For twice the Nyquist aperture size the SNR is 7.85 dB and the BER is 0.0027.

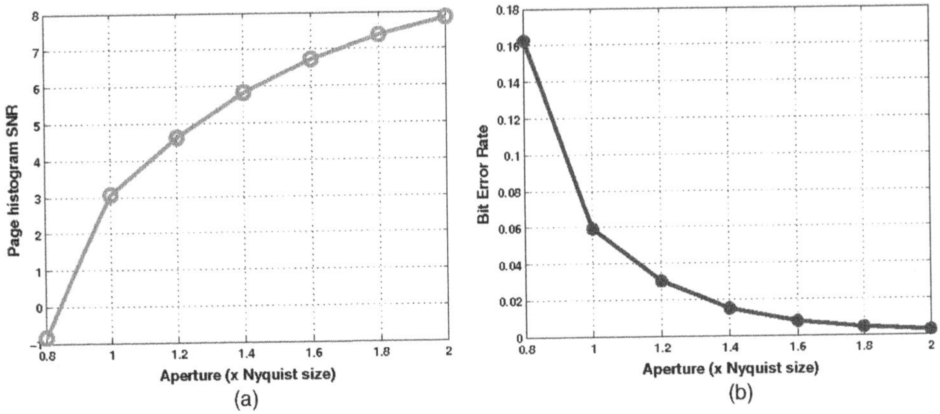

Figure 10.10 *Range of areal apertures (× Nyquist size): (a) page histogram SNR; (b) bit error rate*

10.2.4.6 Optical Noise Variance

Optical noise is modeled as a circularly symmetric complex Gaussian noise. The Gaussian noise random variable with zero mean and same variance is added to the real and imaginary parts of the inverse FT of the data page. The SNR and raw BER degrade drastically beyond an optical noise variance of 10^{-2} as shown in Figure 10.11. SNR decreases from 3.53 to –2.06 dB and the corresponding raw BER increases from 0.0497 to 0.1978 as the optical noise variance goes up from 10^{-2} to 10^{-1}. The change in SNR and BER is not significant up to an optical noise variance of 10^{-3}.

10.2.4.7 Camera Fill Factor

The camera fill factor is simulated similarly to the SLM fill factor. The data page is Fourier transformed and the sinc function with the camera fill factor is point-wise multiplied and the inverse FT is taken. The page histogram SNR increases steeply for areal camera fill factors from 20 to 40% but beyond that the increase is not so rapid. A consistent trend is also seen in the raw

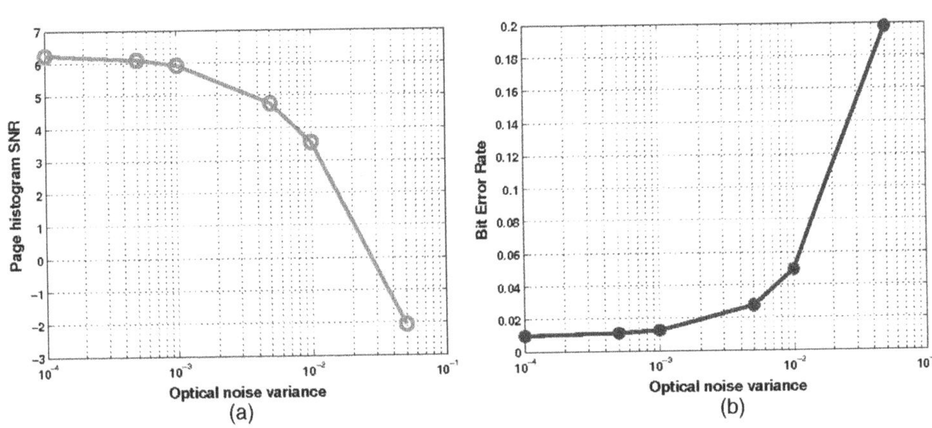

Figure 10.11 *Range of optical noise variance: (a) page histogram SNR; (b) bit error rate*

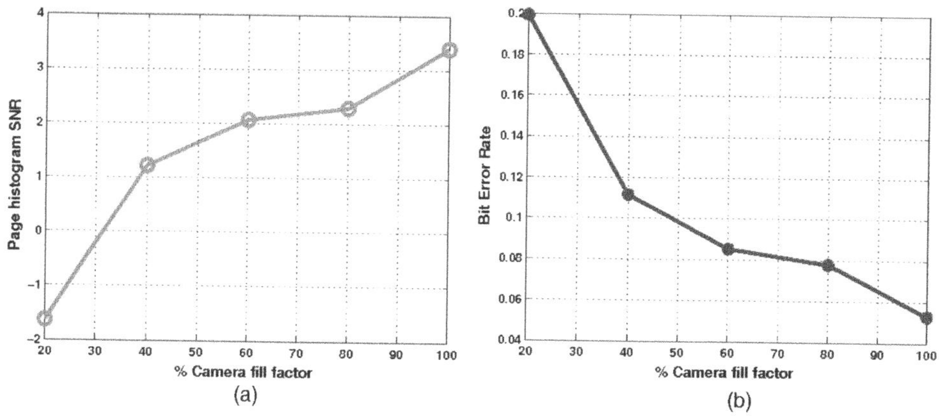

Figure 10.12 *Range of areal camera fill factor percentage: (a) page histogram SNR: (b) bit error rate*

BER. It is about 0.1992 at a camera fill factor of 20% and decreases to 0.1116 for 40% fill factor. The BER improves to up to 0.0529 for 100% camera fill factor with the corresponding SNRs being –1.63, 1.21 and 3.39 dB for 20, 40 and 100% fill factors, respectively (Figure 10.12).

10.2.4.8 Electronic Noise Variance

The magnitude squared operation is performed after the camera fill factor. The electronic noise is then added to the data page. It is modeled as a Gaussian distribution with mean 0.43 and the variance is varied to see the change in SNR and BER. The mean value of the Gaussian distribution is obtained by observing the output histograms of the ones and zeros of the real data page. The peak value of the simulated data page histograms falls at approximately the same output quantized value as in the real data page when we use this mean value for the distribution. Values of the data which fall below zero are truncated to zero, as intensity values cannot fall below zero. For noise levels greater than 10^{-3}, the SNR and BER degrade significantly. SNR drops from 3.05 to 1.58 dB when noise variance increases from 10^{-3} to 5×10^{-2}. The corresponding BER increases from 0.0593 to 0.1002 as shown in Figure 10.13.

10.2.4.9 Dark Noise Variance

The dark noise percentage is the percentage of standard deviation of the uniform distributed noise variable. The SNR and BER do not vary significantly as dark noise standard deviation increases from 1 to 10% as shown in Figure 10.14. However, beyond 10%, the increase in BER is almost linear with decrease in SNR. The SNR for 30% dark noise standard deviation is 0.69 dB and the corresponding BER is 0.1295.

10.2.4.10 Quantization Effects

For our simulator a uniform quantizer was employed. A uniform quantizer is one that has a fixed step size, i.e. the continuous range of values is partitioned to levels of equal spacing and values in each spacing are assigned to a discrete level. The effect of quantization was observed by changing the number of output quantization bits from 2 to 10 (Figure 10.15). It

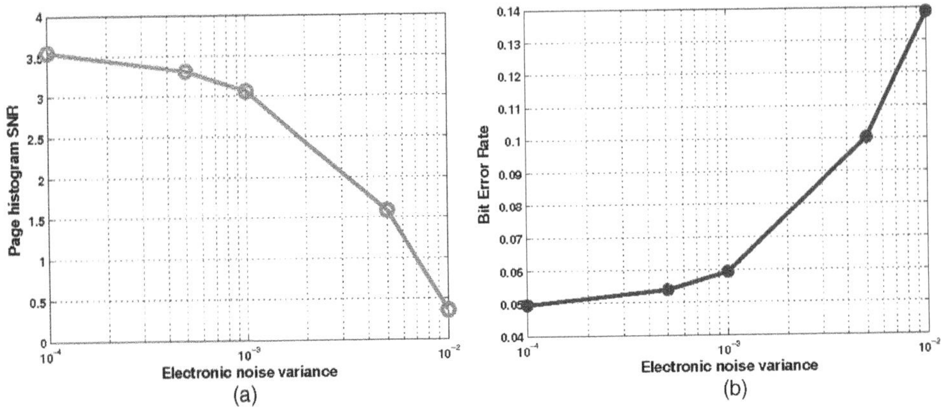

Figure 10.13 *Range of electronic noise variance: (a) page histogram SNR; (b) bit error rate*

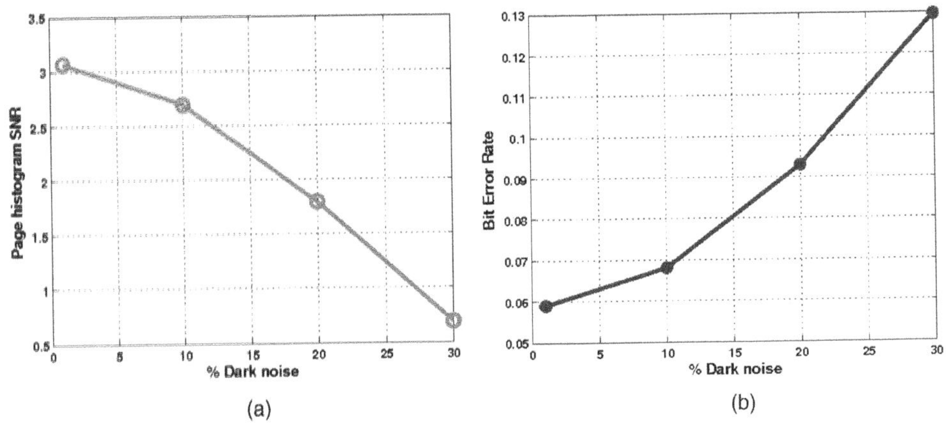

Figure 10.14 *Range of dark noise variance percentage: (a) page histogram SNR; (b) bit error rate*

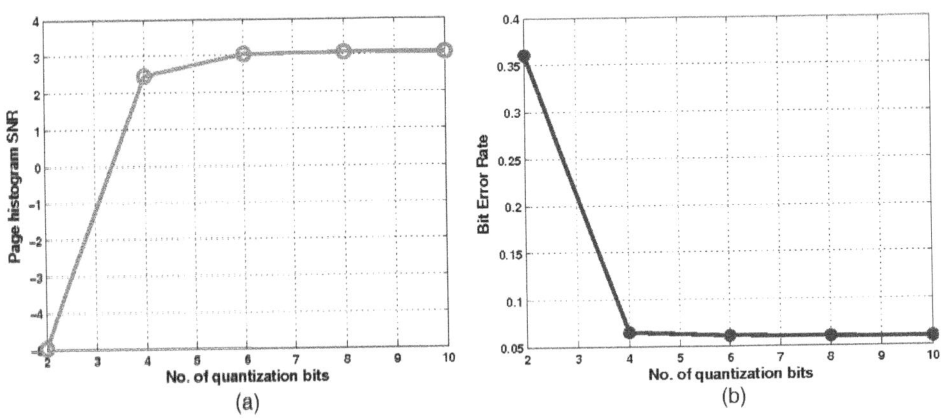

Figure 10.15 *Range of quantization output bits: (a) page histogram SNR; (b) bit error rate*

was observed that the number of output bits has to be at least 4 to get an output SNR of about 2.45 dB and a raw BER of 0.0648. For the case of 2 output quantization bits, the SNR was considerably lower. It is –4.97 dB and the BER is 0.3603, which is too high for the decoder to handle. Above 4 output quantization bits, the improvement in SNR and BER is not significant for this SNR calculation method and the adaptive threshold detector. The HROM camera in Chapter 15 was chosen to have 6 bits of resolution. In an actual system 4–5 bits look feasible as this analysis suggests. The fewer the bits per pixel the easier it is to transfer the data of the chip at high speeds.

10.2.4.11 Comparison of Channel Simulator with Real Data

To compare the simulated data with the real data, we set the parameters close to the real system. The real data pages were provided by InPhase Technologies in 2004 on an old pixel-matched system. The channel simulator parameter values used are similar to those mentioned earlier except for the following: frequency plane aperture = 1.21 × Nyquist size, optical noise variance = 0.023, optical noise mean = 0, electronic noise variance = 10^{-5}, electronic noise mean = 0.43. We vary the optical and electronic noise variance to match the histogram SNR of a real data page, which was about 3 dB. We simulated data pages with dominant electronic noise variance and the resulting histogram is shown in Figure 10.16(a). A page simulated with equal optical and electronic noise has the histogram shown in Figure 10.16(b). An optical noise variance dominant data page has the histogram shown in Figure 10.16(c). All these three simulated data pages have the same page histogram SNR of about 3 dB. However, the histogram of the optical noise variance dominant data page [Figure 10.16(c)] matches better with the real data histogram shown in Figure 10.16(d). The real data were taken with a Micron camera that had a known problem with its quantization which accounts for the spiky nature of the curve. The envelope of the curve is still valid.

For pixel-wise comparison, we observe the NRMSE value and the corresponding scaling constant α for 60 simulated data pages compared with the real data pages. As seen in Figure 10.17 the NRMSE is about 30% and the corresponding α is close to 1, indicating that additional scaling of simulated data pages does not lead to any improvement.

It was also observed that the NRMSE is not uniform across a data page. The NRMSE of 32 × 32 sized blocks in a page is shown in Figure 10.18(a). The darker blocks denote lower NRMSE. Blocks with NRMSE $\leq$10% are shown in Figure 10.18(b). (Black blocks are the blocks with NRMSE $\leq$10%.) Clearly, near the page center the NRMSE is lower and in the edges it is higher, this is because along the optical axis the data page does not have much distortion and away from it there is more distortion. This is not captured in our simulator.

10.2.4.12 Summary

In this section, a physical channel model and simulator was presented. This simulator was used to model the potential impairments to a HDS system. Using quality metrics such as channel SNR, page and blockwise histogram SNR, NRMSE, raw and decoded BER, channel performance was investigated as potential system parameters were varied. Finally, a comparison of the simulated data pages with the data pages obtained from the real system was done. It was observed that the NRMSE is about 30% and the scaling constant α is close to 1.

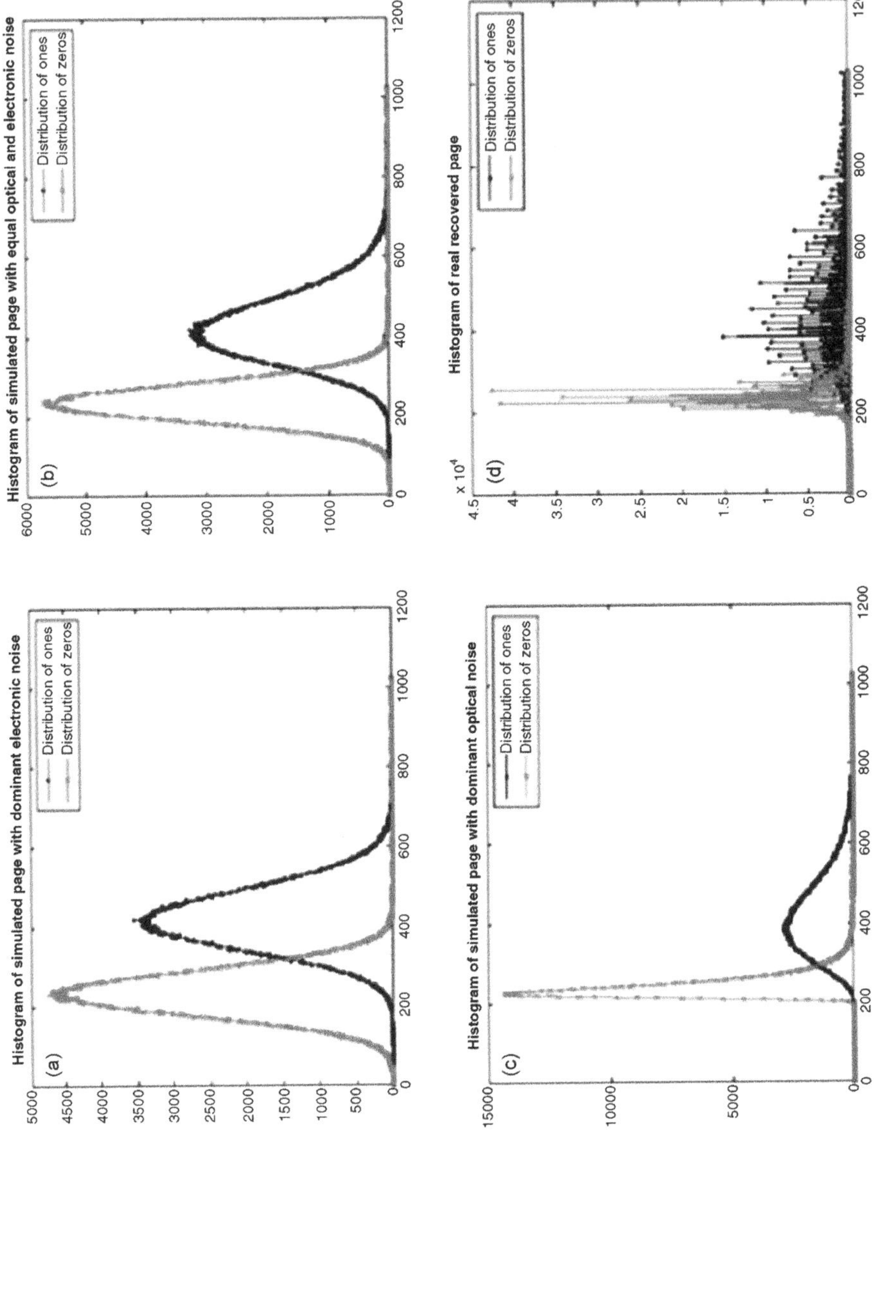

Figure 10.16 *Histograms of: (a) electronic noise variance dominant simulated data page; (b) simulated data page with equal electronic and optical noise variance; (c) optical noise variance dominant simulated data page; and (d) real data page*

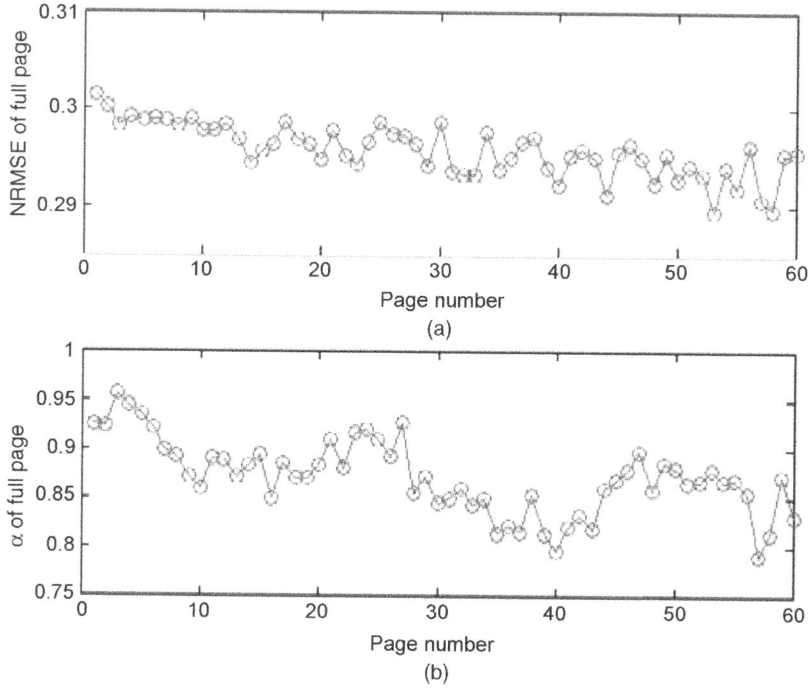

Figure 10.17 (a) NRMSE and (b) scaling constant α for 60 data pages

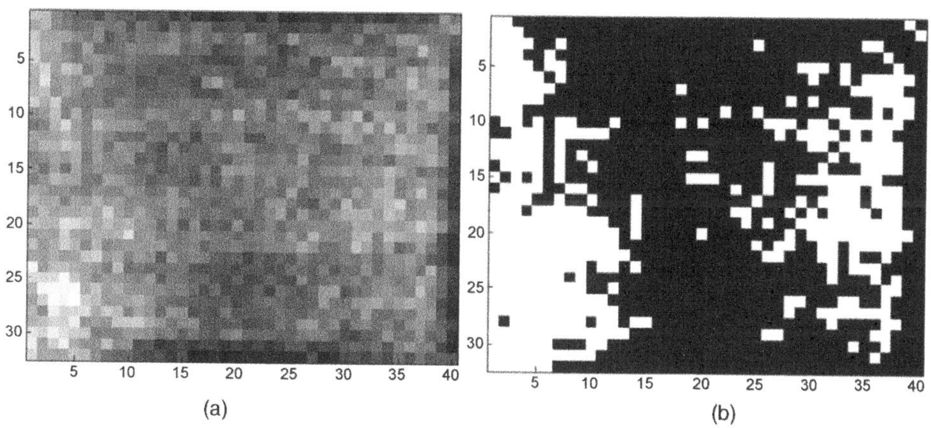

Figure 10.18 (a) Blockwise NRMSE and (b) black blocks in the page with NRMSE ≤10%

10.3 Channel Identification

10.3.1 Introduction

Channel identification is of interest as it enables us to simulate the underlying physical channel for analyzing advanced equalization and detection techniques. The assumed linear

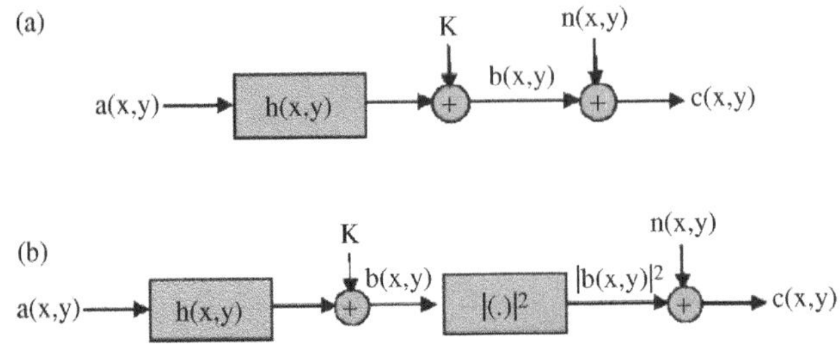

Figure 10.19 *(a) Biased linear channel model and (b) nonlinear channel model*

and nonlinear channel models for channel identification and equalization are shown in Figure 10.19.

In the biased linear model [Figure 10.19(a)] the input $a(x, y)$ is convolved with the impulse response $h(x, y)$ and a constant K gets added to the convolution output. The result is represented by $b(x, y)$. To this a Gaussian distributed noise variable $n(x,y)$ gets added. The output is represented by $c(x, y)$. On the other hand, in the nonlinear model [Figure 10.19(b)], after the addition of the constant K, the output $b(x, y)$ is magnitude squared. The Gaussian noise gets added after that. We use this notation throughout this chapter. This model is not an accurate model as we have not included the optical noise that gets added before magnitude squaring. In reality, the FT of the input gets recorded and during read back the inverse FT is obtained by a lens arrangement, whose real and imaginary components are corrupted by the real and imaginary components of the circularly symmetric complex Gaussian noise. All this is not captured in this model. However, this is a simple model which is useful for simulation of the underlying channel with reasonable accuracy (about 20% NRMSE). Focusing on the linear model, the output $c(x, y)$ is given by:

$$c(x,y) = \sum_{p=-1}^{1} \sum_{q=-1}^{1} h(p,q)a(x-p,y-q) + K + n(x,y) \tag{10.13}$$

By minimizing the mean squared error between output $c(x, y)$ and the channel input $a(x, y)$, we can solve for the linear least squares value of $h(x, y)$:

$$\begin{aligned}
&\arg\min_h \|c-Ah\|^2 \\
&\nabla_h (c-Ah)^T (c-Ah) = 0 \\
&-2A^T c + 2A^T Ah = 0 \\
&h = (A^T A)^{-1} A^T c
\end{aligned} \tag{10.14}$$

In the above equations, the vector **c** is the vectorized form of a block of pixels of the channel output page. The matrix **A** is the matrix whose rows are the vectorized form of the

3×3 neighborhoods of pixels of the corresponding block in the channel input page and its last column is a vector of ones. $\hat{h}$ is the estimate of the vectorized form of the 3×3 matrix of the channel impulse response and the last element is the constant K.

Considering the nonlinear model, the output $c(x, y)$ is given by:

$$c(x,y) = \left| \sum_{p=-1}^{1} \sum_{q=-1}^{1} h(p,q)a(x-p,y-q) + K \right|^2 + n(x,y) \quad (10.15)$$

By minimizing the mean squared error between output $c(x, y)$ and the channel input $a(x, y)$, we can iteratively obtain $h(x, y)$ and the constant K:

$$\arg\min_h \left\| c-(a^T h)^2 \right\|^2$$

$$\frac{de^2}{dh} = \nabla_h \left[c-(a^T h)^2\right]^T \left[c-(a^T h)^2\right]$$

$$\frac{de^2}{dh} = 2\left[c-(a^T h)^]\right](-2a^T h)a$$

$$h(n+1) = h(n) - p\frac{de^2}{dh} \quad (10.16)$$

In the above equations, c is a scalar corresponding to a pixel of the channel output page. The vector $\mathbf{a}$ is the vectorized form of the 3×3 input data page, with the last element being 1, the central pixel of which corresponds to the corresponding pixel in the channel input page. $\hat{\mathbf{h}}(n)$ is the estimated value of the vectorized form of the 3×3 matrix of the channel impulse response with the last element of that vector being K, at the nth step of the iterative least mean squared (LMS) algorithm. Also, μ is the iteration step size. We investigated various step sizes for the real data pages supplied by InPhase Technologies and finally used a step size of 0.05. The initial value of $\hat{\mathbf{h}}$ is the linear least squares solution described in Equation 10.14, except that the square root of $\mathbf{c}$ is used instead of $\mathbf{c}$.

10.3.2 Comparison of Linear and Nonlinear Channel Identification

The channel impulse response was calculated for the linear channel model and the nonlinear channel model as stated by Equations 10.14 and 10.16, respectively. The kernels for the channel impulse response differ for the blocks in the page center and the page corners. We investigated block sizes of 32×32, 64×64 and 128×128. To simulate a recorded page we convolve the kernels with the corresponding blocks of the input page and add the constant K. We also do magnitude square afterwards for the nonlinear model. As the block size reduces, the NRMSE between the real and simulated page also reduces. It is about 16–17% for block size of 32×32, 19–20% for block size of 64×64 and 20–21% for block size of 128×128. The improvement in NRMSE with nonlinear channel identification is about 0.5% with 32×32 sized blocks and 1% with 64×64 sized blocks and 128×128 sized blocks.

For 64×64 sized blocks, the kernel for channel impulse response for the block in the page center is shown in Figure 10.20(a) and for the block in the page corner is shown in

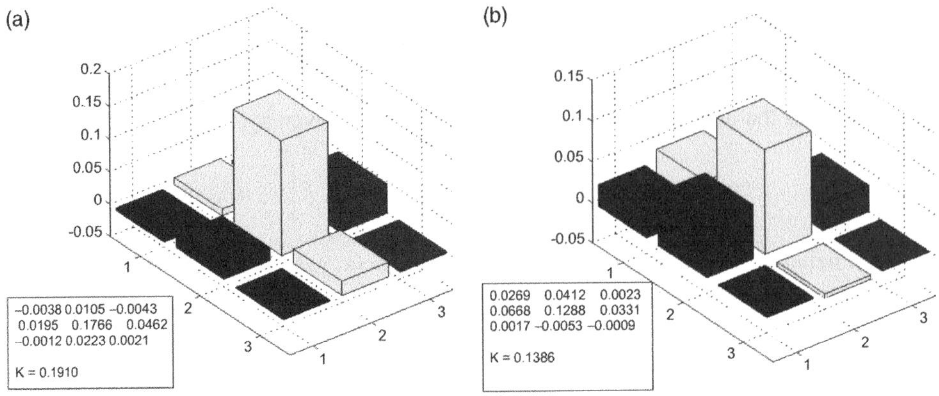

Figure 10.20 *Linear channel model space: (a) estimated MMSE kernel for center block of the page; (b) estimated MMSE kernel for corner block of the page*

Figure 10.20(b). They have different values as shown. The 3×3 channel impulse response kernel for the page center has 91% of the energy at the center value, whereas the kernel for the page corners has only 67% of the energy at the center value. Clearly, there is more ISI in the page corners. This is because the optical aberration is more as we move away from the optical axis. Also, a slight misalignment gets amplified towards the corners.

In the nonlinear channel identification, the channel impulse response kernel for the block in the page center is shown in Figure 10.21(a) and for the block in the page corner is shown in Figure 10.21(b). Clearly, they are different. The 3×3 channel impulse response kernel for the page center has 77% of the energy at the center value; the kernel for the page corners has 75% of the energy at the center value. From Figure 10.21 we can observe that the kernel for the page center has a high value in one of the other

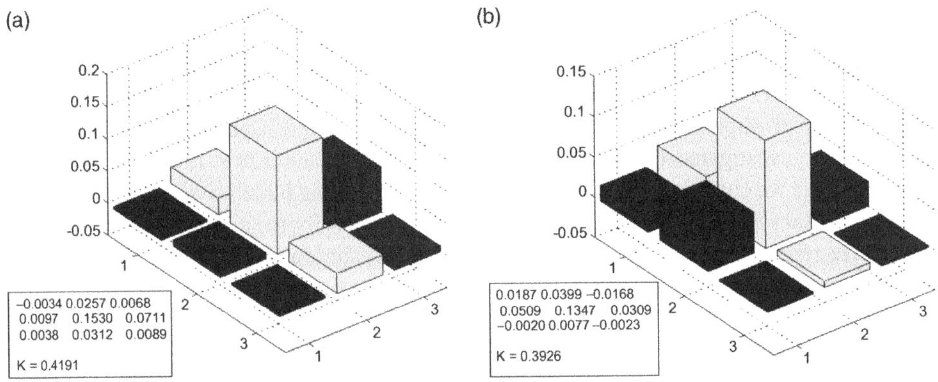

Figure 10.21 *Nonlinear channel model space: (a) estimated MMSE kernel for center block of the page; (b) estimated MMSE kernel for corner block of the page*

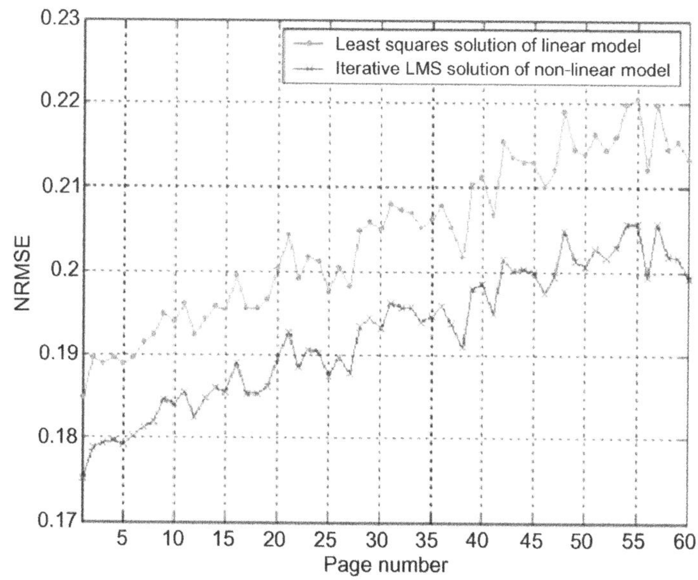

Figure 10.22 *NRMSE between 60 real data pages and simulated pages using linear model versus nonlinear model*

positions other than the center. However, the kernel for the page corner has three other high values other than the one at the center. Thus, the ISI is more spread in the corners.

Comparing the real pages with the simulated pages using the linear versus nonlinear model can be done using the NRMSE metric. The NRMSE between the real pages and the pages simulated using the linear model ranges between 18.5% and 22% for the 60 real data pages. For the pages simulated using the nonlinear model it is between 17.5% and 20%. This is shown in Figure 10.22. There is an obvious increasing trend in NRMSE as the page number increases. This is because of the varying quality of the holograms recorded first (in this case page 60 is recorded first) due to changes in the diffraction efficiency of the holograms. They were recorded with varying diffraction efficiency to see what the effect would be on the analysis.

10.4 Simple Channel Models

While the physical model above is useful for better understanding of the channel and making trade-offs, it is rather slow. It is useful to be able to quickly generate a lot of data that are similar to what comes out of the actual channel. As an example, InPhase would provide ECC code developers a simpler model so they could test the performance of their codes. We started doing this with a version of the amplitude model.

10.4.1 Amplitude Model

In this model, the intensity read signal from a camera pixel is modeled as:

$$I_A = \left|(g \cdot a)*h_A + n_i + jn_q\right|^2 + n_e \tag{10.17}$$

where g is the 2D Gaussian amplitude illumination profile, a is the data (random 1 or 0), h_A is the pixel spread function in the amplitude, $n_o = n_i + jn_q$ is the optical/scatter noise modeled as additive and circularly symmetric complex Gaussian with zero-mean and variance σ_o^2, and n_e is AWGN. The magnitude squaring is on each 2D sample. Roughly the model was set so that the power in the signal term was 10 times the power in the n_o term.

This model is fairly accurate, however it does not model the distribution of zeroes very well. The coherent noise generated using this method gives a distribution similar to a χ^2 distribution with one degree of freedom as seen in Figure 10.23. By regressing

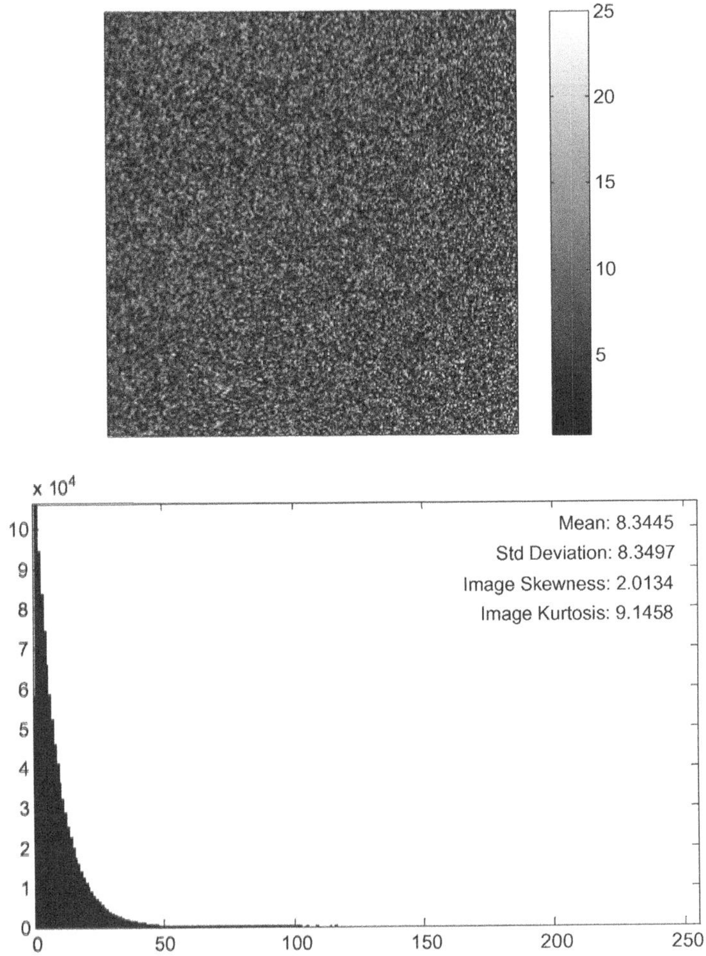

Figure 10.23 Noise image created using circularly symmetric complex Gaussian

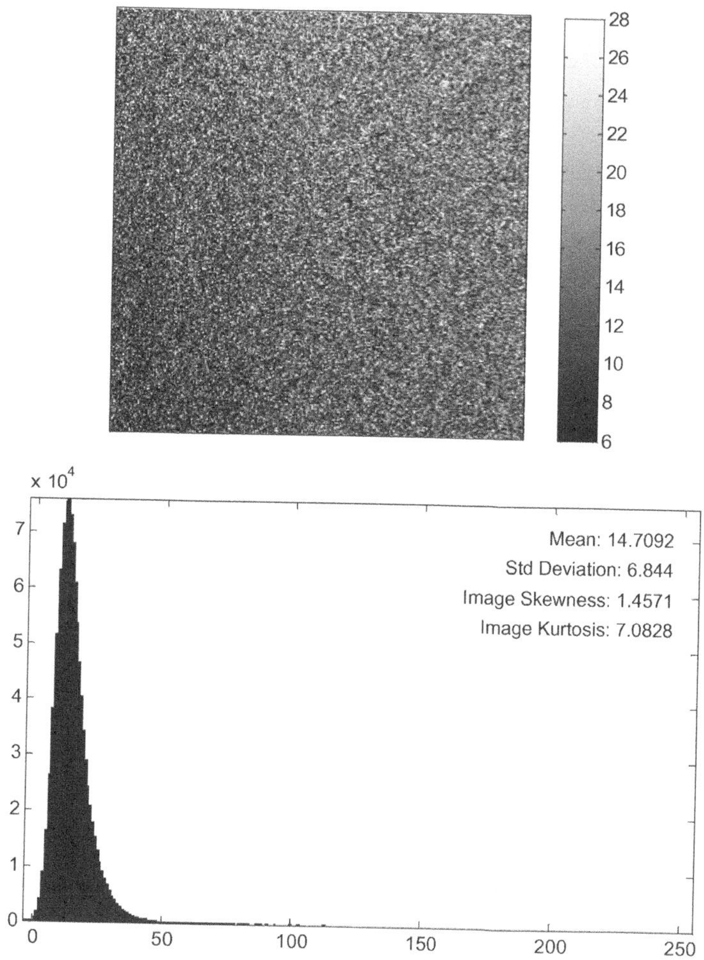

Figure 10.24 *Experimental noise image obtained from InPhase Technologies DVT drive*

actual data as seen in Figure 10.24, it appears that the coherent optical scatter is best fit using a χ^2 distribution with 3.6 degrees of freedom (Figure 10.25). The reason for this can be explained when examining the physical nature of the noise in the holographic system. Because the polytopic filter limits the spatial distribution of scattering centers that will affect the recovered image, the speckle size of the coherent noise can be estimated using simple Fourier optics. Also, because the polytopic filter in the holographic drive is 1.2 times larger than the linear Nyquist size required by the SLM pitch, the noise speckles will be ~0.83 the size of a data pixel or smaller. Getting an accurate noise model can be accomplished by averaging over the maximum speckle

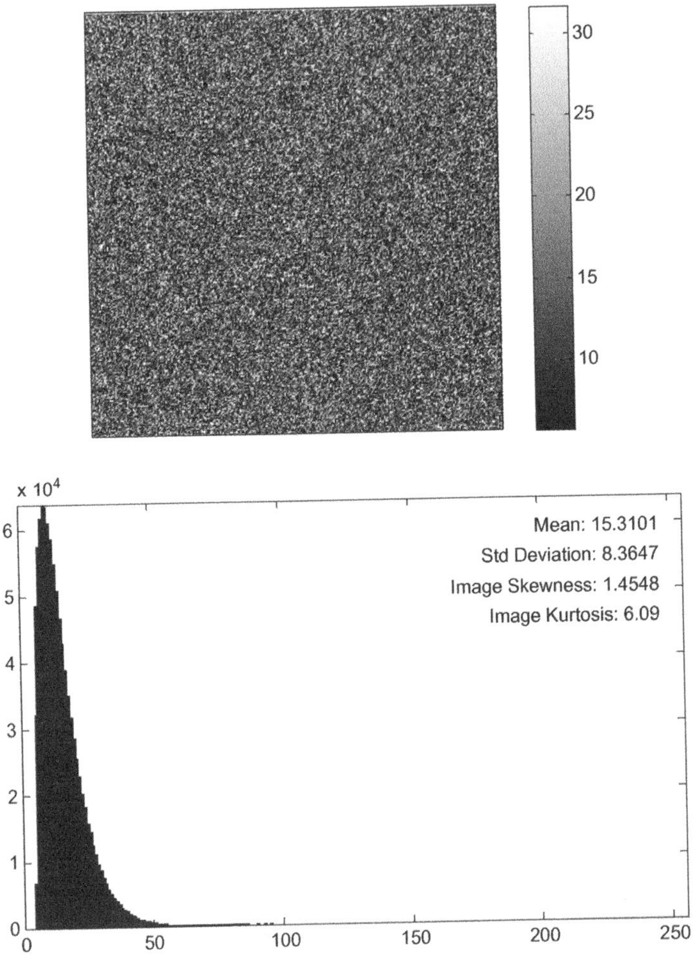

Figure 10.25 *Noise image generated using a χ^2 distribution with 3.6 degrees of freedom*

size and integrating the speckles correctly over the data pixels as shown in Figure 10.26. This method gives the physically correct zero-mean field with the pixel noise statistics seen by the holographic drive.

Acknowledgements

We sincerely thank Dr Lakshimi Ramamoorthy and Professor V.K. Vijaya Kumar for working with us on HDS, and for their permission to use the physical channel and channel identification material that resulted from that collaboration. At InPhase we wish to thank Adrian Hill, Kevin Curtis, Martin Pane, Srini Gopalaswamy and Li Ni.

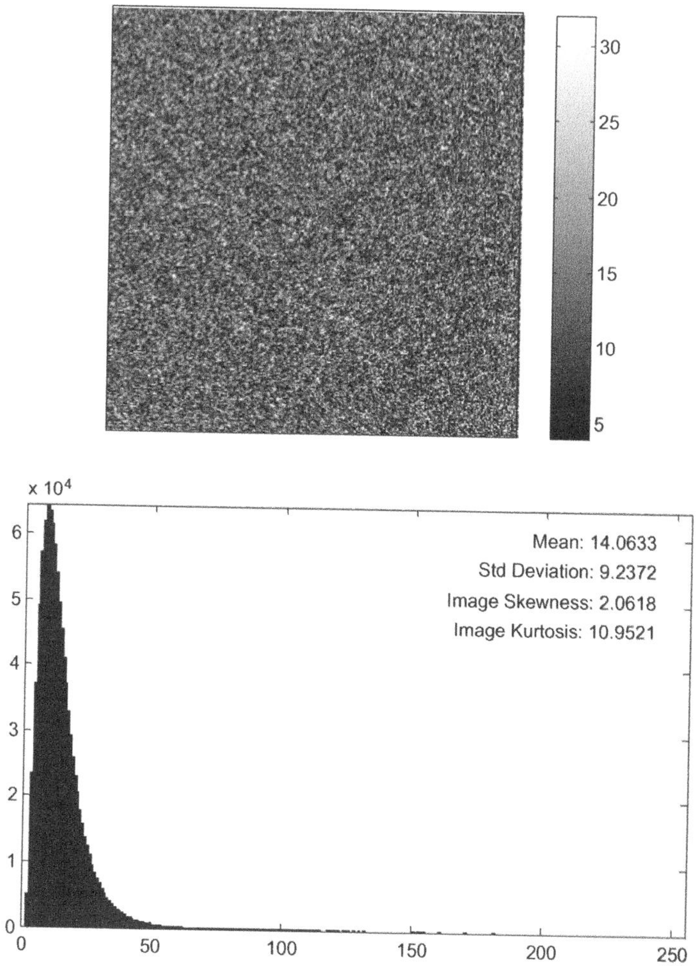

Figure 10.26 *Noise image generated averaged and integrated noise speckles*

References

1. L. Ramamoorthy, Advanced Coding and Signal Processing for holographic data storage systems, PhD Thesis, Carnegie Mellon University, Pittsburgh, Pennsylvania (2008).
2. L. Ramamoorthy, S. Nabavi and B.V.K. Vijaya Kumar, Physical channel model for holographic data storage systems, *Lasers and Electro-Optics Society, The 17th Annual Meeting of the IEEE*, Puerto Rico (2004), Vol. 2, pp. 997–998.
3. C. B. Nurckhardt, Use of random phase mask for the recording of Fourier transform holograms of data masks, *Appl. Opt.*, Vol. 9, pp. 695–700 (1970).
4. V. Vadde, 2-Dimensional equalization, detection and coding for volume holographic data storage systems, PhD Thesis, Carnegie Mellon University, Pittsburgh, Pennsylvania (1999).

11

Data Channel

Adrian Hill, Mark Ayres, Kevin Curtis and Tod Earhart

11.1 Overview

In an idealized binary holographic system, each bit of user data corresponds to the state of a single pixel in a holographic page, which would then be recorded with perfect fidelity in the recording medium. Upon reading the hologram, the reconstructed image of each original hologram pixel would impinge on a single camera pixel, without misalignment, blur, noise, or distortion. Each camera pixel's intensity would correspond to the original binary value of the input data; '1' for a bright pixel, and '0' for a dark one.

In practical devices, of course, none of these ideal conditions are met. The previous chapters describe many of the limitations of the physical channel that result in finite fidelity of the recovered image, including blur, noise, varying local intensity, crosstalk, distortion and misalignment. As a consequence of the limitations of the practical channel, several distinct techniques are required to record and recover data without error; including physical and logical formatting of the data, signal detection and processing, error detection, and error correction. These techniques are the subject of this chapter.

The data channel requirements for a storage device are stringent: on a 300 GB disk, the user expects that there will be zero bit errors, which means that the bit error rate (BER) must be less than 4×10^{-13}. Since this threshold must be achieved over many disks, the effective system BER must be considerably lower (around 10^{-18}) than even this number.

The InPhase professional drive uses an oversampled detection scheme in which the value of each recorded SLM pixel is estimated from multiple detector pixels. After resampling, multiple levels of error correction codes (ECCs) are decoded to preserve the integrity of the data. Within a data page, a Low Density Parity Check (LDPC) code provides powerful error

Holographic Data Storage: From Theory to Practical Systems Kevin Curtis, Lisa Dhar, Adrian Hill, William Wilson and Mark Ayres

correction capabilities for instances where a page is only partially recoverable. A second level Reed–Solomon code provides 'Chapter' (a group of pages) level protection, allowing the data to be recovered even when whole pages cannot be recovered. Finally, a third level 'Anthology' code (also Reed–Solomon) allows the data to be recovered even when entire chapters cannot be recovered with the inner two levels. The overall error correction scheme will be discussed in more detail later in this (book) chapter.

11.2 Data Page Formatting

While conventional optical storage devices read data one bit at a time, holographic data storage (HDS) allows many kilobytes of user data to be retrieved in parallel from each hologram. For example, the InPhase professional drive SLM has 1216 × 1216 pixels, which is more than 1.4 million pixels. The majority of these pixels are used for ECC-encoded user data, with the remainder being used for alignment and synchronization marks, bar code headers and a border around the page. Figure 11.1(a) shows an example data page displayed on the SLM. The features on the page will be described in the remainder of this section.

In order to accommodate misalignments between the reconstructed hologram image and the camera sensor, the camera sensor is considerably larger than the size of the image that lands on it. Features, such as sync marks are included in the page format, so that the position of the hologram image within the sensor area can be determined. These misalignments arise from a number of sources, of which interchange between drives is the most significant. In the sensor X direction (tangential, relative to the disk), the page location can shift ±23.5 camera pixels as the internal drive temperature changes, and because of optical and mechanical differences between drives. In the sensor Y direction, the page can shift ±18 camera pixels due to radial disk tilt changes, and optical and mechanical differences between drives. Also, the magnification of the image can change by ±30 camera pixels because wavelength changes are used to compensate for temperature changes. Rotation differences of

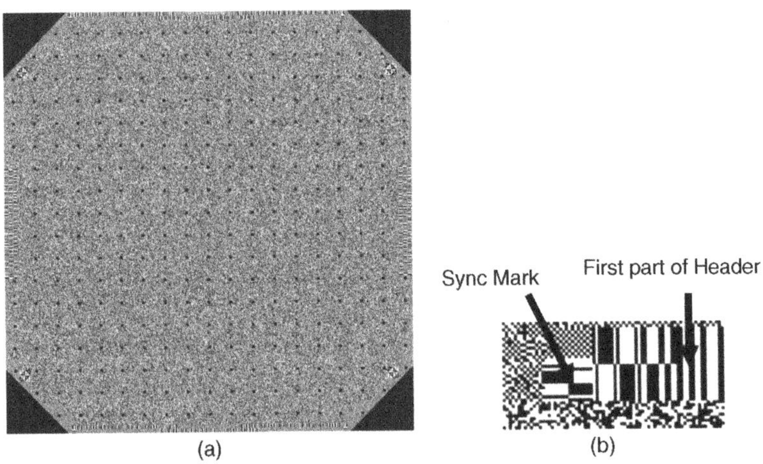

Figure 11.1 (a) Example data page with reserved blocks left blank and showing headers on all four sides of the page. (b) Close-up of left end of header (bar code) and sync mark

±7 camera pixels at the edges of the image arise because the SLM is aligned to its frame to within a tolerance of $\pm0.5°$.

The data page corners are clipped to make the required optical field more circular [Figure 11.1(a)]. The resulting field is 12.94 × 12.94 mm. To allow for misalignment, the optical path includes an aperture with a diameter of 16.21 mm, which is equivalent to 1515 pixels. The camera sensor is an array of 1696 × 1710 pixels, with 8 μm pixel pitch, which results in a field of 13.568 × 13.68 mm. The SLM pixels are 10.7 μm in size, resulting in a nominal linear oversampling ratio of 1.3375.

11.2.1 Sync Marks

The holographic image detected during recovery is not perfectly aligned, and hence sync marks are required to locate the page image on the sensor. Four composite sync patterns are located along the top and bottom edges of the data page. Each composite pattern consists of two horizontal sync marks and one vertical sync mark. The sync patterns are 32 pixels wide and 24 pixels high. Their relatively large size allows their position to be determined by computing the cross-correlation between the camera image and the known sync pattern. To reduce holographic correlation noise build-up within the recording medium, the left and right sync patterns on a single page are mirror images of each other, and on alternate pages, the sense of each pixel in the sync pattern is inverted (dark pixels on one page are light on the next, and so on).

Figure 11.1 shows the location of the sync marks and headers (described next) on the page, together with the details of the sync patterns.

The first stage of processing a hologram consists of finding the positions of these sync patterns, which determine the coarse alignment information for X-Y misalignment, image rotation and image magnification. Image magnification or demagnification occurs when there is a change in temperature (and hence in wavelength) between writing and reading, or when the recovering drive has different optical characteristics (e.g. lens focal lengths) than the recording drive. Image distortion that is not a pure magnification or demagnification may also occur if there are large temperature changes (more than a few degrees centigrade) between writing and reading; the reserved blocks (Section 11.2.3) are used to compensate for these distortions.

11.2.2 Headers (Bar Codes)

Headers are located in the areas between the inner two sync patterns at the top and bottom of the page. Each header contains two simple bar codes, each with a payload of 96 encoded bits. The data bits are encoded with a simple differential modulation scheme, with each bit being represented by a single block 12 pixels high and four pixels wide which is either dark in the first two columns and light in the second two columns, or vice versa. The large block features of the header make it robust to blur, noise and distortions. As with the sync marks, the binary sense of each single block is inverted on alternate pages to avoid inter-page correlation build-up.

The data contained in the header include page format version information, the seed for the selection of the reserved blocks (described next), as well as page and chapter numbers. To further increase robustness, there is a 32 bit cyclic redundancy check (CRC) for error detection, and error correction is provided by a (26, 20) Reed–Solomon code that has a symbol size of 5 bits. Two copies of the Reed–Solomon parity are encoded – one between the first and second sync marks, and the other between the third and fourth sync marks. The same data are encoded in headers at the top and bottom of the page, providing further redundancy in retrieving the information. Header decoding is attempted as each pixel row is transferred from the detector. When the decoded header bit pattern demarked by the best candidate horizontal sync mark patterns produces a valid CRC, the header data are extracted and the horizontal sync mark positions are validated. After one or more successful header decodes, the locations of the horizontal sync marks are used to delineate the possible positions of the vertical sync marks, which are then found by separate cross-correlations. Finally, the positions of four sync marks (two horizontal and two vertical) are transformed to produce the coarse X-Y misalignment, image rotation and image magnification.

11.2.3 Reserved Blocks

After coarse alignment to the sync marks has been performed, precise local pixel alignment is measured using the reserved blocks that are distributed in a grid throughout the page [1]. Reserved blocks are 8×8 pixel blocks containing predetermined patterns designed for accurate alignment measurement as will be discussed in the section on oversampled detection.

The reserved blocks are spaced 64 pixels apart horizontally and vertically, which results in 19 columns and 18 rows of reserved blocks. Of the 342 possible reserved block locations, only 320 are actually used so that reserved blocks do not interfere with the bar codes and sync marks. Overall, the reserved blocks represent an overhead of approximately 1.6%. In Figure 11.1, the reserved blocks are shown as regularly spaced black blocks, though in reality they appear as pseudo-random data.

The page format utilizes a total of 64 different reserved block patterns. To avoid inter-page correlation build-up, these patterns are distributed differently from page to page, and in addition, the sense of the reserved blocks is flipped on alternate pages.

In addition to providing local fine alignment estimates, the reserved blocks are also used for calculating the signal to noise ratio (SNR) and other metrics across the page.

11.2.4 Border Regions

The corners of the page are clipped by 378 pixels (see Figure 11.1) to make the optical field more circular. Also, an eight-pixel border is reserved on all edges of the page, resulting in an effective page size of 1200×1200 pixels. Overall, the field is 12.94×12.94 mm. The extra border allows for some mechanical misalignment in the drive. Pixels corresponding to the border regions are filled with a pseudo-random high frequency pattern, which serves to DC balance the SLM device without generating noise in the channel.

11.2.5 Data Interleaving

The InPhase professional drive uses a LDPC code within each page. (Characteristics of LDPC codes will be discussed in more detail later in this chapter.) Each LDPC code word consists of 2044 bytes of user data together with a four-byte Cyclic Redundancy Check (CRC). Using a code rate of $^1/_2$ results in an overall codeword size of 16 384 bits (4096 bytes). A single data page contains 38 complete code words.

The bytes of a single code word are not written sequentially to rows of pixels across the page. Instead all the code words are spatially interleaved together. This interleaving [2–5] dramatically improves the channel performance by reducing burst errors, which LDPC codes are not designed to correct. Without interleaving, data pages are prone to burst errors because limitations of the optical channel, such as dust, lens aberrations, shadows, and camera or SLM defects tend to be localized in extent on the page. Even without specific defects, nonuniformities in the optical system lead to regions of higher and lower SNR within a page (for example, the center of the page typically has higher SNR than the edges). Distributing each code word over the whole page provides more uniform performance from the LDPC code over all the code words.

The spatial interleaving of code words over a holographic page is analogous to the process in conventional optical storage where code words are interleaved to disperse the bits of a code word across the surface of a disk, to make the disk less susceptible to scratches and other localized defects.

Many interleaving schemes are possible within a two-dimensional (2D) data page – for example see Figure 11.2. In this illustrative scheme, a single bit from each code word in turn would be placed in a raster scan from left to right and top to bottom of the page, starting in the upper left corner, skipping over reserved blocks and any other reserved locations. This

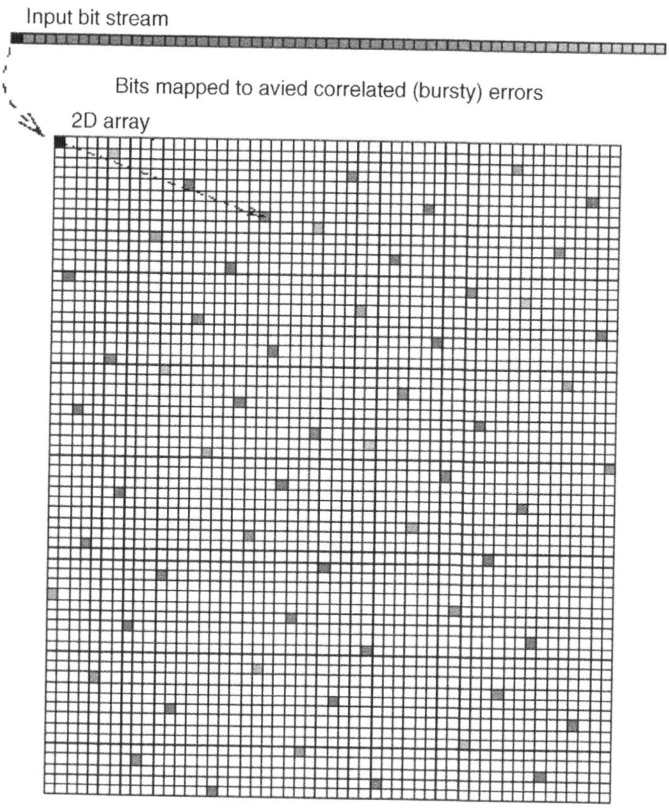

Figure 11.2 *Diagram of bit dispersal method*

would be followed by a second bit from each code word, and so on. Once all bits from the code words have been assigned pixel locations, the remainder of the page is filled with a pseudo-random sequence of bits. This helps maintain the DC balance over the page.

11.2.6 Modulation

In order to achieve a DC-balanced data page, the user data are XORed (exclusive ORd) with a pseudo-random bit sequence, which yields approximately equal distributions of 1s and 0s. Having equal distributions on all pages simplifies the scheduling calculations that define the exposure times used to record the holograms. (The other structures on the page, discussed in the previous sections, are designed to also have the equal distributions of 1s and 0s.)

In the InPhase drive, the pseudo-random bit stream is generated in hardware using a Linear Feedback Shift Register (LFSR), using the primitive polynomial:

$$G(x) = x^{32} + x^{22} + x^{21} + x^{20} + x^{18} + x^{17} + x^{15} + x^{13} + x^{12} + x^{10} + x^{8} + x^{6} + x^{4} + x + 1$$

The LFSR is initialized according to the page header seed, and the sense of the pattern is inverted for odd-numbered pages in order to preserve DC-balancing in the case where many sequential pages are written with the same pattern. This provides DC balance for every pixel, which is not otherwise statistically guaranteed for repeating data because of the limited size of the LFSR seed. In the more typical case of random data, every pixel will balance, on average, over a sufficiently large number of pages.

In the literature there are several papers on other types of modulation for 2D data channels, including HDS data channels. Of these, sparse encoding [5–7] is one of the more common approaches. Sparse encoding changes the distribution of 1s and 0s to make 0s more prevalent. In an on/off modulation scheme, 0s are represented by the absence of a diffracted signal, and therefore do not use *M/#*. Thus, sparse coding can potentially increase the storage capacity and SNR of a system. However, the trade-off is that sparse encoding results in a lower data transfer rate.

Other modulation code characteristics include: avoiding pixel patterns (such as a zero surrounded by all ones) on the page that are likely to generate lower SNR, balancing the on and off pixels, handling defects within a page, differential signaling, spatial frequency equalization, and block modulation encoding to improve SNR or correct errors [8–22]. All such coding techniques that restrict the allowable modulated patterns necessarily have a code rate of less than one. Gray scale encoding, on the other hand, can transmit at a code rate greater than one by encoding more than two signal levels per pixel [13].

In addition, much work has been done on equalization for holographic channels [23–37]. Many of these techniques can be integrated into the oversampled detection process described in Section 11.4.

11.3 Data Channel Metrics

As hologram pages are read from the drive, several parameters are calculated in real time for each page. These parameters provide the servo system with information that may be used to correct misalignments, initiate laser tuning, and so on. The calculated parameters include the familiar SNR as well as total page intensity, intensity centroids, and local summed intensity in four regions near the center of each edge of the page.

11.3.1 Signal to Noise Ratio

Not only are many definitions of SNR possible but a wide variety have been used, even within HDS. Worse yet, the term SNR is often used without definition. However, the most widely used definition is probably [38]:

$$\mathrm{SNR} \equiv \frac{\mu_1 - \mu_0}{\sqrt{\sigma_1^2 + \sigma_0^2}} \tag{11.1}$$

where μ_1 and μ_0 are the means, and σ_1 and σ_0 are the standard deviations, of the detected 1s and 0s, respectively. The numerator represents the distance separating the received binary symbols, and is thus a better proxy for true signaling strength than average power per bit. The denominator is the RMS sum of the two noise standard deviations, and so represents the standard deviation of the total population, assuming that the symbol probabilities are equal.

While Equation (11.1) represents a useful and justifiable performance metric, the noise powers in the denominator do not actually affect detection independently, as is tacitly assumed by this formulation [39]. For an unaugmented binary channel, bits would be detected by simple thresholding; for example, the bit is a '1' if the pixel value is above an optimal threshold and a '0' otherwise. With ideal additive Gaussian white noise (AGWN), raw BER is determined from the statistics of the signal:

$$\mathrm{BER}_{raw} = \frac{1}{2}\,\mathrm{erfc}\left(\frac{Q}{\sqrt{2}}\right) \tag{11.2}$$

where erfc(...) is the complementary error function, and

$$Q \equiv \frac{\mu_1 - \mu_0}{\sigma_1 + \sigma_0} \tag{11.3}$$

This equation arises simply by integrating the tails of the Gaussian distributions that lie on the wrong side of the threshold. As the argument, Q, is functionally related to raw BER, this expression, rather than Equation (11.1), is used as the SNR metric.

Whereas Equation (11.2) is guaranteed to be monotonic with BER, Equation (11.1) can move in the 'wrong' direction in some instances. For example, for $\mu_1 = 1$, $\mu_0 = 0$, and $\sigma_1 = \sigma_0 = 0.2$, the SNR of Equation (11.1) is 3.54, and the BER according to Equation (11.2) is 6.21×10^{-3}. However, if the noises change to $\sigma_1 = 0.24$ and $\sigma_0 = 0.15$, the BER improves to 5.17×10^{-3} but the SNR of Equation (11.1) actually decreases slightly to 3.53.

Raw BER is determined from the *resampled* pixels since the oversampled detector pixel values are not associated with a single identifiable data pixel. (Resampling is discussed in detail in the next section.) The distributions of the 1s and 0s are certainly not Gaussian, but they do not reduce to any easily represented probability distribution function (PDF) either. The variation in the resampled 1s and 0s arises from a combination of residual inter-symbol interference (ISI; that is, neighboring pixel crosstalk that is not removed by the linear resampling process), coherent optical scatter noise, SLM reflectivity variation, and thermal detector noise, among other factors. Resampling involves the scaled summation of 16 separate noisy detector pixel values (see Section 11.4), and so is driven towards a normal noise distribution by the Central Limit Theorem. Qualitatively, the distributions usually appear approximately Gaussian, and Equation (11.2) predicts actual raw BER reasonably accurately. Furthermore, since the pixel values are 'linearized' by taking the square root of the detected irradiance, the resampled pixels will have units of amplitude rather than energy, so the 20 $\log_{10}$ expression for conversion to decibels is appropriate rather than 10 $\log_{10}$. Thus, the formula used to compute SNR in decibels is:

$$\mathrm{SNR} = 20\log_{10}\frac{\mu_1 - \mu_0}{\sigma_1 + \sigma_0} \tag{11.4}$$

Equation (11.4) is used for SNR throughout this book.

Local SNR calculations are made for each of the reserved blocks on the page. These calculations are performed after resampling, at which point there is a single pixel in the resampled image for each original '1' or '0' bit in the reserved block.

The overall hologram SNR is not computed by simply averaging the individual reserved block SNRs on the page because there is no mathematical justification for such averaging, whether the raw ratios or values in decibels are used. However, averaging BERs is valid because the overall BER of two sets of data is simply the mean BER of the two sets.

Under the Gaussian distribution assumption of Equation (11.2), SNRs may be converted to BERs and averaged:

$$\mathrm{BER}_{avg} = \frac{1}{2N}\sum_{i=1}^{N}\mathrm{erfc}\left(\frac{Q_i}{\sqrt{2}}\right) \tag{11.5}$$

where Q_i is the (raw) SNRs for the individual reserved blocks. Then, the aggregate SNR may be determined from the inverse formula:

$$Q_{avg} = \sqrt{2}\,\mathrm{erfc}^{-1}\left(2\,\mathrm{BER}_{avg}\right). \tag{11.6}$$

This method for combining local SNR samples into whole-page SNRs tends to produce aggregate SNRs less than the arithmetic mean of the constituent SNR samples, reflecting the fact that in practice noisy samples dominate performance. The same method of calculating an overall page SNR can be used for computing an aggregate SNR for a group of holograms.

11.3.2 Centroid Calculation

An intensity centroid represents the 'center of brightness' of an image, and is commonly calculated in image processing. In the data channel, it is used for feedback in the wobble servo system. The normalized first-order moment along an axis defines the centroid along that axis. For example, the centroid, $\bar{x}$, in the x direction of an image $I(x, y)$ is given by:

$$\bar{x} = \frac{\sum_x \sum_y xI(x,y)}{\sum_x \sum_y I(x,y)} \tag{11.7}$$

The y centroid is computed analogously. The use of intensity centroids for wobble alignment is discussed in Chapter 14.

11.3.3 Intensity Metrics

The total page intensity is simply calculated by summing the intensities of all the pixels on the camera. The pixel intensities are summed in hardware as the pixel values are streamed off the camera sensor. Summed intensities are also calculated for four rectangular regions on the page, one placed midway along each edge of the image. The purpose of these four regional intensities is described in detail in Chapter 14.

11.3.4 Signal to Scatter Ratio

The signal to scatter ratio (SSR) differs from the preceding metrics in that it is not used directly by data channel algorithms while reading from the medium. However, it is a key metric for write scheduling (determining the exposure time for each hologram in a stack) to ensure that the holograms can be recovered with approximately the same SNR. Its importance stems from the fact that coherent scatter noise is the principal noise source in the channel.

The SSR is the ratio of the diffracted hologram intensity to the intensity of reference beam light scattered by the medium in the absence of holograms. Figure 11.3 shows the dependence of SSR on reference beam angle. The plot is derived from a measurement of the power incident on the detector as a function of the reference beam angle. The dark noise of the detector is first subtracted, and then the plot is normalized so that the mean signal level far from the hologram peaks is unity. Then the SSR of each hologram is determined by the level of its peak.

The importance of the SSR value is evident from the plot in Figure 11.4, which shows the empirical relationship between SSR and SNR as the diffraction efficiency is varied. For SSR values less than approximately 15:1, scatter is the dominant noise source and the SNR increases strongly as the SSR is increased. Above an SSR of 15:1, other noise sources (such

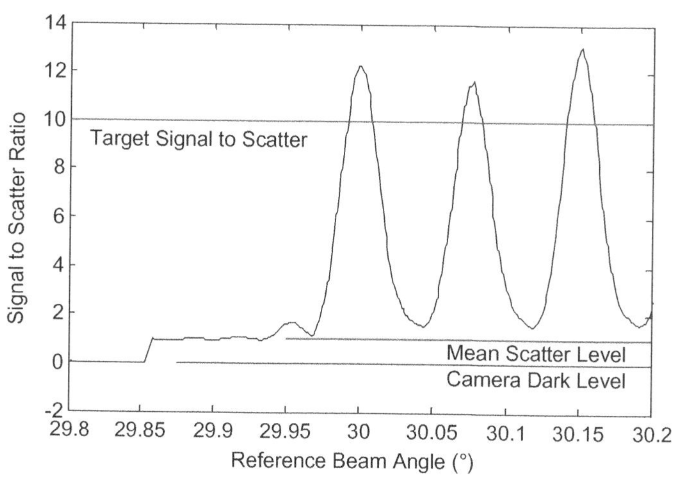

Figure 11.3 *Signal to scatter ratio measurement using power incident on camera versus reference beam angle. The vertical axis is scaled so that camera dark level = 0 and mean scatter level = 1*

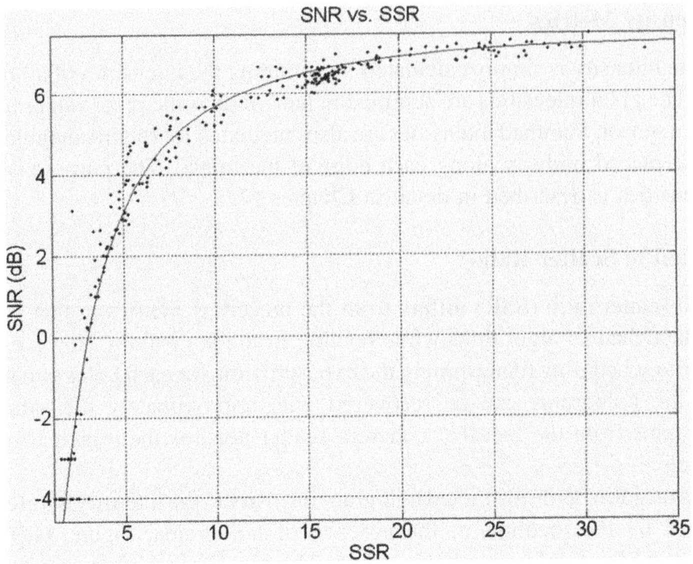

Figure 11.4 *Plot of signal to scatter ratio (SSR) versus signal to noise ratio (SNR)*

as image blur, which is largely determined by the polytopic filter size) begin to dominate, and limit the achievable SNR. The curve indicates that the holograms should be recorded with at least an SSR of 10:1 (corresponding roughly to the knee on the curve in Figure 11.4) so that small changes in diffraction efficiency do not significantly degrade the SNR. This target SSR is shown as the upper horizontal line in Figure 11.3.

11.4 Oversampled Detection

11.4.1 Introduction

High storage densities require high numerical aperture (NA) lenses, operating near the diffraction limit over a wide field. A pixel-matched system, where each SLM pixel is imaged onto a single camera pixel, is simple conceptually but difficult in practice. One difficulty arises from temperature changes – wavelength changes can be used to compensate for the temperature changes, but the wavelength adjustment also results in a system magnification change, which in turn means that a zoom lens would be required to adjust the image size to restore pixel-matching. In addition, pixel matching can only be achieved in a system with negligible field distortion, which is difficult in high NA lens systems.

Pixel-matched systems also impose stringent requirements on allowable image distortions from media dimensional changes (for example, shrinkage) and misalignment. Small errors in the mechanical registration of the holographic medium tend to produce displacement, rotation, and distortion of the holographic image without otherwise degrading its quality. Compensating for these errors with dynamic image alignment and magnification using microactuators leads to prohibitively large, expensive and slow storage systems. For all these reasons, pixel-matched systems are not commercially viable.

Although methods have been proposed for digitally restoring the fidelity of slightly misaligned images [24,40,41], realistic tolerance models indicate that image shifts of up to several hundred micrometers may be encountered in practice. Algorithms based on Nyquist-rate oversampling of the irradiance pattern [42] are prohibitive to implement for mega-pixel data pages because of the large detector bandwidth, power, cost and processing requirements.

Burr [43] proposed an alternative approach for pixel-matched systems that introduces a deliberate image magnification error into a system operating near the pixel-matched imaging condition, thus implementing sub-Nyquist rate oversampling. Such a configuration has the advantages of maintaining a low detector bandwidth, while also shifting the burden from preventing detector misalignment to measuring detector alignment.

This section describes InPhase's expansion of this approach, which includes a practical system for alignment measurement, together with an enhanced resampling process [44].

It is advantageous to limit the spatial bandwidth of the object beam to only slightly higher than the Nyquist frequency of the data pattern. An aperture in a Fourier plane (such as a polytopic aperture) can be used to band-limit the data beam, thereby also minimizing the size of the holograms in a system with a Fourier transform geometry.

The data pattern may contain at most one cycle (one 'on' and one 'off' pixel) per pair of adjacent data image pixels, so the minimum Nyquist frequency of the optical field of the object beam is twice this or one sample per pixel. However, since the spectrum of the irradiance pattern is the autocorrelation of the spectrum of the optical field, the Nyquist frequency of the detectable signal is doubled to a minimum of two linear samples per pixel. As a result, Nyquist sampling requires at least four detector elements (2×2) per image pixel, where the Nyquist rate is defined with respect to the actual irradiance pattern impinging upon the detector. Prior efforts at commercialization of HDS [43,45] have typically employed oversampling at or above the Nyquist rate, presumably on the basis of this observation. However, frequencies above the one-sample-per-pixel rate of the underlying optical field are not information bearing, so the possibility of sub-Nyquist data recovery is not precluded.

11.4.2 Resampling Process

In order to resample an image, the location of every data pixel image impinging the detector pixel grid must be measured to an accuracy not worse than a few percent of the pixel size (i.e. a few hundred nanometers). The resampling process first measures the alignment of local fiducial marks, interpolates to find the alignment of each pixel, and then uses this information to select the correct filter coefficients for resampling each data pixel (Figure 11.5).

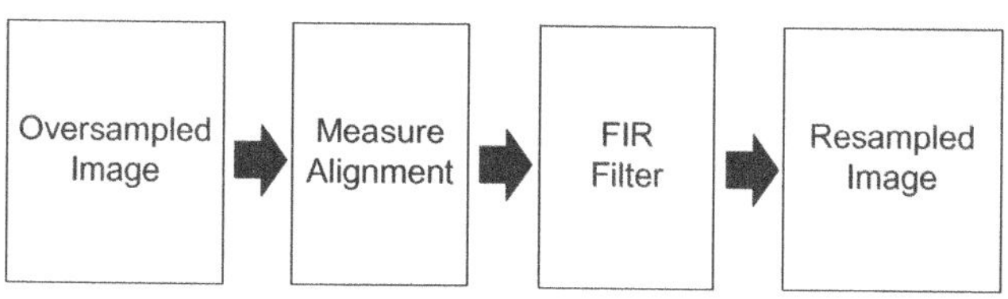

Figure 11.5 *Flow diagram for oversampled detection process*

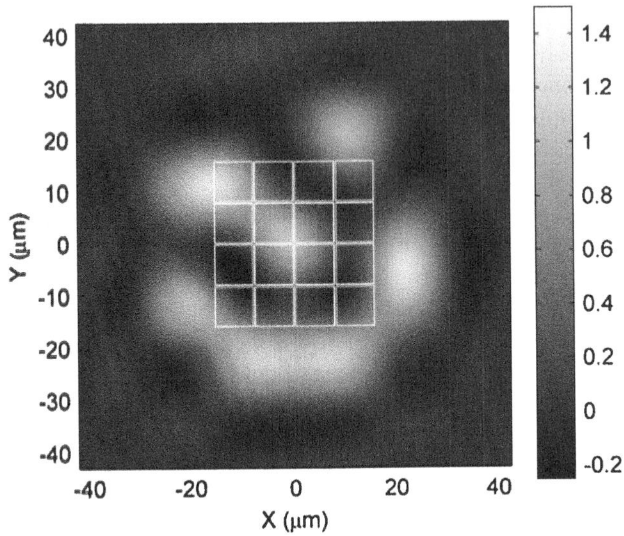

Figure 11.6 *Simulated data pixel image neighborhood (real part of complex amplitude)*

An estimate of the state of each data pixel is obtained using only the detector elements in a 4×4 local window corresponding to the position of the data pixel image on the detector (see Figure 11.6). Furthermore, this resampling function is linear, excepting that a 'linearization' exponent of 0.5 (square root) is applied to each detected intensity value to partially compensate for the nonlinear coherent addition of the data pixel images [46].

A computer model was developed to simulate 5×5 neighborhoods of data pixel images impinging on the 4×4 detector element window, including diffraction effects, pixel fill factors, and coherent optical noise. The model simulated the complex optical field impinging upon the detector on a 256×256 sampled grid. The simulation is highly oversampled spatially, with 32×32 samples representing each SLM pixel. Optical propagation was effected by composing an array representing the optical field of a single SLM pixel at the SLM plane, and Fourier transforming the array to compute the field at the polytopic aperture. A square aperture transmission function is applied multiplicatively. This transmission function had width:

$$D_p = 1.08 \frac{\lambda f}{\Delta_{pix}} \tag{11.8}$$

where λ is the wavelength, f is the Fourier transform lens focal length, and Δ_{pix} is the SLM pixel spacing. The function has the effect of limiting the imaging resolution of the optical field to 1.24 times the Nyquist rate of the data pattern. (The factor of 1.24 was empirically chosen to slightly over-resolve the image, providing margin.) The band-passed optical field pattern is then inverse-Fourier transformed to generate the image field function for a single SLM pixel impinging on the detector. During the simulation, neighborhoods of SLM data patterns are built up in image space by superimposing shifted versions of this archetypical pixel image. (This architecture was adopted rather than Fourier-propagating entire pixel neighborhoods, so that pixel image functions incorporating aberrations and other higher order effects could be easily imported into the simulation.) Finally, the signal detected by each of the 16 detector elements is determined by taking the magnitude-squared of the complex field function and summing it over the active area of each individual detector element.

A set of 16 linear coefficients is required in order to estimate the state of the central data pixel image from the 16 detected values. However, a single set of coefficients is not sufficient because the 4×4 detector window may have an arbitrary fractional alignment to the data pixel image. Instead, coefficient sets are generated in steps of 6.25% of the detector element spacing, for δ_X and δ_Y. This results in a table of $16 \times 16 = 256$ different sets of 16 coefficients. Symmetry of the coefficients in both δ_X and δ_Y allows for the simulation of only the alignments $\delta_X, \delta_Y = 0,\ldots,0.5$ and generation of the other three quadrants of the table by symmetry.

Coefficient sets are derived over 2048 different binary states of the neighborhood for each fractional alignment case. Because it is impractical to simulate all 2^{25} possible states of the 5×5 SLM pixel neighborhood, a strategy is used where the inner 3×3 neighborhood (including the pixel of interest and its nearest neighbors) is swept through all 512 possible combinations four times, and the outer ring of pixels is assigned values randomly. This strategy has been shown empirically to provide good coefficient convergence properties.

The simulated detector values are used to determine resampling coefficients that satisfy the minimum mean-squared error (MMSE) criterion. Each 16 element coefficient vector, $\boldsymbol{w}$, is selected so that:

$$\hat{\boldsymbol{d}} = \boldsymbol{I}\boldsymbol{w} \tag{11.9}$$

where $\|\hat{\boldsymbol{d}} - \boldsymbol{d}\|^2 = \sum_i \left(\hat{d}_i - d_i\right)^2$ is minimized.

$\boldsymbol{I}$ is the 2048×16 matrix of linearized, simulated detector values; and $\boldsymbol{d}$ and $\hat{\boldsymbol{d}}$ are 2048 element column vectors representing the binary states of the central SLM pixel in each neighborhood, and their resampled estimates, respectively. The solution, given by [47]:

$$\boldsymbol{w} = \left(\boldsymbol{I}\boldsymbol{I}^T\right)^{-1} \boldsymbol{I}^T \boldsymbol{d} \tag{11.10}$$

is evaluated with an internal Matlab function. Since the coefficients are selected to produce the MMSE estimate of the binary state of the central data pixel image (as opposed to, say, the irradiance value), the resampling process effectively removes much of the ISI. Finally, the SNR metric from Equation (11.3):

$$Q \equiv \frac{\mu_1 - \mu_0}{\sigma_1 + \sigma_0} \tag{11.3}$$

is used to evaluate performance (μ_1 and μ_0 are the means, and σ_1 and σ_0 are the standard deviations, of the resampled 1s and 0s in $\hat{\boldsymbol{d}}$).

The oversampling ratio itself is of primary importance. Figure 11.7 shows the effect of the oversampling ratio (defined as the data pixel image spacing divided by the detector element spacing) on SNR for three local alignment cases. For these simulations, pixel areal

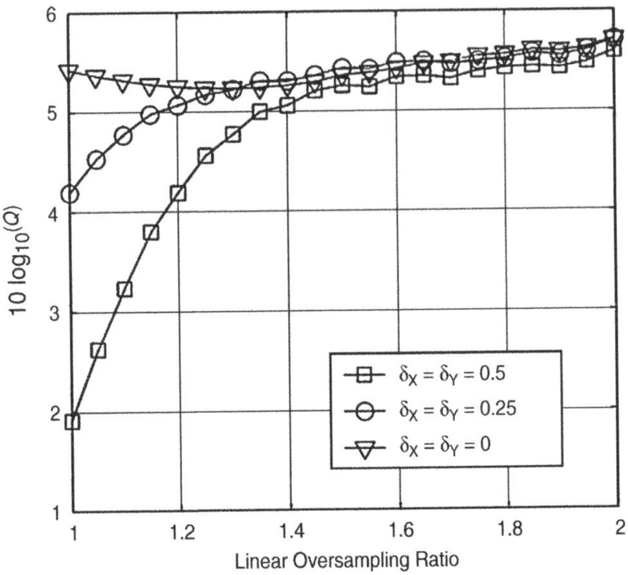

Figure 11.7 Simulated effects of oversampling on SNR for various local misalignments

fill factors for both the detector and the SLM were set to 90%, and coherent pseudorandom noise with power 1/20th of an 'on' pixel power was added. The steep improvement for the poorly aligned case ($\delta_X = \delta_Y = 0.5$, where the center of the data pixel image lands between the corners of four detector elements) as the oversampling ratio increases from unity demonstrates the potential of this approach for relatively small oversampling ratios. Similarly, the flatness of the curve near the Nyquist rate of the irradiance pattern (oversampling ratio ≈ 2) demonstrates the low marginal utility of increasing the ratio to that level.

The choice of an oversampling ratio hinges on a trade-off between performance, cost and the complexity of the system. These factors will not be quantified here, but it is important to note that the total number of detector elements required increases as the square of the oversampling ratio.

The efficacy of this resampling method was first demonstrated with an early test unit using readily available components that included an SLM with 12 μm pixels and a detector with 9 μm pixels, resulting in an oversampling ratio of 4/3. Several performance results for this test unit are shown later in this section.

The resampling approach described here differs substantially from the method of Burr [43]. In that method, an iterative decision feedback approach is used to remove both the coherent and linear crosstalk contributions of a known data image pixel to its unknown neighbor. The process is repeated on an unknown neighbor of the now-known pixel. The method achieves its best performance when multiple passes across the image are used (leftwards, rightwards, downwards and upwards). The present method requires only a single pass, and is not subject to the error propagation issues of the decision feedback method.

11.4.3 Alignment Measurement Method

The resampling method requires submicrometer position information for each and every data pixel. Alignment is derived from the reserved blocks (8×8 SLM pixel data-like pseudorandom patterns, see Section 11.2.3) that are distributed throughout the recorded data page on a grid at a spacing of 64 SLM pixels in both the x and y directions. The reserved blocks serve as fiducials for alignment measurement with a nominal areal overhead of approximately 1.6%.

Reserved block position demodulation is accomplished through a covariance pattern-matching operation at a matrix of integer candidate positions within the detected image. The covariance matrix is calculated between the linearized detector values and a target pattern representing a resampled version of the interior 6×6 pixels of the binary reserved block pixel pattern. The actual pixel patterns in the reserved blocks are selected by computer search such that the covariance between the central 6×6 pixels and any of the other eight edge-bordering 6×6 pixel sub-blocks is exactly zero:

$$\sum_{i'=2}^{n-1}\sum_{j'=2}^{n-1} p(i',j')p(i+i',j+j') = \begin{cases} (n-2)^2, & i,j = 0,0 \\ 0, & |i| \leq 1, |j| \leq 1 \end{cases} \tag{11.11}$$

where $p \in \{-1, +1\}$ is the DC-free binary reserved block pattern. This property ensures that the covariance values nearest the actual reserved block position will reflect the fractional reserved block alignment, rather than noise created by the matching of random patterns. Figure 11.8 shows an example of such a reserved block pattern along with its modified autocorrelation.

Another property imparted by this condition is enhancement of the accuracy in measuring the value of the peak. Since the covariance values neighboring the peak are free of random pattern matching noise, summing these values yields an accurate measurement of the true peak strength. This property is not necessary for the present purpose of alignment measurement but it will be very important for quadrature homodyne detection (see Chapter 12). Synthesized patterns with other desired autocorrelation properties may be considered for further performance improvements [48], although non pattern-dependent noise appears to dominate accuracy in current systems.

The resampled target pattern is computed by choosing the largest square grid of detector elements that will fit within the 6×6 reserved block center, and then simply integrating the binary pattern over the smaller detector pixel areas. No attempt is made to account for optical blur or pixel fill factors in this simplified resampling step. This facilitates variable

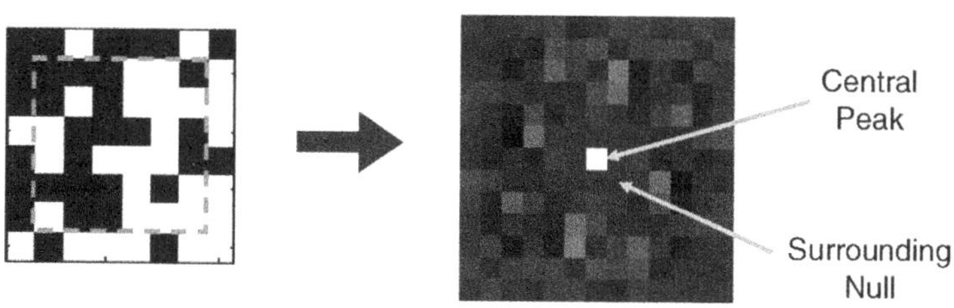

Figure 11.8 *Example of a reserved block with modified autocorrelation, showing an isolated peak*

oversampling ratio testing – the same reserved block patterns can be easily used at any ratio – but further performance gains may be possible by optimizing both the reserved block patterns and target pattern resampling for a specific ratio.

It is useful in practical systems to allow for image displacement tolerances of up to several hundred micrometers. Calculating a covariance matrix of this size would be computationally demanding. Furthermore, much of the search area would contain uncontrolled data patterns that could randomly produce strong correlation peaks. For these reasons, it is highly desirable to augment the technique with a preliminary coarse alignment measurement. Coarse x and y alignment is determined by locating horizontal and vertical bar codes written in the page image borders. The accuracy of the coarse alignment, together with tolerances for image rotation and distortion, permit the individual reserved blocks to be located to within ± 6 pixels, resulting in 13×13 element covariance matrices.

Once the covariance matrix is calculated, the maximum value is identified. The fractional part of the reserved block position is determined by linearly interpolating between the maximum and its immediate neighbors. Figure 11.9 is a plot of the simulated demodulated position demodulation error as a function of coherent noise power for the 4/3 oversampled system. A typical reserved block pattern is demodulated at a diagonal series of 26 positions spanning ± 1.5 detector pixels, and the RMS position measurement error is determined. The process is repeated for various coherent noise levels. The y-intercept of $\sim 2\%$ error indicates the noise-free performance limit determined by suboptimal reserved block pattern or target selection.

The demodulation process is performed on all of the reserved blocks within the data image, producing a grid of reserved block positions. Figure 11.10 shows the experimentally determined displacement from nominal of a typical grid, illustrating a combination of image shift, rotation, and demagnification. The grid is filtered to fill in any missing or anomalous

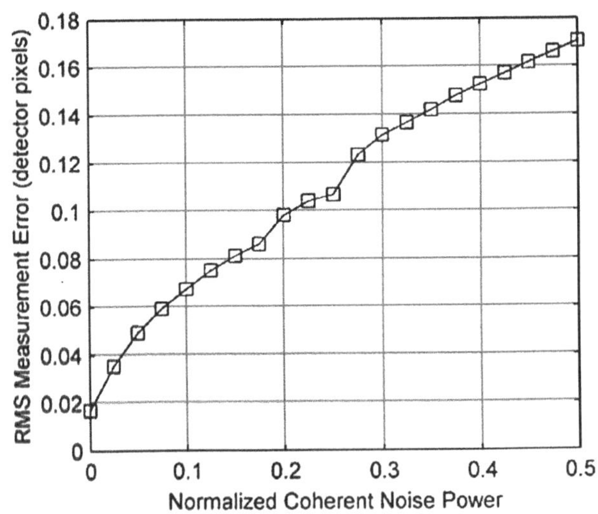

Figure 11.9 Simulated reserved block demodulation measurement error versus coherent noise power (normalized to 1 = 'on' pixel)

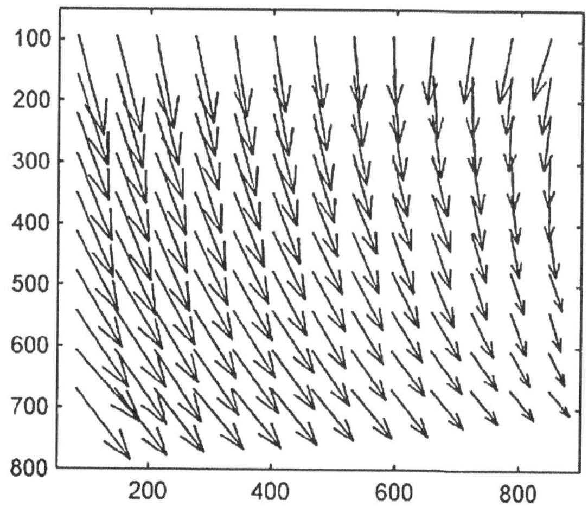

Figure 11.10 Experimental data page reserved block displacement grid (arrow lengths 64 times displacement)

values created by localized image defects. Finally, the filtered grid is linearly interpolated to produce a map of the position of every data pixel image upon the detector. The integer part of the data pixel image position is used to select the 4 × 4 detector window for resampling, and the fractional part (δ_X, δ_Y) determines which coefficient set is used. Since the resampling process involves only the local alignment of individual data pixels, slowly varying distortions such as those illustrated in Figure 11.10 are inconsequential.

Alignment mismeasurement sensitivity is a critical parameter for any processing algorithm based on alignment. Figure 11.11 shows the SNR of a single experimental holographic image, resampled repeatedly while introducing a deliberate error in the x component of the alignment vector. The plot shows a negligible performance decrease for a 5% measurement error, indicating that the 5% resolution of the resampling coefficient table is sufficient.

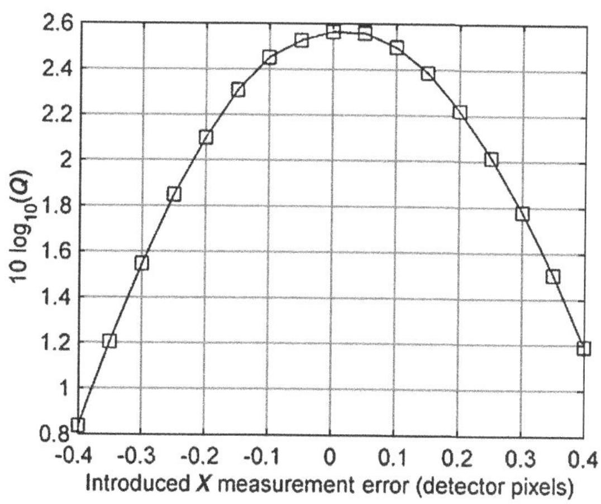

Figure 11.11 Resampling sensitivity to alignment mismeasurement

Hardware implementation of the full algorithm is complex. Because the alignment information required by the resampling step is localized, there is an opportunity for pipelining the alignment measurement and resampling stages. A primary issue is the need to access the large coefficient look-up tables, possibly by several parallel resampling units. Options for mitigating this problem include exploiting the coefficient symmetries to reduce the table size, and interpolating among coefficient values in a coarser table. The InPhase drive performs the correlation calculations to estimate local misalignment in parallel in hardware, using 19 hardware processors (one per reserved block column on the page).

11.4.4 Experimental Results

The effectiveness of the method was first validated experimentally by comparing holographic data pages recovered using a pixel-matched detector with recoveries of the same holograms with the detector replaced by a 4/3 oversampled detector. Stacks of 100 angularly multiplexed holograms were recorded using a 532 nm laser and a 720×720, 12 µm pitch data page, with a 92% pixel fill factor. The data were Fourier transformed and filtered with an aperture with a width 1.08 x Nyquist size (roughly 0.95 mm square). The filtered, Fourier transformed data were then imaged into 1.5 mm thick InPhase Technologies Tapestry HDS3000 media. The resultant holograms were read out using a 19 mW beam in a phase conjugate configuration [49]. Under nominal conditions the oversampled detector had performance equal to that of the pixel matched detector (see Figure 11.12).

A slightly smaller data page (688×688 pixels) was used to test the recovery method's tolerance to detector gross misalignment. The smaller data page was chosen to give a border of 24 pixels (~0.22 mm) between the centered data page and the edge of the detector. Using this data page, tolerance targets of ±0.2 mm displacement in the transverse directions, ±0.4 mm in focus, and ±0.25° of detector rotation were met without SNR penalties. The oversampled detector also achieved a higher SNR with weaker holograms than the pixel matched detector. This result was obtained by varying the recording exposure time to control the strength of the holograms. While performance was equivalent with high diffraction efficiency holograms, oversampled detection was more robust at lower diffraction efficiencies as shown in Figure 11.13. It should be noted that differing electrical

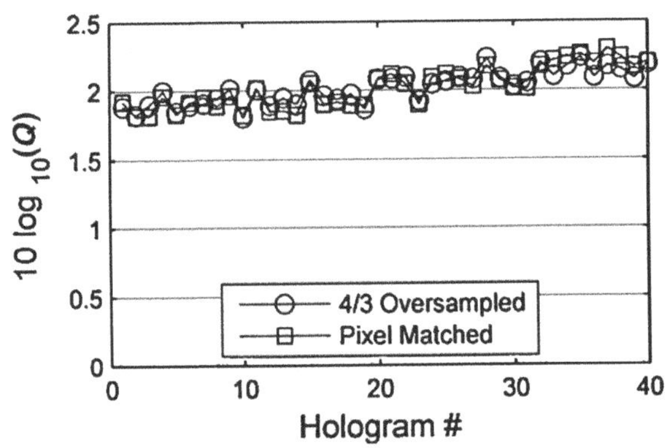

Figure 11.12 Experimental SNR for angularly multiplexed holograms recovered with both pixel-matched and oversampled detectors

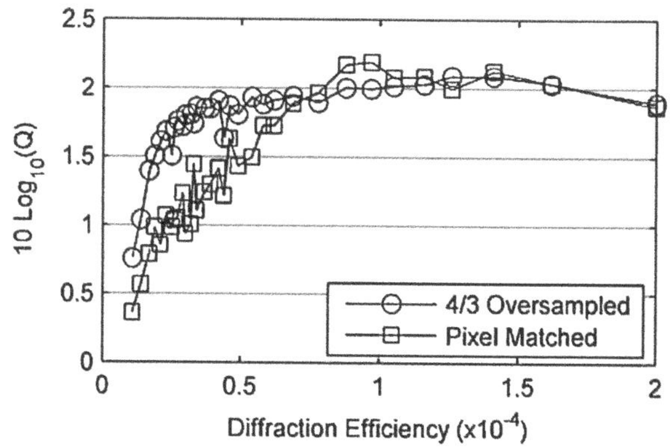

Figure 11.13 *Experimental SNR versus diffraction efficiency*

characteristics and pixel fill factors between the detectors could also have affected these results.

As a further demonstration of the effectiveness of the recovery method, stacks of holograms were stored in Tapestry HDS3000 media, using a combination of angular and polytopic multiplexing. The 720×720 pixel data pages were stored in a 3×5 grid of angularly multiplexed stacks with 80 holograms per stack (total of 1200 holograms). The distance between the stacks was 4% larger than the linear polytopic aperture size (roughly 1.0 mm stack pitch) resulting in a raw bit density of 26.8 Gbits in.$^{-2}$. (The density was limited principally by the small 4/3 oversampled detector size). The two detectors again performed equivalently.

11.5 Page Level Error Correction

The page SNR in a holographic system is lower than the raw SNR in either traditional optical or magnetic storage drives, which leads to a different selection of ECC at the page level. In this low SNR regime, iterative codes such as Turbo codes or LDPC codes perform significantly better than traditional ECCs such as Reed–Solomon codes used in CDs and DVDs [50–55]. These iterative codes use information-preserving 'soft decision' inputs, as opposed to the binary 'hard decision' inputs required by Reed–Solomon codes.

This section provides a description of the page-level ECCs used in the InPhase HDS system, including a discussion about the number of bits needed to represent the soft decision inputs to the decoder, as well as some practical results.

11.5.1 Log Likelihood Ratio

Hard decision ECC schemes require binary valued inputs - that is, either a '0' or '1' bits. In HDS, a simple threshold detection scheme could be used to make the assignments - all pixel intensities greater than some threshold would be mapped to '1' bits, and all intensities below that threshold would be mapped to '0' bits. The task of the ECC decoder is then to correct errors in those decisions or assignments without the benefit of knowing how bright or dim the original pixel actually was.

For LDPC and Turbo codes, each input is expressed in terms of a tentative ('soft') decision about how likely (probable) it is that a given bit is a '1'or a '0'. Even without knowing the details of the ECC implementation, it is reasonable to expect that the overall error correction capability of a code taking soft decision inputs will be greater than that of a hard decision input code, simply because more information is being supplied to the decoder. This expectation is borne out in practice.

Qualitatively, we expect that bright pixels are likely to correspond to a '1', dark pixels are likely to correspond to a '0', and that we are unsure about mid-brightness pixels. Further information about the likely value of a data pixel may be inferred from the brightness of neighboring pixels or detector elements, or by constraints of the modulation code. The probability, considering all factors, that a given bit is a '1' or a '0' may be expressed as a log likelihood ratio (LLR). In a binary communications channel, the LLR is given by [56]:

$$\mathrm{LLR}(r) = \log\left[\frac{p(0|r)}{p(1|r)}\right] \tag{11.12}$$

where $p(0|r)$ is the conditional probability that the bit is a 0, given a received signal, r, and $p(1|r)$ is the conditional probability that the bit is a 1, given the same received signal, r. The 'log' function is the natural logarithm.

In the InPhase holographic data channel where there are equal numbers of 1s and 0s recorded, the LLR is also given by:

$$\mathrm{LLR}(r) = \log\left[\frac{p(r|0)}{p(r|1)}\right] \tag{11.13}$$

Typical recovered hologram images have slowly varying intensity patterns across the page. For this reason, the equations for LLR are applied locally over small square regions of the resampled page, termed 'tiles', to avoid blurring the probability density functions for the 1s and 0s.

Using Gaussian distributions for the probability density functions for the 1s and 0s within a tile:

$$\mathrm{pdf} = \frac{1}{\sigma_j\sqrt{2\pi}}\exp\left[\frac{-(x-\mu_j)}{2\sigma_j^2}\right] \tag{11.14}$$

with $j = 0$ or 1, Equation 11.13 reduces to:

$$\mathrm{LLR}(I) = \log\left(\frac{\sigma_1}{\sigma_0}\right) - {}^1\!/_2\left(\frac{I-u_0}{\sigma_0}\right)^2 + {}^1\!/_2\left(\frac{I-u_1}{\sigma_1}\right)^2 \tag{11.15}$$

where I is the resampled pixel value, μ_1 and μ_0 are the estimated means of the resampled 1 and 0 pixel values within the tile, respectively, and σ_1 and σ_0 are the estimated standard deviations. Since the original pattern of 1s and 0s in a tile is not known, pixels with values greater than or equal to the mean value of the tile are assumed to be 1s and the rest to be 0s. The standard deviation of the 1s and 0s are similarly calculated over these respective populations. (The median, instead of the mean, pixel value might seem a more natural threshold for mapping pixels to 1s or 0s. In practice, the additional computation complexity does not result in a significant improvement in overall channel performance.)

As written, Equation (11.15) may give LLR values from $-\infty$ to $+\infty$, corresponding to 'definitely a 1' to 'definitely a 0'. In practice, the range of the LLRs must match the number of bits used to represent the LLR at the decoder input. For example, if a codec requires a 4-bit LLR, the LLR values in Equation (11.15) are limited to be in the range -8 to $+7$.

If the standard deviations in Equation (11.15) are equal ($\sigma_1 = \sigma_0$), then the equation reduces to a simple linear expression. In practice, the standard deviations are not the same, and the full equation must be implemented in either a field-programmable gate array (FPGA) or an application-specific integrated circuit (ASIC).

Equation (11.15) might theoretically be improved by incorporating measured distributions that are a closer approximation to the real distributions than Gaussians. Note that such a distribution would not reduce to a simple analytical form even under simplifying noise assumptions. In practice, more complex distributions yield only slightly better channel performance at the cost of considerably greater implementation complexity.

11.5.2 Page Level ECC

In the low SNR regime typical of holographic storage, LDPC codes can come much closer to the Shannon limit than Reed–Solomon or Golay codes. The error correction capability of LDPC codes increases as the codeword size increases. The general shape of the plot of SNR versus BER for an LDPC decoder [56] is shown in Figure 11.14. Above a threshold or knee, the BER decreases rapidly in the waterfall region. The location of the knee depends on the code design (length, rate, and parity check matrix). The sharpness of the knee, and the steepness of the waterfall region, is dependent on the code design, number of iterations used in decoding, implementation details, and the uniformity of the SNR of the page. If the page is very nonuniform, the waterfall is less steep because the various code words may have different noise distributions. Below a certain BER, LDPC codes exhibit an error floor where increases in SNR lead to minimal BER improvements. The location of the error floor is also determined by the details of the code design and implementation. In practical decoders, error floors from 10^{-6} to less than 10^{-12} are typical. For a well-designed code, determining

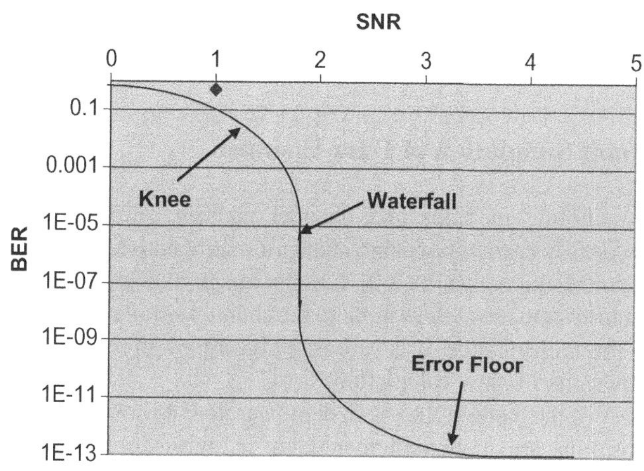

Figure 11.14 *Example of LDPC performance curve*

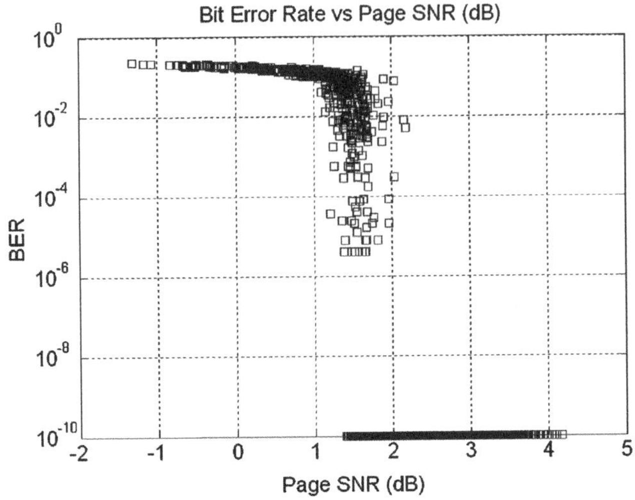

Figure 11.15 *Trellisware decode performance on holographic data pages*

the position of the error floor is time-consuming because the BER must be measured empirically.

The InPhase drive uses a rate $^1/_2$ LDPC code, designed by Trellisware [57] and implemented in an ASIC. The decoder knee is close to 1.5 dB and the error floor is below 10^{-15}. A 32-bit CRC and 16 352 user bits are combined with parity bits to result in a code word consisting of 32 768 bits (4096 bytes). Figure 11.15 shows a plot of BER versus SNR for the Trellisware decoder for data pages from a holographic drive recorded at a density of 350 GB in.$^{-2}$. The apparent error floor at 10^{-10} is not real, and simply reflects the need to plot 0 bit errors as some value on a log axis (there are less than 10^9 data bits on each data page).

The error correction capability of an LDPC decoder is dependent on the number of iterations used during decoding. More iterations (resulting in slower throughput) are required for lower SNR pages. This decoder has an early stopping feature in which the decoder stops iterating once it detects that there are no remaining bit errors on the page.

11.6 Fixed-Point Simulation of Data Channel

Developing oversampling methods and channel metrics, such as those described in Section 11.4, is typically carried out using standard mathematical software packages such as Matlab, Octave or Mathematica [58–60]. Camera images that have been captured in a real holographic storage system are used as input to the channel simulations. At the initial stages of development, the convenience of developing algorithms in a simulation environment outweighs the downside of slow calculations.

Once the efficacy of the methods has been demonstrated in simulations using real camera images, the algorithms are then implemented in a high-level programming language, typically C++, and integrated into the software that controls a prototype storage drive. A key reason for developing a C++ version is to achieve sufficient processing speed so that an optical engineer can modify some physical aspect of the drive (for example, move a component) and see the effect of that change on hologram SNR and BER within a subjectively short time. At InPhase, the implementation of the alignment and resampling algorithm in Matlab has an execution time of several tens of seconds, compared with less than 1 s in the C++ software, running on a standard desktop computer. The calculations in the simulation environment and in the C++ implementation just mentioned are performed using standard floating point libraries.

For a working commercial drive, processing one page per second is still far too slow. To achieve a realistic transfer rate, the algorithms must be implemented in an FPGA or an ASIC. Typically, the algorithms are coded using a hardware description language (HDL), such as Verilog or VHDL.

In FPGAs and ASICs, floating point arithmetic blocks are generally not available, or are prohibitively expensive. Consequently, the floating point version of the algorithms must be recast using fixed-point data representations, which are readily implemented with integer arithmetic blocks that are characteristic of FPGAs and ASICs. In doing so, attention must be paid to the range of values that must be represented at each stage of the channel, and the precision with which those values must be stored.

In fixed-point arithmetic [61,62], floating point values (with whole number and fractional parts) are stored in ordinary integer registers with an implied binary point. For example, an 8-bit register can store plain unsigned integer values from 0 (bits 00000000) to 255 (bits 11111111). Implicitly, there is a binary point (equivalent to a decimal point in ordinary decimal arithmetic) to the right of the least significant bit. If instead, the 8-bit byte is treated as a fixed-point number with five integer bits and three fractional bits, the implicit binary point is between the third and fourth bit from the right. The range of this fixed-point value is now from 0 (00000000) to decimal 31.875 (11111111) where the 1-bits from left to right represent the values $2^4+2^3+2^2+2^1+2^0+2^{-1}+2^{-2}+2^{-3}$. We denote this fixed-point representation of unsigned values as $U(i,f)$, where i is the number of integer bits to the left of the binary point, and f is the number of fraction bits to the right of the binary point. The present example is in fixed-point format $U(5, 3)$.

The 8-bit register, with the same numbers of integer and fractional bits, treated as a signed 2s complement fixed-point number, $S(5,3)$, can store decimal values in the range –16.0 (bits 10000000, or -2^4) to 15.875 (bits 01111111, or $-2^4 * 0+2^3+2^2+2^1+2^0+2^{-1}+2^{-2}+2^{-3}$).

The overall range of values that can be stored in a fixed-point word is principally determined by the number of bits assigned to the integer part, i (Table 11.1). The precision with which a value is stored is determined by the number of bits assigned to the fractional part, f.

There is no requirement that the fixed-point value fits in only 8 bits, and standard FPGAs typically allow integers of up to 32 bits. However, the more bits associated with each value, the

Table 11.1 *Range of values for 2s complement signed and unsigned fixed-point values*

	Minimum	Maximum
Unsigned: $U(i,f)$	0	2^i-2^{-f}
Signed: $S(i,f)$	-2^{i-1}	$2^{i-1}-2^{-f}$

greater the amount of silicon real estate (gates) required to perform arithmetic on those values, so there is a trade-off between cost and total number of bits at each stage of the channel.

The first task in developing a fixed-point implementation of a data channel is determining the minimum number of integer and fraction bits required to represent each physical quantity at each stage of the data channel. This task is best carried out by simulation, comparing the data channel's SNR and BER in a fixed-point simulation with those metrics obtained with a full floating-point simulation. Using too few integer bits may lead to overflow or underflow in the calculations, and consequent errors. The fewer fraction bits there are, the greater the quantization noise that is introduced by the approximate representation of the actual values in the fixed-point representation. These considerations are largely absent in floating point calculations where an IEEE standard 64-bit 'double' floating value can represent values up to $\pm 10^{308}$, and as small as 10^{-308}.

For simulations in C++, fixed-point numbers are represented by a fixed-point class we developed at InPhase to emulate fixed-point arithmetic with different integer and fraction bit counts. Using C++ operator overloading in this class, the same source code implementing the read channel algorithms can be compiled to either operate on floating-point or fixed-point data types with only a single C++ *typedef* change. There is no need to have two separate implementations of the channel algorithms, which has the advantage of not having to keep two code versions synchronized as changes to the channel model are made. Since the fixed-point simulation is bit exact, it also provides a tool for verifying the FPGA and ASIC implementations.

The simulations using fixed-point values are slower than using the native floating-point data types. For this reason a representative set of real hologram images is used as input to the simulations. The images are chosen to cover a range of hologram quality so that the overall SNR and BER vary systematically over the set. The remainder of this section shows two illustrative examples of the results of fixed-point simulations.

One of the first operations in the data channel is linearization, where the square root of each intensity value is computed, as a precursor to the resampling calculations. The input to this calculation is the intensity value for each pixel, which is simply an 8-bit count. This count is clearly an unsigned value, and so is its square root. The number of bits required for the integer part of the square root result is simply half the number of bits of the input, in this case, four. Simulation answers the question of how many fraction bits should be allocated for the linearization result.

Figure 11.16 shows how the finite precision (quantization) of different fixed-point representations of the linearization output affects the final resampled hologram SNR. The x-axis is the hologram index in a set of 55 camera images that give a range of resampled hologram SNRs between 0 dB and 6 dB when a full floating point implementation of the resampling method is used (the uppermost line in Figure 11.16). The lower lines in the graph show the degradation of SNR for the same 55 camera images as the number of fraction bits in the fixed-point linearization result is reduced from 4 down to 0. The results for $U(4, 4)$ are indistinguishable from the full floating point calculation, so a single byte can be used to represent the linearization output in the fixed-point representation. If no fraction bits were used (the lowest line in Figure 11.16), the overall cost to the data channel performance is at least 1.5 dB for all of the camera images.

As a second example of fixed-point simulation of the data channel, Figure 11.17 shows the page BER dependence on the number of fraction bits used to represent the resampling

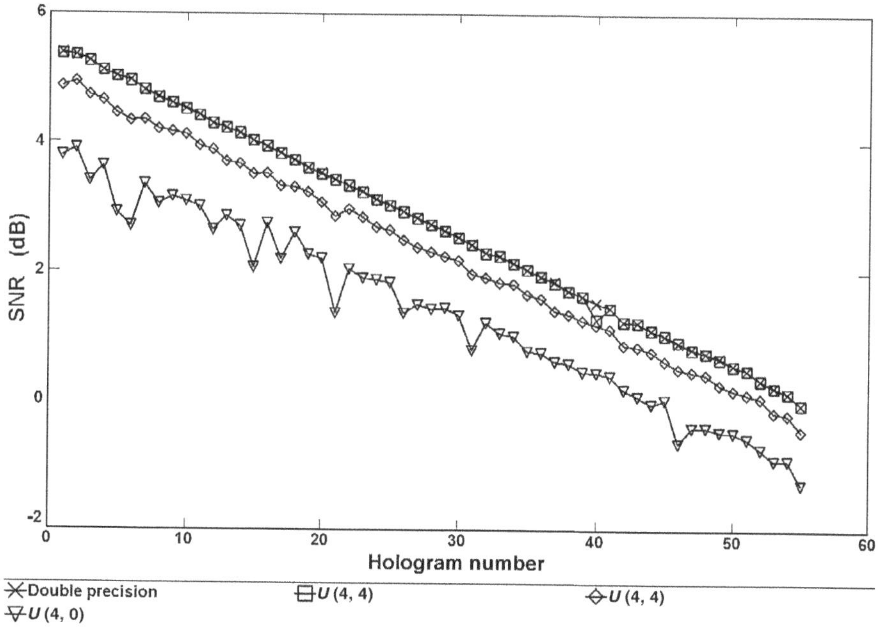

Figure 11.16 Plot of SNR for different quantization in the linearization stage

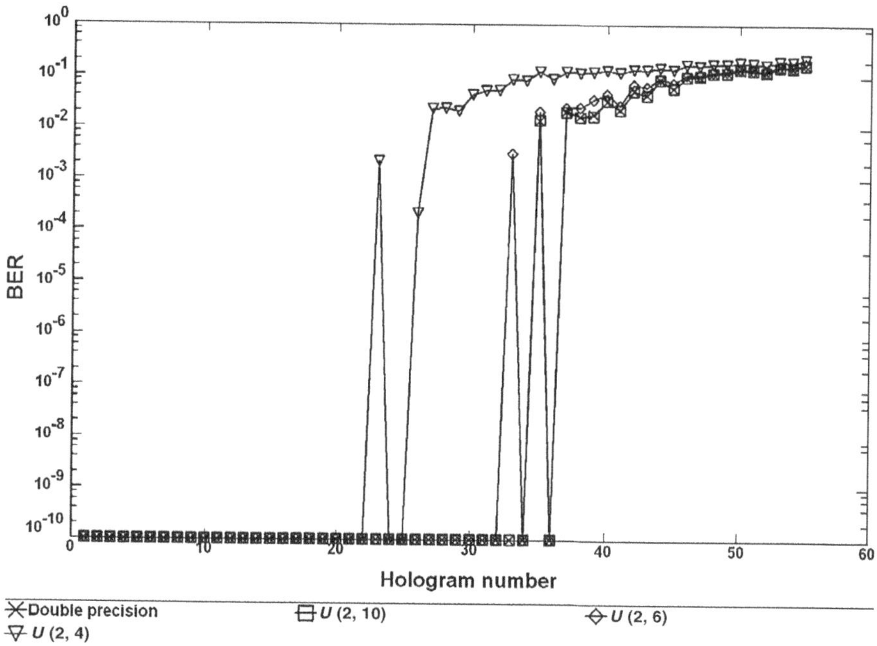

Figure 11.17 Plot of BER for various fixed-point representations of the resampling coefficients compared with the double precision (shown as black squares) implementation

coefficients. (Pages with zero bit errors are plotted as values of 10^{-10} so that a log axis can be used.) Here the BER of individual pages increases from the floating point simulation values when there are less than 8 fraction bits in the fixed-point representation. In the extreme, with only 4 fraction bits, the quantization noise results in a nonzero BER even when the hologram SNR is over 3.5 dB. From this graph, it is clear that the resampling coefficient values cannot be represented by less than 12 bits without introducing additional errors.

11.7 Logical Format

11.7.1 Introduction

All storage devices require a method for protecting and organizing data on a piece of media (the 'logical format') so that the correct data are always retrieved reliably and error-free by a host system. This section describes the concepts, structure, and components for a logical format developed specifically for HDS.

A new logical format was required for HDS because the basic unit of storage is a data image which is a 2D array of light and dark pixels with light pixels usually representing 1s and dark pixels usually representing 0s. An image can have thousands or millions of information symbols. This is in stark contrast to current conventional storage technologies that store and read back data one symbol at a time, either magnetically or optically. These conventional devices may read back data in parallel on multiple tracks, but at most, this represents a few tens of symbols at a time. The logical format must also efficiently accommodate other physical write and read rules particular to holographic storage, such as the need to pre-cure and post-cure recording sites.

The format was designed to be compatible with conventional storage devices to simplify integration of the holographic drive into existing storage systems with existing software. The SCSI standard was used at the interface with multiple layers of formatting employed to support not only multiple levels of ECC but also to provide the flexibility to change the SCSI personality of the drive giving it the ability to emulate multiple storage device types such as optical WORM (write-once read-many), LTO (linear tape-open) and other types of tape drives, and DVD-Recordable, among others.

Figure 11.18 shows the structure of data in the holographic drive logical format. For writing to the media, the drive receives user data from the host computer in logical blocks as dictated by the SCSI specification. The default logical block size for the system is 4096 bytes but a large range of sizes is supported. Before any other steps are performed, a CRC is attached to each logical block to provide error detection. On reading from the drive, this CRC can detect silent data corruption in the drive electronics, and detect the overall error rate of the channel, down to bit error rates as low as 10^{-22}.

After the CRC is added, the data are arranged into chapters (groups of pages), and ECC pages consisting of Reed–Solomon error correction parity are added. This is an 8 bits per symbol code, with the length and redundancy being firmware programmable. In addition to the ECC, other chapter information is added that describes the user data such as compression map (if data compression is turned on), copyright protection of data, data type, and logical-to-physical address mapping information.

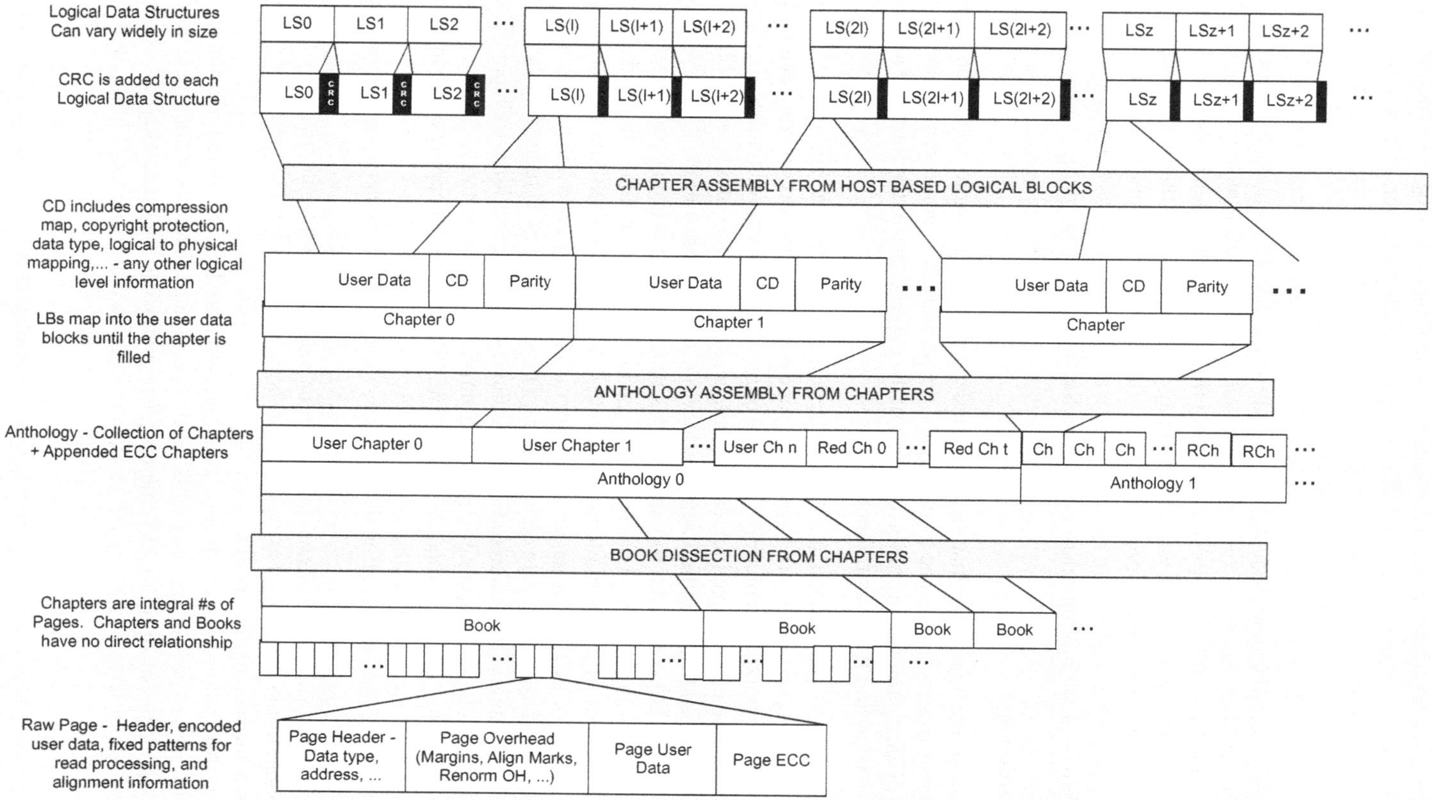

Figure 11.18 *Data structure at different stages in the system*

The chapters are then arranged into anthologies. The anthology is a logical structure that contains another level of Reed–Solomon ECC that spans multiple chapters. The code rate of this Reed–Solomon code is also firmware programmable, and allows the data to be retrieved, even if entire chapters on the media are unrecoverable.

Once the anthology level has been assembled, the final step is to form the data into pages and books, which are the units required by the physical recording channel. A book (or 'stack' in some references) is a group of pages that are multiplexed at the same spatial location on the medium. A book is logically distinct from a chapter, which is also a group of pages. When a book is ready to be written, each page in sequence is streamed through the write channel hardware which composes and then writes each 2D page image. At this stage, the LDPC code is added and the data stream is interleaved, modulated, and augmented with borders and physical formatting information (see Sections 11.2 and 11.5).

11.7.2 Terminology

In order to make the logical format easier to understand, and to avoid confusion with overloaded terms from other storage technologies, the analogy of a public library has been adopted to provide a naming convention for most of the components of the format. The key levels of the format and their basic functions are:

1. Page: A data page is the smallest unit that can be written to media (one hologram), and is essentially an array of 'on' or 'off' pixels. The page has a well-defined structure that may be changed to support different array sizes, encoding, and recovery techniques. The key components of the page format include a layout, header, data areas and reserved blocks, data modulation and ECC coding and interleaving, and randomization.
2. Chapter: This next level of the format provides protection for missing pages and also provides the mapping between host-based logical blocks and user data within a chapter. A chapter consists of a variable length of contiguous pages of user data followed by a variable length of ECC parity pages. A chapter may be any size up to 255 pages and can cross book boundaries (books are defined next). The amount of ECC redundancy can be adjusted to provide a variable amount of protection for lost pages within the chapter. The logical block to chapter mapping is provided through a structure found at the end of a chapter called the 'chapter directory' or CD. The CD provides a mapping of the logical blocks and their sizes within and across chapter boundaries. The CD also provides copyright and security information that prevents unauthorized access to the data in the chapter.
3. Book: A book is a stack of pages recorded in the same location in the media by angle multiplexing. Books are located using track and book addresses for disk media, or x- and y-addresses for rectangular media.
4. Anthology: An anthology is a collection of chapters that are ECC protected. This level of ECC is provided to protect against media defects or damage that occurs after recording. An anthology consists of up to 255 chapters. The number of user data chapters versus redundant chapters is variable, based on data protection and overhead requirements.
5. Bookcase: A bookcase is analogous to a write session in other storage technologies. It is the smallest amount of data that can be written within a specific time period called a 'write session'. The medium within the area of a bookcase is fully cured (i.e. optically inert) after recording. A bookcase also contains a card catalog structure, which is a map

of the usage of every book in the bookcase. The card catalog includes information about how full each book is, if the book contains data or not, what kind of data it contains, and the starting chapter and logical block numbers for books that contain data. This, combined with the chapter directory, provides the drive with a method for locating specific chapters and logical blocks when they are requested through a read command. The bookcase may also include a drive emulation layer (DEL) structure to allow logical block mapping and other conversions necessary to allow the drive to emulate other types of storage devices with different characteristics.

6. Partition: A partition consists of one or more bookcases. A partition contains a single informational data type [for example, user data, library map (LM), or manufacturing data] and is written at a single specific areal density. Partitions allow the medium to look and act like multiple volumes of media. They also allow different data types to be written with different densities and redundancy so that more important data (such as the LM) can be written in a more robust manner than other types of data. Each partition is defined by a partition descriptor (PD) structure that is located within the LM. This structure provides information on the start and end book addresses of the partition, the data type and recording mode used, whether the partition is empty or full, whether the data can be appended to, what chapters and logical block addresses have been written to it, and how to find the card catalog structures within the partition. In addition, partitions can be linked together to create longer, virtual partitions containing the same types of information.
7. Library map: The LM is the main data structure describing the medium. This is the first structure that is read by the drive when a disk is loaded, to ensure that no unwritten areas of the medium are read. The LM may be located at a specific address on the medium or located in an auxiliary memory (such as flash or radio frequency identification, RFID) that is included in the media cartridge assembly. If the LM is too large for the auxiliary memory, only the first part of it may reside in memory, which then points to a full version of the LM on the medium. In most cases, there are at least two current copies of the LM written to medium as well as previous copies to allow robust recovery of the LM. The LM includes information about the format generation, media type, geometry, the system type it was written with, and the medium status (empty, formatted, appendable, write protected, full). In addition, each PD is appended to the LM to provide a full mapping of the media. Optionally, card catalogs (CCs) may also be included with the LM. Both the LM and the PDs may have copy protection and security keys to control and limit access to each partition or the entire media.

Other constructs and concepts beyond the library hierarchy presented within the logical format are:

1. Logical block: A logical block is the standard data unit transferred between a host and most storage devices. Logical blocks are grouped together within the user portion of chapters and written as a group. The format supports multiple logical block sizes as well as variable logical blocks, and can even support different logical block sizes on the same medium. Logical blocks may cross chapter boundaries and may be smaller or larger than pages or chapters. The CD structure within the chapters keeps track of the location and size of the logical blocks. As for nearly all other storage device types, logical blocks are addressed sequentially from 0 to N within a user data partition or logical grouping of user data partitions. If the medium is split into multiple partitions

representing multiple logical volumes, the logical block addressing may be restarted for each logical volume.

2. Bad media mapping: Using any of several algorithms capable of detecting medium defects, the medium may be scanned during manufacturing time, at recording time, or both, to locate any defective areas. As the medium is written, the drive will record which book locations are bad due to defects, or drive or servo errors, in the card catalog, so that these areas will not be read out during a subsequent read. The data from these bad book locations are rewritten in the next available location. Also, through the use of anthology recovery or other algorithms, written data may be detected as bad or deteriorating. These data may be relocated somewhere else on the medium and the relocation addresses recorded in the bad map partition so that the data recovery reliability may be improved. The format is defined with extra capacity to allow for a certain number of books to be marked bad due to media defects or servo/laser unsafe errors.
3. Format generations: A format generation fully describes the system, media type, and format levels used when reading and writing a specific article of media. The format generation can be thought of as a reference key defining how a specific drive type can read and write a specific media type. A format generation defines the media management method used for writing including the usage of partitions, bookcases, anthologies, books, chapters, and pages. It also defines the revisions of the LM, PD, CC, CD, and page format. In addition, the format generation defines the system and media formulation and geometry and which versions of them are compatible with the specified format generation. Thus, each time a feature is added at any level, or new systems or media types and formulations are introduced, one or more new format generations are defined to describe them. This mechanism allows the logical format to easily evolve and take advantage of new technologies, developments, algorithms, and market requirements while still maintaining a linkage and compatibility mapping across widely varying holographic storage product families and markets.

On reading from the media, the LM and card catalog allow for the logical block addresses (LBAs) to be translated into physical addresses on the media. Under normal usage, the page level ECC is the only error correction used for a large majority of books. If there are no errors in the chapter being read back, the chapter parity pages do not have to be read back, which improves transfer rate. If errors are detected in the chapter data, the chapter level ECC is used to recover the bad pages. If there are still errors in the recovered data stream after chapter ECC has tried to recover the pages, then physical retries are performed to try and improve the recovered SNR of these pages. Only if reading again fails to recover the pages will anthology level ECC be used. As for chapter level recovery, if anthology ECC is not needed, the associated pages need not be read out. The use of physical retries and anthology will negatively affect read transfer rate.

Acknowledgements

Over the years, many people have contributed to the data channel described here. We would like to thank Tom Richardson, Martin Pane, Charles Stanhope, Will Loechel, Chris Pruett, J.J. Kane, Li Ni, Brian King and Dan Soo.

References

1. D. W. Woods and K. W. Malang, Micro-positioning movement of holographic data storage system components, US Patent 7,116,626, October 2006.
2. T. J. Richardson, Method for modulating data for storage in page-wise memory, US Patent 6,081,912, June 2000.
3. W.-C. Chou and M. Neifeld, Interleaving and error correction in volume holographic memory systems, *Appl. Opt.*, Vol. 37, No. 29, pp. 6951–6968 (1999).
4. M. Blaum, J. Bruck and A. Vardy, Interleaving schemes for multidimensional cluster errors, *IEEE Trans. Inform. Theory*, Vol. 44, No. 2, pp. 730–743 (1998).
5. L. Ramamoorthy, Advanced Coding and Signal Processing for holographic data storage systems, PhD Thesis, Carnegie Mellon University, Pittsburgh, Pennsylvania (2008).
6. L. D. Ramamoorthy and B. V. K. Vijaya Kumar, Sparse modulation codes for channel with media saturation, *Opt. Data Storage Conf.*, Waikoloa Hawaii (2008), paper ThC4.
7. B. M. King and M. A. Neifeld, Sparse modulation coding for increased capacity in volume holographic storage, *Appl. Opt.*, Vol. 39, No. 35, 2000, pp. 6681–6688 (2000).
8. A. Hoffnagle, *et al.*, Modulation coding for pixel-matched holographic data storage, *Opt. Lett.*, Vol. 22, pp. 639–641 (1997).
9. M.-P. Bernal, G. W. Burr, H. Coufal and M. Quintanilla, Balancing interpixel crosstalk and detector noise to optimize areal density in holographic storage systems, *Appl. Opt.*, Vol. 37, pp. 5377–5385 (1998).
10. J. Ashley and B. Marcus, Constant-weight/lowpass modulation codes for holographic recording, Research Report RJ-10089 (91905), IBM Research Laboratory, San Jose, California (1997).
11. J. Ashley and B. Marcus, Two-dimensional lowpass filtering codes for holographic storage, *IEEE Trans. Commun.*, Vol. 46, pp. 724–727 (1998).
12. G. W. Burr, *et al.*, Noise reduction of page-oriented data storage by inverse filtering during recording, *Opt. Lett.*, Vol. 23, pp. 289–291 (1998).
13. G. W. Burr, *et al.*, Gray-scale data pages for digital holographic data storage, *Opt. Lett.*, Vol. 23, pp. 1218–1220 (1998).
14. J. F. Heanue, *et al.*, Channel codes for digital holographic data storage, *J. Opt. Soc. Am. A*, Vol. 12, pp. 2432–2439 (1995).
15. A. Vardy. *et al.*, Conservative arrays: multidimensional modulation codes for holographic recording, *IEEE Trans. Inform. Theory*, Vol. 42, No. 1, pp. 227–230 (1996).
16. J. J. Ashley, *et al.*, Two-dimensional low-pass filtering codes, *IEEE Trans. Commun.*, Vol. 46, No. 6, pp. 724–727 (1998).
17. J. Liu, *et al.*, Effect of channel coding in digital holographic data storage, *Jpn. J. Appl. Phys.*, Vol. 38, pp. 4105–4109 (1999).
18. T. Kume, S. Yagi, T. Imai and M. Yamamoto, Digital holographic memory using two dimensional modulation code, *Jpn. J. Appl. Phys.*, Vol. 40, pp. 1732–1736 (2001).
19. E. Hwang, K. Kim, J. Kim, J. Park and H. Jung, A new efficient error correctible modulation code for holographic data storage, *Jpn. J. Appl. Phys.*, Vol. 41, No. 3B, pp. 1763–1766 (2002).
20. V. Vadde and B. V. K. V. Kumar, Parity coding for page-oriented optical memories with intra-page intensity variations, *Opt. Lett.*, Vol. 24, pp. 546–548 (1999).
21. T. C. Lee, Differential readout holographic memory, US Patent 3,720,453, March 13, 1973.
22. B. M. King and M. A. Neifeld, Low-complexity maximum-likelihood decoding for shortened enumerative permutation codes for holographic storage, *IEEE J. Selected Areas Commun.*, Vol. 19, No. 4, pp. 783–790 (2001).
23. V. Vadde and B. V. K. V. Kumar, Channel modeling and estimation for intrapage equalization in pixel-matched volume holographic data storage, *Appl. Opt.*, Vol. 38, pp. 4374–4386 (1999).
24. P. Yoon, E. Hwang, B. Kang, J. Park and G. Park, Image compensation for sub-pixel misalignment in holographic data storage, *ISOM Tech. Digest*, pp. 114–115 (2004).
25. J. F. Heanue, K. Gurkan and L. Hesselink, Signal detection for page-access optical memories with intersymbol interference, *Appl. Opt.*, Vol. 35, pp. 2431–2438 (1996).

26. S. Nabavi and B. V. K. V. Kumar, Comparative evaluation of equalization methods for holographic data storage channels, *ISOM/ODS Tech. Digest* (2005), paper TuB8.
27. L. Ramamoorthy and B. V. K. V. Kumar, Two-dimensional equalization and error correction using low density parity check codes for holographic data storage, *Jpn. J. Appl. Phys.*, Vol. 45, pp. 1305–1310 (2006).
28. M. Keskinoz and B. V. K. V. Kumar, Efficient modeling and iterative magnitude-squared decision feedback equalization (DFE) for volume holographic storage channels, *Proc. ICC 2001*, Vol. 9, pp. 2696–2700 (2001).
29. V. Vadde and B. V. K. V. Kumar, Partial response equalization for grayscale volume holographic data storage, *ISOM/ODS Tech. Digest*, pp. 97–99 (1999).
30. S. Nabavi and B. V. K. V. Kumar, Application of linear and nonlinear equalization methods for holographic data storage, *Jpn. J. Appl. Phys.*, Vol. 45, pp. 1079–1083 (2006).
31. M. Keskinoz and B. V. K. V. Kumar, Discrete magnitude-square channel modeling, equalization and detection for volume holographic storage channel, *Appl. Opt.*, Vol. 43, pp. 1368–1378 (2004).
32. M. Keskinoz and B. V. K. V. Kumar, Application of linear minimum mean squared error equalization for volume holographic data storage, *Appl. Opt.*, Vol. 38, pp. 4387–4393 (1999).
33. M. Keskinoz and B. V. K. Vijaya Kumar, Application of linear minimum mean-squared error equalization for volume holographic data storage, *Appl. Opt.*, Vol. 38, pp. 4387–4393 (1999).
34. J. F. Heanue, K. Gurkan and L. Hesselink, Signal detection for page-access optical memories with intersymbol interference, *Appl. Opt.*, Vol. 35, No. 14, pp. 2431–2438 (1996).
35. K. M. Chugg, X. Chen and M. A. Neifeld, Two-dimensional equalization in coherent and incoherent page-oriented optical memory, *J. Opt. Soc. Am. A*, Vol. 16, No. 3, pp. 549–562 (1999).
36. X. Chen, K. M. Chugg and M. A. Neifeld, Near-optimal parallel distributed data detection for page-oriented optical memories, *IEEE J. Selected Topics Quantum Electron.*, Vol. 4, No. 5, pp. 866–879 (1998).
37. J. Ashley, M. Blaum and B. Marcus Coding and signal processing for holographic data recording, Research Report RJ-10013 (89104), IBM Research Laboratory, San Jose, California (1998).
38. Q. Gao and R. Kostuk, Cross-talk noise and storage capacity of holographic memories with a $LiNbO_3$ crystal in the open-circuit condition, *Appl. Opt.*, Vol. 37, pp. 929–936 (1998).
39. A. He and G. Mathew, Nonlinear equalization for holographic data storage systems, *Appl. Opt.*, Vol. 45, pp. 2731–2741 (2006).
40. L. Menetrier and G. W. Burr, Density implications of shift compensation postprocessing in holographic storage systems, *Appl. Opt.*, Vol. 42, pp. 845–860 (2003).
41. G. W. Burr, and T. Weiss, Compensation for pixel misregistration in volume holographic data storage, *Opt. Lett.*, Vol. 26, pp. 542–544 (2001).
42. S. Redfield, Tamarack optical head holographic storage, in *Holographic Data Storage*, H. J. Coufal, D. Psaltis and G. Sincerbox,eds. Springer-Verlag, New York, pp. 343–357 (2000).
43. G. W. Burr, Holographic data storage with arbitrarily misaligned data pages, *Opt. Lett.*, Vol. 27, pp. 542–544 (2002).
44. M. Ayres, A. Hoskins, and K. Curtis, Image oversampling for page-oriented optical data storage, *Appl. Opt.*, Vol. 45, No. 11, pp. 2459–2464 (2006).
45. H. Horimai, X. Tan, and J. Li, Collinear holography, *Appl. Opt.*, Vol. 44, No. 13, pp. 2575–2579 (2005).
46. V. Vadde, and B. V. K. V. Kumar, Channel modeling and estimation for intrapage equalization in pixel-matched volume holographic data storage, *Appl. Opt.*, Vol. 38, pp. 4374–4386 (1999).
47. S. M. Kay, *Fundamentals of Statistical Signal Processing: Estimation Theory*. Prentice Hall, Upper Saddle River, New Jersey, Ch. 8, 1993.
48. D. Calabro, and J. K. Wolf, On the synthesis of two-dimensional arrays with desirable correlation properties, *Info. Ctrl*, Vol. 11, pp. 537–560 (1968).
49. K. Anderson and K. Curtis, Polytopic multiplexing, *Opt. Lett.*, Vol. 29, pp. 1402–1404 (2004).
50. Z. Li and B. V. K. V. Kumar, A new class of quasi-cyclic LDPC codes based progressive edge growth graph, *IEEE Asilomar Conf. Signals, Systems and Computers*, Pacific Grove, California (2004), pp. 1990–1994.

51. H. Hayashi and K. Kimura, Low-density parity-check coding for holographic data storage, *Jpn. J. Appl. Phys.*, Vol. 44, No. 5B, pp. 3495–3498 (2005).
52. H. Pishro-Nik, N. Rahnavard, J. Ha, F. Fekri and A. Adibi, Low-density parity-check codes for volume holographic memory systems, *Appl. Opt.*, Vol. 42, No. 5, pp. 861–870 (2003).
53. M. A. Neifeld and J. D. Hayes, Error-correction schemes for volume optical memories, *App. Opt.*, Vol. 34, pp. 8183–8191 (1995).
54. K. Curtis, Holographic data storage, *CMRR Review*, San Diego, California (2005).
55. E. Hwang, *et al.*, An efficient 3D error correction schemes for holographic data storage, *ISOM Tech. Digest*, (2004), paper TuD6.
56. P. Sweeney, *Error Control Coding – From Theory to Practice*. John Wiley & Sons Ltd, Chichester, 2002.
57. Trellisware Technologies, http://www.trellisware.com.
58. Matlab, The MathWorks, http://www.mathworks.com.
59. Octave, Free Software Foundation, http://www.gnu.org/software/octave/index.html.
60. Mathematica, Wolfram Research, http://www.wolfram.com.
61. R. Yates, Fixed point arithmetic: an introduction, Digital Signal Labs, http://www.digitalsignallabs.com, 2006.
62. W. Sung and K.-I. Kum, Simulation-based word-length opimization method for fixed-point digital signal processing systems, *IEEE Trans. Signal Process.*, Vol. 43, No. 12, pp. 3087–3090 (1995).

12

Future Data Channel Research

Mark Ayres and Kevin Curtis

12.1 Introduction

The previous chapters have summarized work done on data channels for holographic storage, particularly that which directly relates to the commercialization of holographic data storage (HDS). In this chapter, promising work on homodyne detection and other research directions will be presented. These techniques are clearly not ready for product development but represent areas that can improve either the physical recording process or the performance of the channel. Homodyne detection in particular could significantly change the physical data channel for the better and is the main focus of this chapter. Homodyne detection enables the use of phase quadrature multiplexing, which doubles the address space of an HDS device. Homodyne detection additionally permits the use of phase modulation, which reduces noise and improves signal uniformity at the fundamental holographic level. Much of the material was originally published in Ayres [1] and Ayres and Curtis [2].

12.2 Homodyne Detection

12.2.1 Introduction

Coherent detection techniques have been used in communications channels for over a century [3]. The super-heterodyne receiver, the ubiquitous circuit that brings radio-frequency signals down to manageable electrical bandwidths, was invented in 1921 [4].

Holographic Data Storage: From Theory to Practical Systems Kevin Curtis, Lisa Dhar, Adrian Hill, William Wilson and Mark Ayres

Coherent detection enables the reception of signals modulated in phase and frequency, and generally improves the reception of signals modulated in amplitude when compared with direct detection. Digital data modulation schemes employing phase and frequency changes are referred to as *phase shift keying* (PSK) and *frequency shift keying* (FSK), respectively, whereas schemes employing amplitude changes are referred to as *amplitude shift keying* (ASK). Additionally, coherent detection can amplify the signal directly in the optical domain, thereby not only decreasing the required detector integration time but also boosting the signal level above the thermal noise of the detector so as to approach the shot noise limit.

Coherent detection is performed by combining the signal with a reference signal, called the *local oscillator*, and then detecting the intensity of the mixed signal. The coherent superposition includes a cross term that varies in intensity according to the phase difference of the two constituents. Generally, the local oscillator may differ in frequency from the signal carrier, and the process is known as *heterodyne detection*. The intensity of the cross term will vary in time at a rate determined by this difference (beat frequency), and the frequency, as well as the phase, may thereby be determined. If the local oscillator has the same frequency as the signal carrier, then the process is also known by the more specific term, *homodyne detection*. Figure 12.1 shows a block diagram of a generalized single channel optical coherent communication channel, such as a fiber optic link. The photodetector produces a photocurrent proportional to the intensity of the coherent superposition of the two optical fields. Neglecting inter-symbol interference (ISI):

$$i_{PD}(t) \sim |E_L|^2 + |D[t/T]E_C|^2 + 2D[t/T]|E_L||E_C|\cos[(\omega_C - \omega_L)t + (\varphi_C - \varphi_L)] \qquad (12.1)$$

where the signal is composed of data sampled at period T, modulated onto a carrier, i.e. $E_S(t) = D[t/T]E_C e^{j(\omega_C t + \varphi_C)}$, mixed with the local oscillator, $E_L\, e^{j(\omega_L t + \varphi_L)}$. In the homodyne case, the carrier frequency is equal to the local oscillator frequency, i.e. $\omega_C = \omega_L$. The beat frequency $|\omega_C - \omega_L| \ll \omega_C, \omega_L$ is low in the heterodyne case so that $i_{PD}(t)$ is within the electrical bandwidth of the photodetector and subsequent demodulator.

Coherent optical detection was not developed for many years after the advent of coherent radio wave detection owing mostly to the lack of coherent optical sources. However, there is currently great interest in optical coherent detection, especially for fiber optic communications systems [5–7]. In some respects, coherent optical detection is analytically simpler than coherent radio-frequency detection because optical detectors detect intensity (i.e. photon absorption only) whereas radio wave antennae detect field (i.e. mixed photon absorption and emission) [8].

The different variants of coherent optical detection include most, but not all, combinations of modulation scheme (e.g. ASK, FSK, or PSK) with either detection method

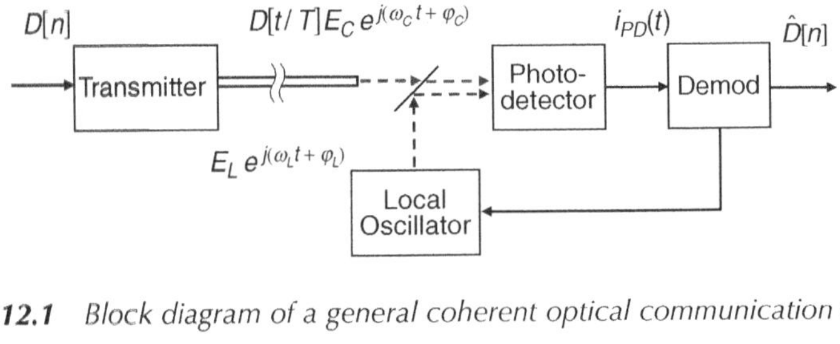

Figure 12.1 *Block diagram of a general coherent optical communication channel*

Table 12.1 *Ideal coherent detector sensitivity in photons per bit for binary transmission at a bit error rate of 10^{-9}. Reprinted from Optical Fiber Communications by L. Kazovsky, S. Benedetto and A. Willner, Copyright (1996) Artech House Inc*

	ASK	FSK	PSK
Asynchronous heterodyne	40	80	—
Synchronous heterodyne	36	36	18
Homodyne	18	36	9

(heterodyne or homodyne). Other technical differences may affect performance, for instance the difference between synchronous heterodyne detection (where the local oscillator is harmonically phase-locked to the signal) and asynchronous heterodyne detection (where it is not). In either case, the local oscillator must be carefully frequency-controlled in order to produce a suitably slow difference frequency, indicated by the feedback path in Figure 12.1. Homodyne detection is always synchronous by design.

Table 12.1 shows the ideal fiber optic coherent detector sensitivity expressed as the minimum required number of average photons per bit in order to achieve a bit error rate (BER) of 10^{-9} or better (taken from Kazovsky *et al.* [19], p. 292). The derivation of these figures will not be reproduced here but they arise by comparison of the photon count probability of the Poisson-distributed source with the binary signaling distance afforded by the respective detection method.

Not shown is the sensitivity for direct detection of an ASK-modulated signal under equivalent circumstances. Such a figure would depend heavily on detector type, specifications, and operating temperature but would likely run in the range of at least several thousand photons per bit. Avalanche photodiode detectors (which are not currently attractive for the large detector arrays required in page-oriented HDS) are dominated by their excess noise factor, and other detector types are dominated by thermal noise.

The enormous sensitivity improvements that may be achieved have motivated fiber optic system designers to explore coherent detection despite its considerable complexity and competing technologies. The incremental improvements in performance gained by going from a relatively simple asynchronous heterodyne system to the best performer – homodyne PSK – are unlikely to warrant the considerable cost and complexity of a fully phase-locked local oscillator laser for fiber optic communications. However, this difficulty is considerably alleviated in HDS. The same laser used to generate the probe beam for holographic reconstruction provides a ready source of coherent photons for the local oscillator, and prompts development of a superior homodyne system rather than heterodyne detection. The task of controlling the phase difference between the signal and the local oscillator could then in principle be achieved by adjusting the optical path length of one or both of the beams rather than phase-locking a laser to generate a local oscillator from scratch.

On the other hand, page-oriented HDS presents a special challenge because of the large number of parallel channels – often over a million – that must be detected. In general, the phase of the holographic signal carrier varies across the page so that a single-channel variable phase retarder could not perform the path length adjustment even if a suitable feedback signal were available. The phase-lock requirement becomes an adaptive optic problem wherein the wavefront of the local oscillator must be carefully matched to the holographic carrier wavefront over the whole image. Figure 12.2 is a conceptual illustration

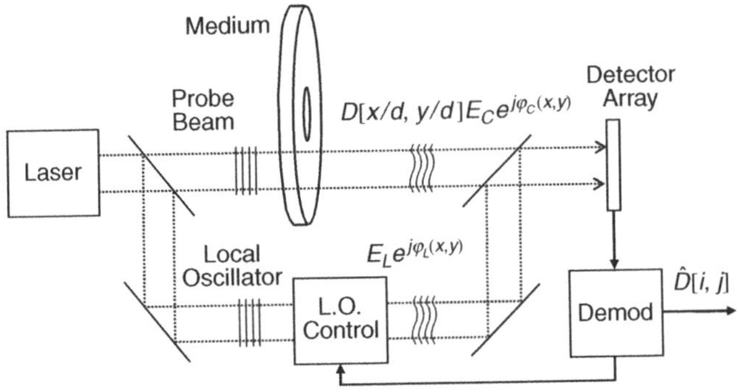

Figure 12.2 *Block diagram of coherent detection in a HDS device*

of this architecture. The signal beam, $E_S(x, y) = D[x/d, y/d]E_C e^{j\varphi_C(x,y)}$ (again neglecting ISI), is modulated by the two-dimensional data spatial light modulator (SLM) with a pixel spacing of d. Mechanical and environmental tolerances cause the signal beam to be modulated by the unknown carrier wavefront, $\varphi_C(x, y)$. Feedback provided by the detector array is used to similarly modulate the local oscillator portion of the laser beam with wavefront, $\varphi_L(x, y)$, in a manner that permits an estimate of the data, $\hat{D}[i, j]$, to be recovered from the detected intensity pattern.

12.2.2 Local Oscillator Generation

The obstacles hinted at in the handful of experiments performed by these researchers highlight the primary difficulty of coherent detection in a page-oriented HDS device. High storage density requires a high numerical aperture (NA) signal beam containing perhaps a million separate field modes, each of which may respond differently to perturbations that affect its total optical path length. Interferometric path-length matching is difficult enough for two single-mode fields, let alone across the entire field of a high-bandwidth holographic signal beam. Recently others have pursued this path [9]. Conceptually, generating a local oscillator suitable for HDS requires measuring the actual carrier wavefront difference, $\Delta\varphi(x, y) = \varphi_C(x, y) - \varphi_L(x, y)$, and then using a phase modulator of suitable bandwidth and resolution to drive this difference to zero over the whole field. However, as we shall presently argue, this process may be performed algorithmically using a simple switchable phase retarder and the method of *quadrature homodyne* detection, rather than physically using a complex multi-channel adaptive optic device.

Figure 12.3 shows a possible method for introducing a collimated local oscillator beam into the signal path in InPhase Inc.'s HDS architecture. All elements in the figure are common to InPhase's current direct-detect ASK architecture except the nonpolarizing beam splitter (NPBS), the switchable retarder, and the local oscillator beam itself. The phase SLM replaces an amplitude-modulating SLM, although with the use of suitable polarization optics, the same actual SLM device might be used. Note also that the local oscillator beam may derive from the same source used to illuminate the SLM during a write operation, which has identical requirements for beam size and collimation. Since the signal and local oscillator beams must be mixed with the same polarization, a NPBS must be used. This means that wasted light from both beams will be emitted from the unused port on the bottom

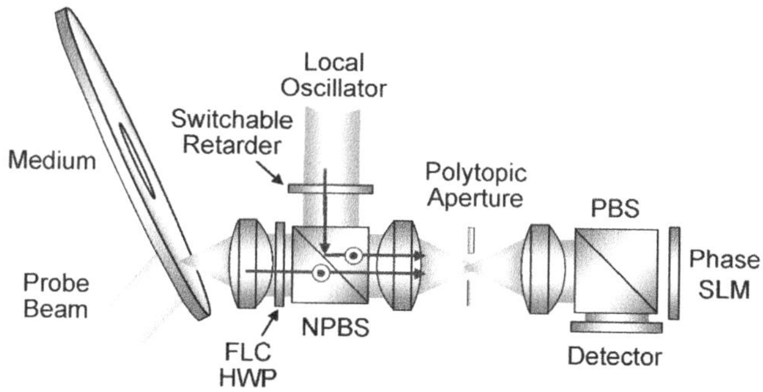

Figure 12.3 *Method for mixing a local oscillator beam with the signal beam in InPhase Inc.'s phase-conjugate HDS architecture. FLC HWP, ferroelectric liquid crystal half wave plate*

of the NPBS, but this effect can largely be mitigated by selecting a weakly reflecting beam splitter (perhaps $\sim$10%) so that plentiful local oscillator power is wasted in preference to the weak diffracted signal.

The switchable retarder is a simple one-channel device that transmits the local oscillator in one of two phase states that have an absolute optical path length difference of approximately 90°. This might be accomplished, for example, with a ferroelectric liquid crystal (FLC)-based variable retarder. The two phase states are not generally sufficient to bring $\Delta\varphi(x, y)$ to zero at *any* point in the image, let alone at every point across the whole image. Nevertheless, this binary control is adequate for the purpose of quadrature homodyne detection.

12.2.3 Quadrature Image Pairs

Using the apparatus of Figure 12.3, two different images of a hologram may be recovered. Assuming the system is reasonably mechanically stable (which is required for write operations in any case), and that the two images are taken within a short time so that thermal variations do not change the optical path lengths, the two images will bear a quadrature relationship to each other as described below. Since the signal and the local oscillator are propagating along the same optical axis, the difference wavefront $\Delta\varphi(x, y)$ will contain only relatively slowly varying components. However, the actual value of $\Delta\varphi(x, y)$ is uncontrolled, and in probability it will drift randomly between $+\pi$ and $-\pi$ with a more or less uniform distribution.

The principles of quadrature homodyne detection are not strictly dependent on the method of oversampled detection and resampling. Nevertheless, our goal here is to present a practicable implementation of the quadrature homodyne method, and image oversampling addresses many of the issues that must be solved in any page-oriented HDS system. Henceforth, the features of oversampled detection will be presumed during this development. The presence of known reserved blocks will be assumed, and the signal beam field at the detector will be written to include the effects of ISI:

$$E_S(x, y) = E_C e^{j\varphi_C(x,y)} \sum_i \sum_j D[i, j] h(x-di, y-dj) \tag{12.2}$$

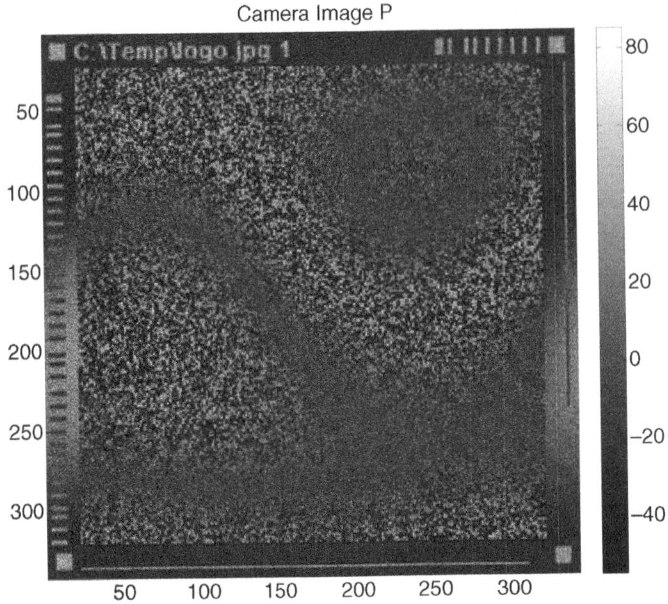

Figure 12.4 *'P' quadrature image simulation of a PSK-encoded hologram recovered with the local oscillator in the first (0°) phase state showing fringes indicative of the* $\Delta\varphi(x, y)$ *pattern*

where $h(x, y)$ is the impulse response (coherent pixel spread function) of the imaging system. $D[i, j]$ is the binary two-dimensional data pattern, which takes on the values (–1, +1) for PSK modulation.

Figures 12.4 and 12.5 are simulated detected images of the same hologram recovered with the switchable retarder in each of the phase states using a hypothetical simulated $\Delta\varphi(x, y)$ function and a moderate amount of pseudorandom coherent noise. The irradiance patterns at the detector are given by:

$$\begin{aligned} I_P(x,y) &= |E_L|^2 + |E_S(x,y)|^2 \\ &\quad + 2|E_L||E_C|\cos[\Delta\varphi(x,y)] \sum_i \sum_j D[i,j]h(x-di, y-dj) \\ I_Q(x,y) &= |E_L|^2 + |E_S(x,y)|^2 \\ &\quad - 2|E_L||E_C|\sin[\Delta\varphi(x,y)] \sum_i \sum_j D[i,j]h(x-di, y-dj) \end{aligned} \tag{12.3}$$

We shall refer to the image of Figure 12.4 with the local oscillator in the 0° phase state as the *P image,* and Figure 12.5 with the local oscillator in the 90° phase state as the *Q image.* Together, the two detected images constitute a *quadrature image pair*. The holographic images of Figures 12.4 and 12.5 consist of 256 × 256 SLM pixels (the small page size was chosen for clarity), and are oversampled with a 4 : 3 detector oversampling ratio as detailed

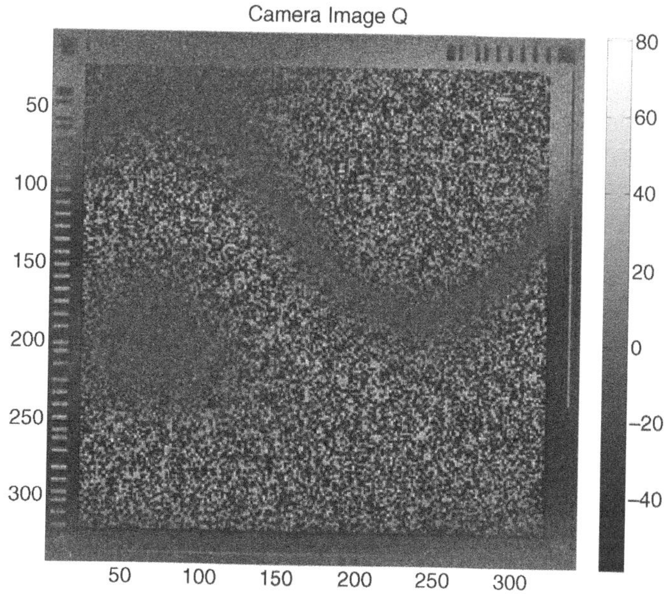

Figure 12.5 'Q' quadrature image simulation of the hologram of Figure 12.4, now detected with the local oscillator in the second (90°) phase state

in Chapter 11. The simulation uses a local oscillator with intensity 100 times that of the hologram, which has the effect of amplifying the signal by a factor of 20. Inspection of the images reveals regions of high and low contrast. The high contrast areas are further divided into regions where the image is inverted e.g. where the page border is light – and regions that are not inverted. In the noninverted regions, $\cos(\Delta\varphi) > 0$, and the state of the pixels may be directly determined from their intensity, which includes the information-bearing $2|E_L||E_C|\cos[\Delta\varphi(x,y)]\sum_i\sum_j D[i,j]h(x-di, y-dj)$ term. In the inverted regions, $\cos(\Delta\varphi) < 0$, and the state of the pixels may also be determined from their intensity. Only in the low-contrast regions where $\cos(\Delta\varphi) \approx 0$ are the data unrecoverable. However, because of the 90° phase change of the local oscillator, the entire fringe pattern has also shifted by 90° so that the regions that are low contrast in the P image are in high contrast in the Q image, and vice versa. Thus, in principle, a quadrature image pair contains all of the information needed to recover all the data in the hologram.

12.2.4 Estimating Phase Difference $\Delta\varphi(x, y)$

A quadrature image pair also contains all the information necessary to recover the carrier wavefront difference, $\Delta\varphi(x,y)$. The two images in quadrature convey the state of both sidebands of the wavefront function. One method would be to measure the contrast ratio in small regions of both images, thus yielding estimates of the terms containing both $\cos(\Delta\varphi)$ and $\sin(\Delta\varphi)$. However, this approach would be sensitive to local variations in the intensity of both E_L and E_C, and the polarity of the image would be indeterminate. Instead, we propose a better method based on the known reserved block patterns introduced in Chapter 11.

Recalling the image alignment measurement method, a cross correlation operation is performed between a portion of the detected image, m, and a target pattern, p, representing a DC-free version of what the middle of the over-sampled reserved block looks like at the detector.[1] Each element of the cross correlation matrix represents a dot product of the target pattern with a block of the detected pattern, or $m \cdot p = |m||p|\cos(\theta)$, where $\cos(\theta)$ is the correlation coefficient between the target and detected patterns. Where the reserved block is detected in a noninverted region of the image, we will have $\cos(\theta) \approx 1$, and where the reserved block is in an inverted region $\cos(\theta) \approx -1$, i.e. there will be a *negative*-going peak in the cross correlation matrix. Since $|p|$ is a constant, we have $m \cdot p \sim \pm|m|$. Furthermore, the sign is determined by the sign of the correlation peak.

In a real, arbitrarily aligned oversampled system, the cross correlation peak will extend over two or three samples in both the x and y directions. Just as an accurate estimate of the position of the reserved block may be found by interpolating among the samples constituting the peak, an accurate estimate of the magnitude of the peak may be found by summing these samples. The null surrounding the peak of the modified autocorrelation effectively isolates the $|m|$ peak magnitude component.

A computer code was developed to implement this demodulation process on the simulated images. Figures 12.6 and 12.7 show normalized cross correlation peak strength

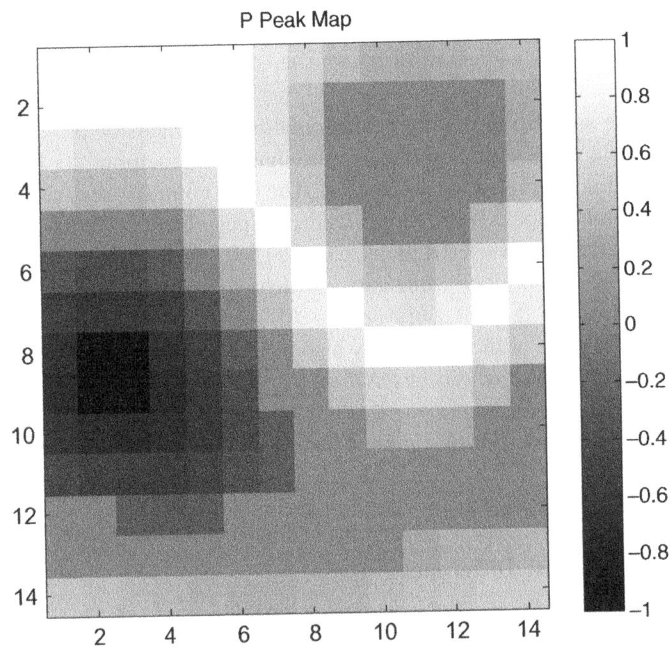

Figure 12.6 *Cross correlation peak strength map, $X_P(i, j)$, for the P image of Figure 12.4*

[1] Recall that a 'linearized' image – the square root of the detected irradiance – was used so that the quantities manipulated were proportional to the optical field. No such linearization is required for homodyne detection since the detected irradiance is already proportional to the signal field.

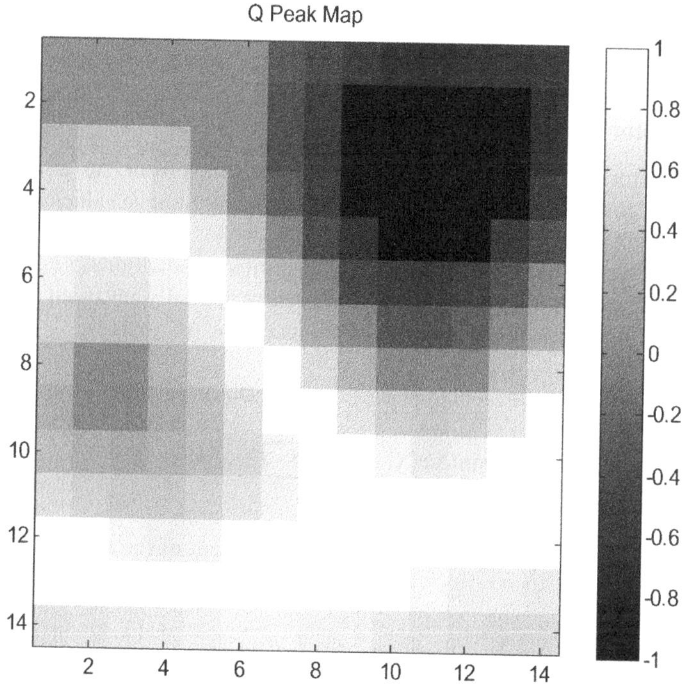

Figure 12.7 *Cross correlation peak strength map, $X_Q(i, j)$, for the Q image of Figure 12.5*

maps for the simulated P and Q images of Figures 12.4 and 12.5, respectively. These images each contained 196 reserved blocks arranged in a 14×14 grid (the density of reserved blocks was increased to a quarter of the page for illustration purposes). Comparing Figure 12.4 with Figure 12.6, it is apparent that the regions of Figure 12.4 that are in high contrast and noninverted show large positive peak strength values (approaching $+1$) in Figure 12.6. Similarly, inverted regions with high contrast have large negative peak strength values approaching -1.

The Q image of Figure 12.5 and peak strength map of Figure 12.7 show a similar relationship. Together, the two peak strength maps represent quadrature projections of the difference wavefront, which may be estimated by the expression:

$$\Delta\hat{\varphi}(i,j) = \tan^{-1}[X_Q(i,j),\ X_P(i,j)] \tag{12.4}$$

where $\tan^{-1}(a, b)$ is the four-quadrant arctangent.

The peak strength maps are sampled at the locations of the reserved blocks. In subsequent calculations, it will be necessary to provide an estimate of the difference wavefront (or equivalently, the peak strength maps) at every detector pixel location. Presuming the wavefront is adequately sampled by the reserved block pattern, this may be accomplished by upsampling, or interpolating the wavefront map to positions between the reserved blocks. In the subsequent numerical implementation of quadrature homodyne detection, this function was performed with a simple bilinear interpolation function of the peak strength maps.

12.2.5 Quadrature Image Combination

For a given detector pixel, knowledge of the difference wavefront, $\Delta\varphi(x,y)$, at the pixel location makes it possible to combine the P and Q quadrature image values and produce an estimate of the signal as follows.

For the normal case where $|E_L| \gg |E_C|$, the $|E_S|^2$ terms of Equations (12.3)(12.3) vanish by comparison. Furthermore, the $|E_L|^2$ terms represent spatially invariant DC offsets, so removing the DC components of the images produces the AC images:

$$\begin{aligned} \tilde{I}_P(x,y) &\approx +2|E_L||E_C|\cos[\Delta\varphi(x,y)]\sum_i\sum_j D[i,j]h(x-di,y-dj) \\ \tilde{I}_Q(x,y) &\approx -2|E_L||E_C|\sin[\Delta\varphi(x,y)]\sum_i\sum_j D[i,j]h(x-di,y-dj) \end{aligned} \tag{12.5}$$

From these, two separate estimates of the real component of the signaling field may be obtained:

$$\begin{aligned} \hat{E}_{SP}(x,y) &= |E_C|\sum_i\sum_j D[i,j]h(x-di,y-dj) = \frac{\tilde{I}_P(x,y)}{2|E_L|\cos[\Delta\varphi(x,y)]} \\ \hat{E}_{SQ}(x,y) &= |E_C|\sum_i\sum_j D[i,j]h(x-di,y-dj) = \frac{-\tilde{I}_Q(x,y)}{2|E_L|\sin[\Delta\varphi(x,y)]} \end{aligned} \tag{12.6}$$

Each of these estimates contains singularities where the denominator goes to zero. Furthermore, as the denominators shrink, the quality of the estimates degrades due to effective noise amplification. The quadrature relationship between the signals guarantees that both will not so degrade in the same place, and in fact the aggregate signaling strength remains constant. The two estimates may be linearly combined:

$$\hat{E}_S(x,y) = \alpha(x,y)\hat{E}_{SP}(x,y) + [1-\alpha(x,y)]\hat{E}_{SQ}(x,y) \tag{12.7}$$

where $0 \le \alpha(x,y) \le 1$ is a spatially varying coefficient that minimizes the noise in the combined estimate. Assuming that the noise fields N_P and N_Q in the P and Q images are independent and identically distributed, the total noise power in the combined estimate is given by:

$$I_N(x,y) \sim \left|\alpha\frac{N_P(x,y)}{\cos[\Delta\varphi(x,y)]}\right|^2 + \left|(1-\alpha)\frac{N_Q(x,y)}{-\sin[\Delta\varphi(x,y)]}\right|^2 \tag{12.8}$$

Differentiating with respect to α to find the minimizing value produces:

$$\begin{aligned} \alpha_{\min} &= \cos^2[\Delta\varphi(x,y)] \\ 1-\alpha_{\min} &= \sin^2[\Delta\varphi(x,y)] \end{aligned} \tag{12.9}$$

Combining Equations (12.7), (12.8) and (12.19) results in:

$$\hat{E}_S(x,y) \sim \cos[\Delta\varphi(x,y)]\tilde{I}_P(x,y) - \sin[\Delta\varphi(x,y)]\tilde{I}_Q(x,y)$$
$$= \frac{X_P(x,y)}{\left[X_P^2(x,y)+X_Q^2(x,y)\right]^{1/2}}\tilde{I}_P(x,y) + \frac{X_Q(x,y)}{[X_P^2(x,y)+X_Q^2(x,y)]^{1/2}}\tilde{I}_Q(x,y) \quad (12.10)$$

In this final form, it can be seen that the estimate simply combines the AC P and Q images in proportion to their local cross correlation peak strength. The negative sign of the peak will restore the intensity pattern to positive polarity in the inverted regions, and the denominator will normalize local variations in image intensity. Local variations in the local oscillator intensity will, however, still affect the quality of the $\tilde{I}_P$ and $\tilde{I}_Q$ AC filtered versions of the image. In practice, it may be advantageous to select a filtering function that optimally rejects low frequency local oscillator variations while passing high frequency signal components rather than simply subtracting the DC component.

Figure 12.8 shows the image produced by combination of the images in Figures 12.4 and 12.5 showing high-contrast noninverted data throughout. The image shows some artifacts from $\Delta\varphi$ estimation errors (as is typical) but is nevertheless of very high fidelity.

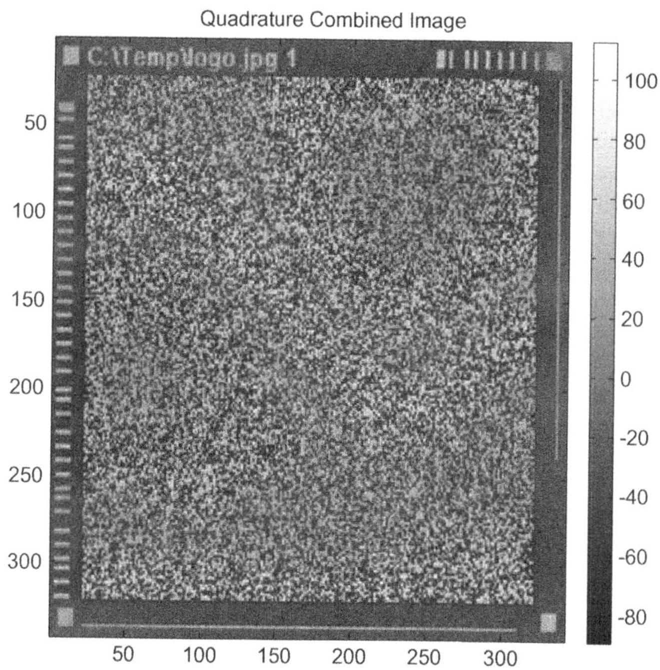

Figure 12.8 *Image produced by combination of the quadrature image pair shown in Figures 12.4 and 12.5*

Quadrature image recombination produces a replica of the holographic image appearing as though it had been recovered using a local oscillator phase-matched with $\Delta\varphi(x, y) = 0$ across the whole image, thereby producing algorithmically that which would have taken an advanced adaptive optic element to produce physically. The price of this advantage is the need to make two separate exposures two recover a single hologram, though phase quadrature multiplexing (described below) does allow for data recovery at a net rate of one exposure per hologram. The optical amplification provided by coherent detection will more than compensate by sharply reducing the actual cumulative exposure time but the increased electrical bandwidth required for the detector and other components will increase cost and complexity.

12.2.6 Quadrature Image Resampling

The recombination process reconstructs the continuously varying field magnitude at whatever resolution the detector happens to sample it. For the case of over-sampled detection, the recombined image is not a direct binary representation of the PSK phase symbols but rather an asynchronously sampled grayscale version of the SLM image. The recombined image may be subsequently resampled using the technique of Chapter 11.

The step of alignment measurement or the 'quiver' method involves precisely locating the position of the known reserved blocks by interpolating the position of the peak within a target pattern cross correlation matrix. This is computationally similar to the cross correlation calculation used to generate the X_P and X_Q cross correlation peak strength maps of Figures 12.6 and 12.7, and it is possible to combine the operations in order to avoid duplication. In addition to finding the peak strengths, the peak position may also be determined by finding the two-dimensional centroid of the values constituting the peak. In the case of negative-going peaks, the values must be inverted for the centroid operation.

This process will yield *two* separate reserved block displacement vector grids (quiver maps), whereas only one is required for resampling. Furthermore, each grid will contain noisy, potentially inaccurate values for reserved blocks that land in low contrast regions. The two grids may be combined into a single high-quality grid using a method similar to the image combination method of the previous section. In fact, a very simple method could be used – for each reserved block, simply select the displacement vector corresponding to the strongest absolute peak strength. However, in the quadrature homodyne computer code developed for this research, a slightly more sophisticated method was employed. The displacement vectors were combined in *proportion* to their corresponding absolute peak strength:

$$\vec{q}(i,j) = \frac{|X_P(i,j)|\vec{q}_P(i,j) + |X_Q(i,j)|\vec{q}_Q(i,j)}{|X_P(i,j)| + |X_Q(i,j)|} \tag{12.11}$$

where $\vec{q}(i,j)$, $\vec{q}_P(i,j)$, and $\vec{q}_Q(i,j)$ are the combined, P image, and Q image displacement vector grids, respectively. In a further refinement, the computer code discards any displacement vector with associated peak strength less than one-third of the other value. This heuristic helps reject vectors taken from very low contrast regions where false cross correlation peaks are sometimes identified.

After the alignment is determined, the combined homodyne image is resampled. The resampling coefficients were optimized for optical field detection by using a homodyne detection model for the minimum mean squared error (MMSE) coefficient training targets. Compared with direct detection, homodyne ISI is more linear (since the linear optical field superposition is detected rather than intensity), but there is more of it (because the optical field pixel spread function is wider than the field-squared pixel spread function for direct detection). These factors somewhat complicate prediction of the performance improvement due to homodyne detection.

12.2.7 Coherent Noise Linearization

Lowering the diffraction efficiency of the holograms allows for more to be stored, so designers tend to be driven to operate at a diffraction efficiency that is only marginally higher than the probe beam scattering efficiency. Scatter, as characterized by the *bidirectional transmission distribution function* (BTDF) of the recording medium, is a critical parameter that is carefully minimized by formulation, process control (especially cleanliness), and recording practices.

Scattered probe light mixes coherently with the holographic signal, and moreover is of comparable spatial bandwidth since it issues from the same Fourier extent. It cannot be reduced by lengthening the exposure time during playback.

Coherent noise degrades performance more than additive noise of equivalent power. For direct detection, the irradiance at the detector is given by the interference equation:

$$\begin{aligned} I(x,y) = |E_S(x,y)+E_N(x,y)|^2 &= |E_S(x,y)|^2+|E_N(x,y)|^2 \\ &+2|E_S(x,y)||E_N(x,y)|\cos\varphi_{S-N}(x,y) \end{aligned} \tag{12.12}$$

where $E_S(x,y)$ and $E_N(x,y)$ are the signal and noise fields, and $\varphi_{S-N}(x,y)$ is the phase difference between them. In the third (cross) term, the signal field is multiplicatively modulated by the noise field and the random phase factor, greatly increasing the variance of the combined irradiance.

For homodyne detection, all three fields interfere:

$$\begin{aligned} I &= |E_L+E_S+E_N|^2 = |E_L|^2+|E_S|^2+|E_N|^2 \\ &+2|E_L||E_S|\cos\varphi_{L-S}+2|E_L||E_N|\cos\varphi_{L-N}+2|E_S||E_N|\cos\varphi_{S-N} \end{aligned} \tag{12.13}$$

where the explicit dependence on (x, y) has been omitted for brevity. In the preferred case, the local oscillator is much stronger than either the signal or the noise, i.e. $|E_L| \gg |E_S|$ and $|E_L| \gg |E_N|$, so terms not involving E_L become negligible by comparison:

$$I \approx |E_L|^2+2|E_L||E_S|\cos\varphi_{L-S}+2|E_L||E_N|\cos\varphi_{L-N} \tag{12.14}$$

Although the local oscillator has amplified the noise along with the signal, it does not amplify the troublesome cross-term. We should expect that the variance of the irradiance is substantially reduced, and performance improved. Furthermore, the noise is now linearly additive. If the noise field has a Gaussian magnitude distribution with uniform random phase (a plausible model – since field is additive, the sum of scattered fields obeys the central limit theorem), and if the noise has a flat spectrum with bandwidth comparable with the signal (also reasonable if the noise sources are uniformly distributed scatterers within the Fourier plane) then the process may be well-modeled as additive white Gaussian noise (AWGN). By contrast, direct detection of the same field leads to a χ^2 irradiance probability density function for each binary state, which complicates analysis and optimization of the downstream channel.

We may estimate the performance improvement by finding the signal to noise ratio (SNR) for the two cases. The SNR was approximated in the direct detection case by previous investigators by finding the mean and variance of a pixel [10]:

$$\begin{aligned} E[v] &= a^2 + 2\sigma^2, \\ \sigma_v^2 &= 4\sigma^2(a^2 + \sigma^2) \end{aligned} \tag{12.15}$$

where a is the signal field magnitude, σ^2 is the variance of both the real and imaginary components of the complex noise field and v is the detected intensity. Considering only 'on' pixels and normalizing to $a = 1$, the investigators then defined an ex post SNR metric:

$$SNR \equiv \frac{E[v]}{\sigma_v} = \frac{1 + 2\sigma^2}{2\sigma\sqrt{1+\sigma^2}} \approx \frac{1}{2\sigma}(\sigma^2 \ll 1) \tag{12.16}$$

We can calculate the equivalent SNR for the homodyne case by noting that the real parts of the signal and noise fields are detected as intensity:

$$\begin{aligned} E[v] &= a \\ \sigma_v^2 &= \sigma^2 \end{aligned} \tag{12.17}$$

So

$$\mathrm{SNR} = \frac{1}{\sigma} \tag{12.18}$$

representing an approximately 3 dB improvement.

12.2.8 Simulation Results

Performance of quadrature homodyne detection was evaluated by applying the algorithm to simulated images. Tests were performed comparing PSK quadrature homodyne, PSK homodyne [i.e. with an ideal $\Delta\varphi(x, y) = 0$ local oscillator], ASK homodyne, and ASK direct detection. The data page was 752×752 SLM pixels including a 16 pixel border, and contained 11×12 reserved blocks on a 64 pixel-spacing grid (~1.6% overhead). The simulated detector page was over-sampled by 4 : 3, producing a 1003×1003 image. The simulated point spread function was established by the 1.124 Nyquist filtering aperture, and the images contained a slight (~0.1%) rotation component to make the resampling coefficient usage approximately uniform. All parameters except for the encoding and detection method were kept identical.

In the homodyne trials, the local oscillator was set to 100 times the power of the signal image. For quadrature homodyne, two waves of tilt in the x direction were also applied. Figure 12.9 summarizes the results of the simulation. The x-axis represents the ex ante signal to noise power ratio, which was varied by adding coherent noise to the 'noise-free'

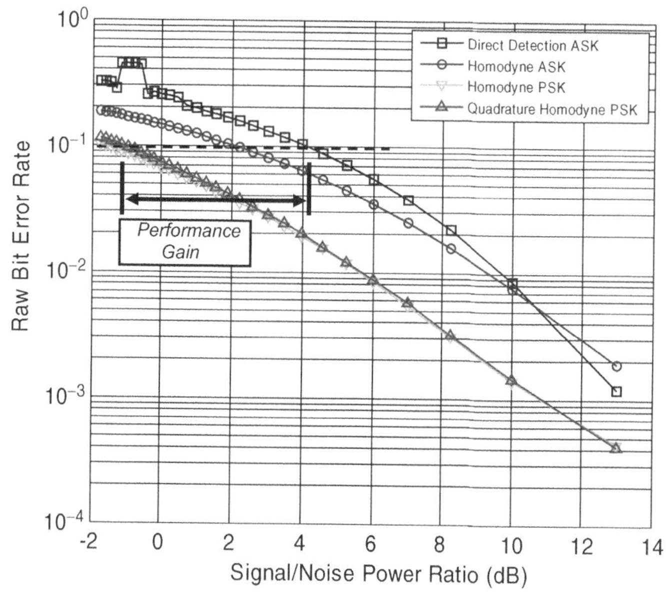

Figure 12.9 *Raw bit error rate versus signal to coherent noise power ratio for the various detection schemes. (The discontinuity in the direct detection curve is caused by alignment algorithm failures in the low SNR regime)*

simulated images. The *y*-axis is the raw BER determined by simple binary threshold detection of the resampled image.

The performance gain due to homodyne detection depends on the signal to noise regime of the system. High-density direct detection currently requires very weak, noisy holograms and heavy error correction in order to tolerate worst-case raw BERs as high as approximately 0.1. At 0.1 BER (the dashed horizontal line in Figure 12.9), homodyne detection of the same ASK signal provides a 2 dB improvement over direct detection, which is somewhat less than the 3 dB figure estimated earlier. This is probably related to the differing performance of the MMSE resampling coefficients where direct detection enjoys a narrower impulse response, and the plot shows that the homodyne advantage is eliminated at higher SNRs. However, this could only be exploited by a device that writes holograms far stronger than the scatter noise floor.

PSK homodyne detection provides almost 4 dB of further improvement over ASK homodyne, slightly more than the 3 dB figure predicted by simple theory. As described above, the quadrature homodyne curve approaches the ideal PSK homodyne curve in performance; although it begins to diverge somewhat in low SNR conditions where local oscillator phase estimates begin to degrade. Nevertheless, quadrature homodyne improves upon direct detection by about 5 dB.

Five decibels represents an enormous improvement for a channel that starts out at only a few decibels. Clearly designers will have the option to write weaker holograms, yielding higher storage densities and shorter write exposures, and will be able to read them with reduced read exposure times. Error correction coding overhead will probably be reduced, leading to further improvements.

It should also be pointed out that most of the system parameters – including the Nyquist aperture size, the oversampling ratio, and the reserved block patterns and spacing – have been optimized for the existing direct detection architecture. Other coherent detection parameters – such as the strength of the local oscillator and the resampling coefficients – have not been as thoroughly analyzed. It is possible that future optimization of all of these parameters specifically for quadrature homodyne detection will yield further improvements and trade-off opportunities.

Further benefits of PSK modulation derive from the fact that PSK-modulated signal beams do not contain the large intensity inhomogeneities of ASK-modulated beams (e.g. the DC Fourier component). PSK signaling also dramatically reduces intra-signal modulation noise compared with ASK signaling [1]. These features in turn eliminate the need for the phase mask subsystem (or another technique) required to mitigate the deleterious holographic effects of ASK modulation, thus greatly reducing the cost, size and complexity of the resulting device.

12.2.9 Phase Sensitivity Issues

While the quadrature homodyne method does not require virtually perfect wavefront matching like the prior methods, there are limits to the wavefront error that may be tolerated. Since the algorithm operates by sampling the $\Delta\varphi$ carrier difference wavefront at the reserved block positions, it follows that $\Delta\varphi$ must be at least Nyquist-sampled by the reserved blocks, i.e.:

$$BW_{\Delta\varphi} < \frac{1}{2\Delta_B} \tag{12.19}$$

where $BW_{\Delta\varphi}$is the x or y bandwidth of $\Delta\varphi$ and Δ_B is the corresponding spacing of the reserved blocks.

By this metric, the simulated images used to generate the results of Figure 12.9 could tolerate as many as five waves of tilt without decreasing the reserved block spacing. However, as is common with sampling systems, performance was empirically found to degrade as the Nyquist limit was approached (the quadrature homodyne simulations instead used two waves of tilt). Still, the ability to tolerate multiple waves of aberration evinces the feasibility of the entire method. An HDS system must, in any case, be able to generate a probe beam with a wavefront tolerance of a fraction of a wavelength, and the local oscillator should be able to be delivered to similar standards. It is the hologram itself that will impart most of the carrier wavefront difference. Every planar component of the hologram must phase align with the probe over at least a large portion of its physical extent, so it is doubtful that the carrier aberration over the entire complex hologram could exceed a 'few' waves when Bragg-matched.

An experiment was performed in order to test this intuition empirically. An ordinary ASK-modulated, direct detected HDS tester was modified to introduce a collimated local oscillator beam into the signal path in a configuration similar to Figure 12.3. Instead of a nonpolarizing beam splitter cube, a 10% reflective pellicle beam splitter was used, and there was no switchable retarder. A data hologram was reconstructed, and the direction

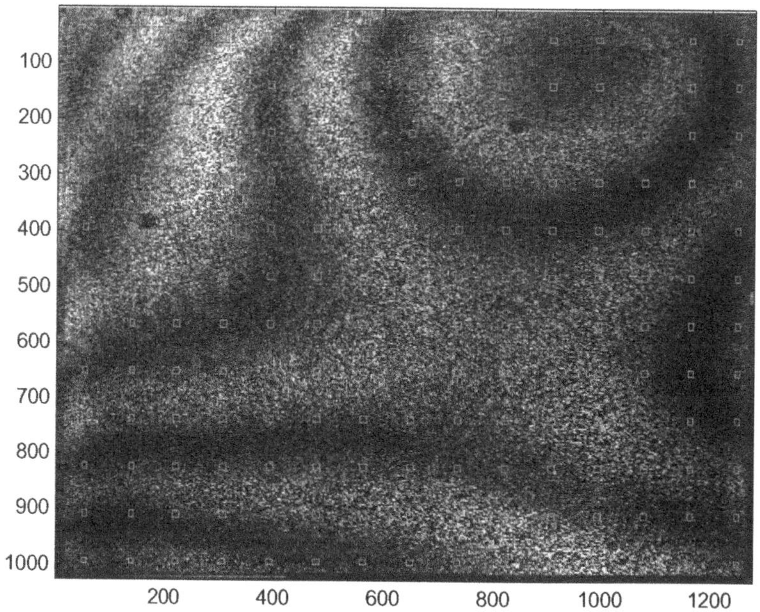

Figure 12.10 *Experimental carrier difference wavefront pattern*

and collimation of the local oscillator was adjusted to minimize the fringe frequency seen at the detector. A typical detector image is shown in Figure 12.10. The approximate positions of the reserved blocks have been indicated with the grid overlay. The irregular fringe pattern does indeed show about three waves total peak-to-valley excursion.[2] Most of the fringe pattern is adequately sampled by the reserved blocks but there are a few regions of high slope that are very close to the Nyquist rate (e.g. the lower left corner).

The results of Figure 12.10 are promising but more work will be required in order to establish bounds on the magnitude and bandwidth of $\Delta\varphi$, both experimentally and analytically. Options to further develop the system include decreasing the spacing of the reserved blocks (perhaps changing their size), or optimizing the optics to improve the flatness of the hologram carrier wavefront. The lenses in the system used for Figure 12.10 were designed without regard to this parameter, and it is possible that the astigmatic pattern is due to either the lenses or the tilt of the medium rather than the hologram itself.

12.2.10 Local Oscillator and Hologram Alignment

While the overall flatness of the hologram carrier wavefront may be adequate, there is still the issue of degenerate 'aberrations' such as tilt and collimation that must also be matched between the two beams. Since the hologram stands in a Fourier relationship with the image

[2] Note that since the data are ASK-modulated, the bright regions correspond to $\Delta\varphi \approx 0$ and the dark regions are 'inverted' with $\Delta\varphi \approx \pm\pi$.

(at least in the architecture of interest here), mispositioning the hologram in a transverse direction will cause a linear phase component (tilt) in the hologram carrier wavefront. Mispositioning the hologram longitudinally (along the optical axis) will decollimate the hologram carrier wavefront, creating a spherical ('zone plate') fringe pattern when mixed with a collimated local oscillator. In order to keep the total fringe bandwidth below the prescribed limits, it will be necessary to carefully control the medium position relative to the read optics in all three dimensions. The medium position actuators (required for spatial multiplexing in any case) must be capable of very fine positioning resolution, and a third actuator providing motion in the height direction must be added. The final resolution required for these positioners will depend on specific system parameters and tolerance trade-offs but is liable to end up in the one wavelength range since one wavelength of transverse shift causes one fringe of tilt in a Fourier system operating at $f/\# = 1$.

Alternatively, these three degrees of freedom may be adjusted by providing fine tilt and collimation control of the local oscillator over a small range, or by some other combination of optical and medium microactuators. In either case, feedback must be provided in order to tell the HDS drive controller how to adjust the actuators. However, within a small range near the ideally aligned position, the fringe pattern itself can provide this information. By decomposing the $\Delta\varphi$ pattern into its dominant linear and quadratic components, the controller can determine both the magnitude and direction required to adjust all three actuators to the optimal position. A single P/Q image pair would suffice if the fringe pattern is adequately sampled but otherwise a search algorithm that analyzes the fringes at different actuator positions might be required to detect aliasing. Since the three actuated degrees of freedom allow the controller to dial out the dominant linear and quadratic components of *any* fringe pattern regardless of their original cause, only higher order components of the carrier wavefront aberrations are ultimately of concern.

12.2.11 Adaptive Homodyne Detection

The quadrature homodyne detection method was developed to be practical with components that are available today or at least from components that could be built using existing technology. However, with the addition of one key component, a system could be built that creates a fully adaptive local oscillator wavefront in order to match the hologram carrier wavefront in the traditional, physical manner. That key component is a fast SLM with the capability to modulate the phase of each pixel in many fine steps, rather than just two states ('gray-scale' phase modulation) [11]. Currently, gray-scale phase SLMs are generally based on twisted-nematic liquid crystals that are too slow to sustain frame rates of interest for commercial HDS. FLCscan be fast but the fast versions typically switch between only two metastable phase states. No attractive technology appears to exist currently but it seems that such a device is likely to be developed within a few years. Analog FLCs and microelectromechanical systems are two possible technologies [12,13].

Figure 12.11 illustrates a possible method for incorporating a gray-scale phase SLM into the present architecture. Compared with the base line direct-detection architecture, this system requires only the addition of two quarter-wave plates (QWPs) and a partially reflecting surface.

During a write operation, the data are written normally, with the only consequence of the new components being that some small portion (perhaps ~10%) of the signal beam photons

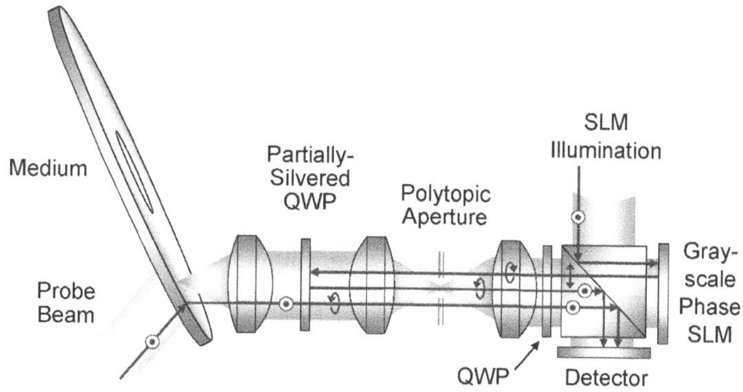

Figure 12.11 *Possible architecture for adaptive homodyne detection*

are retro-reflected by the partially silvered surface, and thereby wasted. During a read operation, the SLM illumination source remains on (instead of switching off, as it does for direct detection), and the required local oscillator wavefront pattern is written onto the SLM. A portion of the modulated light is reflected by the partially silvered surface, which is located in the focal plane shared by the objective lens and the first relay lens. The SLM-modulated light mixes with the holographic signal in the same polarization state and propagates to the detector where it forms another image of the field at the SLM (the total distance being eight focal lengths). (Another alternative would be to place the partially silvered surface within the polytopic aperture, resulting in a 4F path.)

Since the SLM can effectively impart an arbitrary wavefront upon the local oscillator, it is possible to create the $\Delta\varphi = 0$ condition across the whole image and thereby recover the data with a single hologram exposure. However, it will still be necessary to determine the required wavefront in the first place. This might be accomplished by finding $\Delta\varphi$ for an initial hologram by taking P and Q images using the SLM to effect the 90° phase change in the local oscillator. Once established, the wavefront could presumably be used to recover all of the holograms multiplexed at a given location. In the (likely) event that the required wavefront actually evolves slowly as both the recording and recovery conditions change, the controller could employ adaptive algorithms to make updates in the wavefront. For example, if the controller deliberately introduced a small ($\ll 90°$), alternating phase offset in the wavefront used to recover sequential holograms, it could then compare the cross-correlation peak strengths between *neighboring* holograms, and regionally adjust the wavefront phase slightly in the direction that makes the peak stronger. Such a 'wobble servo' would have minimal impact on the signal quality, but would allow the wavefront to be adapted slowly while maintaining single-exposure recoveries. Alternatively, control patterns modulated in quadrature ($\pm\pi/2$ phases) could be written into the reserved blocks so that the deliberate phase offsets are not required.

The modulation capabilities of a gray-scale SLM also opens up possibilities for gray-scale data modulation (more than two states) and/or phase precompensation. Holographic storage currently operates in a low SNR regime where denser than binary modulation is not advantageous but the improvements provided by coherent detection could change that assessment.

12.3 Phase Quadrature Holographic Multiplexing

In addition to the benefits described above, coherent detection techniques such as quadrature homodyne detection have the potential to directly increase the storage capacity of HDS devices. Since homodyne detection allows for the recovery of both the amplitude and phase of the holographic signal, it in principle transmits two channels of information where direct detection provides only one. This principle is exploited in other communications technologies by coding techniques such as quadrature amplitude modulation (QAM) which simultaneously transmits two independent data channels on a single carrier. Because sinusoids separated by 90° in phase are mutually orthogonal, the real and imaginary parts of a complex index grating represent independent degrees of freedom in the holographic address space. These parts correspond to real and imaginary components of the complex amplitude in the reconstructed wavefront, which may in principle be detected as independent data pages using a coherent detection technique.

12.3.1 Phase-Quadrature Recording

Thus, a method of *phase quadrature holographic multiplexing* would double the holographic address space (and, typically, the storage capacity) of an HDS system provided some method can be found for writing independent holograms in phase quadrature. This can in fact be accomplished by using elements that are already present in a device employing quadrature homodyne detection (see Figure 12.12).

Figure 12.12 illustrates InPhase's standard Tapestry® architecture for recording angle/polytopic multiplexed holograms, with the addition of a single element: a switchable retarder has been placed in the path of the beam that illuminates the SLM. For quadrature multiplexing, the switchable retarder must be capable of at least two phase states that are separated by ±90° in path length. The requirements for this device are essentially identical to those for the switchable retarder used for quadrature homodyne detection (see Figure 12.3), and indeed, the same physical element could be employed for both purposes. As with the quadrature homodyne detection architecture, the switchable retarder may optionally be placed in the reference beam path instead of the signal beam path.

Recording quadrature-multiplexed holograms thus becomes a simple matter of performing two exposure operations with the same reference beam. For the first exposure, the

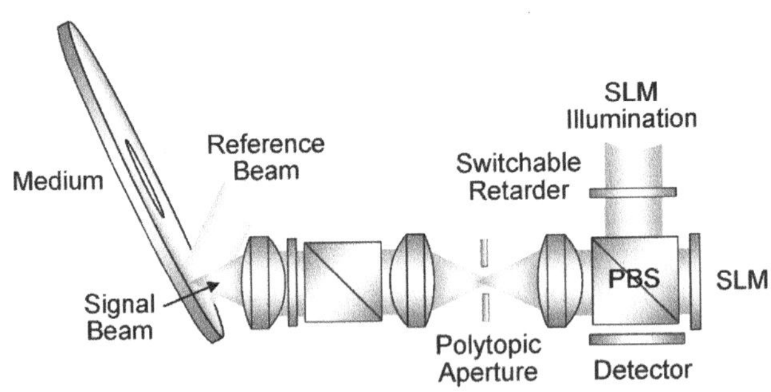

Figure 12.12 *Possible architecture for recording phase quadrature multiplexed holograms*

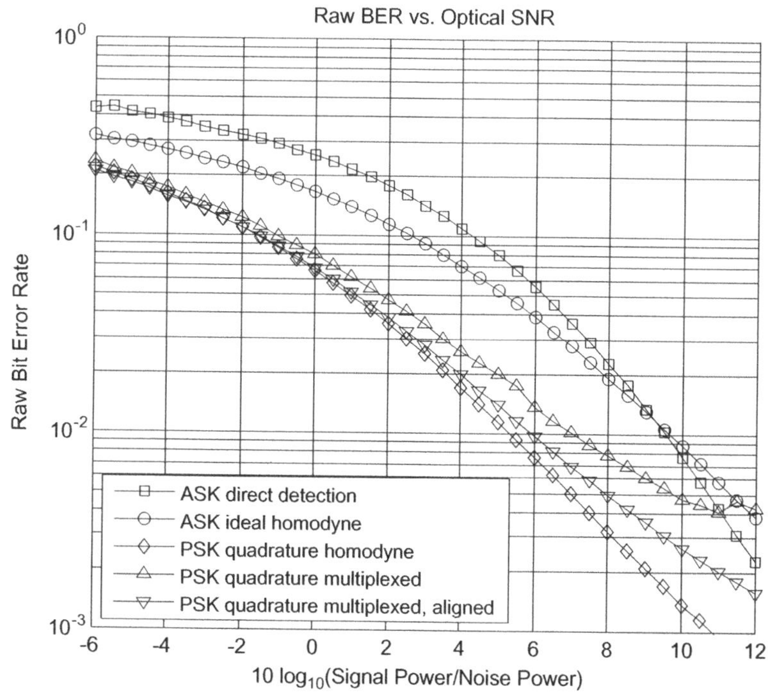

Figure 12.13 *Raw bit error rate versus optical SNR for the various cases including phase-quadrature multiplexed holograms*

switchable retarder is placed in one state (say, the 0° state); and a holographic data page (call it the *P* page) is composed on the SLM. For the second exposure, the switchable retarder is switched to the other (90°) state and another data page (the *Q* page) is written to the SLM. So long as the optical paths in the device remain stable over both exposures, the two holograms will be recorded in phase quadrature with respect to each other. This is to say that the refractive index gratings of each and every Fourier component of the respective holograms will differ in phase by ±90°. The two holograms will thus remain entirely spatially orthogonal so long as the data modulation scheme does not employ any quadrature components. This condition is met for the preferred case of binary PSK modulation wherein the *P* hologram pixels will occupy the 0° and 180° phase states, and the *Q* hologram pixels will utilize the 90° and 270° states. The stability requirements for achieving this are similar to the requirements for recording a single hologram over the extended exposure interval, so generally the practice of recording the pair in quick succession will suffice to ensure the relationship.

Recording quadrature-multiplexed holograms sequentially in this manner also carries the possible advantage of requiring only one reference beam change for every two holograms recorded. In a system employing angle-multiplexing, for example, this may halve the number of galvanometer step-and-settle operations, potentially increasing the achievable recording rate.

Recording rates could be improved even further by recording the two holograms in parallel in the same exposure. This could be accomplished by using an SLM capable of modulating all four quadrants of phase (i.e. at least four levels of gray-scale phase) so that both pages can be represented at the same time. The switchable retarder of Figure 12.12 would not be required in this case. In addition to recording speed, such a technique would improve *M/#* efficiency by a factor of $\sqrt{2}$.

12.3.2 Phase-Quadrature Recovery

Both holograms may be recovered by using a coherent detection method to isolate the two phase quadratures. For quadrature homodyne detection, recovery of quadrature-multiplexed holograms will produce quadrature image pairs similar to those of Figures 12.4 and 12.5, excepting that there will be no low-contrast 'gray' fringes visible. This is because the two holograms are intermixed such that the Q hologram occupies the low-contrast regions of the P hologram, and vice versa. There are several possible variations for recovery using quadrature homodyne detection. In the most conceptually simple method, each hologram is recovered independently using a pair of read exposures, so that four exposures are taken in total. However, it is immediately obvious that this is redundant, and both holograms may be recovered from a single quadrature pair of exposures by applying the recovery algorithm twice – once using the known reserved block patterns from the P hologram, and once using the Q hologram patterns. Further computational efficiency can be achieved by detecting only the P reserved block patterns. The P hologram is then recombined normally, and the Q hologram is recombined from the same information by:

$$\hat{E}_Q(x,y) = \frac{-X_Q(x,y)}{\left[X_P^2(x,y)+X_Q^2(x,y)\right]^{1/2}}\tilde{I}_P(x,y) + \frac{X_P(x,y)}{\left[X_P^2(x,y)+X_Q^2(x,y)\right]^{1/2}}\tilde{I}_Q(x,y) \tag{12.20}$$

The Q hologram may thusly be recombined and subsequently resampled using 'quiver' information wholly from the P hologram. In this form, it is apparent that the quadrature homodyne recombination process merely represents a change of phase basis. The holograms are recorded in an original phase basis wherein the real and imaginary parts encode the P and Q holograms, respectively. They are recovered in an arbitrary, uncontrolled phase basis that varies spatially over the image(s), and then computationally restored to the original basis by the quadrature recombination method.

12.3.3 Reserved Block Equalization

Because the P and Q holograms are generally detected in superposition, there will be considerable crosstalk in the reserved block cross correlation operations unless measures are taken to prevent it. Correlation of the Q hologram reserved block with the corresponding P hologram reserved block target pattern will contribute noise to the X_P and X_Q peak strength maps, thus degrading the estimate of the local oscillator difference wavefront, $\Delta\varphi(x,y)$, producing poor recombined images. Peak positions will also be impacted, leading to poor resampling. These problems may be substantially alleviated by judicious choice of reserved block pattern pairs occupying the same location in the P and Q images. One method for accomplishing this relies on a generalization of the reserved block pattern design principle of Chapter 11. Under that principle, binary reserved block patterns are selected that contain an isolated peak in their modified autocorrelations – i.e. the values of all eight locations neighboring the central peak are zeros. This principle reduces the introduction of pattern dependent noise into the interpolated location of the peak used for resampling. It also sharply reduces the introduction of pattern dependent noise into the X_P and X_Q peak strength maps used for quadrature homodyne detection.

For the following simulation results, the reserved block patterns for the P and Q images were subject to the following additional constraints: the cross correlation between the binary 8×8 reserved block pattern of the P image and the inner 6×6 part of the corresponding pattern for the Q image must contain zeros at the inner 3×3 locations. Similarly, the inner 3×3 of the cross correlation between the Q reserved block and the inner 6×6 of the P reserved block must also be zero. These conditions were selected by computer search from a large pool of pseudo-random candidate block patterns.

12.3.4 Simulation of Phase-Quadrature Decoding

Figure 12.13 shows the performance of the various detection and modulations schemes under discussion, now including the detection of phase-quadrature multiplexed holograms using quadrature homodyne detection. The simulation was performed under conditions similar to those of Figure 12.9, excepting that the page format has been changed to the 1200×1200 format of the Tapestry® drive, and the size of the polytopic aperture has been increased to 1.2 Nyquist. The curves for ASK direct detection and PSK quadrature homodyne detection substantially replicate the curves of Figure 12.9, reflecting the nearly 6 dB performance improvement in the regime of interest. As hoped, the curve for quadrature-multiplexed holograms tracks the curve for detection *without* the quadrature-multiplexed hologram, reflecting a less than 1 dB performance degradation due to crosstalk from the quadrature channel.

The PSK quadrature multiplexed, aligned curve in Figure 12.13 shows performance when the cross correlation peak strength maps are used from the noise-free case, but the rest of the channel operates on the noisy image. This case was intended to determine how much degradation was due to *P-Q* crosstalk from poor carrier wavefront estimation versus how much was due to other factors. Since this case affords considerable improvement, it is likely that nontrivial performance gains might be available from improvement to wavefront estimation and interpolation.

Note also that the curves of Figure 12.13 (and Figure 12.9) are plotted in terms of equal signal power (i.e. diffraction efficiency). However, PSK-modulated holograms are likely to be nearly twice as efficient as ASK in terms of *M/#* efficiency, so their performance could be up to 3 dB better than indicated in Figure 12.13 in an *M/#*-limited situation.

12.3.5 Summary of Improvements

A second-generation HDS device employing the techniques of this chapter would enjoy a direct doubling of capacity due to the doubling of address space from quadrature multiplexing. This might even be achieved without increasing the *M/#* of the medium owing to the improved *M/#* efficiency of PSK modulation. Additionally, the improvement in SNR will allow for a further increase in capacity due to an increase in the error correction code rate. According to the Shannon–Hartley theorem, the capacity of a noisy digital channel is bounded by:

$$C = B\log_2\left(1 + \frac{S}{N}\right) \tag{12.21}$$

where C is the capacity, B is the bandwidth, and S and N are the signal and noise powers, respectively. Applying this to HDS, quadrature multiplexing increases B by a factor of two. Improving S/N by values corresponding to a 5 dB improvement from, say, 2 to 7 dB (2 dB being the decoding threshold for the Tapestry® drive) results in approximately another doubling of C. Thus, it is possible that these techniques could result in over 1 TB of disk capacity without any of the myriad other possible design improvements. The following benefits would also accrue:

- Reference beam angle change (galvanometer step and settle) operations are reduced by half for both recording and recovery.
- Disk move (book-to-book seek) operations are similarly halved.
- Phase mask and associated optics are eliminated.
- Backwards compatibility with first generation disks should be practical.
- Recording exposure times reduced by approximately half (PSK modulation) or more (parallel recording).
- Recovery exposure time reduced by $10\times$ or more (homodyne gain).

12.4 Other Research Directions

Other future research might be focused on incorporating better detection techniques with oversampled detection. Examples are Viterbi or partial response maximum likelihood (PRML) detection [14–16], which are computationally difficult in two dimensions, or Viterbi-like detection algorithms such as multi-strip detection [17,18]. These more complex detectors could potentially improve the performance of the resampling channel by allowing controlled, reversible ISI (blur) while shrinking the size of the holograms. In addition, using empirical log likelihood ratio distributions instead of assuming Gaussian distributions appears to improve the channel performance by 0.2–0.3 dB. Modulation codes that remove bad patterns might also be worthwhile but could be difficult to implement in the context of this channel.

Acknowledgements

We would like to thank Professor B. V. K. Vijaya Kumar at Carnegie Mellon University, and Professors Paul Siegel and Jack Wolf at UCSD for their collaboration with InPhase funded by an Advanced Technology Program by National Institute of Standards. We also wish to thank Marcus Marrow, Martin Pane, Srini Gopalaswamy and Li Ni.

References

1. M. R. Ayres, Signal modulation for holographic memories, PhD Thesis, University of Colorado, Boulder, Colorado (2007).
2. M. R. Ayres and K. Curtis, Homodyne detection of holographic pages, *J. Meeting of the Opt. Data Storage and Int. Symp. on Opt. Memories*, Waikoloa, Hawaii (2008), paper TuB4.
3. R. A. Fessenden,Wireless signaling, US Patent 706,740, August 12, 1902.
4. R. M. Gagliardi and S. Karp, *Optical Communications*, 2nd edn. Wiley-Interscience, New York, 1995.
5. Y. Yamamoto, Receiver performance evaluation of various digital optical modulation-demodulation systems in the 0.5–10 μm wavelength region, *IEEE J. Quant. Elec.*, Vol. 16, pp. 1251–1259 (1980).
6. F. Favre, L. Jeunhomme, I. Joindot, M. Monerie and J. C. Simon, Progress towards heterodyne-type single-mode fiber communication systems, *IEEE J. Quant. Elec.*, Vol. 17, pp. 897–906 (1981).
7. R. A. Linke, High-capacity coherent lightwave systems, *J. Lightwave Tech.*, Vol. 6, pp. 1750–1769 (1988).
8. M. C. Teich, Laser heterodyning, *J. Mod. Opt.*, Vol. 32, pp. 1015–1021 (1985).
9. M. Hara, *et al.*, Linear reproduction of a holographic storage channel using the coherent addition of the optical dc components, *Jpn. J. Appl. Phys.*, Vol. 47, pp. 5885–5890 (2008).
10. C. Gu, G. Sornat and J. Hong, Bit-error rate and statistics of complex amplitude noise in holographic data storage, *Opt. Lett.*, Vol. 21, pp. 1070–1072 (1996).
11. N. Konforti, E. Marom and S. -T. Wu, Phase-only modulation with twisted nematic liquid-crystal spatial light modulators, *Opt. Lett.*, Vol. 13, pp. 251–253 (1988).
12. K. Bauchert, S. Serati and A. Furman, Advances in liquid crystal spatial light modulators, *Proc. SPIE*, Vol. 4734, pp. 35–43 (2002).
13. J. I. Trisnadi, C. B. Carlisleand R. Monteverde, Overview and applications of Grating Light Valve™ based optical write engines for high-speed digital imaging, *Proc. SPIE*, Vol. 5348, pp. 52–64 (2004).
14. S. Nabavi, B. V. K. Vijaya Kumar and J.G. Zhu, Two-dimensional generalized partial response equalizer with conventional Viterbi detector for patterned media, *Int. Communications Conf. (ICC)*, Glasgow (2007), pp. 6249–6254.
15. T. Okumura, *et al.*, Two-dimensional PRML with constant-weight constraint for holographic data storage, *Int. Symp. on Opt. Memories*, Singapore (2007), paper We-J-P14.
16. J.-S. Lee, *et al.*, An efficient balanced code using Viterbi and section division for holographic data storage, *Int. Symp. on Opt. Memories*, Takamatsu, Japan (2006), paper Th-I19.
17. M. Marrow, Detection and modeling of 2-dimensional signals, PhD Thesis, University of California, San Diego, California (2004).
18. S. Nabavi and B. V. K. Vijaya Kumar, Iterative decision feedback equalizer detector for holographic data storage systems, *Proc. SPIE*, Vol. 6282, pp. OT1–OT8 (2006).
19. L. Kazovsky, S. Bendetto and A. Willner, Optical Fiber Communication Systems, Artech House Inc., (1996).

13

Writing Strategies and Disk Formatting

Kevin Curtis, Edeline Fotheringham and Paul Smith

13.1 Introduction

Every storage technology uses a number of techniques for improving the recorded signal quality or rate of recording. Shaping the pulses, run length limiting modulation codes are examples of these techniques used in surface recording to improve the recording process for better recovery of the information. These techniques are called writing strategies and they typically are primarily determined by the media properties. In holographic recording, the media requires a number of writing strategies for recording to the media adequately. In addition, how to record a disk (formatting) requires dealing with the media properties successfully. This is particularly true when the holograms partially overlap in the media. The key writing concept for holographic recording is using the media volume as uniformly as possible. This means conditioning the light to be more uniform as well as using a recording process that uses the media spatially uniformly. The simplest example is changing the writing exposure energy per hologram (scheduling) to yield holograms of equal diffraction efficiency or ideally equal signal to noise ratio (SNR). Formatting the disk is complicated by media time limitations on how long the media can sit after pre-curing the media or after writing the media. In addition after post-curing the media, the edges of the cure beam are not suitable for recording and some space and time constraints must be applied for the next write session. These cure neighbor rules are very important to successful formatting of holographic media. This chapter deals with the strategies for writing and formatting a holographic disk.

Holographic Data Storage: From Theory to Practical Systems Kevin Curtis, Lisa Dhar, Adrian Hill, William Wilson and Mark Ayres

13.2 Media Consumption

13.2.1 Introduction

Storing holographic information at high bit density requires that a recording architecture be optimized in order to take full advantage of the limited media dynamic range. It is important to maximize the volumetric bit density (number of bits per unit volume). When the SLM dimensions are fixed, maximizing the volumetric bit density is done by minimizing the volume exposed by a page. This can be achieved by increasing the numerical aperture (NA) of the data beam's focusing (storage) lens and also by centering the Fourier plane in the middle of the media. Because phase conjugation is used on read-out, a high NA of 0.65 can be used. This is a higher NA than what a standard imaging system can easily achieve. Focusing into the media is complicated by the fact that the Fourier plane contains a DC signal that is as much as 10^5–10^6 times greater in intensity than the data. The media cannot record such a huge dynamic range signal, so techniques have been developed to apply a random phase function to the data with a phase mask in such a way that the DC signal is dramatically decreased. In this section, we present a technique that utilizes a moving non-pixel matched phase mask that allows us to focus the data beam into the media and therefore maximize the bit density while minimizing the amount of required media dynamic range or cumulative index change (*M*/#). In addition, the high NA storage lens is a special design to reduce the hologram volume.

While reducing the volume is a critical concept to minimizing the media's *M*/# usage, more complex writing strategies are also required. Skip sorting of holograms within a track is used due to the nonlinear scheduling and shrinkage effects of partially overlapping books. This concept can be extended to skip sorting of tracks in the future. Short stack recording fills the volume of the media more uniformly inside an individual book. These writing strategies are explained in this section.

13.2.2 Minimizing the Hologram Size

Minimizing the beam waist is an important criterion to minimizing the volume of the hologram and therefore making the best use of the media. The beam waist is approximately a rectangle with dimensions (w,d) as shown in Figure 13.1. The dark area in the figure is the recording layer and the dark thick lines are the substrate surfaces. The holograms are aligned with the beam waist in the center of the media. The width of the hologram or book of holograms in the direction orthogonal to the multiplexing direction is given by $w = \gamma\lambda f/\Delta$. In this equation, λ is the wavelength of light, γ is the linear dimension of the polytopic filter normalized to the Nyquist length (i.e. $\gamma = 1$ is a Nyquist filter), f is the focal length of the storage lens, and Δ is the pixel pitch of the SLM. In the direction of multiplexing the size is

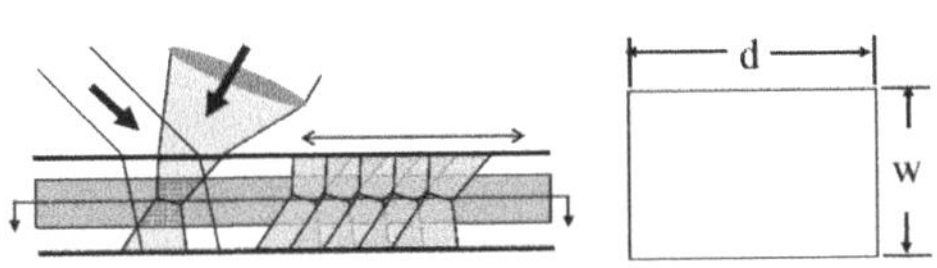

Figure 13.1 Diagram of hologram size with signal and reference beam intersecting at the center of the media

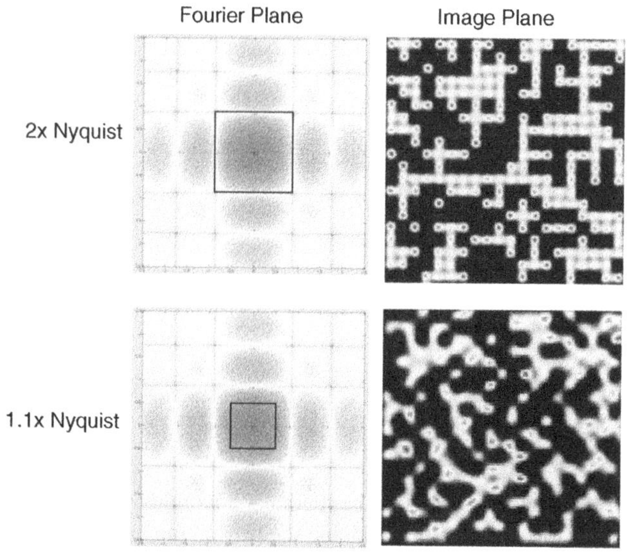

Figure 13.2 *Fourier plane filter and corresponding image*

given by $d = w\cos(\theta) + w\sin(\theta)\tan(\theta + \alpha)$. In this equation, θ is the angle of the signal beam and $\alpha = \sin^{-1}(\text{NA})$, where NA is the numerical aperture used by the signal beam.

The important factors for minimizing the beam waist are: a high NA lens (appears directly in d but also related to f/Δ in expression for w), the Nyquist factor, the wavelength of light, and the angular range used. In practice, it is possible to slightly under-fill the books with the reference, to further decrease the size of the book beyond the (w,d) given.

The Nyquist factor is an obvious place to look for making the book smaller. However, the smaller the polytopic filter the more blurry the image will be and therefore the more inter-symbol interference will be present. Figure 13.2 shows the Fourier transform (FT) plane with the black box the size of the polytopic filter and the corresponding image on the camera (image plane). The top of Figure 13.2 shows the case for a 2x Nyquist filter, and the bottom shows the case for a 1.1x Nyquist filter. Notice the blur in the 1.1x Nyquist case. However, as discussed in Chapter 11, most of this SNR loss can be regained by equalization. The polytopic filter size also relates to tolerances on interchange and alignment. The trade-off to find this best size filter has to consider both effects [1,2]. For these reasons, currently we use 1.3x Nyquist area or a $\gamma = 1.14$.

13.2.3 FT Lens Design

The arguments above are for a perfect lens. However, aberrations can also limit the size of the beam. The storage lens in the professional drive is an isoplanatic high NA lens. It is approximately 0.65 NA with a focal length of 12.395 mm. The isoplanatic lens allows for large assembly tolerances as well as large media alignment tolerances in radial and focus as explained in Chapter 4. The lens consists of three singlet and one doublet lenses all with spherical surfaces. Since the data beam's optical axis intersects the media at an angle of 25°, normally there would be a lot of aberration causing the hologram or FT plane to enlarge. To prevent this effect, one of the elements (the one closest to the media) is intentionally

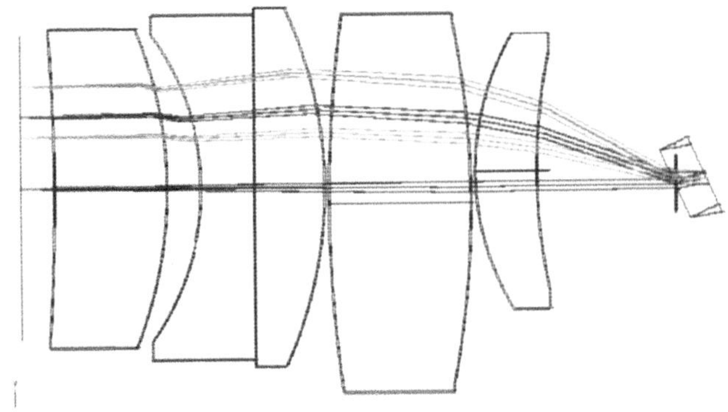

Figure 13.3 *Storage lens with decentered element*

de-centered to compensate for entering the media at an angle as shown in Figure 13.3. By eliminating the aberrations of the beam inside the media, the holograms have the smallest volume in the media. By making the volume smaller, more efficient use of the media *M/#* is achieved. In addition, tighter track spacing can be used without overlapping the beams in the radial direction. By minimizing the aberrations in the media, other localized high intensity areas in the volume of the media are minimized. Any high intensity area (not just the DC spot in the FT plane) can cause media distortion or local saturation and lower the SNR of the holograms. Figure 13.4 shows a plot of the data beam intensity at a location 575 μm into the recording layer. Both plots use the same scale for intensity. The lens with the offset element shown in Figure 13.4(b) has a much smaller footprint. In addition, the central intensity point in the standard lenses is actually much brighter (more of a hot spot) than the central point of

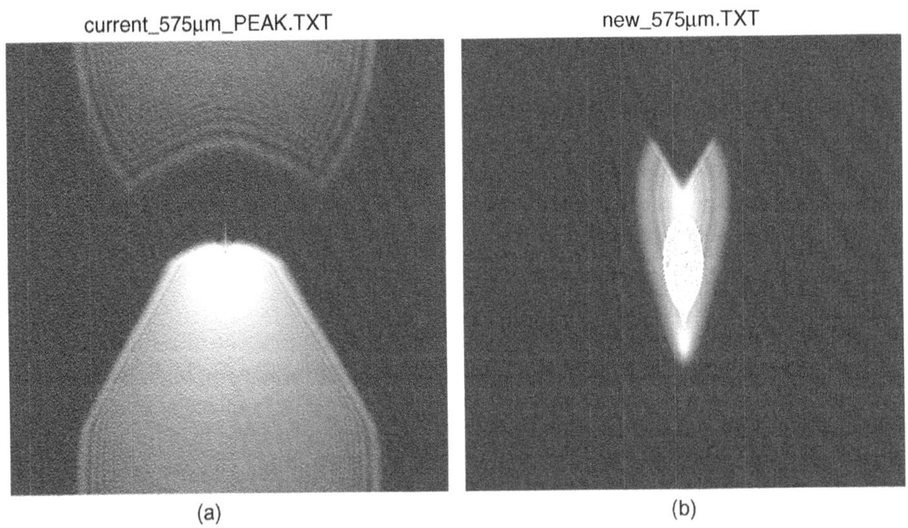

Figure 13.4 *Data beam footprint 575 μm in the media: (a) standard FT; (b) FT with offset lens*

intensity with the offset lens element. The size and uniformity of the data beam using the lens with the offset element is smaller and better for all planes inside the media due to lack of aberrations.

13.2.4 Phase Mask

The use of phase masks or phase plates to make the FT plane more homogeneous and therefore better for recording is well known in holography and goes back to the late 1960s. There are a large number of excellent references on this subject including the use of both digital (discrete levels) and analog phase masks [3–23]. By making the beam more spatially uniform, the average modulation depth is better for all components and the DC component does not locally saturate the media. Saturation can cause serious distortions of the media and significantly hurt the SNR. While a phase mask decreases the effect of the DC signal, there can be residual DC or other intensity nonuniformities at or near the FT plane that lead to a nonuniform media usage. Ideally the phase structure would be incorporated in the SLM [24]. Imaging a pixel matched phase mask is another potential solution, however, this requires positional manufacturing tolerances of about $\pm 0.5\ \mu m$ to match the phase mask to the SLM, and this is unattainable in a commercial product.

InPhase has explored analog phase masks at great length. These masks can be imaged onto the SLM without the need to pixel match. However, these phase masks force us to balance between DC suppression and filtering loss through the polytopic filter. This is because a more aggressive phase mask spreads the FT plane further which is then clipped by the polytopic filter. This can result in what we call rabbit tracks on the holograms from clipping at the polytopic filter (Figure 13.5). Rabbit tracks can also be caused by moving the phase mask too fast during recording.

Figure 13.6 shows the experimental set-up used to study phase masks. The phase mask is set on a linear stage. The phase mask could be held stationary, moved per exposure or moved

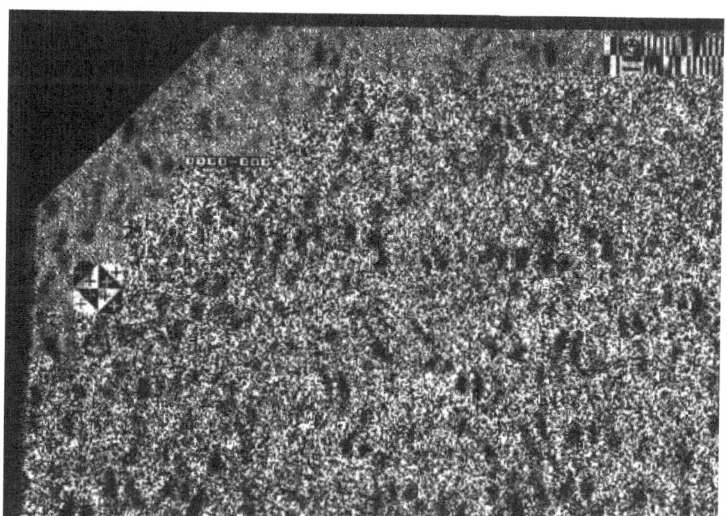

Figure 13.5 Rabbit tracks on page

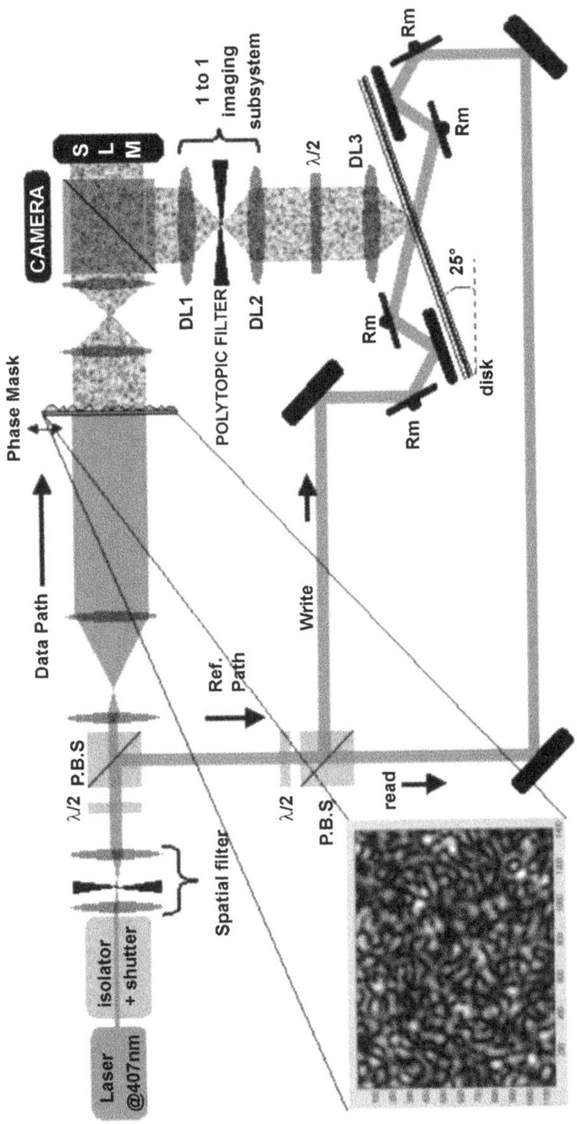

Figure 13.6 Experimental set-up for studying phase masks

continuously to evaluate the effects of the motion. The recovered holograms were read out (using phase conjugate) back to the camera. This set-up used a polytopic filter with an area of 1.2x Nyquist.

The phase mask was designed using several steps. First, a very uniform random phase field must be made as the starting point. Then, a perturbation to this random field is made. Using this new field, the suppression in the FT plane is calculated, and the SNR of the image field after filtering with the polytopic filter is calculated. If both suppression and SNR are improved then perturbation is kept as the new random field. This process is repeated until the best combination of suppression and SNR is found. After this process is complete, the field is checked for phase discontinuities. If any discontinuities are found, a phase unwrapping algorithm [25] is used to smooth the phase at that location. Many phase masks with different phase depths and phase feature sizes were designed, fabricated, and their performance and alignment tolerances experimentally measured. The combination of ~225 μm feature size with a maximum phase depth of 5 waves at 405 nm was found to have acceptable performance and tolerances. These phase masks are very smooth as shown in Figure 13.7. The phase mask results in about 40 dB of DC peak suppression with minimal increase in FT size.

If a stationary phase mask was used to multiplex many holograms in one location, no matter what phase mask was used the SNR of the book would decrease as more holograms were added. Any residual DC signal or local hot spot can cause a nonlinear media response that can lead to a low-pass filtering effect in the FT plane and therefore a high frequency enhancement in the image plane. Figure 13.8 shows actual data from a data page that shows the noise/distortion from a stationary phase mask when multiplexing a lot of holograms in one location. This nonlinear effect limits the number of pages in a book unless the phase

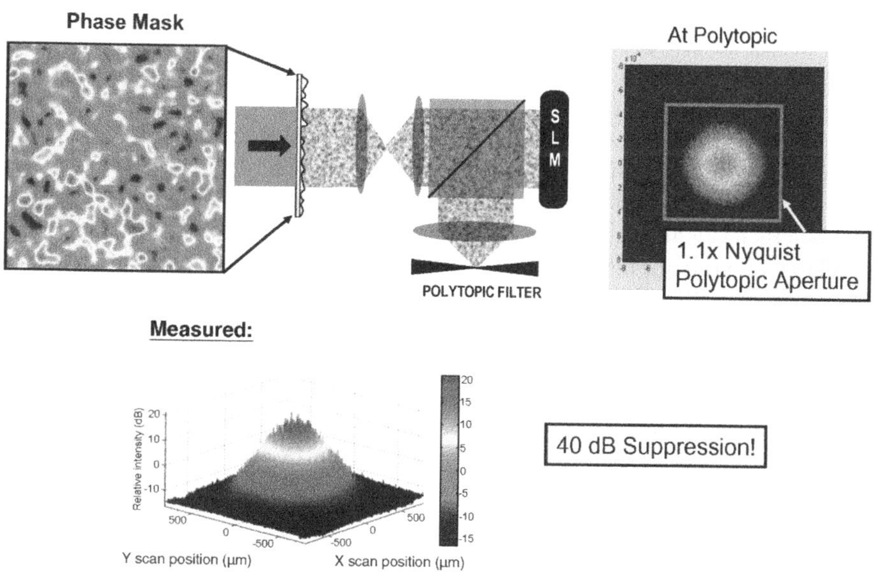

Figure 13.7 Phase mask imaged onto SLM and then passed though polytopic filter. The suppression of the phase mask in the FT plane and the intensity distribution is also shown

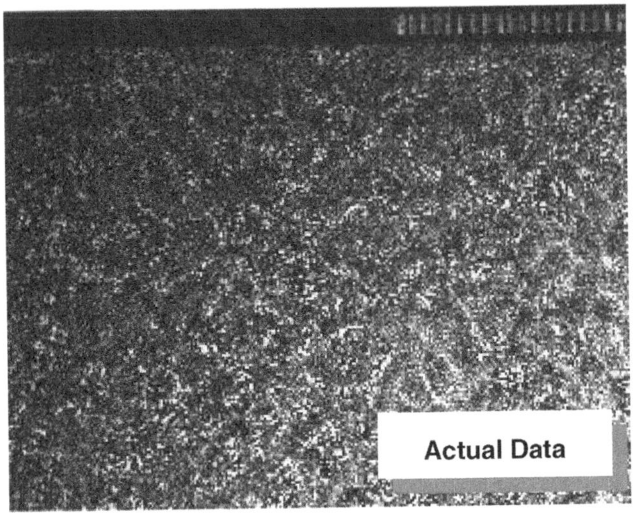

Figure 13.8 Data showing high frequency noise/distortion in image

function is changed. When the phase mask is moved during recording, thereby changing the phase function, the media nonlinearities are not allowed to build-up from book to book and therefore mitigate any high frequency enhancement [26]. Thus, with a moving phase mask, more pages can be added to a book without changing the SNR as long as the spacing between pages is kept the same.

The phase mask is moved linearly while recording. The continual motion cannot be too fast or the hologram diffraction efficiency is decreased and/or rabbit tracks appear. It cannot be moved too slowly or the high frequency noise starts to occur with large books. The speed range depends on how fast you are recording and the length of the exposures. The phase mask described above needs to be moved at least 5 µm per hologram (exposure time plus galvo move/servo overhead time) for good holograms to be recorded. This phase mask must also be moved slower than 20 µm per hologram exposure time in order to not lower diffraction efficiency and start to see rabbit tracks.

In addition to speed, the direction of motion can be important. If the phase mask is moved at just the right speed in the same direction as the reference beam is scanned to multiplex holograms, an interesting phenomenon called 'dark band' can occur. In this case, regions of a page are out of phase with the first sinc sidelobe of the same region of an adjacent page. The destructive interference on read-out of the sidelobe of the adjacent page and the desired page shows up as a dark band in the page. This dark band extends up to one-fifth of the page on the side of the page that is closest to the reference beam (side that has the least Bragg selectivity).

13.2.5 Short Stacking

When looking in detail at the intensity profiles of the light in the media in polytopically multiplexed books, there is a nonuniform usage of the media. Figure 13.9 shows this

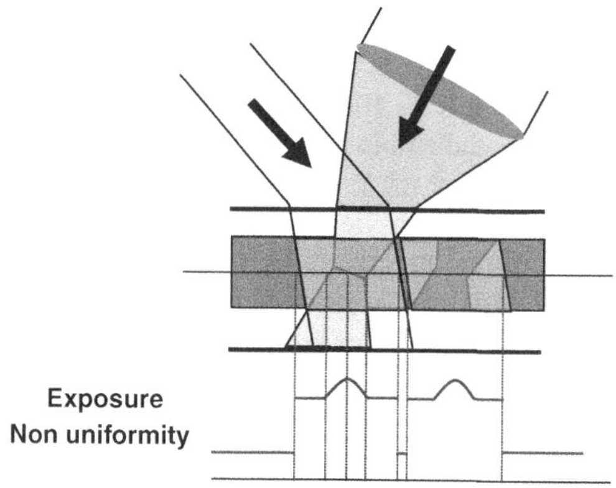

Figure 13.9 Illumination uniformity in polytopically multiplexed books

nonuniform media usage because the intensity is greater where the signal and reference overlap. This nonuniformity can also be seen from book to book when there is a partial overlap of a page or book onto the media with different exposure history. This wasted media but can also build up index changes that distort the holograms and can lower SNR. Short stacking is a technique that utilizes small media movements within a book so that the index variations within a book do not build up and cause unwanted noise. Also, one of the biggest benefits is that it improves the efficiency of *M/#* usage compared with writing all the holograms in the same spot. We get a boost in diffraction efficiency from the extra *M/#* in every short stack.

The idea is to break the book up into sections called stacks. Figure 13.10 shows the concept of short stacking. In this example, instead of multiplexing the entire 340 pages of the

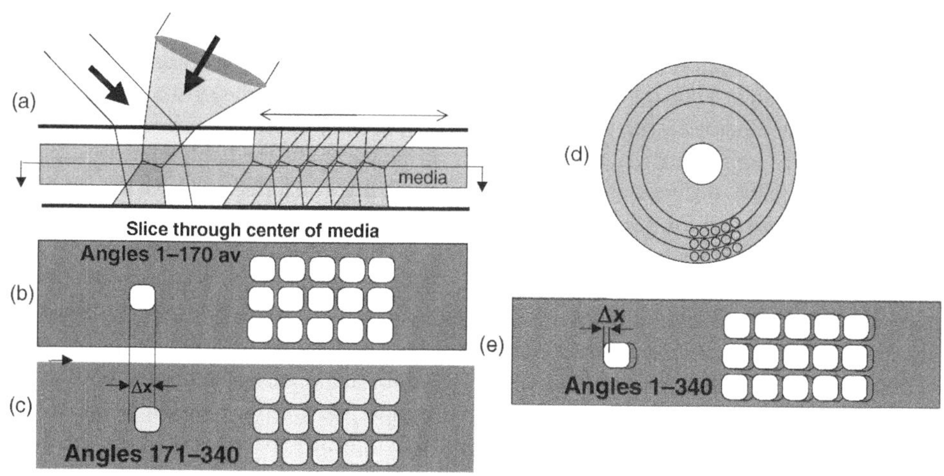

Figure 13.10 (a) Books being recorded polytopically. (b) The first 170 holograms are stored in one location. (c) The rest of the book (pages 171–340) is recorded with a small in track shift. (d) Three tracks. (e) The total book with two short stacks looking down on the media

book in exactly the same place, the pages are broken into two groups. Figure 13.10(b) shows the first short stack for the vantage point of looking down on the media. This short stack is the first 170 pages of the book. In Figure 13.10(c), the media is shifted by Δx, and the rest of the pages are recorded. This is repeated within the tracks. Figure 13.10(d) shows three tracks on the disk with each book having short stacks where the short stacks are shifts along the track dimension. Figure 13.10(e) presents a view of the entire book showing the short stack displacement. This displacement effectively blurs out the intensity profiles shown in Figure 13.9 to make the resulting profile smoother and more uniform. By smoothing out the exposure profiles, the media's bulk index build-up is smoothed out as well. The direction of the shift matters when the data beam does not have uniform intensities at every depth of the media. If one side of the beam tends to be hotter than the other, it is advantageous for *M*/# usage to give it the fresher media with the short stack shift. Typically the short stack displacement (Δx) has to be at least a few micrometers to be effective. Current formats use short stack moves of 215 μm with book pitches of 860 μm. Shifts can be used to design more uniform exposure profiles in the media.

The largest drawback to short stacks is transfer rate. Since disk theta moves are slow compared with galvo moves, transfer rate is improved by having books/stacks as large as possible with minimal number of disk moves. Thus, a compromise between best recording in terms of SNR and transfer rate requirements must be reached.

13.2.6 Skip Sorted Recording Within and Between Tracks

Using the media uniformly requires that tracks/books that overlap are not put down sequentially according to radial/theta address. If you record a book, and move over a polytopic distance and try to record another book, the second book's SNR will be compromised. This is due to two factors. The second book would spatially be in two types of media. One half of the book would be in unused, virgin media, and the other half would be in media that already has had many holograms recorded in it. There are two differences in unused media and media that have many holograms recorded in it – roughly half way up the index change versus exposure energy curve. First, the curve is nonlinear since it describes a media that saturates. Thus the media sensitivities are different in these two regions. The second difference between the two portions of the partially overlapping book is that the virgin media has approximately twice as much shrinkage as the media that is half way to saturation. These two differences result in holograms that are distorted and have lower SNR. Media distortions result in lower SNR as well as peak shapes (angular selectivity curves) that are wider or misshapen.

This phenomenon can occur anywhere where books are partially overlapped. Partial overlap of books can occur along the track or across tracks. Currently we do not overlap across tracks but in the future the track pitch could be reduced. To handle the partial overlap of book along a track, the books are recorded in a fashion analogous to building a brick wall, lying down one brick at a time layer by layer. For example, the first two books are recorded in virgin media separated by the book size. Then the polytopically spaced book is recorded on top of these two books with the center of this book midway between the centers of the previous books. After this the next book in the track is recorded in unused media, and then the polytopic book is recorded. In this way, the polytopically multiplexed holograms are always written in media with a uniform amount of usage. The books that are recorded in

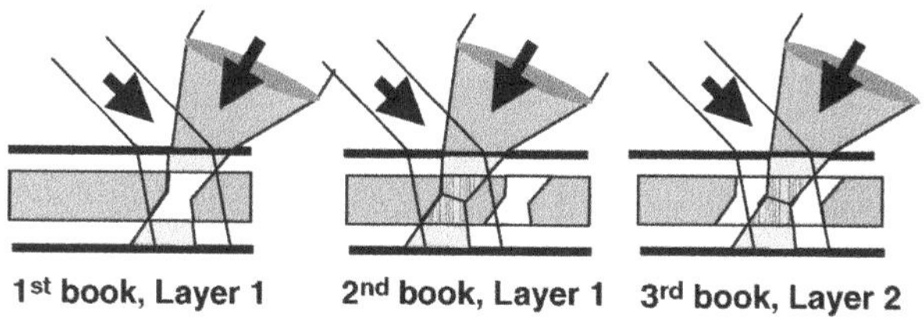

Figure 13.11 *Illustration of skip sorted recording inside a track*

unused material are called layer 1 books. The polytopic spaced books on top of the layer 1 books are called layer 2 books. This allows for all the layer 1 books to have the same schedule and all the layer 2 books to have the same schedule albeit a different schedule for each layer. Figure 13.11 illustrates this kind of recording. Currently we record all layer 1 books in a bookcase (bookcases include at least one cure beam location but can be multiple cure beams) before we record the layer 2 books in that cure beam. This method allows us to take full advantage of the media's diffusion properties for growing the diffraction efficiency of holograms in the dark. See Section 13.4.2 for details on bookcases. This is skip sorted recording [27] inside a track.

Skip sorted recording is also used in data track layout. While data tracks are not overlapping in the radial dimension, shrinkage due to one track extends beyond the physical edge of the track and into the neighboring data track. Any stack that is written in a shrinkage gradient and then read without the same shrinkage gradient will experience an SNR loss. Since the media will ultimately be uniformly (and completely) shrunk in the post-cure process, it is necessary to write all stacks in uniformly shrunk media as well. Figure 13.12 shows the writing sequence of seven data tracks within a larger cure track, and Figure 13.13 shows the shrinkage sequence of the first three data tracks written. The first track is written in unshrunk media, which causes the neighboring media to be shrunk by a small amount with a gradient across its width. The second track is put down beyond the edge of this shrinkage zone, so it is also in unshrunk media. However, the second track puts another shrinkage gradient across the unused media between the two tracks opposite the first track's shrinkage gradient, so the superposition of the two gradients results again in flat media. It is now slightly shrunk uniformly, and the third track may be written without penalty.

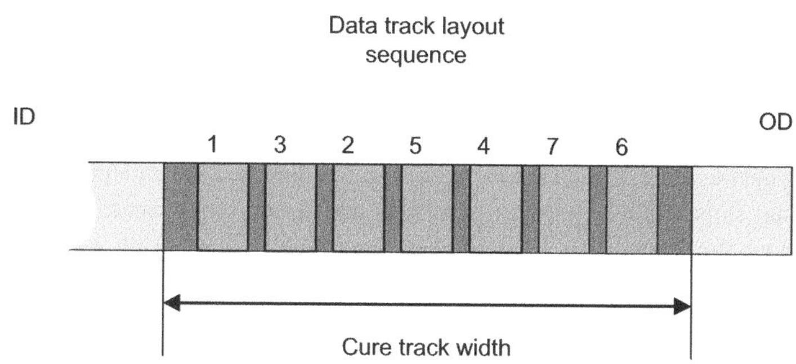

Figure 13.12 *Full data track write sequence, highlighting the skip sort technique. ID, inner diameter; OD, outer diameter*

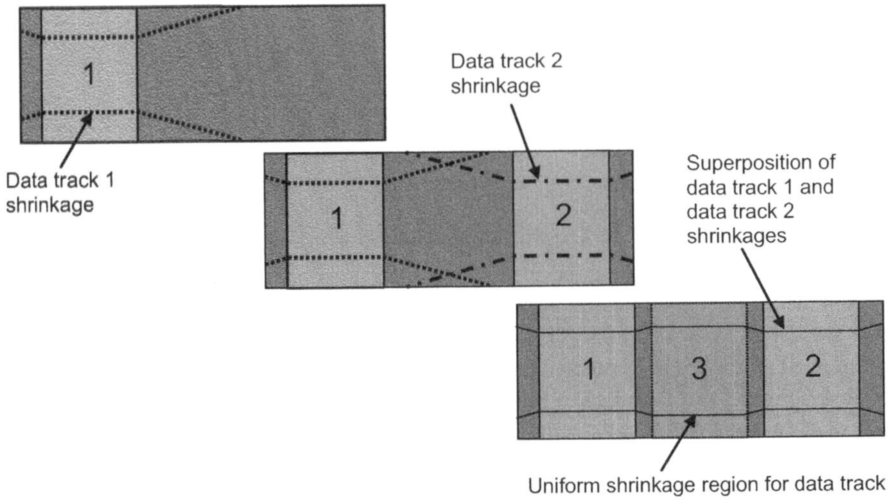

Figure 13.13 *Shrinkage sequence of first three data tracks written within a cure track. Skip sorting of tracks is done in the radial direction (the long axis of the dark rectangle is in the radial direction)*

13.2.7 Angular Scheduling of Holograms in a Book

As discussed in other chapters, InPhase currently records all the pages in a book with the same spacing. However, the actual angular selectivity is a function of the reference beam angle. Currently, a very conservative equal spacing for holograms in the second 'null' of the sinc function for the worst reference beam angle is used. By adjusting the spacing of the holograms with reference beam angle (angular scheduling) the book can hold significantly more pages [28]. Most capacity calculations in the 1990s for angle multiplexing already assume this angular spacing [29,30]. Figure 13.14 shows the Bragg selectivity (first null) of the data pages in the InPhase drive as a function of reference beam angle. Instead of spacing all the holograms at 0.07° (more than second null for worst case reference angle), the correct spacing for that particular reference angle should be used. The second null is a very good spacing for SNR but has to be adjusted across the reference angles. While putting all the holograms in the first null of the angular selectivity function would lower the SNR considerably, it is possible to put some of the data pages in the first null. For example, the spacing between the first hologram and the second could be the first null spacing but the spacing between the second the third hologram could be a second null spacing to reduce the additional noise. This alternating of first and second null spacing could be continued throughout the book. This technique has tentatively been done with little loss of SNR. Therefore, some fraction of the holograms can be angularly scheduled closer than the second null to further increase density.

13.2.8 Angular Fractional Page Interleaving

The Bragg selectivity of a high NA data page is largely determined by the side of the page that is closest to the reference beam because this side has the worst selectivity. Figure 13.15

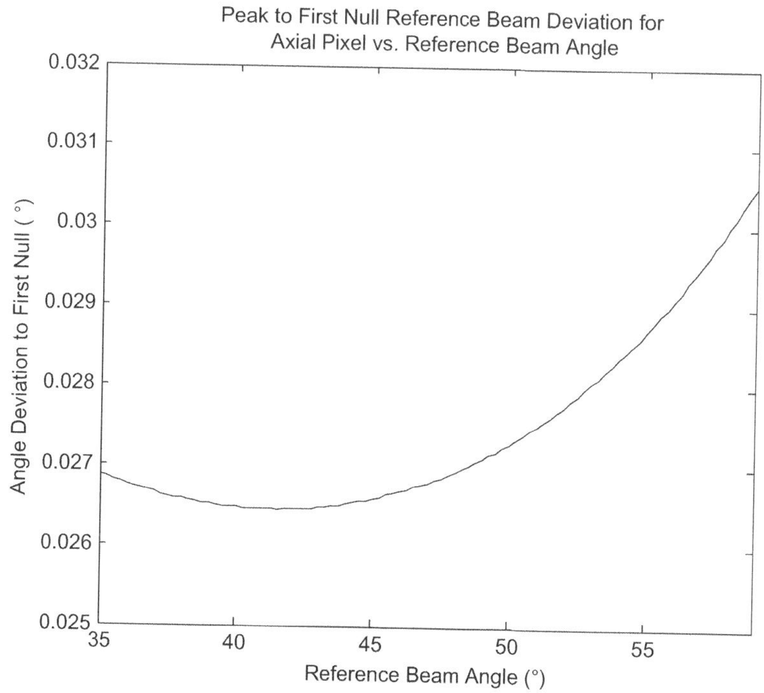

Figure 13.14 Bragg selectivity versus reference beam angle

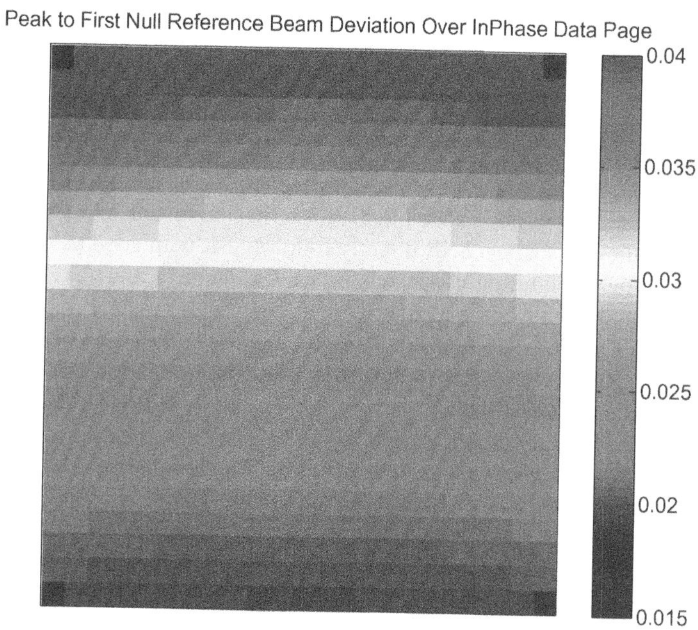

Figure 13.15 Plot of Bragg selectivity for across a data page. The top of the page corresponds to angles closest to the reference beam

plots the Bragg selectivity of a hologram stored at a reference angle of 35° as a function of pixel location across the page. While entire data pages can be scheduled to put them at the overall second null, this is not the best that can be done. The concept of angular page interleaving is to place half pages or fractional data pages in between full data pages scheduled at the second null. The empty or unused part of the fractional data pages has to be the part of the page that has the worst selectivity. The rest of the page has high selectivity and can be accessed by interleaving these fractional pages. As Figure 13.15 shows, the Bragg selectivity varies by a factor of 3 across the page. The top half of the page has twice the selectivity width as the edge of the page. This technique is not currently in our technology roadmap as given in Chapter 3 but it can help extend the roadmap in the future.

13.3 Scheduling and Write Pre-compensation

13.3.1 Introduction

The need to adjust the energy of each hologram in order to get equal diffraction efficiency in photopolymers or other materials has been well known for about two decades [31–34]. This process is called scheduling holograms. The entire write process includes pre-cure, recording holograms, and post-cure. For photopolymers, the media sensitivity is a significant function of media temperature. Fortunately, the *M*/# does not seem to change with temperature, just the sensitivity. In addition, to help compensate for thermal effects the writing should pre-compensate by changing the wavelength with respect to the media temperature. This change in wavelength also changes the media sensitivity as different wavelengths absorb more than others. This section covers scheduling, write pre-compensation, and the thermal effects pertinent to scheduling.

13.3.2 Basic Scheduling

Historically, scheduling has been done to achieve uniform diffraction efficiency. However, in a product the desire is for all the holograms to have the same SNR. This is approximately equivalent to having all the pages have the same signal to scatter ratio (SSR). Since some angles have more scatter or noise than others, this criterion results in a different schedule than equal diffraction efficiency. SNR directly follows SSR with good recording practices due to scatter being the limiting noise source (see Chapters 11 and 13). In order for the exposure schedule to remain consistent between drives, the drives have to be set so that the power at the media is consistent to about 5% and the power beam ratio between reference and signal has to remain constant to within 10%. The best beam ratio was experimentally determined by recording to full density with different beam ratios and looking at the average recording time needed to get the same SNR (Figure 13.16). As seen below, the best ratio is 4.5 to 1 reference beam's power to data beam's power. The optimal ratio highly depends on the intensity profiles of the beams throughout the media. The beam ratio needs to remain constant over wavelength and thermal changes in the drives which for InPhase requires careful polarization control to a tenth wave retardation. Overlap of the reference and data beams also needs to remain good over temperature to less than 100 μm in radial.

To get the basic recording schedule, iterative approaches are used as in Pu *et al.* [31]. Figure 13.17 shows a schedule derived for roughly equal SNR. The time line is the exposure

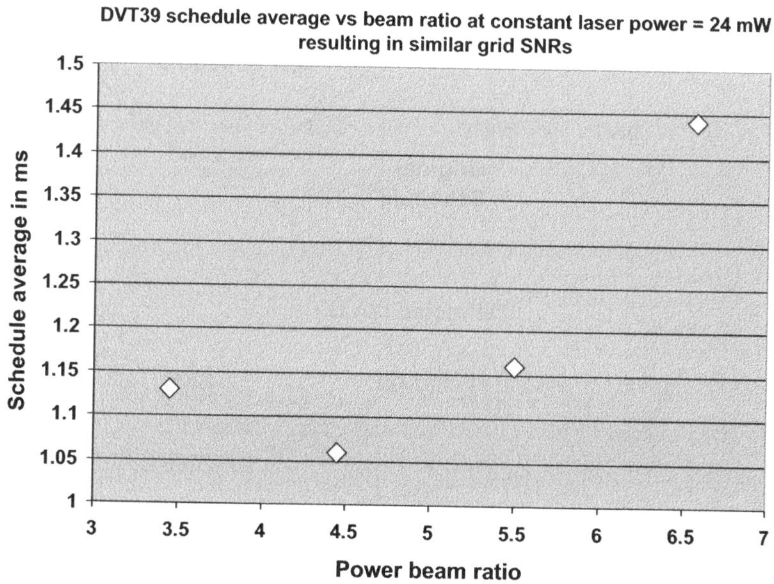

Figure 13.16 Beam ratio versus schedule average

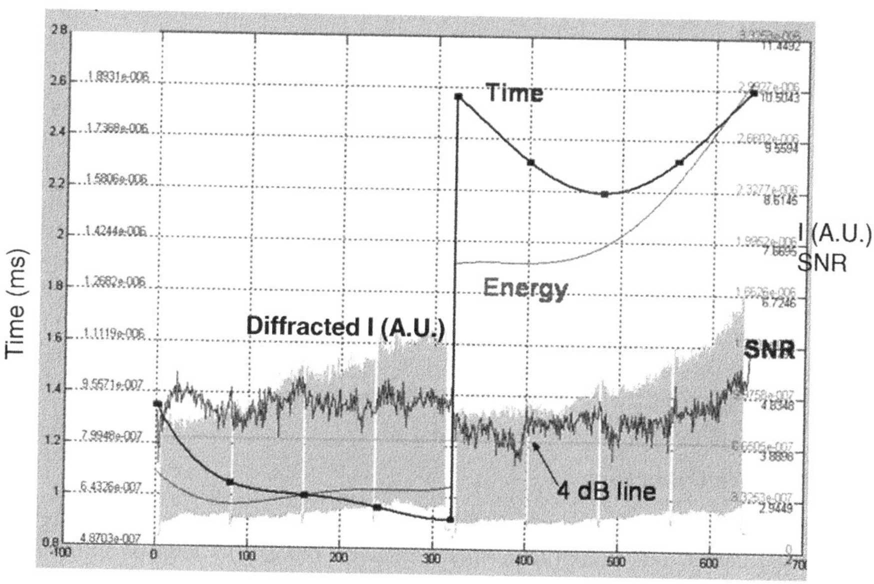

Figure 13.17 Exposure time (s), exposure energy (J), diffracted power (arbitrary units) and SNR (dB) versus hologram number. The horizontal axis shows the layer 1 holograms for numbers 1 to 320 and layer 2 holograms for numbers 321 to 640

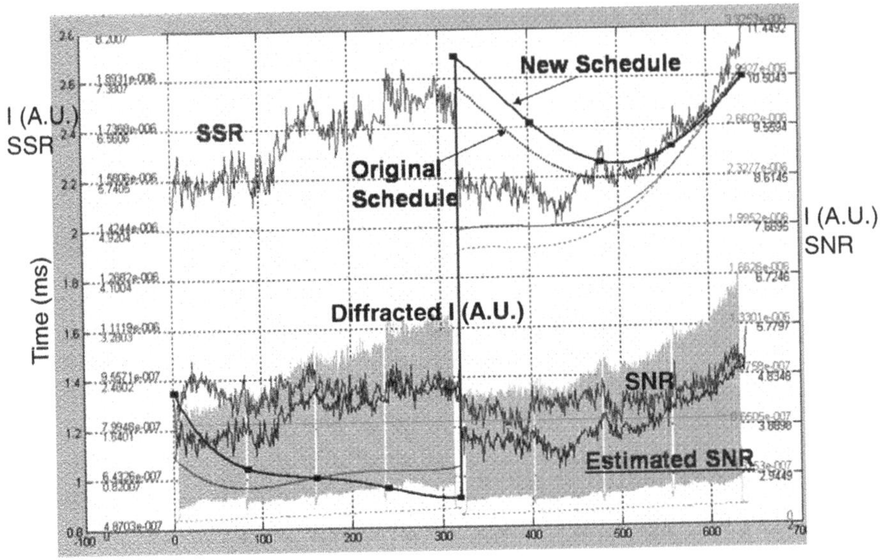

Figure 13.18 New schedule adjusted from the old one after looking at SSR and estimating SNR

time between 1 ms and 2.6 ms. The energy line is the corresponding energy at the media for each hologram. The schedule is shown for both layer 1 and layer 2 books. Due to angle of incidence and other issues, the amount of energy at the media varies with angle. The time schedule does try to deal with this. The top of the gray region is the diffracted power. The SNR line is the SNR of the pages varying between 3.5 dB and 5.5 dB. The SNR variation is larger than desired for this schedule.

Figure 13.18 shows the same graph as Figure 13.17 but the SSR and an estimate of the SNR for the holograms are plotted. The SSR was measured from the data and then used to estimate a SNR. The SSR was measured by using both the baseline scatter from outside the hologram angular area and an estimate of crosstalk from adjunct holograms. The SSR line is the SSR for each hologram which for these holograms is a bit low (~6.5–7). For the estimate of the hologram SNR, we use the SSR versus SNR graph (see Chapter 11). This graph was used to convert SSR into an estimate of the SNR of the holograms. This estimated SNR line is a little lower than the actual SNR line. Looking at the ratio of SNR/SSR, the schedule was adjusted to yield the new schedule line. This new schedule slightly improves the SNR uniformity by improving the SSR uniformity of the holograms.

13.3.3 Pre-cure Calibration

It is necessary to pre-cure every portion of the media before holograms can be written. The reason for pre-cure is that holographic media contains oxygen in addition to the photoinitiator and monomer. When a photoinitiator molecule is excited by a photon of light, it tends to bind to an oxygen molecule rather than starting polymerization. This unwanted reaction will continue to occur until all the oxygen molecules are used up, at which point the photoinitiator will polymerize the monomer and create the hologram.

Variations in the amount of energy necessary to completely bind up all the oxygen are dependent on several factors. If the amount of oxygen present in the media changes, then a different amount of energy will be necessary. Since the media substrates are partially permeable to an oxygen molecule, variations in altitude will change the amount of oxygen and therefore the precure energy. If the amount of photoinitiator changes in each batch of media, it will alter the required pre-cure energy. Also the photoinitiator absorption varies with wavelength. Wavelengths closer to 395 nm have 60% efficiency while 405 nm only has 40% efficiency. Thus changes in the pre-cure diode wavelength drive to drive can cause variations in energy required. In addition, if the diffusion rate of the molecules in the media changes, as with different temperatures or possibly at different humidity levels, it will also need a different amount of pre-cure energy.

The effect of not pre-curing the media enough is that each hologram will have less diffraction efficiency, since the energy is going to binding photoinitiator and oxygen rather than to the hologram. This is generally observed in the holograms that were written first in any location on the disk, since the amount of unbound oxygen decreases with each hologram that is written. The effect of pre-curing the media too much is that some of the energy designated for binding photoinitiator with oxygen goes instead to polymerizing the monomer. As a result, there is not as much monomer available to write holograms. Figure 13.19 shows the results of under pre-curing and using the correct pre-cure on the diffraction efficiency books of holograms. The exact tolerance in pre-cure energy is difficult to measure because the effect of over pre-cure is not easily quantifiable but is estimated to be around $\pm 10\%$.

Due to the difficulty in measuring the effect of over pre-cure, a pre-cure calibration algorithm can use the under pre-cure effect to measure the required precure time. First, a location on the disk is under precured with the cure beam, and a low-density stack of

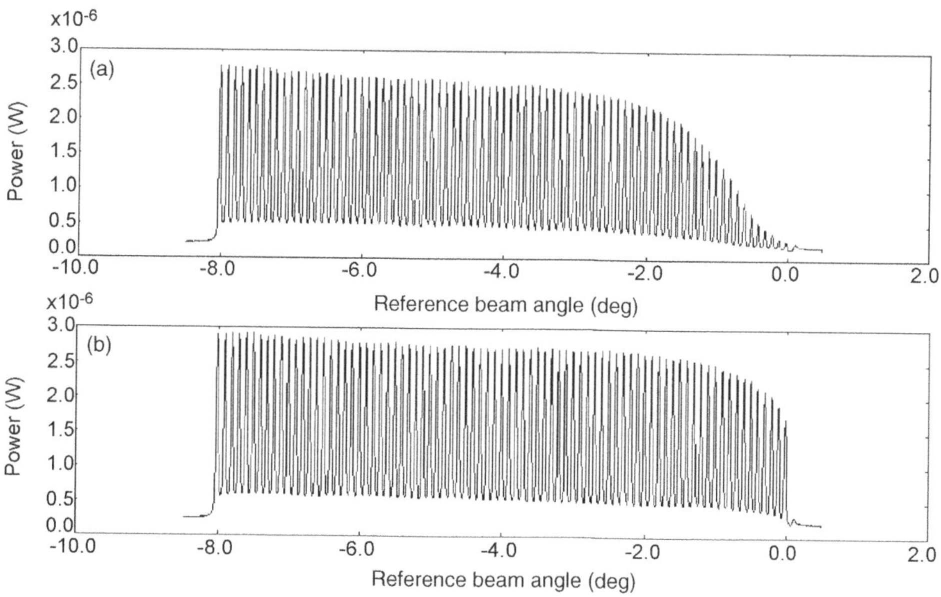

Figure 13.19 *Plots of diffraction efficiency versus angle for a stack of holograms with no pre-cure (a) and with roughly the correct pre-cure (b)*

holograms (200 holograms) is written slightly off center in the cure spot. Once the stack has been written, the same cure spot is pre-cured a second time, resulting in a slightly over-pre-cured location. A second stack of holograms is then written in the same cure spot. After both stacks have been written in the same cure spot, the spot is post-cured, fixing the holograms.

Since the information density is low, it should not be approaching the over-pre-cure limit with lower diffraction efficiency, even though it was purposefully over pre-cured. The difference between the diffraction efficiency of the second stack and the first stack can be used to determine exactly how much more pre-cure was needed in the first stack. If the peak diffraction efficiency plots are normalized and overlaid, shown in Figure 13.20, it is clear that they diverge from one another on the right side of the plot. The location of this divergence, called the pre-cure knee, is directly proportional to the amount of additional energy necessary to correctly pre-cure the stack. Also shown in Figure 13.20 is the difference in diffracted powers between the over- and under- pre-cured stacks. The derivative of the power difference (curve-fitted to remove high-frequency variations) is used to find the pre-cure knee. The pre-cure knee location is defined at the left-most negative slope crossing of −0.0045. In other words, the pre-cure knee is defined by the rate at which the two diffracted power curves diverge from one another. Once the rate of divergence is high enough, it is labeled as the pre-cure knee.

It is theoretically possible to calculate the total amount of energy passed to the media by all the holograms written before the pre-cure knee and convert that into the amount of additional time necessary to pre-cure the first stack of holograms. However, due to nonuniformities in the data and reference beams, and different amount of overlap between

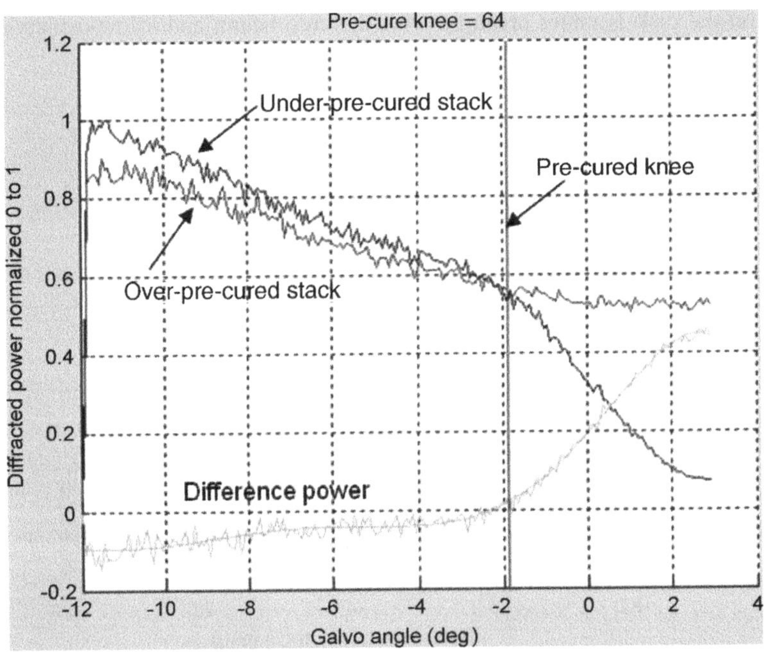

Figure 13.20 *Overlaid diffraction efficiencies demonstrate the location of the pre-cure knee, located at the point of divergence between the two plots*

systems, it is easier to experimentally determine the required additional pre-cure time. The measurement of amount of energy delivered to the media is done on a system on the build line by using different pre-cure times on many different stacks and measuring the location of the pre-cure knee for each one.

With the current 300 GB format, each calibration cure spot used will reduce the total capacity by an average of 605 MB based on the 300 GB format that will be used for product release. Even an unoptimized routine can calibrate the pre-cure in under 20 s.

13.3.4 Write Pre-compensation Process

Holographic gratings written in a polymer based medium are distorted due to temperature changes. The expansion or contraction is different in each dimension due to constraints placed on the media by the substrates. Normally, the distortion can be compensated for by tuning the wavelength of the read-out laser and slightly adjusting the position of the read-out angle. In order to accurately predict the proper read-out wavelength and galvo position, it is necessary to know the exact angle, wavelength, and temperature during the writing stage. Using a pre-compensation technique, it is possible to set the wavelength and writing angle for every hologram based on a chosen temperature so that the read-out angle and wavelength can be predicted without knowledge of the original temperature. In addition, by pre-compensating the wavelength of the recording for temperature changes, the supported operating temperature range of the drive is increased in comparison with the range that can be compensated with recording at a fixed laser wavelength and just compensating on read-out.

13.3.4.1 Wavelength Compensation

The procedure behind pre-compensating the wavelength due to temperature variation is autonomous from the rest of the recording process. In general terms, media temperature is measured and the wavelength is adjusted to an absolute value based on the measurement. Ideally, this wavelength adjustment should be made as often as the temperature changes. The media temperature is measured with an IR sensor that looks at the media location to be written. The difference between the temperatures measured by the sensor and the true temperature at the center of the media is relatively small and has some lag with temperature change.

The correction curve is defined by a center wavelength, a center temperature, and a slope of $-0.2\,\text{nm}\,°\text{C}^{-1}$. This slope is determined by the media properties. The center wavelength and temperature must be fixed for all drives. In the professional product, the wavelength range is 402–408 nm, and the temperature range is 10–40 °C. Therefore, the center wavelength and temperature should be chosen to be 405 nm and 25 °C. In reality, the working temperature of the drive is likely to be slightly warmer than the outside temperature, indicating a working temperature range closer to 15–45 °C with a center temperature of 30 °C. The change is such that increasing temperature needs decreasing wavelength. If the temperature is low during recording, the pre-compensation can use a higher wavelength and this actually increases the range that can be compensated versus recording at a fixed wavelength.

The reality of setting the wavelength due to a temperature shift is somewhat more difficult than the ideal model simply because setting the wavelength likely alters the single-mode operation of the external cavity laser. Therefore, each time the wavelength is significantly adjusted the laser single-mode search algorithm must also be run. However, adjusting the wavelength too much can slow the write process. In order to reduce the number of wavelength adjustments, the wavelength should be kept stable unless the temperature drifts by more than some prescribed amount. The wavelength sensor in the product is accurate to 100–150 pm, which implies that the wavelength should only be adjusted if the temperature changes by 0.8 °C. The specified maximum rate of temperature change is 10 °C h^{-1} or 0.167 °C min^{-1}. At this maximum change rate for temperature drift, the wavelength should only be set once every 4.8 min, corresponding to an insignificant impact on write transfer rate.

If the holograms were written with the proper wavelength pre-compensation, then no knowledge of the writing temperature is needed to predict the proper recovery wavelength. The recovery wavelength is determined from the recovery temperature in the exact same way as the writing wavelength versus temperature.

13.3.4.2 Angle Compensation

Holograms read with a different wavelength than they were written at will be found at different galvo angles than they were written at. The galvo angle pre- and post-compensation procedure is analogous to the wavelength compensation procedure. The difference is that every galvo angle requires a different adjustment line. To start, the center temperature is selected to be the same as the center temperature chosen for wavelength compensation. The center galvo angle at that center angle is chosen to be the uncompensated writing angle, and the slope of the line is given by:

$$\text{Galvo shift slope} = -2.13534 \times 10^{-5}x^2 + 0.000867897x - 0.0258147 \qquad (13.1)$$

where x is the uncompensated writing angle (the galvo angle in Figure 13.21). The galvo compensation assumes that the wavelength pre- and post-compensation is also always being used.

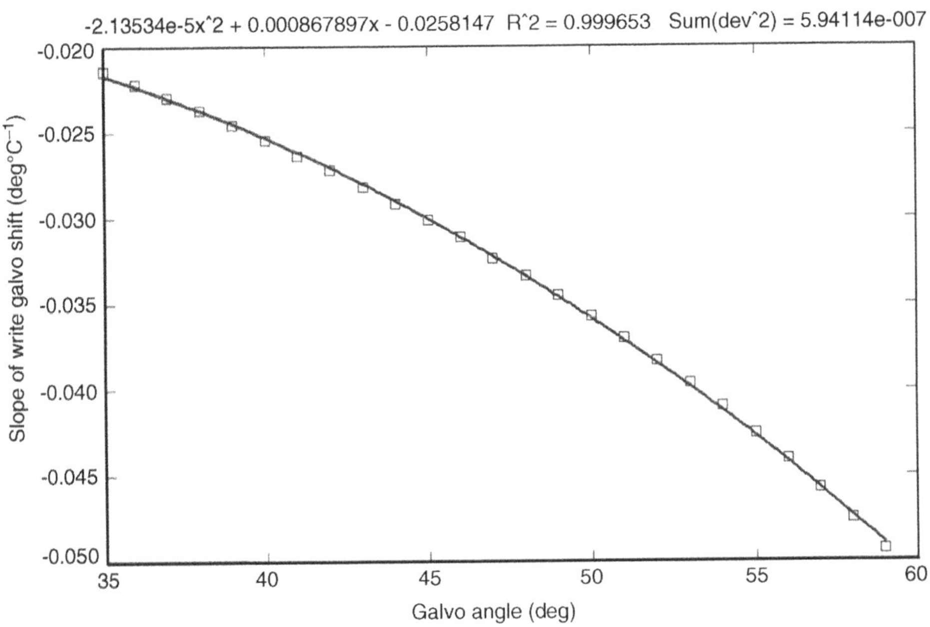

Figure 13.21 *Slope of galvo shift versus galvo location*

Table 13.1 Galvo parameters

A	-9.0392×10^{-12}
B	6.7262×10^{-11}
C	-6.4469×10^{-7}
D	3.0567×10^{-6}
E	-8.6791×10^{-3}
F	1.2904×10^{-1}

Like in the wavelength compensation case, the galvo angles are exactly the same for both writing and reading, and dependent only on the measured temperature. Unlike the wavelength compensation, inaccurate predictions in the read galvo position can highly affect the quality of the hologram. A galvo accuracy of 0.002° requires that the temperature change by only 0.04 °C. At the highest rate of temperature drift, this amount of change can occur in approximately 15 s. It is therefore reasonable to perform the galvo pre- and post-compensation on the fly for every book during both writing and recovery.

The galvo equation can be derived theoretically to a curve that matches:

$$\Delta\phi(\phi, \Delta T, \Delta\lambda) = \left(A\phi^2 + C\phi + E\right)\Delta T + \left(B\phi^2 + D\phi + F\right)\Delta\lambda \tag{13.2}$$

In this equation, the angles are in millidegrees, the wavelength is in picometers, and the temperature is in millidegrees centigrade. Equation (13.2) makes it clear that the angle change comes from mostly the wavelength change but also from the temperature change. The parameters for Equation (13.2) using InPhase's current media are given in Table 13.1.

13.3.5 Thermal Effects on Schedule

The media is a photopolymer that has diffusion as part of the recording mechanism. Diffusion is strongly dependent on temperature. In addition, pre-compensation requires the use of a different wavelength to record the data when at a different temperature. Currently the absorption of the media also varies significantly across the wavelengths used. These two reasons require the schedule to be adjusted or scaled with temperature changes. Figure 13.22 shows plots of measured pre-cure energy and sensitivity (energy to 80%). Both

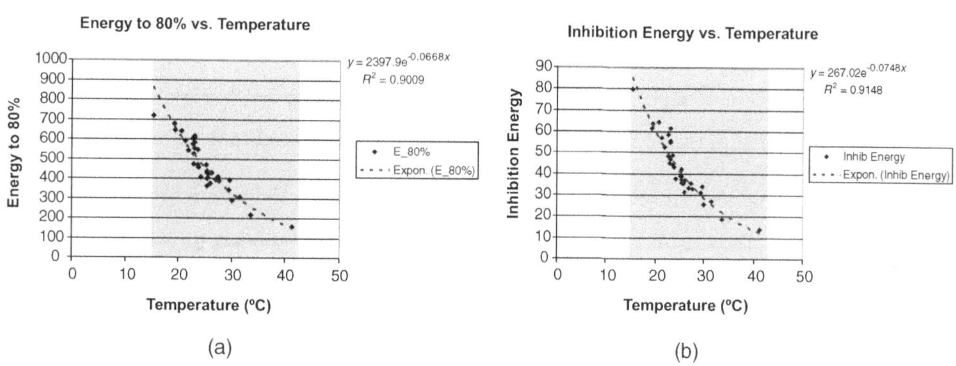

Figure 13.22 *(a) Energy to 80% usage of M/#. (b) Energy required for pre-cure of media versus temperature*

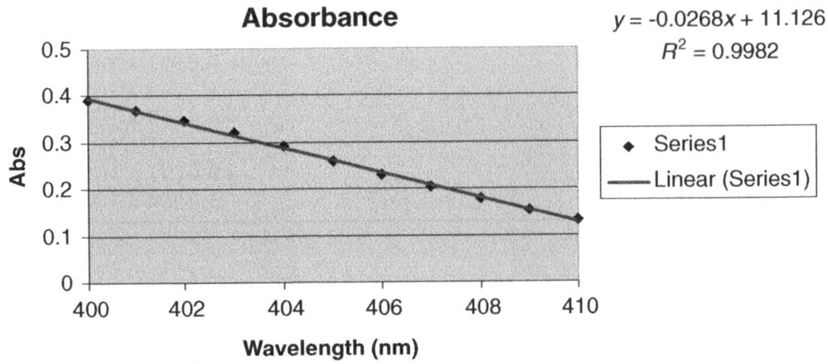

Figure 13.23 *Plot of media absorption versus wavelength*

parameters seem to scale the same versus temperature. Thus the media is much more sensitive at 45 °C than at 15 °C.

Figure 13.23 gives the media absorption versus wavelength change in the wavelength range of interest. This change is also significant over the wavelengths that the drive uses (402–408 nm). Figure 13.24 plots the average exposure time versus temperature considering both wavelength and temperature changes. This is the cumulative scaling of the schedule. The point at 29 °C is the data point for average schedule time and the rest of the curve was generated by scaling as per wavelength and temperature changes. The cure diode does not change wavelength significantly and therefore only media effects scale the post- and pre-cure times. In addition, Figure 13.24 is for constant laser power. The laser naturally puts out more power at cooler temperatures and less power at hotter temperatures and this curve does not take this into account. In addition, the

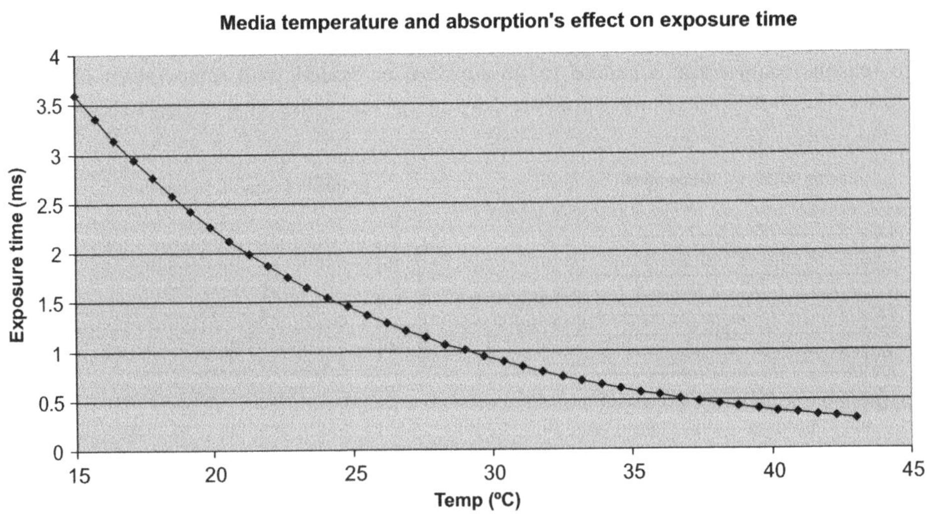

Figure 13.24 *Average exposure time versus temperature for both thermal and wavelength effects*

drive would adjust the laser power to further minimize the potential changes to the transfer rate.

As shown in Figure 13.22(b), the pre-cure energy must also be scaled with temperature. The pre-cure, schedule, and post-cure energies all scale with temperature. In addition to these energies, the media pre- and post-cure time-out limits will also vary with temperature. Fortunately the *M/#* does not vary with temperature so that the media can always support the same density regardless of temperature.

13.4 Media Formatting

13.4.1 Introduction

Media formatting encompasses all that relates to how the data are arranged or written to the storage medium. This section will focus on disks as the storage medium but the general concepts apply to other shapes such as cards. Media characteristics determine many aspects of how to layout the disk. For example, media properties influence the feasibility of multi-session or the ability to stop and restart recording, both key issues when formatting a disk. In addition to data zones, library zones (directory of where to find particular data) and calibration zones need some space on the disk too. Formatting a disk often trades off performance for ease of implementation. For initial products the ease of implementation is of more importance than maximizing the performance of the drive.

13.4.2 Considerations

With InPhase's particular system architecture and media, the data tracks are grouped into cure tracks (the latter as wide as the cure beam in the radial dimension). Cure neighbors are defined as the unit cure areas closest to a given cure location in the tangential and radial dimensions. Bookcases are a group of neighboring cure areas that has to be written all at once. Given this vocabulary, the major considerations for developing a format are: choosing the number of data tracks per cure tracks, defining rules for when sites may be pre- or post-cured, deciding the spacing of bookcases, and taking into account media time-outs (see Section 8.10.3).

Pre- and post-cure time-outs constitute the main media time-outs that affect the format. The first negatively affects the diffraction efficiency of a book's first pages and the second negatively affects the SNR of all the pages of a book. Therefore, recording in an area must take place within a limited time after pre-curing and post-curing must happen within a limited time after data recording. These time limits highly depend on media formulations with longer time-outs affording more format flexibility. Writing speed in conjunction with media time-outs determine how many cure sites may be pre-cured ahead of the actual recording or how many data get written before having to do a round of post-cure. More subtly the time-outs can also depend on the density of recording in some formulations.

Cure rules determine what cure sites can be or must be pre- or post-cured at any time. They are governed by the time-outs mentioned previously and on what book addresses are next in line for recording. We call one important subset of cure rules the 'cure neighbor rules'. The definition of neighboring cure sites may comprise only the closest neighbors in the tangential direction or it may also include the closest sites in the radial dimension.

Depending on this definition of neighbor, cure rules either concern only the cure sites in the current cure track (1D cure rules) or take into account sites in the cure tracks around the one been written in (2D cure rules). Pre- and post-cure may use different rules so that a site's pre-cure neighbors may differ from its post-cure neighbors. This allows more freedom in bookcase shape. The pre-cure neighbor rule enforces that all pre-cure neighbors to one cure site must be pre-cured before any data are written to the site. This guarantees that no book is recorded on only partially pre-cured media. The post-cure neighbor rule makes sure that all post-cure neighbor sites must have all their data written before a site can be post-cured. This rule guards against over pre-curing sites with scattered light from the post-cure of a neighboring site.

All the formats described below have the same pre- and post-cure neighbors. Bookcases are contained in a section of a single cure track for simplicity. Within these bookcases the right distance between cure locations provides uniform stitched curing. The first two format examples space out the bookcases such that 1D cure rules can apply. The third example uses 2D cure rules to pack the data on the disk as tightly as possible. With 1D cure rules, a bookcase is also the smallest write session. This offers the possibility to write one bookcase at a time. With 2D cure rules a writing session extends in both the tangential and radial dimensions offering the possibility of writing data tracks across cure tracks seamlessly. In this case the cure areas are uniformly stitched in radial as well. The 2D cure rules relieve the constraint of having an integer number of data tracks per cure track.

Cure track and data track pitch directly influence the critical capacity number: smaller pitches enable the format to hold more capacity. The SNR of books in data tracks put too close together suffer from extra noise. Data tracks too far apart may not yield the desired capacity but less obviously they can also result in poor archival performance (see Section 8.11). The cure system hardware determines the size of the cure beam at the media which in turn determines the absolute minimum cure track pitch. At this closest distance, two cure track edges overlap such that the intensity across the boundaries is uniform with the rest of the cure track. A format using 2D cure rules needs this 'stitched' spacing of cure tracks. Formats using 1D cure rules impose a different minimum cure track pitch. For those, the closest two cure tracks must accommodate the post-curing of one cure track without pre-curing any part of the other one. Format Examples 1 and 2 (see Section 13.4.3) show this extra spacing between cure tracks. Formats with 1D cure rules are easier to generate, easier to implement in a drive, and easier to work with.

The media's shrinkage affects format too by imposing the recording of dummy books. These books are written but never read. They serve to pre-shrink media so that the user data are recorded in optically flat media. Typically format dummy books appear in zones between fully shrunk and virgin media during the recording process (radial edges of bookcases). Holograms in these zones undergo distortion between the write and read stages, making them unsuitable for user data. Random data fills the dummy books but they are otherwise recorded like any other book. The less a media formulation shrinks, the less dummy books a full density format requires. Different layers of holograms do not necessarily need dummy books in the same locations.

Allowing multi-session recording adds another level of complexity to designing and implementing formats. Multi-session here means feeding the drive one set of data at a time with some unknown amount of time in between each set. Media properties still limit the overall time a user may take to fill up a disk. The easiest path to multi-session uses a format with 1D cure rules and predefined bookcase sizes. A data set is then recorded in however many bookcases it needs and any unused space in the last bookcase opened becomes wasted space. Dynamic allocation of bookcase size and associated dummy books could minimize such wasted space.

Finally, all formats need to put aside some small areas to calibrate writing parameters before recording data to a disk and to record a library map of the user data on the disk. Single cure sites generally provide enough space for these special tasks.

13.4.3 Format Types with Examples

The following examples demonstrate a number of the features described above. Format Examples 1 and 2 have linear cure rules but suppose different write speeds (Figures 13.25–13.27). The first format assumes 5 MB s^{-1} or higher writing speeds while the second works for 15 MB s^{-1} and higher. The higher recording speeds allow longer bookcases and so a fewer number of dummy books. This helps both capacity and transfer rate. The dark gray areas at the end of bookcases represent dummy books. The lighter gray areas indicate where the user data get written but the data tracks are not drawn. Instead the figure exposes the boundaries of unit cure areas. Format Examples 1 and 2 each have two cure sites put aside for recording library map information (dark gray squares) but no calibration sites.

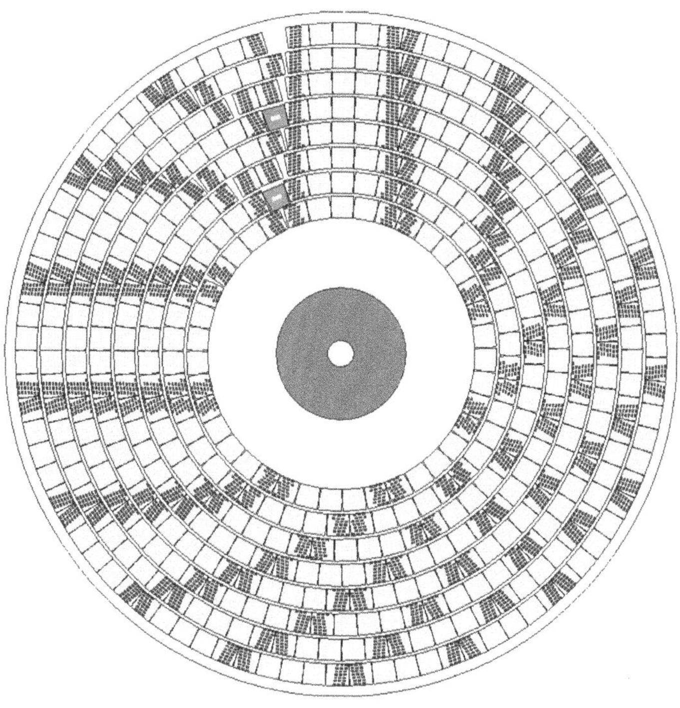

Figure 13.25 Format Example 1

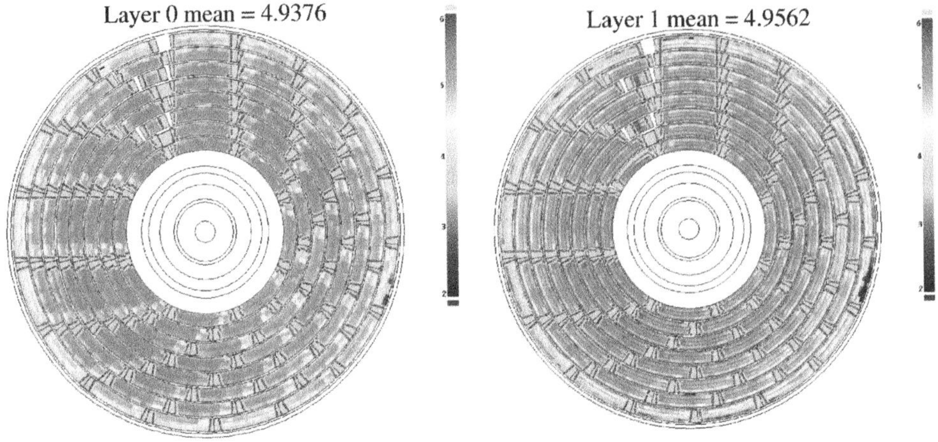

Figure 13.26 *Average SNR maps of the two layers*

Format Example 3 is an example of a format using 2D cure rules (Figure 13.28). We refer to this type of format as a 'wave format' because recording starts in one place (left of the upper most bookcase in the figure) and then propagates in two dimensions at once, washing across the disk. This format requires longer media time-outs and a more complicated

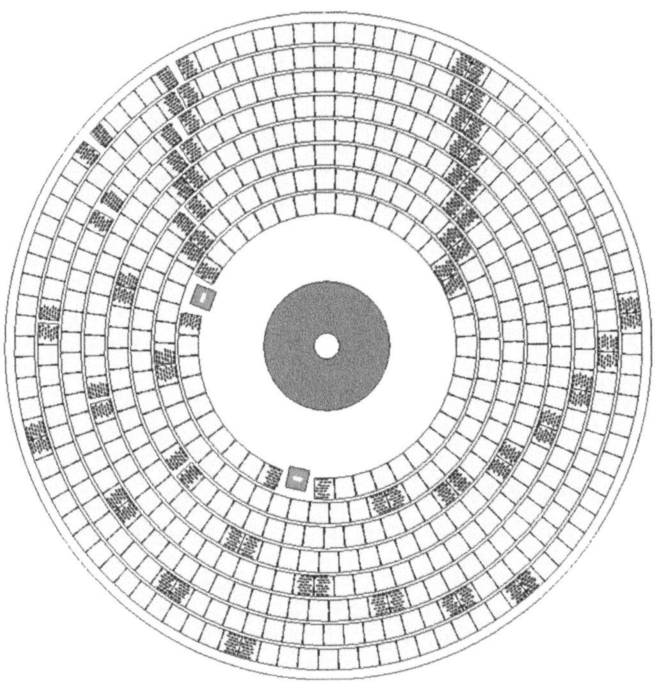

Figure 13.27 *Format Example 2*

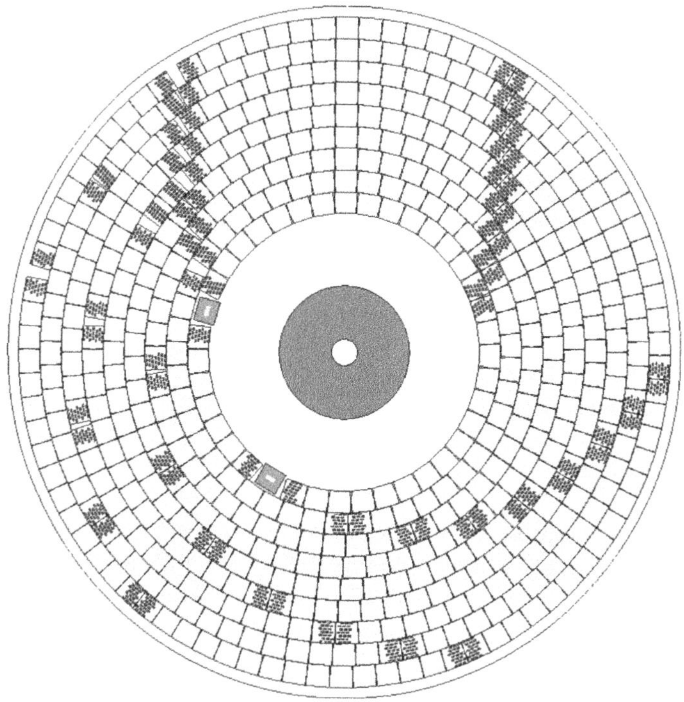

Figure 13.28 *Wave format example*

firmware implementation than the other two but it fits in one more cure track worth of data on the disk, increasing the user capacity from 326 (in Example 1) to 412 GB.

13.4.3.1 Format Example 1

Format Example 1 is shown in Figure 13.25.

1D cure rules
Pre-cure time-out: >14 min
Post-cure time-out: >10 min
Necessary writing speed: $>5\,\mathrm{MB\,s^{-1}}$
Total books: 16524
Dummy books: 1414 (8.5%)
Layers: 2
Tangential book spacing: 860 μm
Radial book spacing: 600 μm
Short stacks: 4
Short stack spacing: 215 μm
Total pages per book: 320
Pages per short stack: 80
Galvo page spacing: 0.075°

Galvo short stack spacing: 0.150
Outer track radius: 62.7 mm (edge = 63 mm)
Inner track radius: 26.0 mm (edge = 25.7 mm)

Cure spots: 496
Tangential cure pitch: 4.45 mm
Radial cure pitch: 4.728 mm
Number of data tracks per cure track: 7
Bookcases: 101
Library cure sites: 2
Library sites: 4

Assumed user bytes per page: 67 512

Total capacity: 326 GB

Figure 13.26 maps the average book SNR of holograms written with Format Example 1, separately for both layers. The drive's writing speed was a little over 5 MB s^{-1}. The SNR scales from a high 6 dB average down to 2 dB average. The overall averages for both layers are slightly less than 5 dB. The edge wedge of the disk was out of specification in this example inducing some corresponding SNR loss at the outer edge.

13.4.3.2 Format Example 2

Format Example 2 is basically the same as Example 1 but designed for 15 MB s^{-1} and above write transfer rates (Figure 13.27). This format and results are presented in reference [35].

1D cure rules
Pre-cure time-out: > 14 min
Post-cure time-out: > 10 min
Necessary writing speed: > 15 MB s^{-1}
Total user books: 16 298
Dummy books: 1120 (6.4%)
Layers: 2
Tangential book spacing: 860 μm
Radial book spacing: 575 μm
Short stacks: 4
Short stack spacing: 215 μm
Number of data tracks per cure track: 7
Total pages per book: 320
Pages per short stack: 80
Outer track radius: 62.712 mm
Inner track radius: 26.012 mm

Cure spots: 551
Bookcases: 40

Library cure sites: 2
Library sites: 4

Assumed user bytes per page: 67 512

Total capacity: 352 GB

13.4.3.3 Format Example 3

Format Example 3 is an example of a wave format (Figure 13.28). It uses the same basic parameters as Example 2.

2D cure rules
Total books: 19 650
Dummy books: 574 (2.9%)
Layers: 2
Tangential spacing: 860 μm
Radial spacing: 600 μm
Short stacks: 4
Short stack spacing: 215 μm
Total pages per book: 320
Pages per short stack: 80
Galvo page spacing: 0.075 °
Galvo short stack spacing: 0.150
Outer track radius: 62.7 mm (edge: 63 mm)
Inner track radius: 26.0 mm (edge: 25.7 mm)

Cure spots: 559
Tangential cure pitch: 4.45 mm
Radial cure pitch: 4.1 mm
Number of tracks per cure track: 7
Number of bookcases: 41

Library cure sites: 2
Library sites: 4

Assumed user bytes per page: 67 512

Total capacity: 412 GB

13.4.4 Format Files

Format files are binary files containing all the instructions that a drive needs to format a disk. These files include all spatial as well as temporal information. Spatially the file defines the layout parameters for the location of each page (angularly, radially and tangentially), library map, dummy book, cure site, and neighbor relationships. Temporally the format file contains all the sequencing information, schedules, media time-outs, diffusion time, and pre- and post-cure times.

Acknowledgements

Many people have contributed greatly to this material. Specifically we wish to acknowledge Alan Hoskins, Friso Schlottau, Ken Anderson, William Wilson, Adrian Hill, Sam Weaver, Cotton Anderson, Dave Beal, David Samuels, J.J. Kane, Rod Harris, Jimmy Wade, Brad Sissom and Oyvind Haehre.

References

1. O. Malki, *et al.*, Optimal aperture size for maximizing the capacity of holographic data storage systems, *J. Meeting of the Opt. Data Storage and Int. Symp. on Opt. Memories*, Waikoloa, Hawaii (2008), paper TuP9.
2. M.P. Bernal, G. W. Burr, H. Coufaland M. Quintanilla, Balancing interpixel crosstalk and detector noise to optimize areal density in holographic storage systems, *Appl. Opt.*, Vol. 37, pp. 5377–5385 (1998).
3. J. T. LaMacchiaand D. L. White, Coded multiple exposure holograms, *Appl. Opt.*, Vol. 7, pp. 91–94 (1968).
4. C.B. Burckhardt, Use of a random phase mask for the recording of Fourier transform holograms of data masks, *Appl. Opt.*, Vol 9, pp. 695–700 (1970).
5. Y. Tsunoda and Y. Takeda, Hologram memory with high quality and high information storage density, *Jpn. J. Appl. Phys.*, Vol 11, pp. 656–665 (1972).
6. C.B. Burckhardt, Fourier transform recording with random phase shifting, Bell Labs, US Patent 3,604,778, September 14, 1971.
7. W.C. Stewart, A.H. Firester and E.C. Fox, Random phase data masks; fabrication tolerances and advantages of four phase level masks, *Appl. Opt.*, Vol. 11, No. 3, pp. 604–608 (1972).
8. W.J. Dallas, Deterministic diffusers for holography, *Appl. Opt.*, Vol. 12, No. 6, pp. 1179–1187 (1973).
9. M. Abe, *et al.*, Phase mask for use in holographic apparatus, Ricoh Co. Ltd, US Patent 3,995,948, December 7, 1976.
10. A. Iwamoto, Artificial diffuser for Fourier transform hologram recording, *Appl. Opt.*, Vol. 19, No. 2, pp. 215–220 (1980).
11. D.G. Esaev, A.A. Lorei and S.P. Sinitsa, Continuous random phase mask, *Sov. Phys.– Tech. Phys.*, Vol. 22, No. 9, pp. 1150–1152 (1977).
12. V.Y. Levin, *et al.*, Fabrication and testing of phase masks for information storage and processing devices, *Soviet J. Opt. Tech.*, Vol. 45, No. 3, pp. 171–174 (1978).
13. A. Verbovetskii, *et al.*, Randomizing phase mask for recording Fourier holograms of paraphase-coded binary data, *Opt. Spectrosc.*, Vol. 45, No. 5, pp. 810–812 (1978).
14. A. Akaev and T. Kerimkulov, New experimental results on the efficiency of a random phase mask in holographic memories, *Opt. Spectrosc.*, Vol. 51, No. 3, pp. 296–299 (1981).
15. Y. Nakayama, and M. Kato, Linear recording of Fourier transform holograms using a pseudo-random diffuser, *Appl. Opt.*, Vol. 21, No. 8, pp. 1410–1418 (1982).
16. V.G. Mityakov and V.B. Federov, Comparison of phase masking methods in holographic recording of binary code transparencies, *Opt. Spectrosc.*, Vol. 57, No. 2, pp. 184–188 (1984).
17. J. Hong, I. McMichael and J. Ma, Influence of phase masks on cross talk in holographic memory, *Opt. Lett.*, Vol. 21, No. 20, pp. 1694–1696 (1996).
18. M.P. Bernal, *et al.*, Effects of multilevel phase masks on interpixel cross talk in digital holographic storage, *Appl. Opt.*, Vol. 36, No. 14, pp. 3107–3115 (1997).
19. Q. Gao and R. Kostuk, Improvement to holographic digital data-storage systems with random and pseudorandom phase masks, *Appl. Opt.*, Vol. 36, No. 20, pp. 4853–4861 (1997).
20. M.P. Bernal, *et al.*, Experimental study of the effects of a six-level phase mask on a digital holographic storage system, *Appl. Opt.*, Vol. 37, No. 11, pp. 2094–2101 (1998).

21. J. Yang, L.M. Bernardo and Y.S. Bae, Improving holographic data storage by use of an optimized phase mask, *Appl. Opt.*, Vol. 38, No. 26, pp. 5641–5645 (1999).
22. M.J. O'Callaghan, Sorting through the lore of phase mask options: performance measures and practical commercial designs, *Proc. SPIE*, Vol. 5362, pp. 150–159 (2004).
23. J. Knittel, *et al.*, Beam-shaping random phase mask for holographic storage, *Opt. Data Storage Conf.*, Buena Vista, Florida (2009), paper TuC4.
24. M.J. O'Callaghan, J.R. McNeil, C. Walkerand M.A. Handschy, Spatial light modulators with integrated phase masks for holographic data storage, *Optical Data Storage Conf.*, Montreal, Canada (2006), p. 208.
25. D.C. Ghigliaand M.D. Pritt, *Two-dimensional Phase Unwrapping, Theory, Algorithms, and Software*. John Wiley & Sons, Ltd, New York, 1998.
26. K. Anderson, E. Fotheringham, S. Weaver, B. Sissomand K. Curtis, How to write good books, *Opt. Data Storage Conf.*, Montreal, Canada (2006), paper TuC1.
27. K. Curtis, *et al.*, Skip sorted recording, US Patent 6,614,566, September 2, 2003.
28. N. Kinoshita, *et al.*, Angular interval scheduling for angle-multiplexed holographic data storage, *J. Meeting of the Opt. Data Storage and Int. Symp. on Opt. Memories*, Waikoloa, Hawaii (2008), paper TuP10.
29. S. Li, K. Curtisand D. Psaltis, Storage density of 3-D disks, *OSA Tech. Dig. Ser.*, Vol. 16, pp. 79–83 (1993).
30. H.-Y. S. Liand D. Psaltis, Three-dimensional holographic disks, *Appl. Opt.*, Vol. 33, pp. 3764–3774 (1994).
31. A. Pu, K. Curtisand D. Psaltis, Exposure schedule for multiplexing holograms in photopolymer films, *Opt. Eng.*, Vol. 35, No. 10, pp. 2824–2829 (1996).
32. K. Blotekjaer, Limitations on holographic storage capacity of photochromic and photorefractive media, *Appl. Opt.*, Vol. 18, pp. 57–67 (1979).
33. D. Psaltis, *et al.*, Adaptive optical networks using photorefractive crystals, *Appl. Opt.*, Vol. 27, pp. 1752–1759 (1988).
34. Y. Taketomi, *et al.*, Incremental recording for photorefractive hologram multiplexing, *Opt. Lett.*, Vol. 16, No. 22, pp. 1774–1776 (1991).
35. K. Curtis, Practical issues of servo, lenses, lasers, drives and media for HDS, *IWHM Irago*, Aichi, Japan (2008), paper 21A1.

14

Servo and Drive Control

Alan Hoskins, Mark Ayres and Kevin Curtis

14.1 Introduction

Perhaps the most promising design for a holographic data storage (HDS) system employs a page-based architecture wherein the Fourier transform (FT) of a data page, created using a high numerical aperture (NA) lens, is interfered with a plane wave reference beam to record holographic fringes near the center of the holographic medium (Figure 14.1). High data density is achieved in the system through angle multiplexing, where successive data pages are stored in the same location by changing the reference beam angle, and through polytopic multiplexing, where neighboring books of angularly multiplexed holograms overlap within the media and undesired reconstructions are filtered out by a polytopic aperture. Polytopic multiplexing and the design of the polytopic aperture are covered in more depth in Chapters 3 and 4. While mathematically the FT operation is shift invariant, this invariance is limited in real optical systems due to finite spatial support and aberrations in the FT lens. Even neglecting these positional tolerances, high NA holographic systems that utilize thick media have unique tolerances on the media tilt, temperature and laser wavelength. This chapter discusses the sources for these tolerances [1], identifies which tolerances can be mitigated with servo systems, and shows that it is possible to optimize some system parameters to increase the tolerances.

In addition to reviewing these tolerances, two servo systems based on the recovered holographic signal are presented to control the reference beam pitch angle and wavelength compensation requirements unique to page-based holographic memory systems. First, a basic algorithm to compensate for pitch misalignment and holographic medium dimension change is reviewed [2], and then the principle alignment method for sequential recovery

Holographic Data Storage: From Theory to Practical Systems Kevin Curtis, Lisa Dhar, Adrian Hill, William Wilson and Mark Ayres

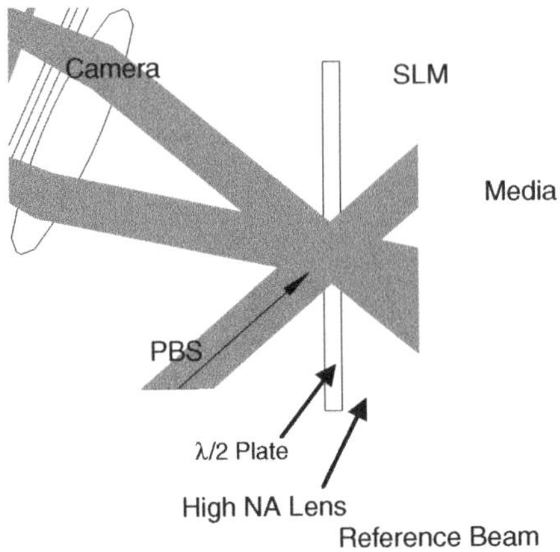

Figure 14.1 *Page-based HDS system using FT geometry and plane wave reference beam*

(wobble servo [3]) for these variables is presented. After these servo concepts have been presented, other basic algorithms and the fundamentals of drive control are reviewed including the logic for the read, write, and interchange operations. While the algorithms presented here are sufficient for a professional drive implementation, other servo principles are possible [4,5]. For example, a disk rotation servo system using a red laser to read marks on the media and radial motion servo system are presented in Chapter 9.

14.2 Holographic System Tolerances

14.2.1 Introduction

This section examines the system signal to noise ratio (SNR) tolerances in regards to holographic medium position, medium tilt and reference beam orientation, wavelength, and wave-front in a page-based system using a phase conjugate geometry. Phase conjugate systems read out the stored holograms using a probe beam propagating in the opposite direction of the recording reference beam. These systems have a large advantage as the recovered pixel wavefronts retrace the same optical path as the recorded pixel wavefronts, and thus lens aberrations are automatically removed and the system is diffraction limited regardless of the system design. To understand the tolerances of these systems, a definition of SNR is chosen that is easy to measure and has a well-defined relationship to the recovered bit error rate (BER; Figure 14.2). SNR degradation due to medium mis-positioning and reference beam wavelength and orientation errors fundamentally arises from several causes: imperfect phase conjugation of the recovered data beam, Bragg mismatch, book to book crosstalk, and clipping by the polytopic filter. Another important cause of SNR degradation is the effective vignetting of the recovered pixel wavefronts by the complex three-dimensional pupil defined by the overlap of the hologram and read reference beam and the wavefront aberrations between the writing and reading reference beams. In order to experimentally measure the various tolerances without entangling the wavefront or beam overlap, oversized reference beams were used with total aberrations less than 50 milliwaves peak to peak and

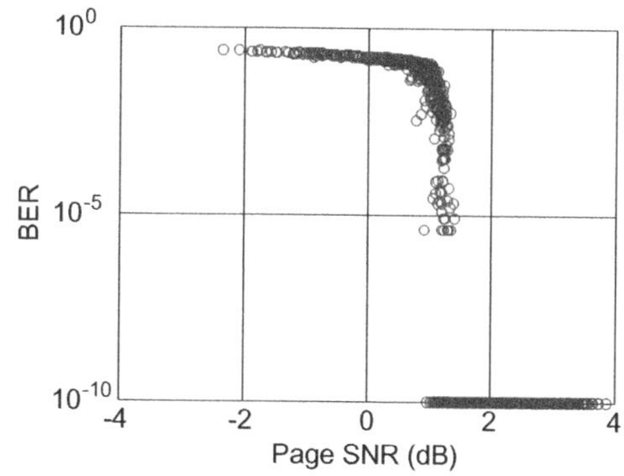

Figure 14.2 *Corrected BER versus page SNR. Pages with 0 bit errors are marked as having a BER of 10^{-10}*

root mean square (RMS) aberrations less than 10 milliwaves. It was found that the medium position tolerances can be very accurately predicted with a k-space-based model.

The first SNR degradation category, imperfect phase conjugation, occurs when the reconstructed pixel wavefronts use a different path through the data optics on recovery due to medium orientation or wavelength-induced magnification. This condition is illustrated in Figure 14.3 for a medium shift in the x direction. Because the pixel wavefronts are using different portions of the lens during recording and reconstruction, significant aberration to the point spread function (PSF) results which in turn reduces the SNR of the recovered data page. This tolerance is entirely determined by the design of the data path optics. In Chapter 4, it is shown that an isoplanatic lens design can greatly increase the tolerances associated with this effect.

In order to understand the SNR degradation due to Bragg mismatch and book-to-book crosstalk an important metric is the signal to scatter ratio (SSR). Coherent scatter tends to be the dominant SNR-limiting noise source for page-based holographic systems. The SSR is the ratio of the hologram diffraction efficiency to the 'scatter efficiency' of the system. The relationship between SSR and SNR is presented in Chapter 11. At data densities greater than 300 Gbits in.^{2}, the diffraction efficiency of stored holograms is typically set such that the SSR of the holograms is 10–15.

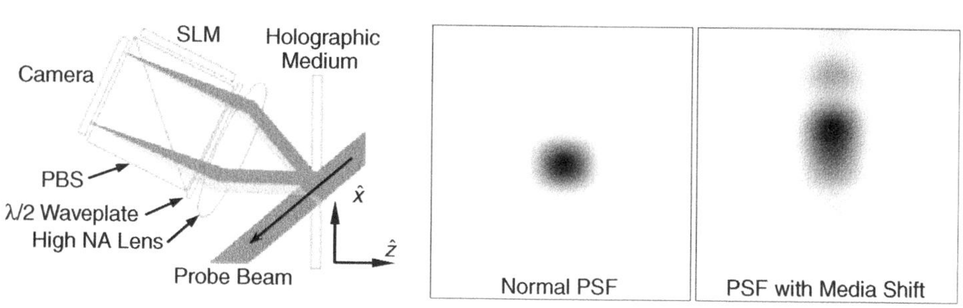

Figure 14.3 *Recovery of data pages using a phase conjugate probe beam. A medium shift in the x direction will cause aberration of the point spread function (PSF) of the system and degrade the SNR of the recovered data page*

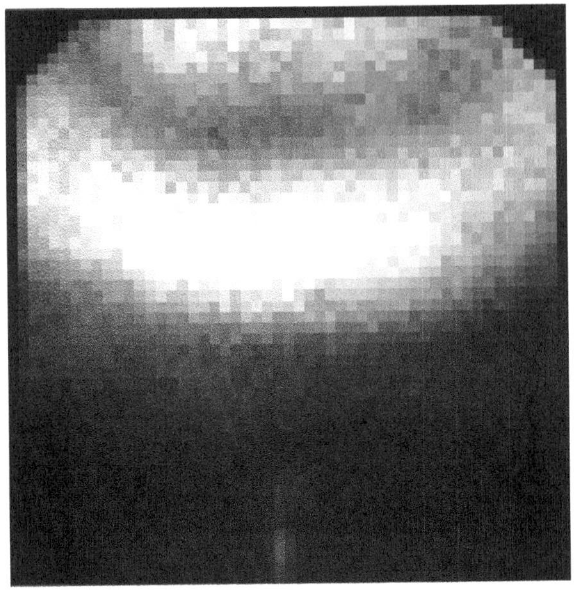

Figure 14.4 Camera image showing recovered page after thermal change

Bragg mismatch occurs when all or part of the hologram is not optimally aligned to the reference beam wavefront and the corresponding loss in diffraction efficiency (DE) attenuates the recovered data signal, degrading SSR and SNR. An example of this effect is seen with temperature changes or wavelength detuning; in this case it becomes impossible to Bragg match the entire data page with a conjugate of the original plane wave reference beam [6–8]. This results in a narrow band of recovered pixels on the camera as illustrated in Figure 14.4. The degree of SNR degradation in this case is fundamentally limited by the thickness of the media, the DE of the holograms, the NA of the FT lens, and the angular address of the data hologram.

Book to book cross-talk occurs when spatial media misalignments (i.e. disk r, θ) cause components of neighboring books to be admitted through the polytopic filter instead of being blocked by it. This effectively results in an increase in the coherent scatter and a corresponding decrease in SSR and SNR. SNR degradation in this case is dependent on the degree of misalignment, the book spacing, and the SSR of the recorded holograms. While this cause of degradation is apparent with minimal system examination, experimentally it is found that with a polytopic filter and book spacing exceeding the filter size only slightly (i.e. by the positioning accuracy), this noise term is not significant as the SNR degradation due to vignetting by the polytopic filter dominates.

Vignetting by the polytopic filter is a clear cause of SNR degradation that happens whenever the hologram is not correctly imaged due to media misalignment. In the InPhase professional drive the polytopic filter, located at the Fourier plane of the spatial light modulator (SLM), is sized to only slightly larger than the Nyquist size needed for resolution of the SLM. Because of this, slight medium misalignments can result in sub-Nyquist sampling and a corresponding decrease in SNR. This type of SNR degradation is a function of the NA of the FT lens and a servo system must be used to minimize it.

The final cause of SNR degradation due to medium mispositioning or reference beam orientation concerns the overlap of the reference beam with the data beam during recording and the probe beam with the hologram during recovery. Because the hologram is a recording of the image of the polytopic filter, changes in alignment and poor overlap can result in sub-Nyquist sampling of the SLM in the medium plane and a corresponding decrease in SNR. In the InPhase professional drive the size of the reference and probe beams is such that the SNR degradation due to poor overlap is negligible. A related effect involves the quality of the reference and probe beams. For perfect hologram reconstruction, the probe beam should be a perfect conjugate of the recording reference beam. Any deviation from this condition results in SNR degradation as partial Bragg mismatch causes apodization of the recovered data beam. Through experimentation it has been found that controlling difference in reference/probe beam focus to less than two waves and all other aberrations to less than 50 milliwaves RMS, the SNR loss is negligible. Because of this and the wavefront tolerances of the InPhase professional drives, this cause of SNR degradation will not be covered in this chapter.

Of these causes of SNR degradation, servo systems can be used to mitigate some while others must be alleviated by system design. Specifically, imperfect phase conjugation and SNR degradation due to overlap and reference beam wavefront deviation between recording and recovery can be mitigated using isoplanatic lenses in the data path and well designed reference beam optics, respectively. SNR degradation due to Bragg mismatch and polytopic vignetting must be minimized with servo systems. The remainder of this chapter will examine the specific tolerances and the algorithms used to control the servo systems.

14.2.2 Experimental and Modeled Tolerances

In this section tolerances for media positioning errors in all directions and rotations are presented for the InPhase professional drive system. Focus data are given from earlier versions of the drive where the camera focal plane and/or media height were adjustable. (These adjustments were removed from later versions as they were determined to be unnecessary.) Tangential shift of the media is also from a previous version of the drive. The reference beam wavefront limits and beam overlap tolerances are given as safe rules of thumb from many collective experimental results. Unless otherwise specified, the tolerances presented were measured for holograms at high recording density ($\sim$350 Gb in.$^{-2}$). This density was used for authenticity but experiments soon showed that book to book crosstalk was not a major factor and identical results can be obtained using lower recording density if the SSR and page spacing was controlled to be identical. In these cases, a desired SSR can be achieved by controlling the diffraction efficiency of the recorded holograms. All experiments were performed using 1.5 mm thick InPhase Tapestry sandwiched between two 1.0 mm thick polyolefin substrates. The allowable SNR roll-off for the experiments was arbitrarily set at 0.3 dB for each type of misalignment.

In addition to the presented experimental data, a ray-based, thick-lens, diffraction model was used to study the media positioning tolerances. This model consisted of a simple geometric model to construct the complex electric field of a SLM pixel in the holographic medium. This field was then interfered with a reference beam field to create holographic gratings. The gratings were modified by temperature effects and medium deformations and a recovered pixel complex field was generated using a phase conjugate probe beam. The recovered field takes into account reference beam wavefront aberrations, wavelength

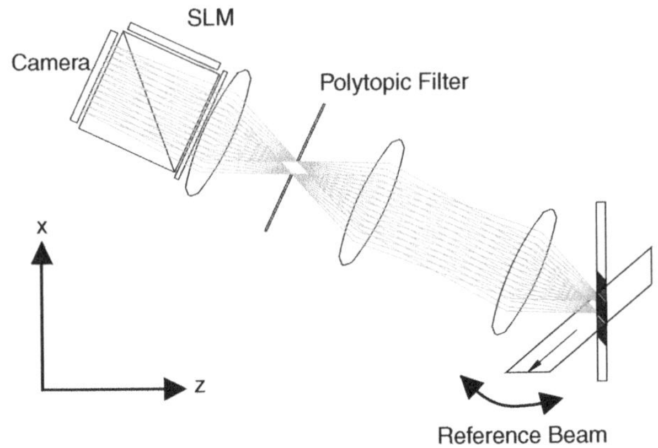

Figure 14.5 System diagram showing coordinates used to measure media shift and rotation tolerances

changes, reference beam angle changes, and beam overlap effects. This field was then used to generate a complex field distribution in the camera plane where an over-sampled camera image was created by integrating the intensity over the pixel areas. Coherent scatter, fixed pattern noise, camera gain nonuniformities, and shot noise were added and the result was extended to a data pattern that is re-sampled to determine the SNR and estimate the BER.

The media positioning tolerances measured were with regards to the coordinate systems shown in Figure 14.5. The tolerances for disk shifts in the x, y and z direction (tangential, radial and axial directions, respectively) were measured for a 0.3 dB SNR loss. Media rotations about the x-, y- and z-axis (radial tilt, tangential tilt and disk rotation) were performed about the disk axis for the z-axis and about the hologram for the x- and y-axes. For angle multiplexing, the reference beam is rotated about the y-axis. These measurements were done using an InPhase Technologies professional drive where possible and on an older testbed version with similar optics otherwise.

The reference beam angle tolerance was obtained by recovering the data holograms while varying the reference beam angle. The SNR degradation caused by this misalignment is due entirely to Bragg mismatch. As the reference beam changes from the optimum angle for Bragg matching the data page, the diffraction efficiency of the page drops into the scatter noise resulting in a SNR roll-off of 0.3 dB in $\pm 0.005°$ (Figure 14.6). This rate of SNR loss does depend on the angular address of the hologram as the Bragg selectivity is a function of reference beam angle. This tolerance is also dependent on the thickness of the media and the SSR of the holograms. Because of these limitations, it is necessary to control the position of the reference beam using a servo system; this is done for sequential reads in the InPhase drive using a wobble algorithm presented later in this chapter.

The SNR roll-off for media shifts in the radial direction (perpendicular to the plane of the reference beam and data path) is dominated by polytopic vignetting and imperfect phase conjugation. On reconstruction the pixel wave fronts use a different path through the data path optics than during recording, resulting in page distortion and pixel blur. The over-sampling detection algorithm is able to compensate for page distortion, but the increase in size of the PSF due to the polytopic vignetting degrades the SNR by 0.3 dB for shifts of

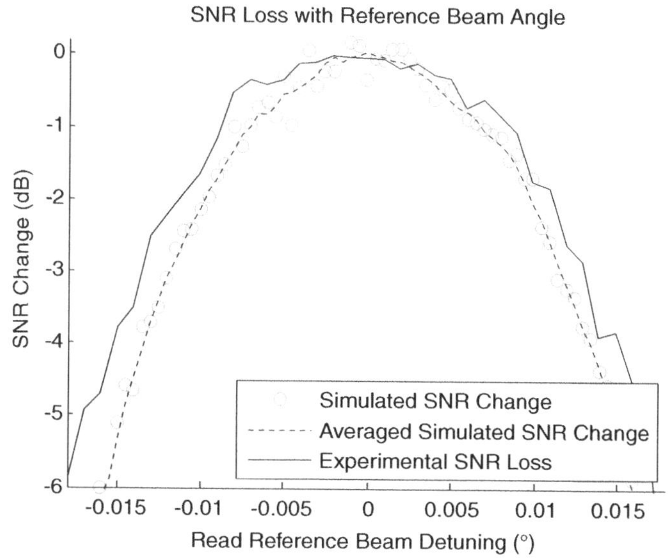

Figure 14.6 *SNR loss versus read reference beam angle detuning. A ±0.005° angular detuning corresponds to a 0.3 dB SNR loss*

±45 μm (Figure 14.7). This vignetting makes it necessary to control the radial positioning of the media with a servo system. While book to book crosstalk should also have an effect for radial shifts, a shift of 45 μm adds an insignificant amount of scatter due to crosstalk at the current recording density and polytopic size (~800 μm square).

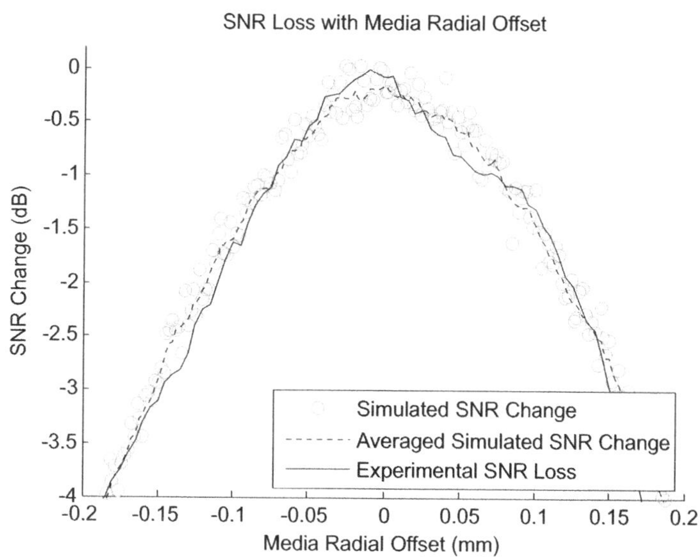

Figure 14.7 *SNR loss versus media radial off-track error. The effects of book to book cross-talk are negligible*

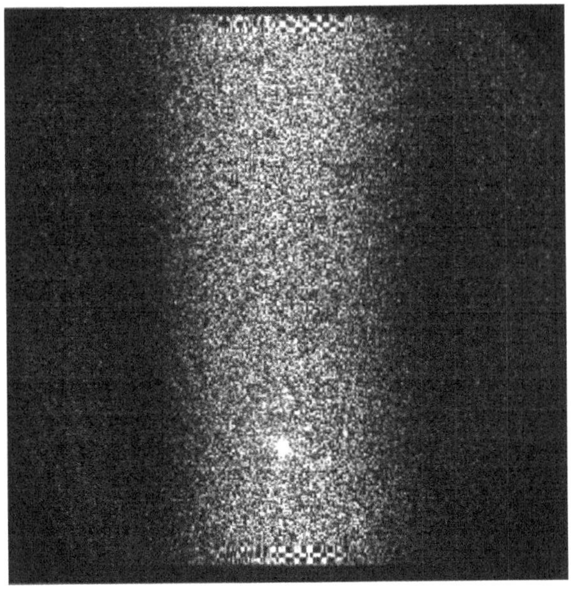

Figure 14.8 *Partial page Bragg mismatch due to media rotation*

The SNR tolerance for a disk rotation error is dominated by Bragg mismatch due to the rotation of the holographic gratings. As the index gratings rotate with the medium, it becomes impossible to Bragg match the entire data page with the plane wave probe beam, resulting in a narrow strip of pixels being Bragg matched (Figure 14.8). This Bragg-based effect dominates the SNR roll-off for all holograms and thus the tolerance does not change appreciably as a function of the track radius. In the InPhase Professional drive, data are written in tracks from 28 to 63 mm from the disk center. Experimental results indicate that a rotation of 0.022° (corresponding to a shift of 10 μm for the innermost track) results in a 0.3 dB loss in SNR (Figure 14.9). Because of imperfect phase conjugation and the polytopic vignetting at larger track radii, this tolerance falls to 0.020° for the outer tracks (corresponding to a shift of 22 μm). In the InPhase Technologies professional drive, the media is encoded with a tracking pattern that is read with a red laser to control media rotation to within $\pm 33 \times 10^{-6}$ °. Like radial shifts, book to book crosstalk did not affect the tolerance as the increase in scatter for a 22 μm shift is negligible.

Tangential shifts of the media (along the x direction in Figure 14.5) is essentially a drive manufacturing tolerance for the InPhase professional drive since the media is tracked in this direction, defining how accurately the disk spindle needs to be placed with regards to the data path optics during drive manufacture. The measurement of this tolerance (Figure 14.10) seems to be asymmetric; media shifts towards the data path optics are insensitive with a 50 μm shift being required for a 0.3 dB SNR loss. Shifts away from the data path result in a 0.3 dB loss in only 20 μm. The measurement of this tolerance was done using a prototype version of the InPhase professional drive and may have been significantly improved by isoplanatic lenses in the data path.

The SNR degradation caused by or axial misalignment of the medium is a result of imperfect phase conjugation. While the effects of this shift should be asymmetric with

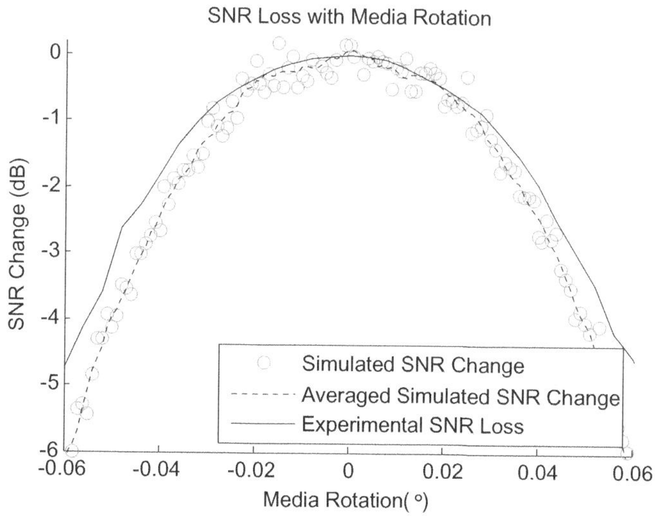

Figure 14.9 *SNR loss versus media rotation. The SNR loss is dominated by the rotation and Bragg mismatch*

direction as lens aberrations tend to worsen toward the edges of the field, the experimental and model results do not support this; Figure 14.11 shows the SNR degradation versus medium axial position. The SNR loss seems to be symmetric with a 50 μm shift in either direction corresponding to a loss of 0.3 dB. While this tolerance is relatively robust, the media must be held to this tolerance during medium interchange from drive to drive and also over the full operating temperature range of the drive. Like tangential and radial shifts, it was found that book to book crosstalk did not affect this tolerance for a data density of 350 Gbit in.$^{-2}$.

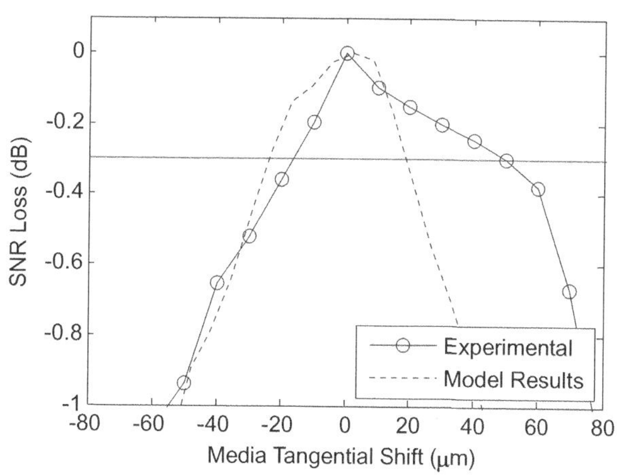

Figure 14.10 *SNR loss veersus media tangential shift*

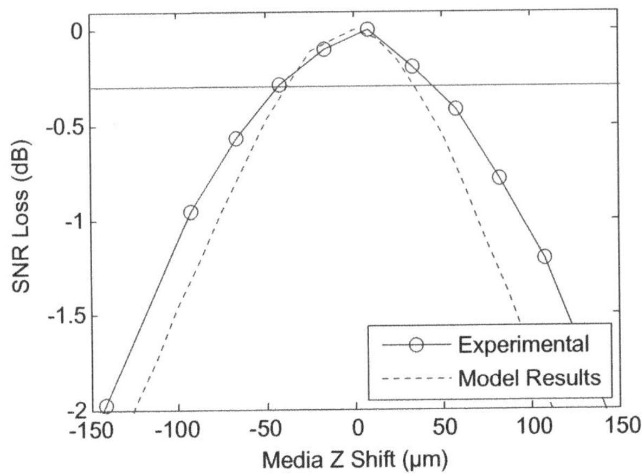

Figure 14.11 *SNR loss versus media focus. A shift of ± 50 μm is required for a SNR loss Of 0.3 dB*

Medium tilt tolerances are primarily dominated by Bragg-based effects, though imperfect phase conjugation can also have a noticeable effect. Medium tilt in the tangential direction (rotation around the *y*-axis in Figure 14.5), is indistinguishable from reference beam angle detuning; the SNR drops by 0.3 dB with a tilt of only ± 0.005° (Figure 14.12). This tolerance is necessarily tight as this is multiplexing direction and holograms are placed as close together as possible. If the reference beam angle is changed to compensate for the media tilt, the tolerance is a function of imperfect phase conjugation; in this case, a tilt of ± 0.125° is needed before the SNR drops by 0.3 dB.

Tilt in the degenerate, or radial, direction (rotation about the *x*-axis in Figure 14.5) is very similar to disk rotation in that only a fraction of the pixels are Bragg-matched with a given probe beam angle due to the rotation of the index gratings. Like disk rotation, these Bragg effects dominate the SNR tolerance resulting in a 0.3 dB SNR roll-off for a tilt of ± 0.015°

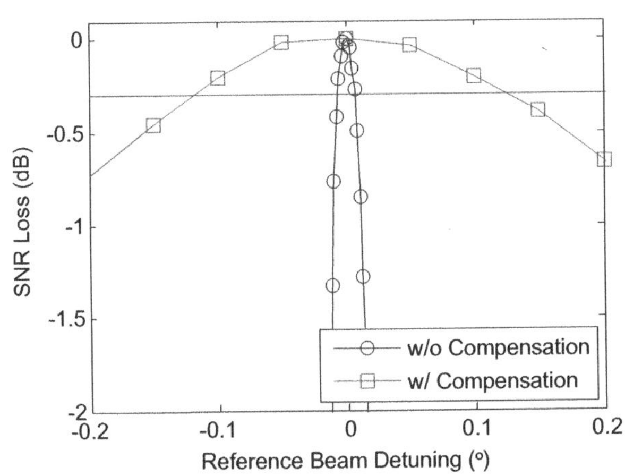

Figure 14.12 *SNR loss versus tangential tilt with and without changing the reference beam angle to compensate*

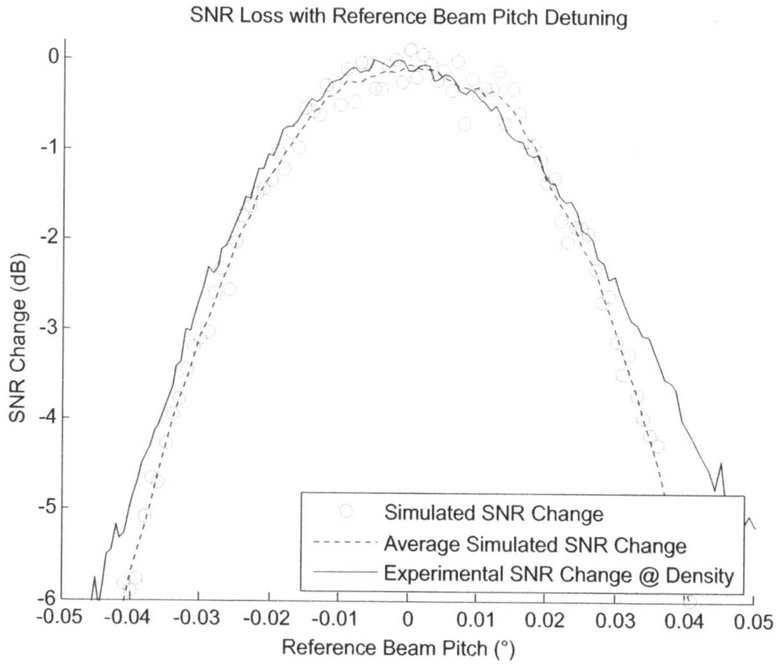

Figure 14.13 *SNR loss versus probe beam pitch on an InPhase professional drive. A tilt of ±0.015° degrades the SNR by 0.3 dB*

(Figure 14.13). Because disk rotation and degenerate tilt produce similar effects in momentum space, it is possible to use one effect to compensate for the other [9]. For this to be effective it is necessary to have isoplanatic lenses in the data path. If this condition is met, the Bragg effects of medium rotation can be mitigated using a probe beam pitch servo system.

Thermal expansion of the holographic disk is anisotropic due to the different coefficients of thermal expansion (CTEs) of medium and substrates. This anisotropy, combined with the dependence of the medium index of refraction on temperature, changes the holographic gratings in such a way that it becomes impossible to Bragg match the entire data page without changing the probe beam wavelength and angle to compensate. Thus, the SNR degradation due to temperature changes and wavelength detuning is caused by Bragg mismatch. Figure 14.14 shows that the wavelength detuning tolerance of the holograms is 0.3 nm for a degradation of 0.3 dB. This SNR degradation corresponds also to an uncompensated temperature change of 1.5 °C. When temperature changes are compensated for using probe beam wavelength and angle changes, the tolerance is limited by imperfect phase conjugation due to the different paths taken by the data pixel wavefronts. Figure 14.15 shows the SNR loss versus compensated medium temperature change. In this case, the faster roll-off of the experimental data is due to other changes in the InPhase drive that are not captured by the holographic model such as laser wavefront and pointing. In addition, decreasing the effective media's thermal expansion increases the allowable thermal change and has been demonstrated using InPhase Tapestry materials.

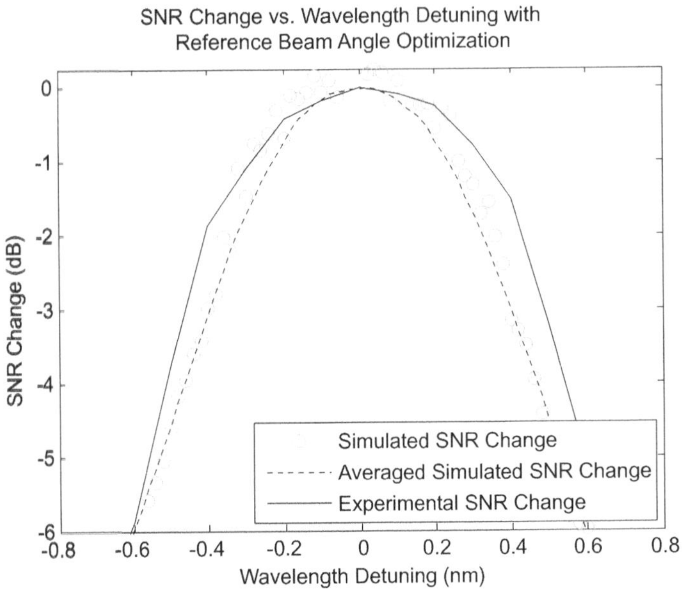

Figure 14.14 *SNR loss wavelength change with reference angle optimization*

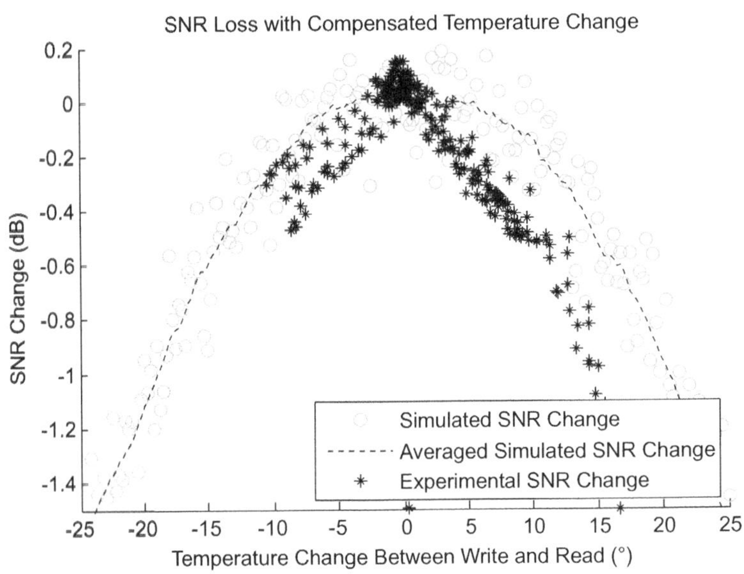

Figure 14.15 *SNR change versus media temperature with wavelength and angle compensation*

14.2.3 Tolerance Summary

The tolerances of the InPhase professional drive which employs a page-based, phase conjugate, Fourier geometry, and a high NA storage lens have been examined. Analysis has shown the major underlying sources of the medium positioning tolerances fall into three categories: polytopic vignetting, Bragg mismatch, and to a lesser degree, imperfect phase conjugation. Experiments and simulations have been used to find the tolerances and to verify their sources. From experimentation the reference wavefront rule of thumb has been to limit the focus error to less than two waves and the remaining aberrations to less than 50 milliwaves RMS. At this level there is negligible SNR loss. In fact, depending on the exact aberrations present in the reference beam, this rule of thumb may be tighter than actually required.

Table 14.1 tabulates the experimental results and the tolerances examined. The first tolerance examined, reference beam angle, is necessarily tight as this is the multiplexing direction and holograms are stored as close together as possible. Because of this and the fact the degradation is due to Bragg mismatch of the entire page, the axis must be controlled by a servo system to mitigate the SNR loss. Medium shifts in the x, y and z directions cause SNR degradation due to polytopic vignetting and imperfect phase conjugation. Imperfect phase conjugation can be mitigated using isoplanatic lens designs but servo systems must be used to properly image the hologram onto the polytopic filter. The final cause of SNR degradation, partial page Bragg mismatch, occurs when the medium is rotated about the x- or z-axis, the medium changes temperature, or the wavelength of the probe beam is different than the reference beam. These tolerances cannot be mitigated using design techniques without capacity trade-offs as they are a function of the media thickness and the system NA; reducing either of these variables decreases the capacity of the system. The exception to this is to decrease the media's CTE which improves the temperature tolerances without affecting system performance. The other theoretical causes of SNR degradation, book to book crosstalk and reference/probe beam overlap, had little effect due to the system design and the data densities of the InPhase professional drive.

14.2.4 Tolerance Analysis

The various tolerances and drive accuracies enumerated above can be used to estimate the expected losses for holograms due to multiple misalignments using a root square sum (RSS) method. This simple analysis makes the assumption that the tolerances are independent and that the SNR roll-off and drive accuracy for each follows a normal distribution. The results of this analysis are shown in Figure 14.16. At the data densities in the current InPhase professional drive, there are roughly 2–3 dB of margin on average after 50 years of accelerated lifetime testing available before chapter and anthology error correction codes (see Chapter 11) are needed to recover the data. As can be seen from Figure 14.16, greater than 99% of the holograms should be recoverable without relying on robust alignment algorithms.

Hitachi and InPhase have also proposed an alternative method [10] that relies on the assumption that the noise sources are independent. In this method, it is first necessary to define the normalized noise of the system using the following formula:

$$\sigma' = \frac{\sqrt{\sigma_{on}^2 + \sigma_{off}^2}}{\mu_{on} - \mu_{off}} \qquad (14.1)$$

where μ_{on} is the mean of the recovered on pixels, μ_{off} is the mean of the recovered off pixels and σ_{on} and σ_{off} are their respective standard deviations. Under optimal conditions, this

Table 14.1 *Tolerances examined and results of measurements*

Condition	Dominant cause of SNR degradation	Measured tolerance (0.3 dB SNR drop)	InPhase professional drive accuracy
Probe beam angle, multiplexing direction	Bragg mismatch	$\pm 0.005°$	$\pm 0.005°$
Medium radial offset	Polytopic vignetting	$\pm 45\,\mu m$	$\pm 5\,\mu m$
Medium rotation (tangential detrack)	Bragg mismatch (partial page)	$\pm 0.020°$	$\pm 0.010°$ (trade-off with tangential offset)
Medium tangential offset	Polytopic vignetting	$\pm 20/50\,\mu m$	$\pm 5\,\mu m$ (trade-off with medium rotation)
Reference beam angle, radial (degenerate) direction	Bragg mismatch (partial page)	$\pm 0.015°$	$\pm 0.005°$
Disk tilt – tangential direction (with probe compensation)	Imperfect phase conjugation	$\pm 0.125°$	$< 0.01°$
Disk defocus (media normal offset)	Imperfect phase conjugation	$\pm 50\,\mu m$	$\pm 20\,\mu m$
Camera defocus	Imperfect phase conjugation	$\pm 50\,\mu m$	$\pm 10\,\mu m$
Wavelength detuning	Bragg mismatch (partial page)	± 0.3 nm	± 0.1 nm
Uncompensated temperature changes	Bragg mismatch (partial page)	$+ 1.5\,°C$	N/A
Compensated temperature changes	Imperfect phase conjugation	$+ 10\,°C$	N/A
Reference/probe wavefront difference (safe rule of thumb)	Data beam apodization	<50 milliwaves RMS	<10 milliwaves RMS
Reference/probe beam overlap	Data beam apodization	Radial $100\,\mu m$ Theta $>100\,\mu m$	N/A

N/A, Not applicable.

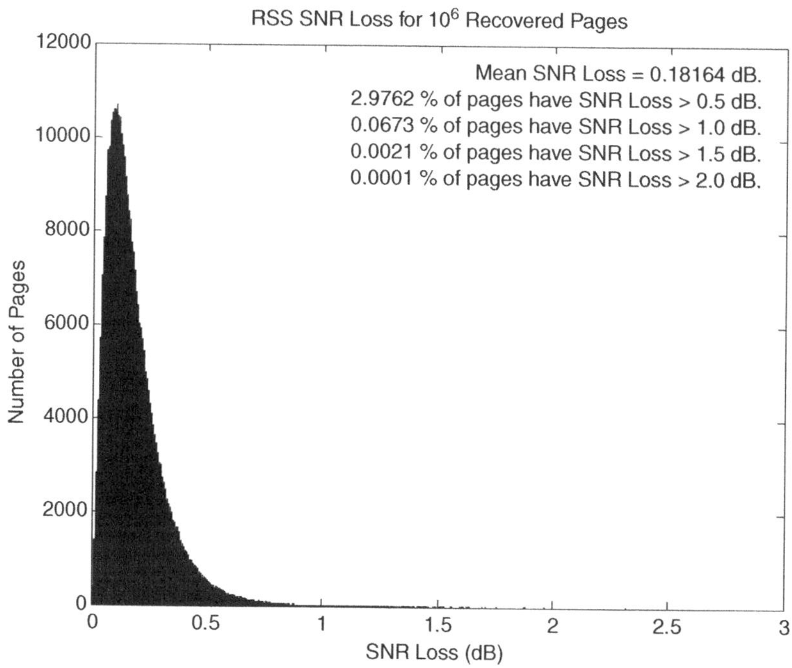

Figure 14.16 *RSS analysis of SNR loss in InPhase DVT holographic drive due to positional tolerances using 10^6 recoveries*

normalized noise, σ'_0, is called the DC noise and the SNR of the recovered data can be defined using:

$$\text{SNR} = 20\log_{10}\left(\frac{1}{\sigma'_0}\right) \tag{14.2}$$

When unique independent stresses are applied to the system, for example the mis-positioning of the medium or probe beam, it is possible to determine the effects on recovered SNR using the formula:

$$\sigma' = \sqrt{\sigma'^2_0 + \sum_i \sigma'^2_i} \tag{14.3}$$

where σ'_i is the unique noise due to each stress applied to the optimal system. Using this method to analyze the monocular architecture presented in 2007 by InPhase Technologies and Hitachi, it was found that the current drive technologies are more than sufficient to recover the holograms without error with typical optical drive tolerances.

14.3 Algorithms

14.3.1 Introduction

Now that the medium and probe beam positioning tolerances are known, servo systems must be designed to align holograms to within the required tolerances. As discussed in the professional drive chapter, the disk positioning in both radial and theta is done extremely accurately. The radial run-out is approximately 4 μm while the disk encoder allows the disk location in theta to be measured to a few nanometers. Referencing the positional tolerances above, these accuracies allow for hologram recovery with negligible SNR loss. While the medium positioning is thus well compensated, the algorithms used to control the probe beam angle, pitch and wavelength have yet to be presented. This section covers the algorithms that are used to set these axes within the required tolerances. First a simple method of determining reference beam pitch and wavelength detuning is reviewed which makes use of the Bragg mismatch of recovered pages. This method is then extended to an iterative, high speed method to align the pitch, wavelength and angle of the probe beam to recover a single hologram from an arbitrary starting alignment. Finally, a wobble servo is presented that dynamically maintains alignment while recovering holograms in sequence using a closed-loop servo without affecting data transfer rate.

14.3.2 Theory of Thermal and Pitch Compensation

In the previous section it was shown that the cause of SNR degradation with probe beam pitch and wavelength detuning is Bragg mismatch of parts of the recovered data page. One convenient method for understanding Bragg mismatch effects is to decompose the recording fields into their respective angular spectrum of plane waves and projecting the resulting components onto the surface of the momentum sphere (the k-sphere formulation). For page-based systems with Fourier geometry and a plane wave reference beam, this decomposition is simple as the storage lens performs a FT of the SLM and the waveform of each SLM pixel is nearly planar in the holographic medium. Because of this, it is only necessary to examine the principal plane wave for each SLM pixel (Figure 14.17). This plane wave, easy to identify as it propagates in the same direction as the pixel's chief ray, is represented in momentum space as a delta function on the surface of a sphere with radius $2\pi n/\lambda$ (where n is the index of refraction and λ the wavelength). When this wave is interfered with the plane wave reference beam in the holographic medium, a periodic modulation will be written into the index of refraction proportional to the intensity of the interference pattern. This index grating can be represented in momentum space by a vector found simply by taking the vector difference of the plane wave propagation vectors as shown in Figure 14.17. The grating vectors are then modified by medium anisotropic shrinkage and tilt or rotation. Explicitly,

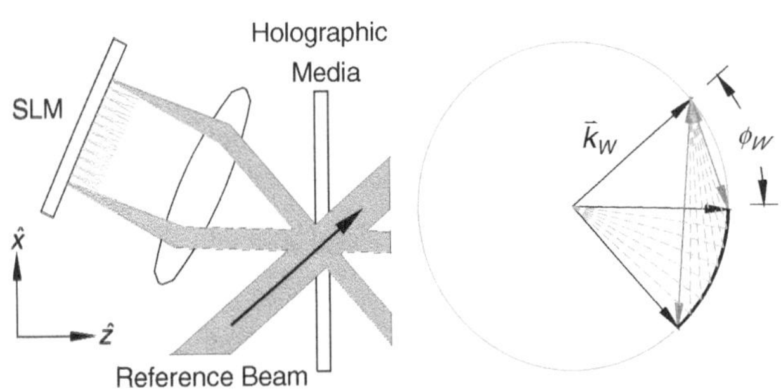

Figure 14.17 Page-based system using a high NA FT lens and its momentum space representation. The angle ϕ_W is the angle of the write reference beam

the plane wave for the pixel in mth row and nth column on the SLM will have the propagation vector $\vec{k}_{m,n}$ and will from a grating that can be represented in momentum space by the vector $\vec{K}_{m,n}$:

$$\vec{K}_{m,n} = \pm \begin{bmatrix} \cos\theta & \sin\theta & 0 \\ -\sin\theta & \cos\theta & 0 \\ 0 & 0 & 1 \end{bmatrix} \begin{bmatrix} 1/x_{SF} & 0 & 0 \\ 0 & 1/y_{SF} & 0 \\ 0 & 0 & 1/z_{SF} \end{bmatrix} \left(\vec{k}_W - \vec{k}_{m,n}\right) \tag{14.4}$$

where $\vec{k}_W$ is the reference beam propagation vector, θ is the angle of rotation about the z-axis, and the variables x_{SF}, y_{SF} and z_{SF} are the hologram volume scaling factors in the x, y and z dimensions due to medium thermal expansion and post recording cure shrinkage. Rotation about the multiplexing axis and degenerate (pitch) axis are ignored as the probe beam orientation can be changed to compensate these rotations.

During read out of the holographic grating, the dependence of the diffraction efficiency on probe beam orientation can be found using:

$$DE \propto \mathrm{sinc}^2\left(\frac{\Delta \vec{k}_{\hat{N}} L}{2\pi}\right) \tag{14.5}$$

where L is the thickness of the holographic medium and $\Delta\vec{k}_{\hat{N}}$ is the distance of the vector sum of the probe beam propagation vector and the pixel grating vector from the surface of the momentum sphere along the direction of the medium normal (Figure 14.18) [11]. When the probe beam orientation is different than the reference beam or the grating has been modified by rotation or thermal expansion, Bragg mismatch occurs and $\Delta\vec{k}_{\hat{N}}$ can be found using the relationship:

$$\Delta\vec{k}_N = \left|\left(\vec{k}_P + \vec{K}_{m,n}\right)\cdot\hat{N}\right| - \sqrt{\left|\vec{k}_P\right|^2 - \left|\left(\vec{k}_P + \vec{K}_{m,n}\right) - \left(\left(\vec{k}_P + \vec{K}_{m,n}\right)\cdot\hat{N}\right)\hat{N}\right|^2} \tag{14.6}$$

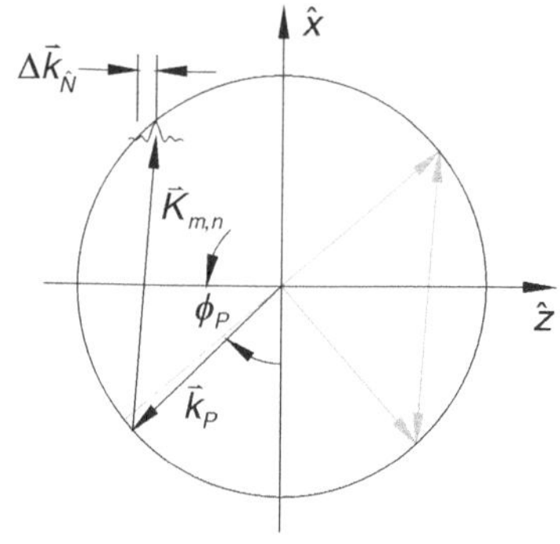

Figure 14.18 Momentum space representation of hologram recovery using a phase conjugate reference beam. Because of the phase conjugate geometry, the + order of the grating is recovered

where $\hat{N}$ is the medium's unit normal vector and $\vec{k}_P$ is the probe beam propagation vector in the holographic medium. This equation can be solved to find the probe beam angle in the multiplexing plane (i.e. the plane containing the data beam optical axis and medium normal vector), $\phi_{m,n}$, where $\Delta\vec{k}_{\hat{N}} = 0$ and the diffraction efficiency is at its maximum. While a closed form solution is possible and has previously been presented [2], it is more productive to examine the equation quantitatively for a pair of separated pixels on the SLM.

If the medium is properly oriented during recovery with no wavelength, bulk index, or volume changes, then the angle of the probe beam will be the same as the angle of the write reference beam. If, however, the medium has been rotated about an axis in the multiplexing plane or has undergone expansion or had a change in its bulk index of refraction, it is impossible to Bragg match the entire data page without changing the probe beam pitch and wavelength to compensate. Without this compensation only a narrow band of pixels will be Bragg-matched with a given probe beam angle (Figure 14.19). As the probe beam angle is changed, this band of pixels will move across the camera as the DE of some pixels increase while others decrease. For any two SLM pixels, the difference in probe beam angles where their relative diffraction efficiency was a maximum will be a function of the pitch and wavelength changes need to compensate the Bragg mismatch.

$$\Delta\phi(\rho, \Delta\lambda) = \phi_{Pixel1} - \phi_{Pixel2} \tag{14.7}$$

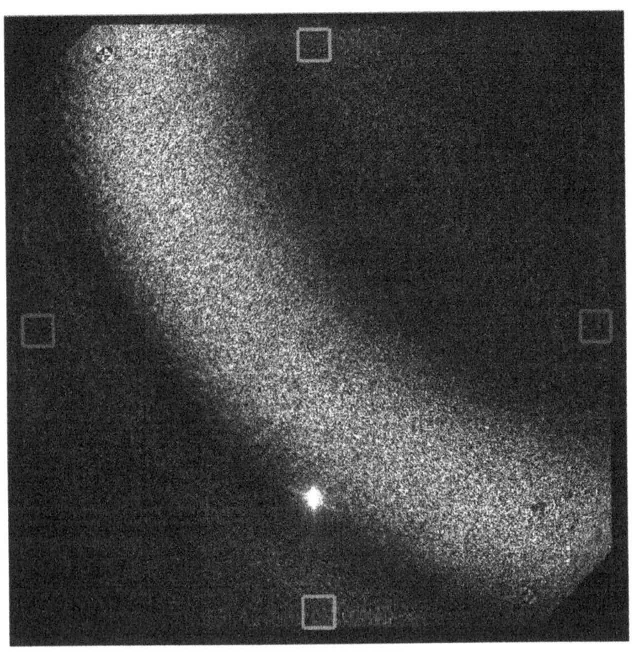

Figure 14.19 Recovered hologram showing partial Bragg mismatch. The boxes are regions on the camera that are monitored while scanning the probe beam angle

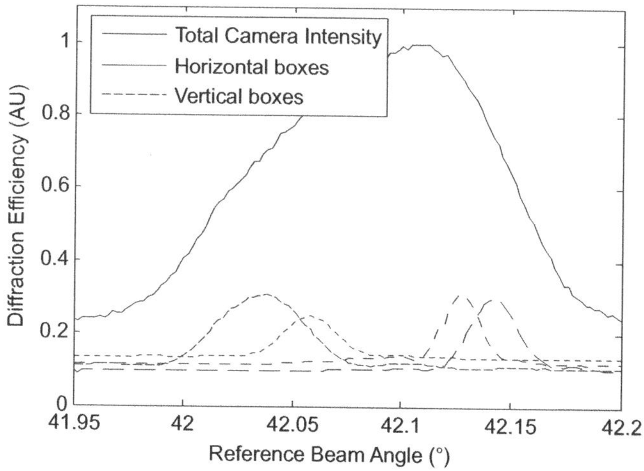

Figure 14.20 *Hologram diffraction efficiency as a function of probe beam angle. The smaller peaks are the intensity reported by the boxes shown in Figure 14.19*

This function can be further simplified by expanding it to first order using a simple Taylor expansion:

$$\Delta\phi(\rho, \Delta\lambda) = \frac{\partial \Delta\phi}{\partial \rho}\rho + \frac{\partial \Delta\phi}{\partial \Delta\lambda}\Delta\lambda \tag{14.8}$$

The partial derivatives of $\Delta\phi$ versus the two variables are constants that can either be solved for using Equation (14.8) or determined experimentally. If solved for, it can be shown that the derivatives are only a function of the system geometry. They are insensitive to the amount of detuning, medium index changes, and to the anisotropy of the expansion. They are simply a function of the angle of the write reference beam and the pair of pixels being examined. This result means that if the angle of the probe beam is scanned while monitoring the DE of specific pairs of pixels (Figure 14.20), the amount of probe pitch and wavelength detuning can be determined simply by solving a system of linear equations.

This result can be simplified even further by choosing specific pixel pairs. For a pixel pair whose chief rays are confined to the multiplexing plane, the function $\Delta\phi$ is dependent only on the wavelength detuning. Conversely, $\Delta\phi$ is independent of the wavelength detuning for a pixel pair whose chief rays are coplanar and in a plane orthogonal to the multiplexing plane. Thus, by monitoring the horizontal boxes in Figure 14.19, the wavelength detuning of the probe beam and the needed angle change can be determined. It should be noted that wavelength changes can result in a significant magnification change, thus in addition to a tunable laser and a method of changing the probe beam angle, an over-sampled detector is required. In a similar fashion but using the vertical boxes in Figure 14.19, the probe beam pitch can be calculated.

While this servo signal can be used to find the correct probe wavelength and pitch it is in generally too slow to be part of a feedback system. However, the same concepts of Bragg mismatch leads to the dither align and wobble servo algorithms presented next.

14.3.3 Dither Align

While the preceding analysis demonstrates a signal that can be used to align probe beam pitch, wavelength, and multiplexing angle, it is not very practical for a drive system for several reasons. First, the bandwidth of the signal is low thus the transfer rate of the drive can be negatively affected. Second, there are many potential errors trying to detect peaks in the regional intensity signals, especially for holograms stored at high density. A more effective method relies on the total camera image and the movement rate of the intensity pattern with small iterative probe beam angle shifts. The dither align algorithm uses this movement rate and the SNR signal of the recovered hologram to set the optimal probe beam wavelength, pitch and multiplexing angle from any arbitrary starting point.

While the intensity pattern movement rate is used to determine the wavelength and pitch, the SNR quality metric is used as the basis for determining the reference beam angle error. SNR is calculated by embedding known data patterns within each hologram (referred to as 'reserved blocks' [12]), and measuring the fidelity of the detected pattern according to a formula given in Chapter 11. SNR for a given hologram as a function of reference beam angle generally has the form of a sharp peak, with SNR reaching a maximum value at the optimal angle, and falling off steeply as the angle deviates from optimal, e.g. approximating:

$$\mathrm{SNR}(\phi) = \mathrm{SNR}_0 - C(\phi - \phi_0)^2 \tag{14.9}$$

where ϕ is the external probe beam angle with respect to the medium normal, ϕ_0 is the optimal read-out angle and C is a constant defining the quadratic peak shape. SNR_0, the peak SNR of the hologram, is not known in advance and indeed varies as a function of write reference angle and medium deformation. An alignment error indicating both the sign and magnitude of the probe angle error, $err = \phi - \phi_0$, cannot be determined from a single SNR sample. However, from the SNR peak model it is apparent that the *derivative* of $\mathrm{SNR}(\phi)$ is proportional to *err*. The read-out beam angle error can thus be determined from the difference of two SNR samples offset in ϕ:

$$err = \phi - \phi_0 = \frac{\mathrm{SNR}(\phi - \Delta\phi) - \mathrm{SNR}(\phi + \Delta\phi)}{4C\Delta\phi} \tag{14.10}$$

where $\Delta\phi$ is a constant read-out beam angle offset.

Using this, the dither alignment proceeds as follows: The three axes – reference beam angle (ϕ), reference beam pitch (ρ) and wavelength (λ) – are initialized to starting values ϕ_0, ρ_0 and λ_0 representing the current normal location of the desired hologram. A dither offset of $+\Delta\phi$ (typically 6 mdeg) is applied and a first holographic image exposure is taken (i.e. at $\phi_0 + \Delta\phi$); then an offset of $-\Delta\phi$ is applied and a second holographic image exposure is taken. The alignment error in angle ϕ is then determined from Equation (14.10). If either of the exposures used for the iteration failed to produce an SNR measurement, then total page intensity, which has a similar quadratic peak, is used instead.

Probe beam pitch and wavelength misalignment are determined by quantifying the apparent motion of the intensity pattern in the dither align image pair. As revealed in the previous section, the vertical component (i.e. in the multiplexing plane) of this motion corresponds to the wavelength tuning error, and the horizontal component indicates a probe beam pitch error. For small errors, where the holograms are close to alignment, the difference in page intensity centroids suffices to reliably detect the motion but this method

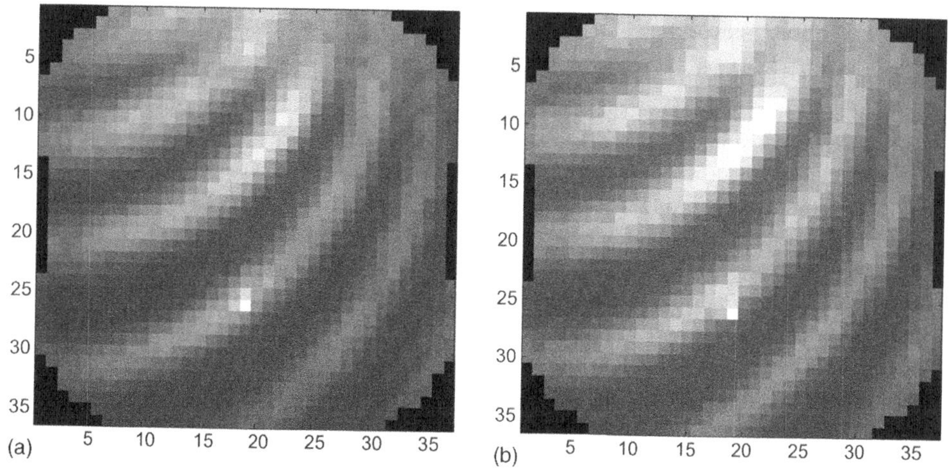

Figure 14.21 *Dither diffraction pattern motion for hologram book with a pitch error of +0.25° and a wavelength tuning error of –3.0 nm: (a) +Δϕ image; (b) –Δϕ image*

is not reliable for poorly aligned holograms. One illustration of this situation is shown in Figure 14.21. Large alignment errors of the reference beam pitch ($err_\rho = +0.25°$) and wavelength ($err_\lambda = -3.0$ nm) have caused fragments of six separate holograms to be reconstructed simultaneously, appearing as the curved, bright bands. A slight upward and leftward movement of the band pattern in Figure 14.21(b) compared with Figure 14.21(a) is apparent to the eye but would not be reliably detected as a difference in the intensity centroid. For this reason, a more complex algorithm based on a modified cross-correlation of the two intensity images was developed.

A cross-correlation operation is performed between reduced resolution versions of the two holographic images (36 × 37, as in Figure 14.21). The DC component of each image is removed, and only the inner 3 × 3 terms of the cross-correlation matrix are calculated, corresponding to correlations between the images shifted by only 0 or ±1 elements. The wavelength and pitch errors are then proportional to the difference between the top and bottom row sums (for λ), or the right and left column sums (for ρ) of the correlation matrix. The method was refined and verified on large sets of image pairs spanning the entire three-dimensional misalignment space in ρ, λ and ϕ.

During each iteration of the dither align algorithm, a pair of dithered images are collected and used to determine the alignment error in ϕ, λ and ρ. All three axes are adjusted to reduce the error, and the process is repeated using the new axes positions as the starting positions. During each iteration, the alignment errors are checked to see if alignment is complete. If a settling criterion is met – i.e. the magnitude of each error sample is smaller than a predetermined threshold for several samples in a row – then alignment is achieved and the process terminates. A final, aligned holographic image exposure ('page check') is taken to qualify the alignment, and to actually recover the hologram data. If the algorithm has settled on the wrong page, then the reference beam angle is adjusted to nominally select the right page, and the algorithm continues iterating in order to settle on that page.

14.3.4 Wobble Servo

Angle-multiplexed holograms are recorded in a volume medium by varying the reference beam angle in small increments. However, upon playback, misalignment, medium shrinkage, thermal expansion, and other distortions cause a displacement of the optimal recovery angles from their recording angles. This displacement typically contains a dominant low-frequency component, along with much smaller high-frequency excursions. The low-frequency component is a characteristic since the holograms are all subject to the same dimensional distortion. Attempting to recover the holograms at the original recording angles would result in errors and/or reduced margin in locations where the displacement is high.

The purpose of the wobble algorithm is to provide an estimate of the optimal probe beam angle, pitch, and wavelength for recovery of the next hologram in a sequence using information available from previously recovered holograms. The available information consists of a scalar quality metric (for example SNR or brightness), the diffracted intensity patterns, and the recovery angles of the previous holograms. The method can be described as using two conceptual steps: (1) derive a feedback error signal which is an estimate of the displacement from the optimal recovery position for each hologram; (2) apply the feedback signal as input to a compensation algorithm that produces one or more axis control commands to be applied for the recovery of subsequent holograms. Probe beam angle, pitch and wavelength can all use this feedback scheme all at the same time. In addition, because it only uses information from previous holograms, the read transfer rate is *not* affected by the servo system. Thus data are recovered from each exposure of the media.

The wobble algorithm relies on the SNR signal in the same manner as the dither algorithm, namely it assumes the SNR peak of adjacent holograms is identical and it recovers with a small fixed offset, $\Delta\phi$, between pages. In order to estimate the probe angle error, *err*, while recovering a sequence of holograms with only one exposure which is near the SNR peak for each hologram, it is necessary that (1) $\Delta\phi$ be small and (2) each *err* sample is calculated from the difference in SNR of *two different* holograms within the sequence, i.e.:

$$err_h \approx \frac{(-1)^h\left\{\mathrm{SNR}_h\left[\phi_h-(-1)^h\Delta\phi\right]-\mathrm{SNR}_{h-1}\left[\phi_{h-1}+(-1)^h\Delta\phi\right]\right\}}{4C\Delta\phi} \tag{14.11}$$

where the subscript denotes the hologram number in the sequence. The nominal read-out beam angles ϕ_h and ϕ_{h-1} should be separated by the true spacing between holograms h and $h-1$ in order to produce the most accurate estimate. Furthermore, the alternating sign factor $(-1)^h$ has been introduced so that even-numbered holograms are always sampled at $-\Delta\phi$ from their estimated nominal positions, while odd-numbered holograms are sampled at $+\Delta\phi$. Thus, an error sample can be generated with each new hologram when the alternating 'wobble' offset $\Delta\phi$ is applied to the sequential recovery angles. Figure 14.22(a) depicts the even/odd sampling and the derivation of the error signal from the difference in SNR. The method is analogous to 'wobble tracking' employed for track following by some optical disk drives [13]. This method has more noise than a traditional servo signal so a recursive least squares (RLS) filter for estimating the angle of the next hologram based on a (weighted) LS fit to past observations is used in lieu of a simpler (e.g.

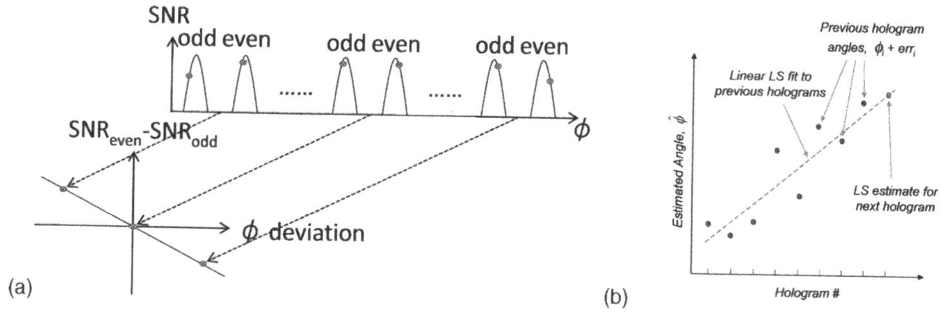

Figure 14.22 *(a) Derivation of error signal. Reproduced with kind permission from Yawara Kaneko, (b) Principle of RLS compensator*

proportional–integral–derivative, PID) control loop. Figure 14.22(b) conceptually illustrates the use of the RLS filter to estimate the location of the next hologram. To initialize the filter, the RLS estimator state may be saved near the beginning of one book (e.g. a dozen pages in) for use at the beginning of the next book.

In addition to the reference beam angle error signal, the imparted wobble produces changes that can be used to determine misalignment in pitch and wavelength. This is possible because the presence of these angular misalignments causes the wobble offset to produce a shift in the best Bragg-matched region of the holographic image. The shift can be detected as a change in the position of the intensity centroid of the detected images.

The principle for pitch measurement is illustrated in Figure 14.23. The wave vectors $\vec{k}_{P1}$ and $\vec{k}_{P2}$ for two read-out beams at slightly different angles are indicated by the arrows, and the locus of the polarization density distribution created by the interaction of a read-out beam with the hologram is indicated by the shaded patch. In a perfectly aligned system, the polarization density patch would lie entirely on the surface of the k-sphere (area in dashed lines), but in the figure a tilt error has been introduced. This is manifested as a clockwise

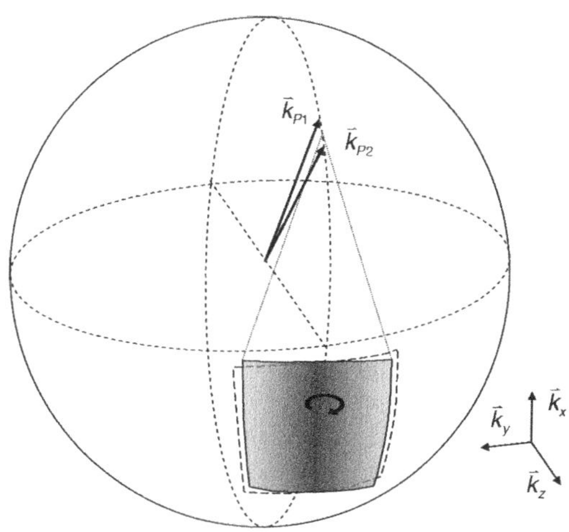

Figure 14.23 *Principle of centroid shift error signal in k-space*

rotation of the patch about an axis parallel to $\vec{k}_x$ passing through the tip of the read-out beam wave vector, as indicated by the circular arrow. The rotation causes the vertical edges of the polarization density patch to separate from the surface of the k-sphere, indicating a Bragg mismatch. This causes diminished DE at the edges of the holographic image, as illustrated by the paler shading.

When holographic exposures are taken at the two angular read-out beam offsets indicated by wave vectors $\vec{k}_{P1}$ and $\vec{k}_{P2}$, the polarization density patch translates up and down as though it was rigidly attached to the tip of read-out beam wave vectors. This causes the line of intersection between the patch and the k-sphere to shift rightwards (for $\vec{k}_{P1}$) or leftwards (for $\vec{k}_{P2}$), and hence the bright, best Bragg-matched part of the image to shift from right to left. This may be detected as a shift in the centroid of the image intensity pattern in the y direction. Conversely, if the rotation is counter clockwise instead of clockwise, then the intensity centroid will shift from left to right instead of right to left, with the amount of centroid shift being a function of the degree of rotation.

In a similar manner, the centroid shifts in response to media rotation, or rotation about an axis parallel to $\vec{k}_z$. In fact, the method cannot actually distinguish between these two rotation components but instead it will indicate a 'zero' alignment error (i.e. no centroid shift) when the hologram is optimally Bragg-matched (though in principle the two rotations could be separated by measuring two holograms as described above). Thus, a small media rotation misalignment can be corrected by a small read-out beam tilt, and vice versa. Furthermore, a combination of small errors in both rotation and tilt can be Bragg-matched by rotating the reference beam slightly out of the plane formed by the $\vec{k}_x$ and $\vec{k}_z$ axes (pitch correction). In a commercial system, the optimal pitch angle setting will change within an angle-multiplexed hologram stack due to medium deformation, out-of-plane errors in the beam steering optics, etc. In the preferred embodiment, media rotation is set to some nominal, invariant value for a hologram stack, and read-out beam pitch is dynamically adjusted in response to the centroid feedback signal in order to optimize the Bragg matching of each hologram.

In yet another embodiment, a centroid shift in the vertical direction (as opposed to the horizontal direction as above) can be used to indicate a wavelength or dimensional mismatch. Again referring to Figure 14.23, a wavelength mismatch would be indicated graphically by changing the radius of the k-sphere and the length of wave vectors $\vec{k}_{P1}$ and $\vec{k}_{P2}$, and an isotropic dimensional mismatch (caused, say, by thermal expansion or contraction of the medium) would be graphically indicated by changing the radius of curvature of the shaded polarization density patch. In either case, the curvature of the k-sphere would no longer match the curvature of the polarization density patch, and so the polarization density patch would separate from the k-sphere at the top and bottom of the patch when best Bragg-matched in a horizontal locus across the middle. Thus, the read-out beam angle wobble will cause the locus of highest intensity to shift up and down, which may be detected as a shift in the vertical centroid of the intensity in the detected images.

In an integrated system, the wobble offset can be used to derive all three error signals simultaneously, which are in turn used to close independent servo loops. Because changes in the alignment of sequential holograms are slowly varying, and because the error signals are relatively noisy, low-gain servo compensators are required, such as the RLS filter described above.

We have demonstrated a system that recovers sequential holograms with less than 0.15 dB average SNR loss per hologram when compared with careful optimal alignment. In the demonstration, the reference beam angle and pitch servos were each closed using a RLS compensator. The wavelength error signal was merely monitored in order to trigger retuning between books for sufficiently high error. (It is not desirable to change the wavelength dynamically because of its interaction with the required read-out angle, and in any case the wavelength should only need to be retuned infrequently in response to temperature changes.) The graphs of the simulated and experimental centroid servo signal are given in Figure 14.24. These plots show centroid shift in pixels versus pitch angle. These curves have the characteristics of a good servo signal in that they cross through the point (0, 0) and can provide magnitude and direction of the required correction.

14.3.5 Other Algorithms

There are several other basic algorithms that are used during interchange, seek, and retry operations. These tend to be brute force algorithms to find specific optimums. For example, a reference beam angle scan may be used to identify the angle of peak hologram diffraction efficiency. From these data the optimal recovery angle for the first hologram can be identified very accurately. Another type of scan operates similarly, but the SNR is measured instead of DE. Other alignment axes, such as the disk radius and angle (r, θ) may be scanned while SNR or DE is measured. These operations may be used to center the polytopic aperture on a holographic image as part of an interchange operation. For example, the disk is scanned in radial while SNR is measured. The peak SNR location can be identified from the scan data and the best radial disk location corresponds to the peak.

14.4 Drive Control

14.4.1 Introduction

The final section of this chapter gives an overview of the algorithms needed to control a holographic drive during the operations of data recording and recovery. As shown earlier in the chapter, high NA holographic data storage devices have unique media alignment tolerances and are sensitive to media deformation resulting from changes in temperature or ambient humidity. These changes can necessitate complex algorithms during recording as well as interchange and compensation during recovery. It is also necessary to monitor and adjust the power used to record and cure holograms due to changing environmental conditions and resulting changes in media sensitivity. This section gives a simplified overview of these fundamental operations.

14.4.2 Record Operation

A simple control flow diagram for a record operation is shown in Figure 14.25. This flow diagram is for illustrative purposes only and is strongly dependent on the writing speed and

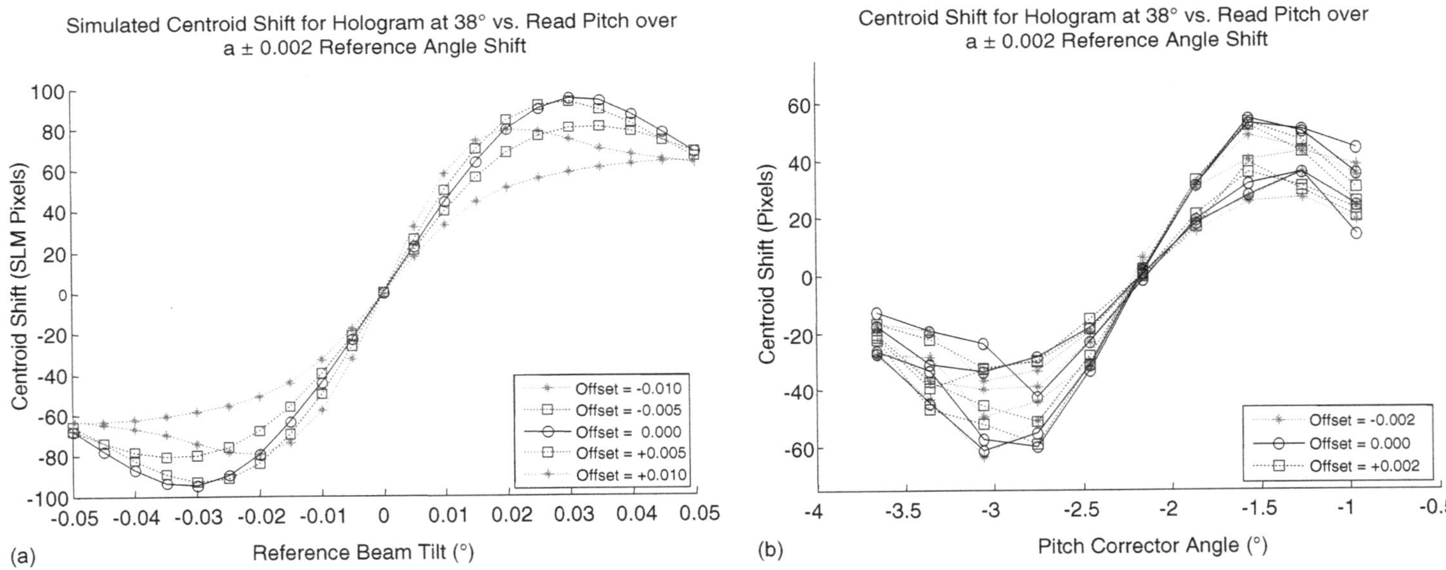

Figure 14.24 *(a) Simulated and (b) experimental intensity centroid shift as a function of pitch error*

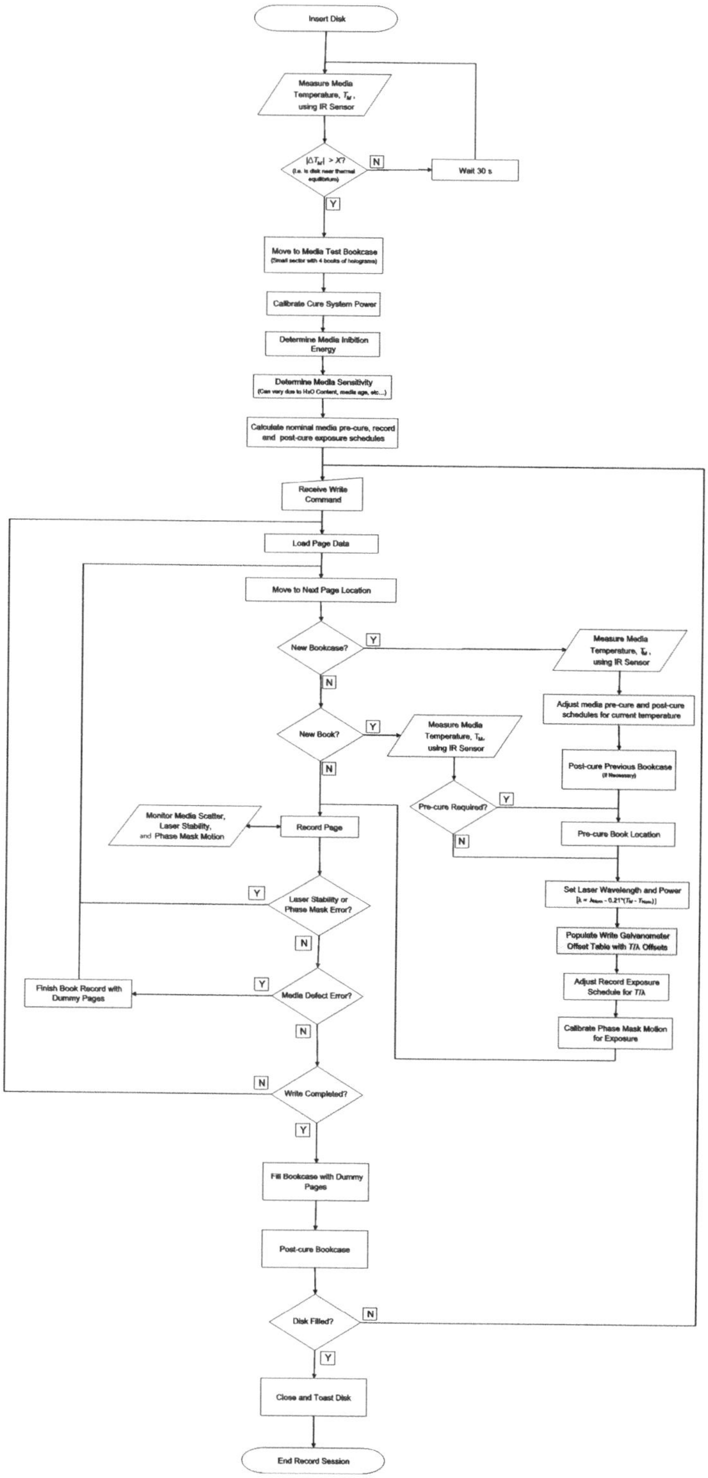

Figure 14.25 *Simple record operation flow diagram*

the holographic medium formulation. The diagram presented is roughly based on the InPhase DVT drive and the InPhase Tapestry 300r® holographic media.

After the media is loaded, the temperature of the media is monitored to ensure that the drive temperature is within its operating range and that the media is near thermal equilibrium. Once this condition is reached, a small area of the media is used to fine tune the exposure times used during the record. This step involves writing several books of holograms (typically 4) to determine the needed energy to overcome the inhibition exposure (pre-cure energy) and to gauge the sensitivity of the media. The media sensitivity is strongly dependent on the temperature of the media but can also be potentially affected by the absorbed H_2O content, the age of the media, and its storage conditions. For media formulations with high *M/#*, this step can potentially be skipped, as it is possible to over pre-cure the media and write holograms stronger than necessary to compensate for any media variations. The trade-off with these formulations is increased shrinkage and associated media deformation caused by the post-cure process. This is not a problem as long as the writing speeds are high enough to maximize the bookcase size. The exposure calibration is also used verify that the recording and cure laser powers are within the specifications required for drive operation. After the calibrations are complete, data recording may be carried out.

As part of the recording process, the media must be pre-cured before writing and each bookcase must be post-cured after they are filled. In addition, at the beginning of each book record, the media temperature is measured and the laser is tuned to the correct wavelength such that the holograms are recoverable at a standard temperature with the system's central wavelength. As part of this pre-compensation, the reference beam angles used to record each page are adjusted so the angle of the probe beam will be at its nominal value with standard conditions. The pre-cure, post-cure and exposure schedules are also adjusted to compensate for the temperature-dependent photosensitivity of the media and the motion of the phase mask is calibrated for the new exposure schedule.

The laser mode stability, phase mask motion, axis servo motors, and media scatter from the reference beam are monitored continuously during the record process. Errors with the mode stability, axis servos or phase mask will cause the pages to be rewritten. Increased media scatter, typically indicating a media defect, will cause the system to write a dummy book in that location (to keep media usage uniform) and will resume writing the data at the next book location.

When all the user data have been written but there is still space in the bookcase for more holograms, the system will fill the bookcase with dummy data and post-cure the bookcase. This step ensures that the media usage is uniform and minimizes media distortion and index structure build up from diffusing monomer. When the entire disk is filled the disk is finalized by writing library books, isolated books of holograms containing the disk map and file allocation tables, and the disk is bleached. The bleaching process involves flood curing the entire disk to ensure that no remaining monomer exists in the media.

14.4.3 Read Operation

A flow diagram for a read operation is presented in Figure 14.26. Again, this flow diagram is simplified to illustrate an actual read operation based on the InPhase DVT drive. When

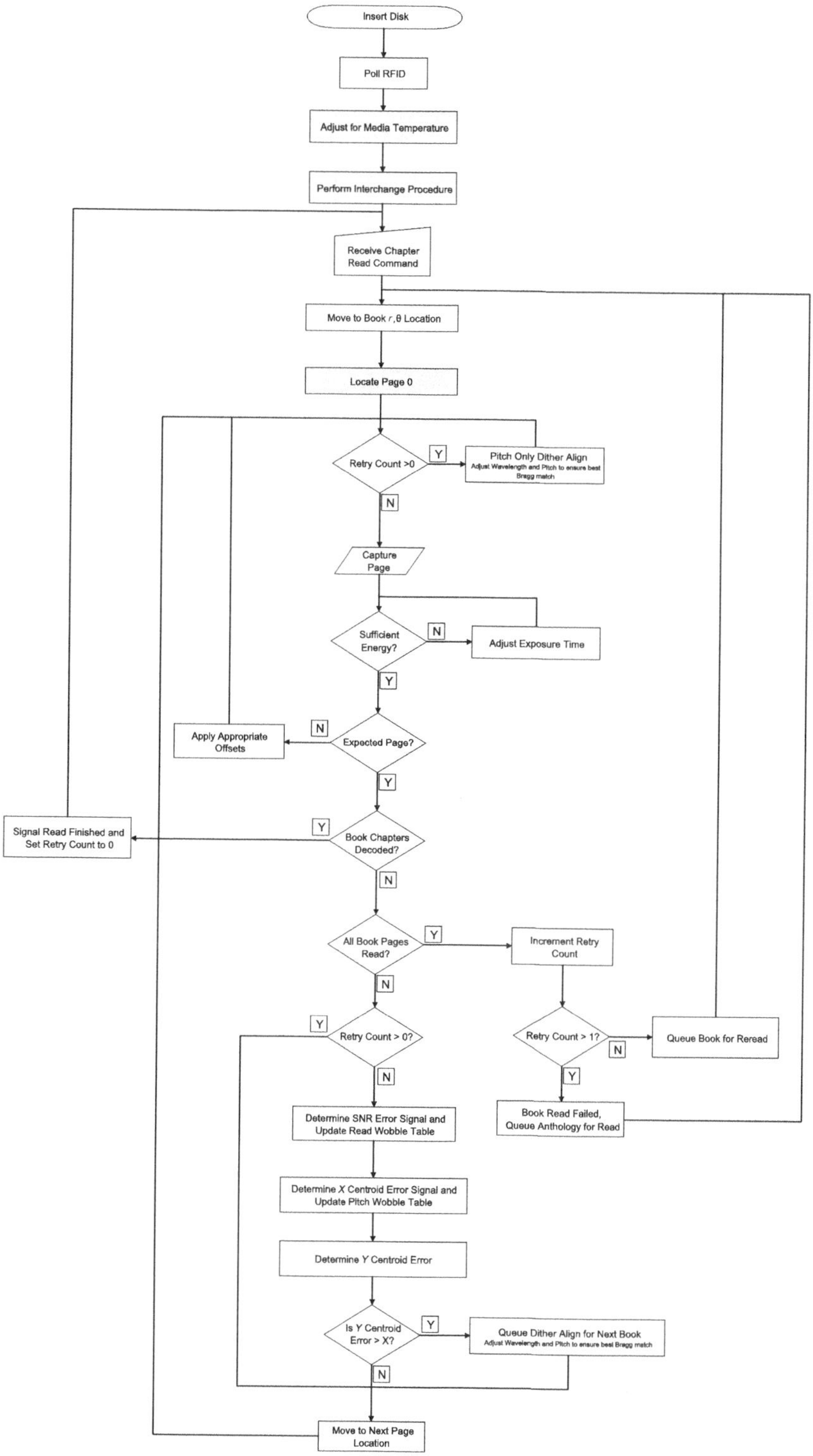

Figure 14.26 *Simplified read operation flow diagram*

a disk in loaded, the drive locates the library map using a radio-frequency identification (RFID) chip in the cartridge. After adjusting the probe beam wavelength and orientation to compensate for media temperature, the drive then uses the library books, which typically have a higher diffraction efficiency and are recorded at lower density that the user data, to perform an interchange procedure. The mechanical differences between drives, run out of the disk spindle, and tolerances of the disk hub change the location of the holograms on the disk (r, θ) and the orientation of the probe beam needed to recover the holograms with each disk load. The interchange procedure, discussed below, is used to correct these offsets.

Once the library books are read and the data locations on the disk are mapped, a read command is issued by the host to retrieve a desired set of logical blocks. These blocks are organized into a set of chapters that will be read to retrieve the user data. While each book can contain several chapters and the user data may be located in only a single chapter, for simplicity's sake the entire book is read starting at page 0. Due to a host of causes, including nonuniform medium distortion, temperature-induced medium expansion, and changes in disk seating, the location of the first page of the book with regards to the reference beam orientation can change from book to book. The algorithm to locate this page, discussed later in this section, depends on the previous book read.

The book is then read a page at a time using the wobble algorithm to track the deviations of the probe beam pitch and Bragg angle from nominal. As discussed in previous sections, the wobble algorithm also tracks the recovered hologram irradiance centroid motion in the plane of the reference beam and data beam and uses this motion to determine if a wavelength adjustment is necessary. Wavelength adjustments typically happen at the beginning of a book read. The exposure is controlled by a servo system to optimize the usage of the image sensor dynamic range. On the occasions where the read fails, the system will verify the book is properly aligned with the polytopic aperture and re-read searching for the correct probe beam orientation for each page. If this also fails, the system will rely on anthology error correction to recover the user data.

14.4.4 Interchange Operation

The interchange operation (see Figure 14.27) starts by moving the media and probe beam orientation to the default library map location. It then attempts to center the book on the polytopic aperture (see Figure 14.28) using a coarse theta scan of the disk while monitoring the recovered hologram intensity. Theta is set to maximize the intensity and a dither alignment is performed to reduce the Bragg mismatch of the hologram being recovered. Once the Bragg matching is optimized, the probe beam angles are set to those expected for the media temperature and wavelength of the probe beam. The imaging of the hologram on the polytopic filter is then fine tuned using radial disk scans and coupled probe beam pitch/disk theta scans while monitoring SNR. In some cases, the interchange routine then might need to calibrate the probe beam pitch and angle offsets to compensate for drive build variations. This involves scanning the recovered library book to determine the correct offset across all the pages. The library book is read to determine the user data locations on the disk. The data in the library book are written in several redundant locations, each recorded by the RFID chip, for cases where the primary library book is unrecoverable.

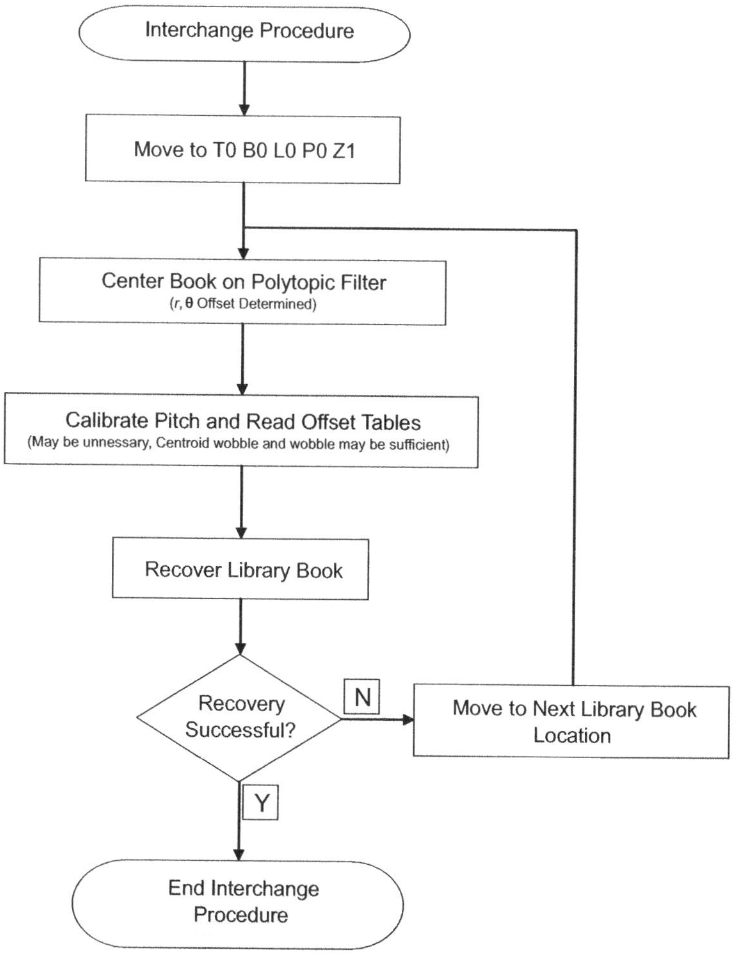

Figure 14.27 *Interchange algorithm*

14.4.5 Locating Page 0

The algorithm to locate the first page in a book being recovered is shown in Figure 14.29. This algorithm depends on what book is being read and whether a previous read attempt was unsuccessful. Since the probe beam orientation can depend on the disk seating and drive variability, experience has shown that for large seeks (large disk motion), it is necessary to search for the first page by changing the probe beam angle while monitoring the recovered image SNR. When the books are being read sequentially, this is not necessary and the location of the first page can be estimated using the location of the previous book recovery. In cases where a book recovery originally failed and is being retried, a dither align is usually necessary to change the probe beam angle, pitch, and wavelength to recover the first page. If a dither align is not sufficient, it is rarely necessary to verify the proper imaging of the hologram onto the polytopic aperture using algorithm to center the book on the polytopic shown in Figure 14.28.

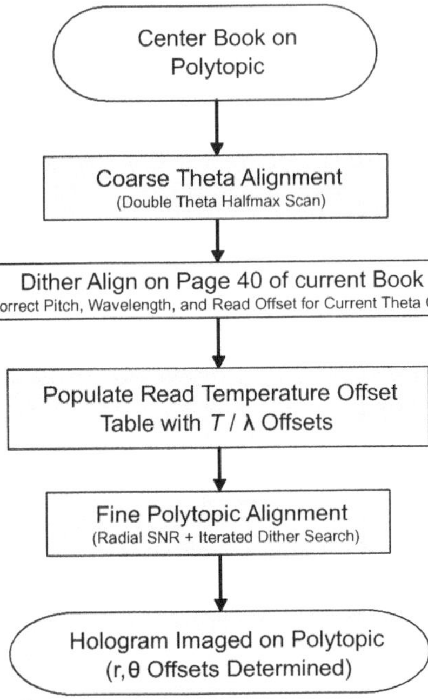

Figure 14.28 *Simple algorithm to center book image on polytopic filter*

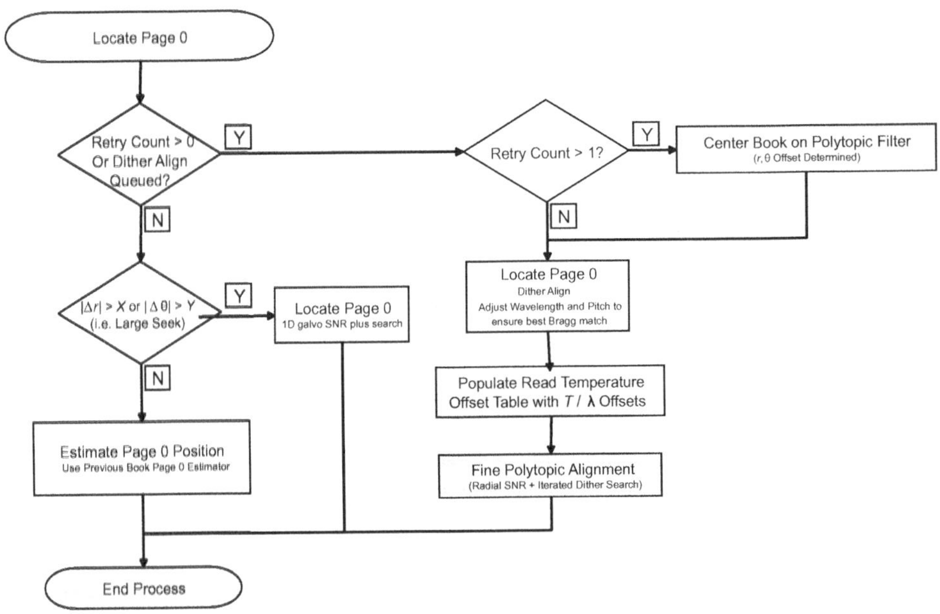

Figure 14.29 *Algorithm to determine probe and medium orientation for the first page of a book*

Acknowledgements

We would like to thank Ken Anderson, Paul Smith, J.J. Kane, Oyvind Haehre, Steve Hughes, Friso Schlottau, David Beal, Ernest Chuang, Edeline Fotheringham, Greg Hertrich and Keith Malang.

References

1. A. Hoskins, *et al.*, Tolerances of a page-based holographic data storage system, *Opt. Data Storage Conf.*, Portland, Oregon, (2007), paper WB2.
2. A. Hoskins, *et al.*, Using Bragg effects to determine media orientation and wavelength detuning in a holographic data storage system, *Int. Workshop on Holographic Memory*, Penang, Malaysia (2007), paper 27 p 13.
3. M. Ayres, *et al.*, Wobble alignment for angularly multiplexed holograms, *Joint Int. Symp. on Opt. Memories and Opt. Data Storage*, Waikoloa, Hawaii, July (2008), paper ThC01.
4. K. Hayashi and N. Kihara, A new tracking servo system for angle multiplexed holographic data storage drives, *Int. Symp. on Opt. Memories,* Takamatsu, Japan, October (2006), paper Th-I-37.
5. E. Hwang, *et al.*, Real-time video demonstration of holographic disk data storage system, *Proc. SPIE*, Vol. 6282, pp. 6282–6285 (2006).
6. A. Hoskins, *et al.*, Temperature compensation strategy for holographic storage, *Opt. Data Storage Conf.*, Montreal, Canada (2006), paper WC4.
7. K. Curtis, *et al.*, Holographic storage medium having enhanced temperature range and method of manufacturing the same, US Patents 6,348,983 and 6,650,447, February 19, 2002 and November 18, 2003.
8. M. Toishi, *et al.*, Temperature tolerance improvement with wavelength tuning laser source in holographic data storage, *Joint Int. Symp. on Opt. Memories and Opt. Data Storage*, Honolulu, Hawaii, July (2005), paper ThE5.
9. Y. Matsumura, *et al.*, Tilt compensation method of two-beam angle multiplexing holographic memory, *Int. Symp. on Opt. Memories*, Takamatsu, Japan, October (2006), paper Mo-D-08.
10. T. Ishii, *et al.*, Tolerance analysis of monocular architecture, *Opt. Data Storage Conf.*, Buena Vista, Florida (2009), post deadline.
11. H. Kogelnik, Coupled wave theory for thick hologram gratings, *The Bell System Tech. J.*, Vol. 48, No. 9, pp. 2909–2947 (1969).
12. M. Ayres, A. Hoskinsand K. Curtis, Image oversampling for page-oriented optical data storage, *Appl. Opt.*, Vol. 45, pp. 2459–2464 (2006).
13. A. B. Marchant, *Optical Recording: A Technical Overview*. Addison-Wesley, Reading, Massachusetts, pp. 180–181, 1990.

15

Holographic Read Only Memories

Ernest Chuang and Kevin Curtis

15.1 Introduction

For traditional optical storage the ability to replicate media very quickly and inexpensively has formed the foundation for the successful launch of platforms. The CD was developed to distribute music, the DVD to distribute movies and Blu-ray to distribute high definition movies. The ability to copy the entire contents from a disk master to a low cost disk copy in a few seconds is a very powerful economic force. Only after media replication and the read only memory (ROM) versions were successful were recordable versions able to successfully come out on the market, leveraging the cost reductions achieved by the ROM versions. If holographic data storage (HDS) is to be the next generation consumer technology, it needs to have the ability to quickly replicate full media to have the same economic advantage as current optical technology.

This chapter details both the technology for compact inexpensive holographic read only memory (HROM) drives required for the consumer market as well as a high speed low cost replication method for the media. We present new technologies such as non Fourier plane filters to implement polytopic filtering and/or Nyquist filtering, which is an enabling technology for any drive design. This massively reduces the complexity and cost of drives, whether read only or recordable. We also developed the unusual concept of asymmetric phase conjugation to simplify the drive optics. In addition to traditional disk media, we developed a small card for mobile content distribution that would be complementary to solid state memory (SSM) formats. This unique platform capitalizes on advantages of holographic storage, such as not requiring fast-spinning media to achieve good transfer rates, to allow new formats for optical storage such as card media to be utilized.

Holographic Data Storage: From Theory to Practical Systems Kevin Curtis, Lisa Dhar, Adrian Hill, William Wilson and Mark Ayres

The goals of this development were:

- A compact drive capable of 4 GB capacity on a small media card with 2 MB s^{-1} transfer rate, backwards compatible to solid state memory.
- Demonstration of the two-step mastering process for high diffraction efficiency masters allowing quick high density replication.
- Prototype tools needed to master and replicate holographic media.
- A method for fast replication from a single master media (under 4 s replication)
- A roadmap to 50 GB on a flash-sized card.
- A low cost reader (~US$20) and low cost replicated media (< US$0.40).

A general comparison of HROM and SSM is given in Figure 15.1, where an up arrow signifies an advantage, a down arrow a disadvantage, and a horizontal arrow signifies an equal or neutral position on that feature. In general, for mobile application SSM has been the default and also only choice, as its small size, power utilized, cost, and ruggedness are very favorable. However, for high volume content distribution, typically new mask sets are needed to fabricate content-specific chips. This technology results in a costly media product compared with traditional optical replication and also incurs significant market delays. Since most of the volume and almost all profits typically come from the media sales, media cost is a very important aspect and could be used to lower drive costs.

With the technology developed and explained below, holographic storage can implement low cost full media replication for disk or card media. This is a significant advantage that until now has mostly been held by other technologies such as SSM or some bitwise volumetric approaches. By pushing the technology towards consumer applications, technologies such as the angular filter can help enable low cost versions of HDS for readers and recorders such as the Monocular implementation. In addition to supporting holographic

ITEM	HROM	SSM ROM
MEDIA CAPACITY ROADMAP	▲	▼
MEDIA SIZE	▶	▶
MEDIA PRICE	▲	▼
PRICE PER GB	▲	▼
MEDIA TRANSFER RATE	▶	▶
MEDIA SECURITY	▲	▼
MEDIA LIFETIME	▲	▼
DRIVE PRICE	▼	▲
DRIVE SIZE	▶	▶
POWER CONSUMPTION	▼	▲
WOW ! (COOL) FACTOR	▲	▼
REPLICATION TURN AROUND TIME	▲	▼

Figure 15.1 *Comparison of HROM with SSM*

storage as the next generation optical platform, perhaps for three-dimensional (3D) movies or ultra high definition movies, by combining with other unique properties of HDS, new formats and products can be realized.

15.2 System Design Considerations

The basic HROM system architecture was driven in large part by the desired reader form factor. For a drive suitable for a portable consumer unit, small size, low cost, low power, and robust operation are all critical features. To achieve the most simple and compact reader, our original design concept was to use lensless phase-conjugate read-out in combination with reflection holograms, as shown in Figure 15.2(a). During recording, the Van der Lugt geometry is used, illuminating a spatial light modulator (SLM) with a converging beam so that the data modulated beam naturally converges to a Fourier plane before entering and being recorded into the media. This would allow the image to be similarly reconstructed without lenses using conjugate read-out. The design is simple, low cost, and keeps all beam delivery and read-out components on the same side of the media to help minimize drive volume.

However, during the course of development there were two major changes to the architectural concept. The first change was the introduction of a Fourier transform (FT) lens. The Van der Lugt geometry offers an elegant system design, but generates only an approximate FT, where the approximation becomes less valid with increasing numerical aperture (NA) of the beam. At high NA, the transform size is larger than a true FT, forcing an enlarged polytopic aperture and corresponding book spacing, and hence a decrease in recording density. Also the short focal length (about 2.4 mm) targeted for the small reader form-factor resulted in a strict depth of focus tolerance of about $\pm 25\,\mu m$. To avoid these limitations, we introduced an isoplanatic lens to perform the transform, specifically designed to produce an accurate FT with wide shift and depth of focus tolerances. This is discussed in more detail later, as well as in Chapter 4.

The second major change was switching from reflection to transmission holograms. Further investigation found that it is impossible to fully compensate anisotropic thermal expansions or contractions of the media in the reflection geometry. Anisotropic thermal expansion arises from the media structure, where a holographic recording material having

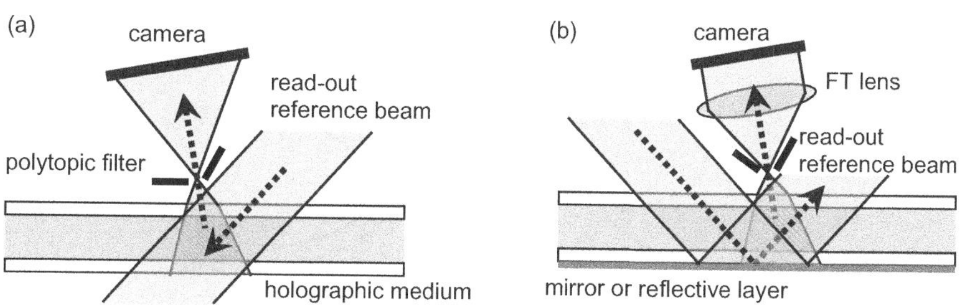

Figure 15.2 (a) Lensless conjugate read-out with reflection holograms. (b) Conjugate read-out using Fourier transform lens with transmission holograms

a relatively high coefficient of thermal expansion (CTE) is bound between substrates having relatively low CTE. During thermal expansion the lateral (x-y) expansion of the holographic material is constrained by the bonds to the substrate material, forcing most of the dimensional change of the material to be in the media thickness (z) direction.

A comparison of the effect of anisotropic CTE on transmission and reflection holograms is illustrated in Figure 15.3, which shows k-space diagrams for the two cases. Figure 15.3(a) shows a representative case of recording in the transmission geometry at a wavelength λ_1 between a plane wave reference beam R_1 and a complex signal beam consisting of components ranging from S_{1A} to S_{1B}. The recorded hologram gratings are represented by a similar range of grating vectors G_1. Under thermal expansion, the media expands disproportionately in the media thickness z direction, which shrinks the grating vectors mainly in the z direction to G_2, represented by dotted lines. In the case of transmission geometry, read-out can be compensated by adjusting the angle and wavelength λ_2 of the read-out reference beam R_2, so that all components of the recovered gratings stay closely Bragg-matched, as shown in Figure 15.3(b).

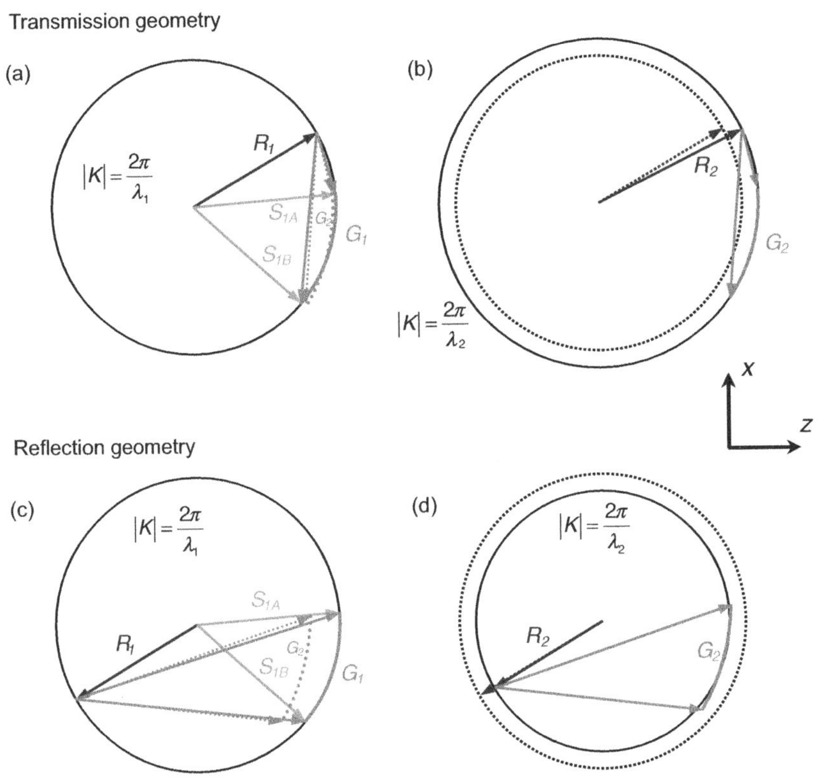

Figure 15.3 *Illustrative examples of effect of anisotropic media CTE on hologram recording and read-out. Transmission geometry: (a) recording; (b) read-out. Reflection geometry: (c) recording; (d) read-out. Changes are exaggerated for clarity*

The reflection geometry case is shown in Figure 15.3(c) and (d). Because of the large z components of the reflection grating vectors, the shrinkage of the vectors mainly in the z direction has two conflicting effects on read-out. The shortening of the vector lengths requires a longer read-out wavelength λ_2 to shrink the k-sphere and keep the gratings close to Bragg match on read-out. However, the curvature of the grating space G_2 is flattened relative to G_1, so it needs a larger read-out k-sphere (shorter read-out wavelength) for a matching curvature. These opposing requirements make it impossible to find a combination of read-out reference beam angle and wavelength that achieves good Bragg match for all hologram grating components. Unless the media CTE anisotropy is strictly constrained or thin media is used to reduce readout sensitivity to Bragg mismatch, recovered hologram integrity will quickly degrade with even small media temperature changes in the reflection geometry. Based on simulations, we estimate that for a reflection geometry version of the HROM reader and media to be operable over a 40 °C temperature range (e.g. 0–40 °C), the media CTE_z needs to be kept under about 155 ppm °C^{-1} if the lateral CTE_{xy} is constrained by the substrates to 70 ppm °C^{-1}.

We therefore chose to design the HROM system architecture for reading transmission holograms with an FT lens as shown in Figure 15.2(b). Data and read-out paths are still maintained on the same side of the media by folding the optical path using a reflective layer applied to the media *after* replication. Because the holograms are not recorded with the reflective layer, hologram diffraction will only occur from the reflected reference beam, avoiding any potential interference from diffraction on the incident pass of the reference.

15.3 Reader Design

Based on the system concept of Figure 15.2(b), we designed a reader [1,2] with the target of achieving a user data capacity of 4 GB on a small media card with transfer rate of 2 MB s^{-1} in as small a drive form factor as possible.

The drive uses a combination of angle and polytopic multiplexing for storing holograms. Using an FT lens with 0.65 NA and a 2.4 mm focal length, the polytopic multiplexing book spacing is about 0.19 mm. This allows an array of 106 × 106 books to fit in a hologram recording area of 22 × 23 mm. The hologram footprint is slightly elongated in one direction due to the beam projection angle. The signal path is tilted 10° off normal to the media with the reference beam angle range spanning 45–55.6° from normal in the opposite tilt direction. With a recording material thickness of 0.7 mm, this angular range allows 54 data pages per book. Each data page holds about 7.4 kB of user data after accounting for overhead, resulting in a total storage capacity of over 4 GB.

15.3.1 Optics and Optomechanics

The optical head of the reader is as shown in Figure 15.4. The head consists of a blue laser diode (LD), lenses and prisms for collimation and beam shaping, a custom miniature mirror actuator to change the beam angle at the media, a phase conjugate Fourier transform (PCFT) lens with a 2.4 mm focal length, a polytopic filter, and a custom CMOS camera. The polytopic filter is a 0.19 mm square aperture of 1.2x Nyquist area located 0.5 mm from the media surface. The laser is centered around a wavelength of 406.7 nm, but has some limited

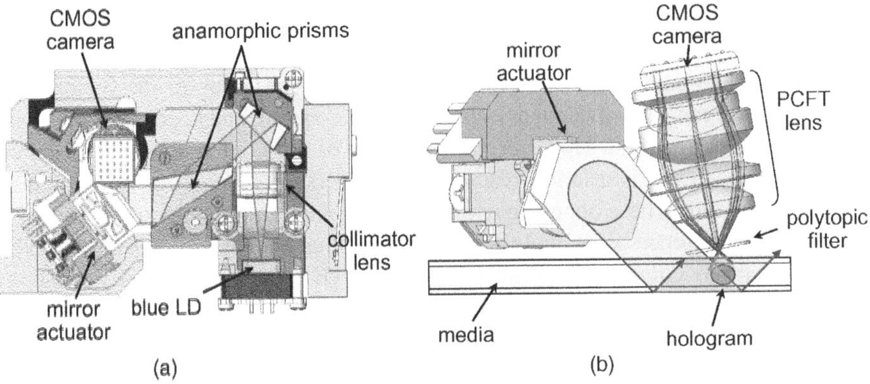

Figure 15.4 Layout of optical head of drive. (a) Top view showing illumination path up to the mirror actuator. (b) Cross-section view showing the illumination path from the mirror actuator to the hologram and the data path from the hologram to the CMOS camera sensor

wavelength tuning ability with temperature. For thicker media (needed to get to 50 GB on a card) better tunability is needed.

The optics consists of two functional parts, an illumination path and a data path. The function of the illumination path is to collect light from the LD, collimate it, shape it into a round beam by the anamorphic prisms, and direct it onto the hologram at the appropriate angle for read-out. The mirror layer in the media is used to reflect the light, folding the optical system and providing another layer of security. The beam is scanned in angle to access the different pages of a book, and the entire optical head is translated in x and y to access different books. The function of the data path is to collect the light from the reconstructed hologram and produce an image of the data page on the CMOS sensor.

The PCFT lens is a custom isoplanatic design and is a key technology component of the HROM system. It is optically 0.71 NA but when used with a 0.65 NA data beam it produces an accurate FT over the entire data field with focus tolerance of over $\pm 100\,\mu m$ and lateral shift tolerance of $\pm 100\,\mu m$. The reader lens shown is a quintet type. We have also designed a simpler triplet type that can be used and that will be discussed in Section 15.12.

15.3.2 Drive Mechanism

The outer drive mechanism that moves the optical head over the media card is shown in Figure 15.5, and was built by Alps Electric. It measures approximately 64×67 mm, with a height of about 1 cm. The head assembly moves over a stationary media card to access different hologram book locations and is controlled by algorithms implemented in firmware. The lead screws have a pitch of 0.67 threads mm^{-1} and are moved by small inexpensive stepper motors allowing an approximate resolution of $5\,\mu m$. Backlash is calibrated and compensated by firmware.

15.3.3 Mirror Design and Control

The custom mirror actuator was also designed and built by Alps Electric. InPhase designed the feedback sensor and the control and calibration electronics. Voice coils are used for the

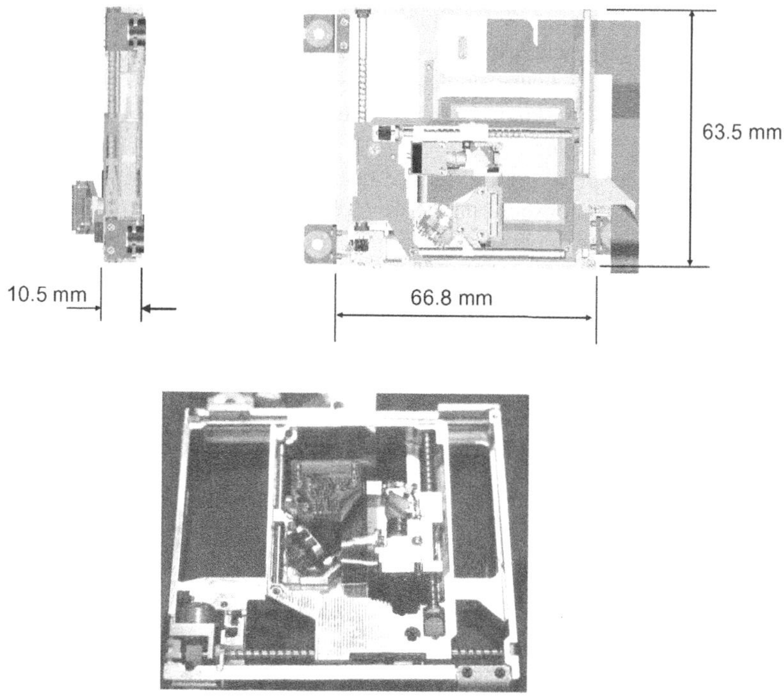

Figure 15.5 Photograph and CAD drawing of HROM reader mechanism with dimensions

tilt actuation, and the feedback sensor consisted of a Hall sensor and a small magnet placed on the yoke of the mirror actuator. Figure 15.6(a) shows the back of the mirror actuator and the voice coils. The size of the actuator and mirror is approximately $8 \times 8 \times 6$ mm. The mirror actuator can be controlled in two dimensions. A large scanning range is needed for multiplexing in one direction but some tilt control is needed in both directions to compensate for system and media tilt as well as alignment tolerances. For the device of Figure 15.6(a),

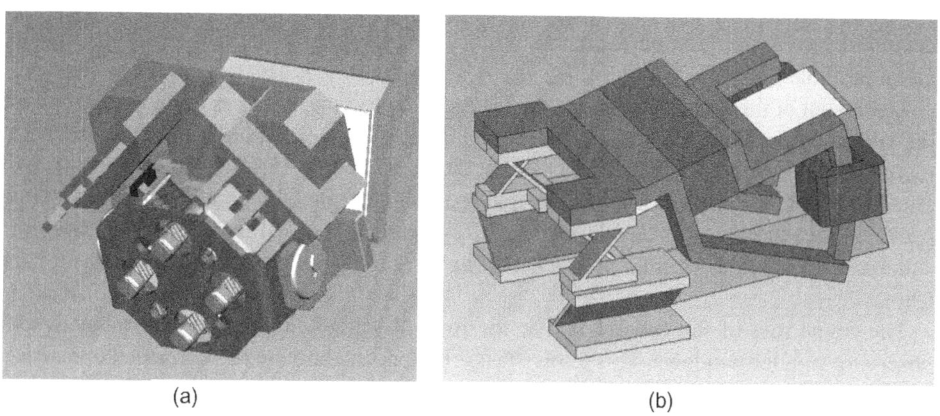

Figure 15.6 (a) Back view of original mirror actuator, and (b) Next generation mirror actuator

the angular range of motion was 8° and 11° for the two principal directions. The angular resolution was about 0.001° with use of the feedback sensor. Because the center of rotation for the mirror is within the actuator, angle changes will result in a shift of the beam at the media, requiring that the reference beam be oversized to ensure full overlap with the hologram over the angular range.

In order to reduce the drive profile, reduce beam size, and improve light efficiency, a next generation mirror actuator shown in Figure 15.6(b) was developed to center the rotation of the reference beam at the hologram position. This had 16° of sweep in the major axis and 3° in the minor axis. More information is available in Chapter 4.

15.3.4 Electronics

In order to control the reader, including many functions such as data channel, actuator controls, laser control, and data transfer to computer, the following prototype electronics were developed. The main printed circuit board assembly (PCBA) includes the following circuitry:

- ARM processor [a reduced instruction set computer (RISC) architecture processor], processor flash, and processor RAM
- buffer RAM (for image data storage)
- FPGA (for actuator control logic, system interface, and camera interface)
- CPLD (for glue/reset logic)
- actuator control circuitry for X/Y, mirror, and laser
- expansion interface for future hardware channel card
- power management and regulation circuitry.

The data channel was on a separate PCBA that attached to this board. Firmware was written that controls the basic drive functions and components, but overall top-level drive control was still done by software on a computer. Figure 15.7 shows the prototype electronics board with reader attached. Figure 15.8 shows a diagram of the electronics.

15.3.5 Camera Development

During the prototyping stage InPhase contracted with Cypress Semiconductor to develop a custom camera chip and package for use in the reader. This custom chip allowed us to develop a small and high performance prototype. The general camera specifications are given in Table 15.1.

The layout of the sensor was carefully designed to push the sensor as close to the edge of the silicon as possible. This was critical for fitting the camera into the optical head with margin for the reference beam to pass on the side. Figure 15.9 shows a diagram and photograph of the shell-case package that was chosen for cost and size reasons. This packaging technology was developed for cell phone cameras and is very inexpensive and suitable for very high volume production. The packaging vendor we used was Xintec in China.

The frame rate of the camera is 500 fps for full frames but can operate faster when observing smaller windows. This allows fast servo optimization to be performed on partial windows before capturing the full data page.

The InPhase professional drive camera uses 8 bit ADC resolution. However, through modeling, 6 bit resolution was determined to be more than enough to give sufficient

Figure 15.7 Photograph of current electronics board and drive attached

accuracy on signal to noise ratio (SNR), so 6 bit ADC resolution was chosen for this custom camera to reduce the internal data flow off the chip. Cost was a driving force in the camera and packaging design. The target was US$1 for the camera in volumes of 1 million units a month after some period of time at this volume.

15.3.6 Power Consumption

Aimed for use in a portable drive, the reader has been designed for the entire system to operate off of a low-voltage battery with power consumption held to around 100 mW. The main PCBA accepts low voltage inputs and makes the necessary conversions for system operation.

Tables 15.2–15.4 show the preliminary estimated power consumption of the drive, based upon conversion from a 3.7 V battery and assuming the mirror actuator from Figure 15.6(b), CMOS camera using buried photodiode array (see Chapter 4), and pulsed operation of the laser. Table 15.2 shows the average drive power factoring in idle time, assuming a usage model with a 20% duty cycle. Tables 15.3 and 15.4 show preliminary estimates of power consumption for the operating and idle cases, respectively.

15.3.7 Data Channel

The data channel is currently implemented in software. It uses our standard over-sampling ratio of slightly greater than 4/3 with the camera pixel pitch being 4.6 μm and the SLM pitch imaged to about 6.44 μm. During data recovery, the first operation is to find the header for the data page. Figure 15.10 shows a photograph of the data page with the header located at the top of the page. This allows for a rough estimate of where the page is located within the camera sensor area. The camera area is deliberately oversized with respect to the data page to allow tolerance margins for imperfect alignment, whether from manufacture, environment effects, or media loading. Using the header location information, the reserved blocks are found and an accurate local estimate of block location is found. This information is used to resample the camera pixels. The pixels are then used to calculate log likelihood estimation

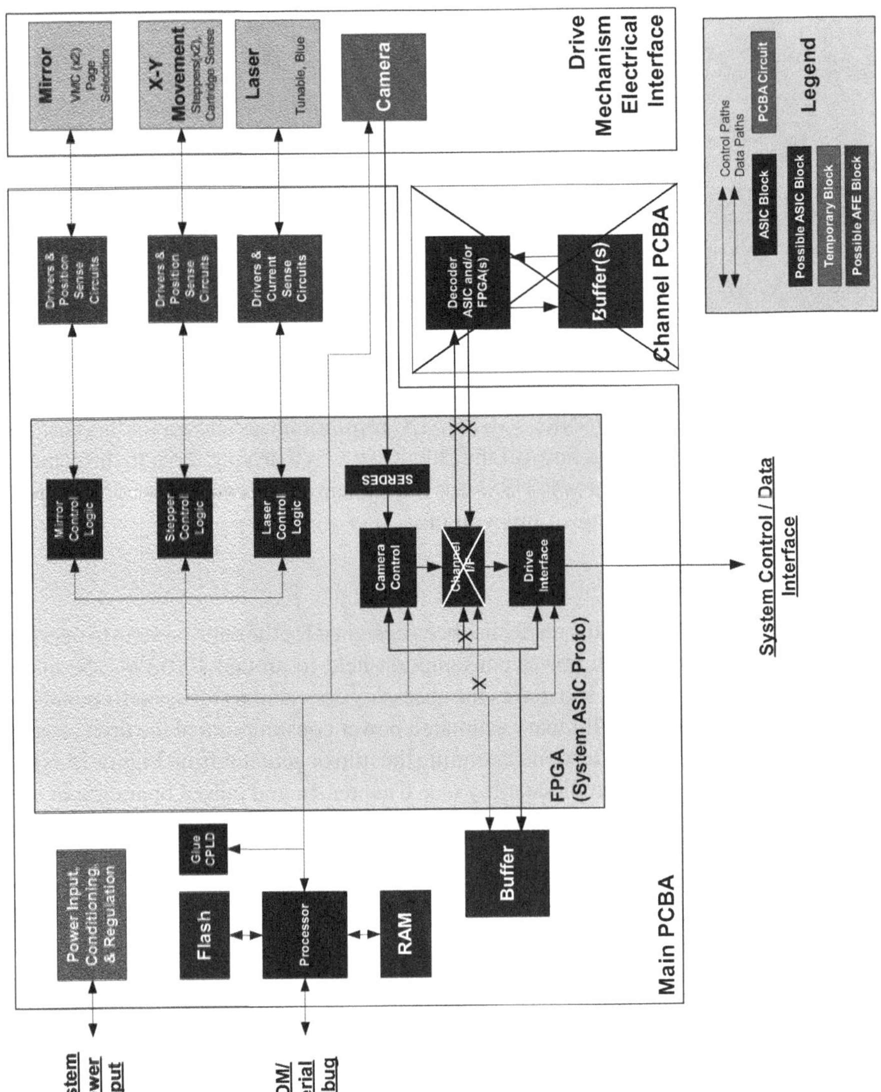

Figure 15.8 *Prototype electronics architecture*

Table 15.1 CMOS camera specification

Parameter	Specifications	Remarks
Pixel architecture	6 transistors	
Pixel size	4.6 × 4.6 μm	
Pixel format	640 × 640	2.94 × 2.94 mm active area
Pixel rate	2160 MHz	Combination of 2 data channels
Frame rate	500 fps	At full resolution, faster with windowing
Shutter type	Pipelined snapshot	Allows integration during readout
Programmable gain	1x to 2x	In 4 steps using SPI programming
On chip ADCs	6 bit	8 on chip at 27M samples per second
Supply voltage	1.8 V	
Interface	SPI	Serial to Parallel interface (SPI)
Package	30 pin BGA	

and then passed to the inner error correction code which is a low density parity check (LDPC) code. See Chapter 11 for more details.

The HROM data page specifics are:

- The data page is 400 × 424 pixels.
- Pixel pitch is 6.44 μm.
- Two barcodes (12 × 224 each) yielding 40 bits of header payload.

15.4 Media Design

The recording media is based on the same blue sensitive material as is used in the InPhase professional drive and also uses the same APO-like material for substrates. Various techniques were developed to minimize the effective thermal expansion of the media to significantly improve the thermal performance and reduce the required laser tuning range.

The general appearance of the media is a small card. A representative photograph of the media card is shown in Figure 15.11. The media format has the following basic characteristics: A frame which includes features (three contact reference points) for referencing the

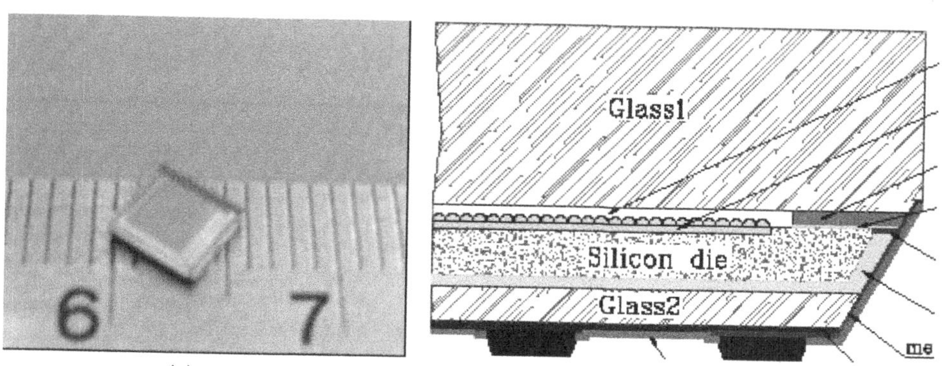

(a) (b)

Figure 15.9 *(a) Photograph of sensor and (b) diagram of package*

Table 15.2 *Average drive power*

$P_{Operation}$ (mW)	P_{Idle} (mW)	Duty cycle	Total power (mW)	Total current (mA)
461	9	0.2	99	27

Table 15.3 *Drive power (maximum operating)*

	Voltage (V)	Current (mA)	Efficiency	Duty cycle	Battery current (mA)	Battery power (mW)
Laser	4.5	50	0.7	0.5	43	161
Camera	3.3	3.6	1	1	3	12
Camera	1.8	10	0.9	1	5	20
ASIC	2.5	20	1	1	14	50
ASIC	1.8	30	0.95	1	15	57
ASIC	1.2	50	0.9	1	18	67
X-Y Actuators (2)	5	200	0.65	0.05	21	77
Mirror Actuator	0.64	40	0.65	0.45	5	18
Battery Voltage	3.7					
Totals					125	461

position of the media to the drive, latching into the drive, and an indent for holding media. It includes two protective substrates with antireflective and scratch resistant coatings on the top substrate that is fixed to the frame, and a mirrored back surface on the bottom substrate. The recording material is between these substrates. A protective cover is placed over the bottom substrate to protect the mirror surface and also to allow for a product label. The prototype media size was 42 × 35 mm with a window area of 28 by 24 mm. The total thickness of the frame was 4 mm with the recording layer 0.7 mm thick and each substrate 1.2 mm thick. Thinner media is used initially for the first product generation to ease Bragg sensitivity for current component limits on laser tuning and servo performance. As technology improves, higher storage capacities can be accommodated in future generations by increasing the recording layer thickness while decreasing the substrate thickness to maintain backwards compatibility.

Table 15.4 *Drive power (idle)*

	Voltage (V)	Current (mA)	Efficiency	Duty cycle	Battery current (mA)	Battery power (mW)
Laser	4.5	50	0.7	0	0	0
Camera	3.3	3.6	1	0	0	0
Camera	1.8	10	0.9	0	0	0
ASIC	2.5	20	1	0.05	1	3
ASIC	1.8	30	0.95	0.05	1	3
ASIC	1.2	50	0.9	0.05	1	3
X-Y Actuators (2)	5	200	0.65	0	0	0
Mirror Actuator	0.64	40	0.65	0	0	0
Totals					2	9

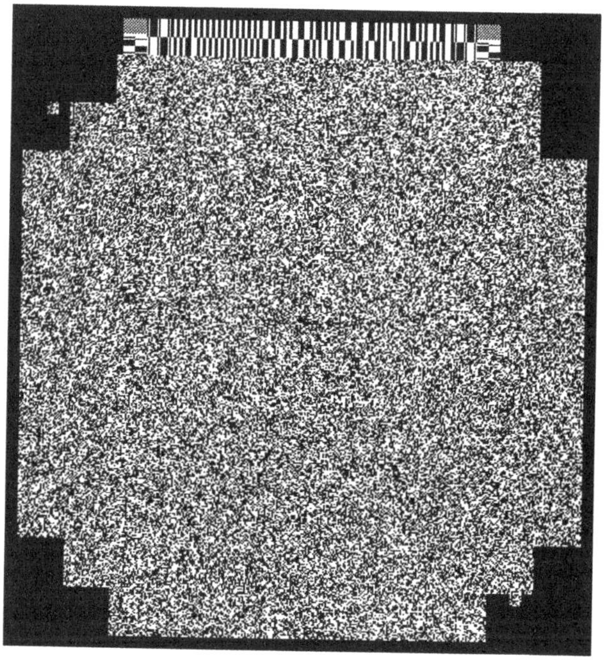

Figure 15.10 HROM data page

15.5 Two-Step Mastering

The distinguishing feature for any ROM format is the ability to create a master and quickly replicate media for distribution. Previous proposals for holographic media replication have included a method for sector-by-sector replication from a holographic master to a target disk using an array of mutually incoherent reference beams [3,4] and also a system based on

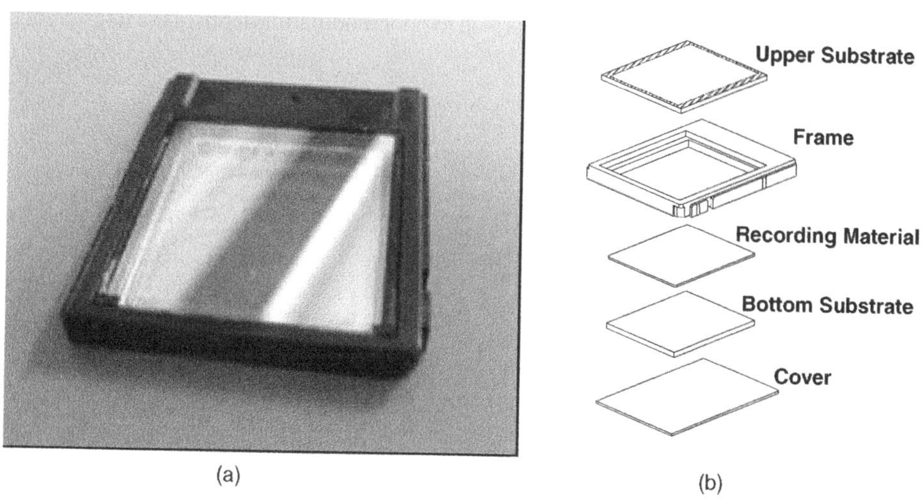

Figure 15.11 (a) Photograph of the media card and (b) diagram of media card components

conventional serial read-out optical disk in which multiple DVD-like patterns are replicated to a target disk from a set of amplitude mask masters [5]. Also proposed was an amplification process using a photorefractive crystal to reduce the loss in diffraction efficiency at density [6].

However, these architectures suffer from the choice of either using a complex process requiring the swapping of multiple masters during replication or using a holographic master in which the holograms may not have sufficient strength for efficient replication, particularly at high data densities. Here we present a two-step mastering process [7] for generating a holographic data master with much higher diffraction efficiency than would otherwise be possible using a standard page-wise recording process.

Replicating holograms from a holographic master to a blank media has previously been demonstrated by reading the master with an appropriate reference beam and positioning the blank media behind the master such that it captures the light interference between the transmitted reference and the diffracted hologram, thus duplicating the holographic data to the new media such as shown in Figure 15.12. In the case of card media containing arrays of angle and polytopic multiplexed books, a single broad plane wave illuminating the master media will simultaneously reconstruct a hologram page that was recorded at that angle from every book, which are then simultaneously copied in parallel to the replicate media. This is then repeated for all reference beam angles used in the master to complete the replication process.

However, this normally presents difficulties in recording a master for high storage densities, because of the low diffraction efficiencies that result from dividing the limited dynamic range of the recording media among a large number of holograms sequentially exposed at the same location, and also by any additional overlap between neighboring book locations. In particular for the HROM system we have described, the architecture dictates that the replicated media must be located 0.5 mm from the polytopic filter plane. This forces the master to be placed on the opposite side of the replicated media from the filter position as seen in Figure 15.12, resulting in an extremely large hologram footprint and high overlap factor between books in the master. Diffraction efficiencies will typically be around 10^{-3} or less, resulting in a large difference between the signal and reference beam intensities incident on the blank media during the replication exposure. Beam ratios around 1:1 are usually optimal, whereas large imbalances lead to poor modulation depth of the interference

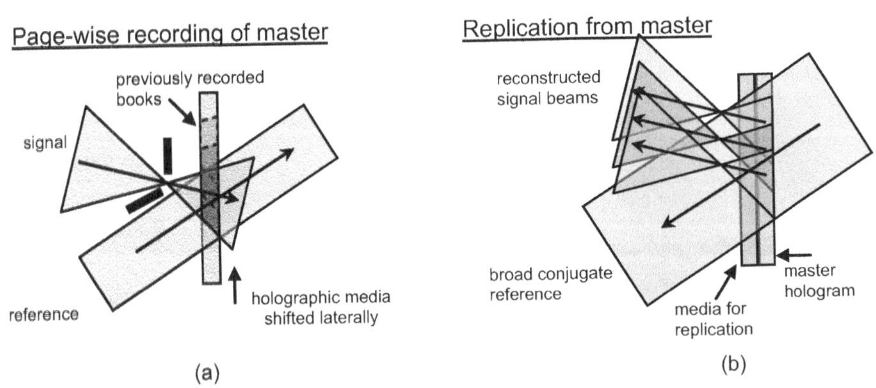

Figure 15.12 (a) Traditional one-step holographic mastering and (b) replication from master

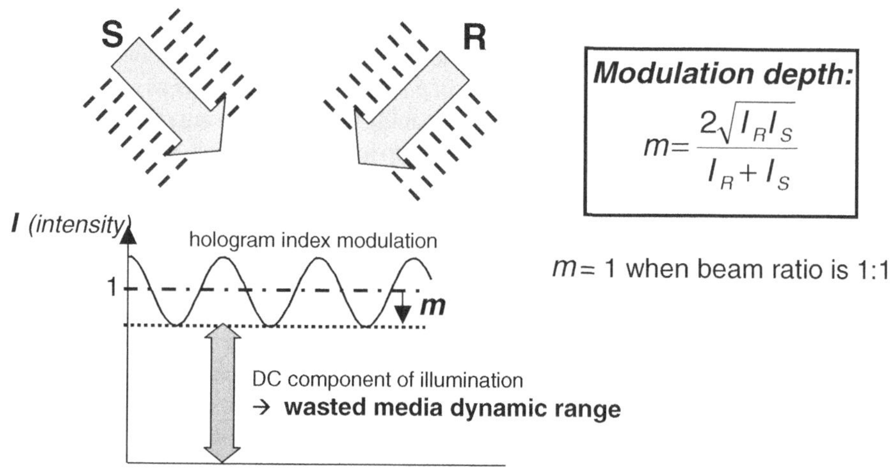

Figure 15.13 *Effect of beam ratio on hologram modulation depth and media dynamic range usage*

pattern, weaker holograms, and considerable waste of the media dynamic range, as shown in Figure 15.13. This ultimately has an impact on limiting the achievable storage density and transfer rate of the system.

To address this issue, we developed a two-step approach to mastering as shown in Figure 15.14. In the first step, all holograms are recorded page by page as usual, but divided among multiple 'sub-master' media. Each sub-master contains all of the book locations but only a subset of the total angular range. For example a target of 30 pages per book could be divided among three sub-masters with 10 pages each. This effectively multiplies the available media dynamic range by the number of sub-masters used, allowing for much stronger hologram recordings. In the second step, all holograms in the sub-masters are

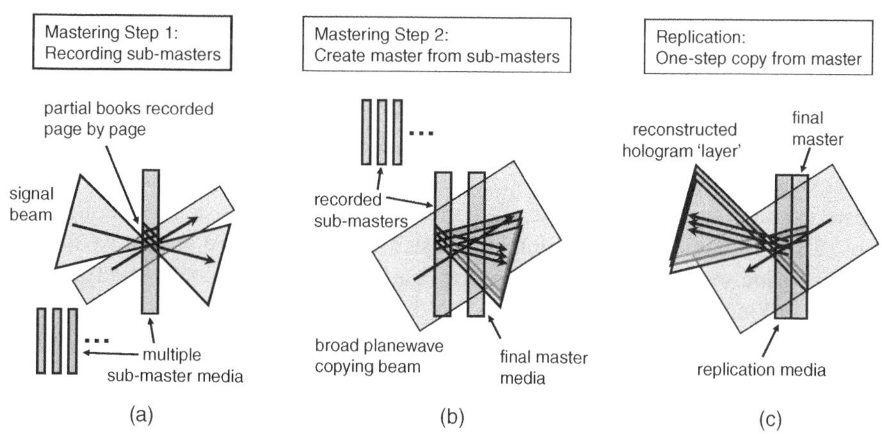

Figure 15.14 *Two-step mastering process: (a) step 1, page-wise recording of multiple sub-masters: (b) step 2, copying from sub-masters to a single final master: and (c) replication from the single master*

combined to a single final master by essentially the same process used for replication, illuminating each sub-master together with the final master with a series of broad plane waves to copy pages from all book locations in parallel. We refer to this as a hologram or page 'layer'. A layer is a figurative description meaning the same page out of all the different books recorded on the media. The term is not meant to imply an actual physical separation of these layers in the media.

This parallel recording by layer eliminates redundant exposures of the same area of the media by the reference beam to record overlapping holograms at the same reference angle, so media dynamic range is used more efficiently. Diffraction efficiencies can actually be amplified from the sub-master to the final master, resulting in a single master media containing all holograms at a much stronger level than would be possible under a conventional one-step process where all pages are sequentially recorded directly into the master. Replication would be performed as described earlier. For the system architecture presented in this chapter, a reflective layer would be applied to the replicated media as a final step, allowing subsequent read-out but no further copying.

Another key advantage of the two-step mastering process is that it allows more freedom in choosing the placement of media at each step. When holograms are recorded page by page, ideally we would like to record with the media at the Fourier plane to minimize the hologram footprint and minimize overlap of neighboring books in order to use the media dynamic range most efficiently. However in the case of one-step mastering, it is impossible for the master and replicated media to share the same space during the replication procedure. So as in the case of Figure 15.12 as was mentioned before, if the replicated media must be close to the Fourier plane and a physical aperture must be present at the Fourier plane during recording of the master, then the master media must necessarily be placed on the side of the replicated media further from the Fourier plane.

In contrast, with the two-step mastering method the intervening step of copying to the master media allows the sub-master and replicate media to share the same space (in different steps) if needed. Sub-master recording can take place with the media exactly at the Fourier plane, which is the most advantageous for the page-wise recording step. A relay lens system is used so that polytopic filtering can be performed at a Fourier plane earlier in the signal path. One-step mastering could also get some partial benefit with a relay lens but the close proximity of the replicate media to the Fourier plane would still prevent recording the master at exactly the Fourier plane.

The gain of using two-step mastering in combination with a relay lens can be illustrated with an example, using the HROM system parameters we have presented thus far. Here we compare the one-step mastering case of Figure 15.12 with the two-step mastering case of Figure 15.14. With one-step mastering, the master media is recorded at least 2.6 mm from the Fourier plane, so that during replication the Fourier plane of the reconstructed holograms will occur 0.5 mm after passing through the 3.1 mm card window (substrate-polymer-substrate) of refractive index 1.5. (Actually it would need to be even further away to account for the full 4 mm card thickness.) At this distance, given the 0.65 NA signal beam divergence, a 5 mm diameter reference beam is needed to cover the signal beam footprint in the master media, where we assume the master media recording material is 1.5 mm thick with 1.0 mm substrates. With the book spacing of 0.19 mm we can calculate a 'book overlap factor', which is the average number of books that overlap at any point in the media, taking both the signal and reference beam spot sizes into account. For this one-step mastering example, the book overlap factor is about 690, using an average of the signal and reference beam volume footprints for simplicity. For the two-step method where sub-masters are recorded at the Fourier plane, the signal beam footprint and required reference beams are much smaller, resulting in a book overlap factor of only 26. We can similarly calculate a 'signal beam overlap factor', which only considers the signal beam overlap disregarding the reference beam. The book overlap factor is relevant for page-wise recording, where both the signal and reference beam illuminations use up the media dynamic range. The signal beam overlap factor is used for read-out, as a measure of how many holograms are reconstructed in parallel with a given read-out reference beam illumination. For the same system parameters, the signal beam overlap factors are about 487 and 16 for the one-step and two-step mastering cases, respectively.

With these overlap factors we can take a closer look at the expected diffraction efficiencies attainable at various steps in the mastering and replication processes. Figure 15.15 summarizes an example comparison between one-step and two-step mastering methods, assuming three sub-masters are used for the two-step method, each piece of media has an *M/#* of 36, and each book is 54 pages.

The diffraction efficiency η per hologram in step 1 is estimated from the usual *M/#* relationship:

$$\eta = \left(\frac{M\#}{M}\right)^2 = \left[\frac{M\#}{(\text{pages per book})(\text{book overlap factor})}\right]^2 \tag{15.1}$$

	One-step mastering	Two-step mastering
Mastering step 1		
Available media *M/#*	1 × 36 = 36	3 × 36 = 108
Book overlap factor	690	25
Signal beam overlap factor	487	16
Diffraction efficiency per hologram	9.3×10^{-7}	5.9×10^{-3}
Diffraction efficiency per layer	0.045%	9.5%
Modulation depth for transfer step	**0.043**	0.59
Mastering step 2		
Available media *M/#*		1 × 36 = 36
M/# limited by modulation depth		21.1
Diffraction efficiency per layer		15.2%
Modulation depth for transfer step		**0.72**
Replication step		
Available media *M/#*	1 × 36 = 36	1 × 36 = 36
M/# limited by modulation depth	1.5	25.9
Achievable diffraction efficiency per layer	**0.081%**	**23.0%**

Figure 15.15 *Comparison of achievable diffraction efficiencies with one-step and two-step mastering methods (assuming 3 sub-masters used)*

where it is assumed that the page-wise recording is done with an optimal beam ratio. The layer diffraction efficiency is calculated as the hologram diffraction efficiency multiplied by the signal beam overlap factor. The modulation depth for the next step of copying by layer can be found from the equation in Figure 15.13, which reduces to $m = 2\sqrt{\alpha(1-\alpha)}$, where α is the layer diffraction efficiency. The modulation depth decreases the usable *M/#* of the blank target media by a factor of $(1-m)$. In step 2 of the two-step method, each layer is recorded in a single holographic recording exposure, so the book overlap factor can be ignored. Step 2 amplifies the layer diffraction efficiencies while recombining all layers to a single master. From the final modulation depths before replication, we can see that in the one-step mastering case over 95% of the originally available replication media *M/#* is being wasted by the poor modulation depth of the master, whereas 72% of the replication media *M/#* is preserved by using the two-step mastering method. The final replicated holograms are over 280 times stronger with the two-step method.

For the sake of simplicity, this brief analysis used many first-order approximations but it is clear that the two-step method offers the potential to create master media with much higher diffraction efficiencies than could be achieved with a traditional one-step mastering method.

15.6 Mastering and Replicating Disk Media

For card media where the holographic books are arranged such that all the reference angles of each book are the same angles for all books in the media, it is easy to see how a broad plane wave can reconstruct a hologram from every book simultaneously in order to replicate data one layer of pages at a time. However, for disk media that rotates, if the reference angles are fixed with respect to the drive's frame of reference, then the reference angles for all the books on the media are not oriented in the same direction in the media's frame of reference. In fact, only books aligned along a single radial spoke would share the same orientation of reference angles. On the opposite side of the disk the reference angles would tilt in the opposite direction.

Some potential ideas for dealing with disk media have been proposed. The use of global master reference beams that are not in the recorder's frame of reference could be used, at the cost of adding considerable complexity to the system. Also the global reference beams would not be usable in a drive to read media, and effectively requires writing twice as many holograms to the media. See Chapter 3 for a description of this type of replication concept for the collinear architecture. The requirement that the total hologram set be divided into 'layers' of nonoverlapping holograms can require a prohibitive number of multiplexed layers in the master, depending on the multiplexing method used. Chuang *et al.* [8] propose this using an imaging system to image master onto replicated media as well. The space bandwidth product of such a lens system is not feasible, and in combination with the problems above this approach is very difficult.

A solution to the disk replication problem was proposed by Sony in Chuang *et al.* [5], where a conical mirror was used to generate a conical recording reference beam centered on and covering the entire media for replication as shown in Figure 15.16(a). This architecture was aimed at storing DVD-like track patterns for bitwise serial read-out with the disk continuously spinning. Different disk patterns would be multiplexed by using a different conical reference angle, requiring a different conical mirror for each disk pattern. Read-out is performed locally using a conjugate reference beam with a matching cylindrical curvature

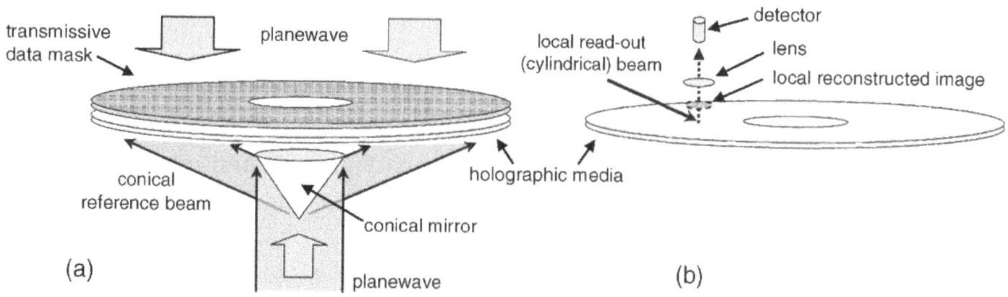

Figure 15.16 Method for (a) replication from data mask using a conical mirror with (b) local bitwise conjugate read-out

and angle to the recording conical beam in that area, where the beam angle selects the disk pattern for read-out.

In addition to the complexity and time of changing conical mirrors during replication, the use of conical mirrors makes it difficult to generate conical beams that cover a typical disk (12 cm diameter) and span a useful angle range. The result was typically angles beyond 60° from normal which are not desirable or usable angles for angle multiplexing many holograms. Typical reference angle ranges are 30–60°. For these desirable angles a conical mirror would have to cover a large inner radial portion of the disk, rendering it impractical.

Here we present a method for replicating disk media quickly (whole virtual layer at a time), by using two new techniques [8]. First we generate the conical beam by use of prisms, as shown in Figure 15.17(a). Such a prism would not obstruct any disk area and can cover the desired range of angles. However, in order to copy a different virtual layer of holograms the prism would have to be changed. This is likely too slow for a commercial process. The second technique is to use a volume holographic media to store all of the required conical waves by angle multiplexing with a plane wave. Then during replication, the conical reference hologram media is used to generate the conical reference needed to replicate that layer by just changing the angle of the read-out plane wave. This conical beam then copies

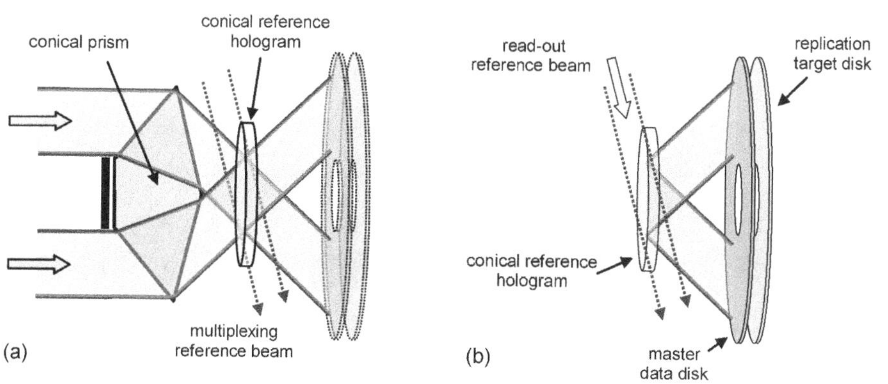

Figure 15.17 (a) Recording holograms of each conical beam reference needed by angle multiplexing. (b) Using conical reference beam hologram to read out master disk and copy to replication disk. Changing angle of the read-out reference beam reconstructs a different conical beam and copies a different virtual layer of the disk

the virtual layer of pages as shown in Figure 15.17(b). This can be done very quickly as for a card but with some loss of light throughput.

It should be noted that the reference beam scan angle in the drive must be aligned to the radial direction of the media for this to work and the holograms need to be small in size. The small size is for compatibility with a consumer version of an angle multiplexed product [9] which needs a very short focal length lens (2–2.5 mm) to maintain low cost and size.

15.7 Sub-mastering System

The sub-mastering system is the system that records the holograms page by page into the sub-master media. We designed a system to record at 406.7 nm with a krypton laser source from Coherent, and also incorporated the capability to read holograms – thus the recorded sub-master media can be checked *in situ* before moving to the mastering system. The sub-master and master media are large 3 in. square glass-media-glass coupons. This size allows for up to four sub-master hologram arrays to be recorded into the same piece of media. The media kinematically mounts into the sub-mastering and mastering tools so that precise alignment with sub-masters can be maintained. Figure 15.18 shows a photograph and CAD drawing of the entire sub-mastering tool.

Figure 15.19 shows a close-up of the SLM, camera, data path optics, and the sub-master media coupon with frame for kinematic mounting. The data path optics consists of the same PCFT lens as in the reader used in conjunction with a relay lens system to locate the Fourier plane in the middle of the recording layer. The SLM has 1216 × 1216 pixels with 10.7 μm pitch and can run at 500 frames s^{-1}. We only use 400 × 424 pixels for the data page. The relay lens magnifies the data page (from 10.7 to 6.44 μm pixel pitch) to the correct size for over-sampled read-out on the small CMOS sensor used in the HROM drive. The reference beam is introduced into the media by a two-galvo mirror relay on the right side. This configuration allows the angle to be changed without the changing the beam location on the media. Below the media is another galvo relay for conjugate read-out of the holograms using the camera next to the polarizing beam splitter and the SLM. Thus, the tool can verify the sub-masters before going to the next step. The media is carried on a very precise high-speed x-y mechanical stage having an accuracy of 1 μm, which is used to translate the media to record

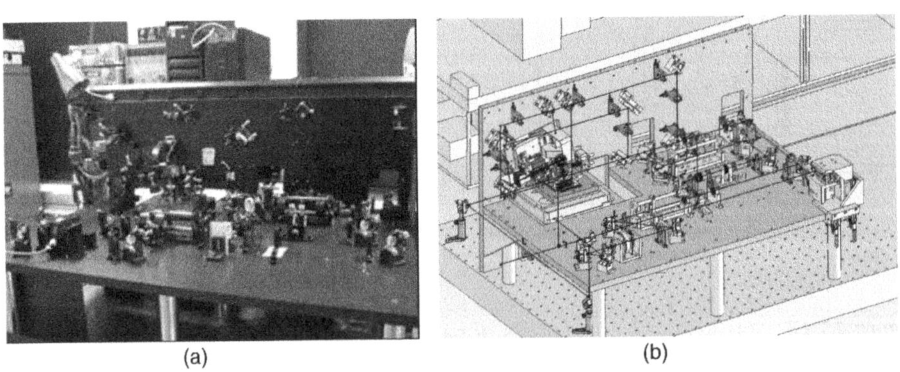

Figure 15.18 *(a) Photograph and (b) CAD drawing of the sub-mastering system*

(a)

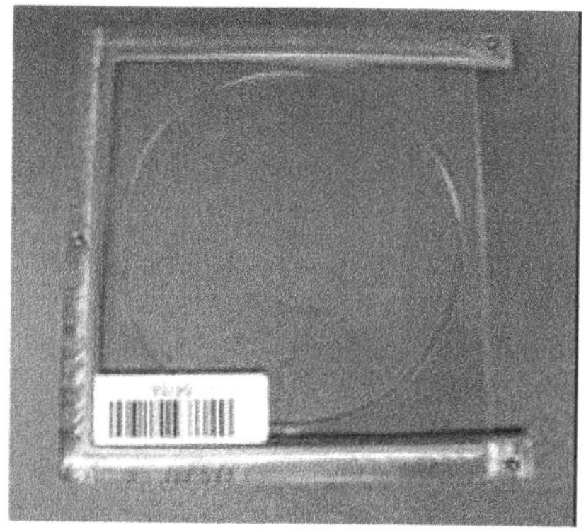

(b)

Figure 15.19 (a) Close-up photograph of sub-mastering system around the data path and media and (b) photograph of sub-master media

multiple books and multiple sub-masters. The entire set-up rests on a floating table for vibration isolation.

The control of these systems use custom InPhase designed electronics for the camera, SLM, shutter, and galvos. The precision stages have their own electronics and firmware that communicate through a standard interface. The overall control is from software running on a computer. The software controls the timing of events, recording format (where the holograms are recorded), recording exposure schedule, and the data channel operation. We recorded $5 \times 5 \times 5$ grids of holograms on sub-master media and read them out on the sub-master station with average SNR as high as 9.5 dB.

15.8 Mastering System

Because the mastering and replication steps are very similar, we designed a station that can perform both tasks with their respective media. The mastering step uses a sub-master media coupon containing up to four sub-master hologram arrays and combines them into one master media coupon. This is the second step in the two-step mastering approach that is needed for high diffraction efficiency mastering. Figure 15.20 shows a photograph and CAD

(a)

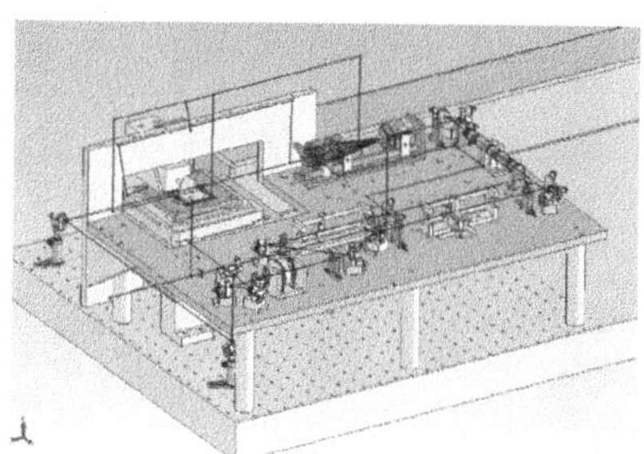

(b)

Figure 15.20 (a) Photograph and (b) CAD drawing of the mastering and replication system drawing of the mastering and replication tool. The system uses a 1 W 406.7 nm krypton laser in order to minimize the recording times. The optics shown are solely to deliver a very uniform plane wave to the media that is large enough to cover the data area of both sub-master and master media. The large plane wave is used to copy an entire layer of holograms in parallel from one media to another. There is a large galvo mirror that changes the angle of the reference beam to select each layer for copying into the blank master media coupon below. It also uses the same precision stages as the sub-mastering system in order to accurately align each sub-master to the master media. At the completion of the transfer process, the master media will hold the combined data pages from all of the sub-masters. Since relatively few exposures are required, many masters can be created quickly for running multiple lines.

15.9 Replication System

The replication operation is performed on the same station as the mastering operation, except that during the replication process the master media holograms are copied into an HROM media card for use in the reader. The x-y stage is used for aligning the master media to the card, but no translation is needed during replication. Having a high power laser is most critical at this step. The goal for replication was a total time of less than 4 s, with the optical recording time being approximately 2.7 s with another 1.0 s for media handling. The target diffraction efficiency of the final holograms is approximately 0.05% to enable the target transfer rate in the reader.

15.10 Margin Tester System

A control margin tester was built that can read either sub-master media coupons or HROM media cards. The laser source is an InPhase-built blue tunable external cavity laser diode (ECLD). The tester also uses the same precision stages as used in the sub-master and the mastering stations. Figure 15.21 shows a photograph and CAD drawing of the margin tester. The ECLD is on the left side of the table and emits light toward the front of the table towards the mirrors at the front, left corner of the table. This system uses the same type of galvo

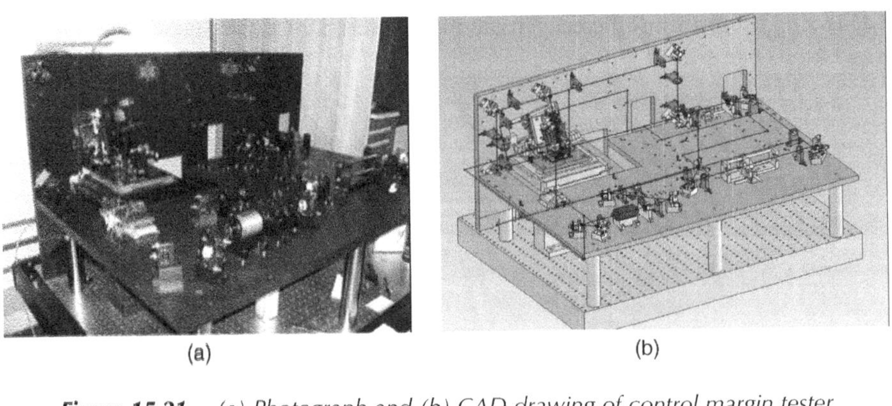

(a) (b)

Figure 15.21 (a) Photograph and (b) CAD drawing of control margin tester mirrors used in the sub-mastering tool for scan control of the reference beams. There are two read-out paths with galvo mirror relays above and below the media to allow read-out from either transmissive media or from HROM cards after the reflective layer has been applied.

The system also has two cameras. The first camera is a commercial camera with nearly identical pixel pitch as the sensor used in the drive. It is interfaced to a computer via an industry standard Camera-Link frame grabber card. The second camera is the custom CMOS sensor developed by InPhase and Cypress Semiconductor and used in the reader. It is interfaced to the computer via the custom electronics built by InPhase. This arrangement allows for various cameras to be tested for performance reading actual holographic data.

15.11 Experimental Results

In one experiment, we recorded two separate sub-master arrays on the sub-mastering station, both arrays having 5 × 5 book locations. One array contained 3 pages per book for a total of 75 holograms, and the other array contained 2 pages per book for a total of 50 pages. Figure 15.22(a) shows a typical sub-mastered, recovered data page and the SNR from all 125

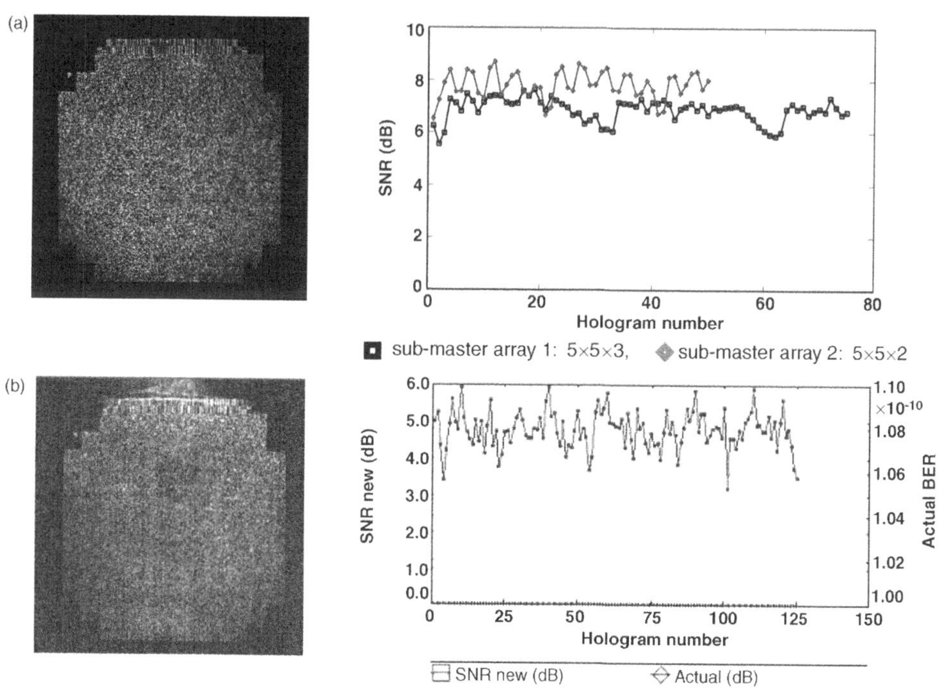

Figure 15.22 *(a) Typical sub-mastered hologram, and SNR from two sub-master arrays, 5 × 5 × 3 and 5 × 5 × 2. (b) Typical replicated hologram and SNR and BER from all 125 holograms*

holograms recovered on the sub-mastering station. Average SNR for all holograms was about 7.3 dB.

Using the mastering station, the two sub-masters were copied to a single master resulting in a 5 × 5 × 5 array of 125 holograms. This master was replicated to a card using the same process, and the card was then aluminum-sputtered by Hitachi Maxell on the back surface to provide the reflection layer. The replicated card was then played back in the reader. Figure 15.22(b) shows a typical recovered hologram from the replicated card as well as the SNR from all 125 holograms. The average SNR for all holograms was about 4.8 dB with no errors during recovery.

The following two sections detail important technologies to simplify and lower the cost of this prototype drive.

15.12 Asymmetric Phase Conjugation

As explained above, the sub-mastering system used the isoplanatic quintet design lens shown in Figure 15.4. We also designed a triplet reader lens that still allowed for holograms recorded with the quintet to be reconstructed with excellent fidelity. This technique is called asymmetric phase conjugation, because the lens used to record the hologram is not the same lens used during reconstruction (see Chapter 4 for more information of asymmetric phase conjugation and lens design). This can be used to simplify the reader optics while still using a near perfect lens for mastering.

In this example of asymmetric phase conjugation, a simple 3-element lens (shown in Figure 15.23) is used to recover holograms written with the quintet. Without media shifts, tilts or rotations the triplet was designed to perfectly phase conjugate the pixel wavefronts recorded by the quintet. With a media axial shift of 80 μm, the lens is still diffraction limited over the lens field (see Figure 15.24). This 3-element spherical lens assembly could be further simplified and replaced by a 2-element aspheric lens assembly without loss of performance. This technique may also allow shortening of the focal length of the InPhase professional drive to allow smaller and cheaper components while still maintaining backwards read compatibility.

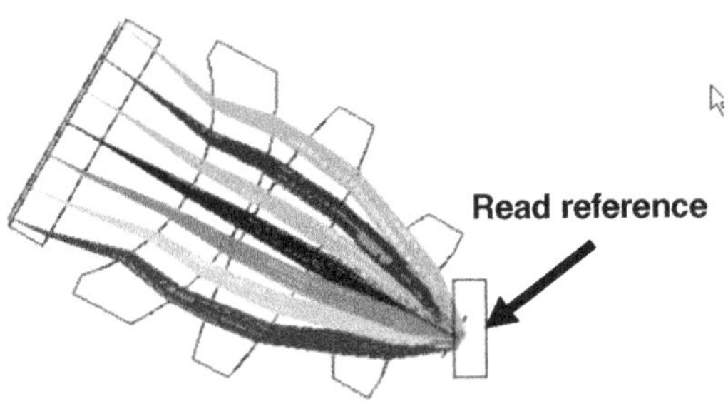

Figure 15.23 Asymmetric read-out using 3-element lens to recover holograms written with the 5-element lens in Figure 15.4

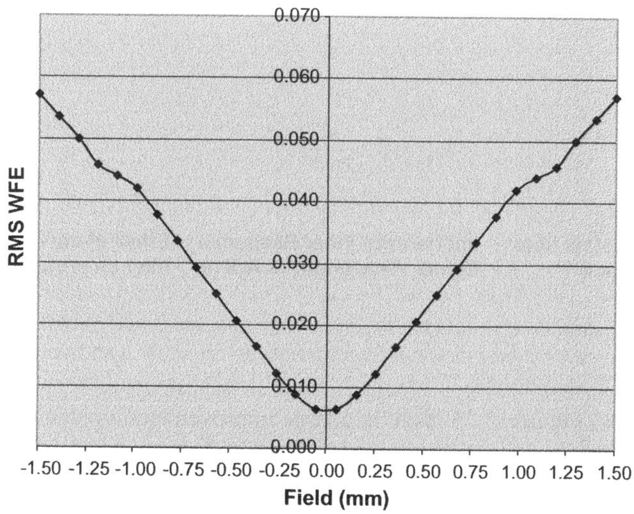

Figure 15.24 *Root mean square wavefront error versus field in asymmetric system with 80 μm axial shift in media*

15.13 Non Fourier Plane Polytopic Filter Designs

In order to filter out adjacent book reconstructions we designed a polytopic filter that is a physical aperture at the beam waist (FT plane). This means that the aperture and beam waist have to be in the drive and outside the media. This makes for a longer optical path (affects drive height), adds additional elements (cost), requires critical alignment or servo (cost), requires a larger reference beam (large book overlap), and wastes energy,cost, media performance, transfer rate). In summary, using a light block at the beam waist has the following issues for either a reader or writer:

- Beam waist is outside of media – optical path is more complex/longer (drive height and cost).
- Beam waist moves with media position => aperture might need to be servo controlled to maintain the position (drive complexity and cost).
- Beam waist outside of media waste light as more undesired pages from other books are reconstructed (transfer rate, power).
- Beam waist outside makes reference beam larger (drive height and cost).
- Beam waist outside media requires more media *M/#* to get a certain density and likely introduces more noise on replication (media cost, cycle time, achievable capacity).

We have designed another type of crosstalk reduction filter, an angular filter. We can filter in angle space because as the undesired books are not on the optical axis of the system, their reconstructed holograms will be at higher angles from the axis than the reconstruction of the desired book of holograms [10]. The preferred embodiment is a thin film filter on a curved surface. The curvature amplifies the angles of incidence of the off axis books. This appears to be a very attractive technology for HROM and consumer versions of the recordable drive such as the monocular architecture. More information is available in Chapter 4.

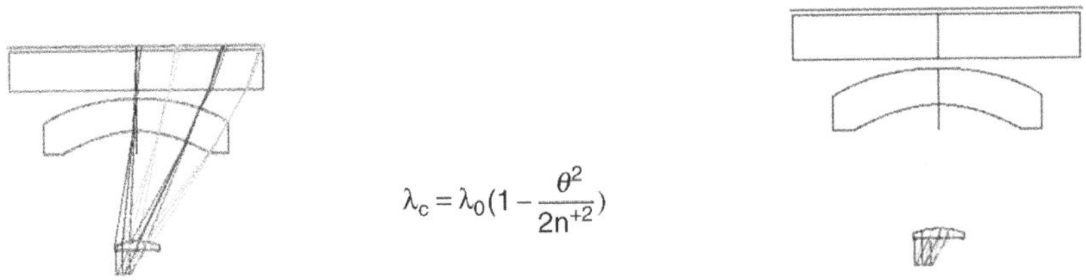

Figure 15.25 *The unfiltered system passing light from adjacent books (crosstalk noise) is shown on the left and the curved filter element filtering out the crosstalk is shown on the right*

The examples in Figure 15.25 result in a large improvement in reducing the size of the reader. The lens designs shown fit in a drive height of 7.5 mm which is 3 mm smaller than the prototype reader with an external filter.

15.14 Cost Estimates

Media cost estimates for high volume manufacturing includes many factors. The retail price estimates are approximately US\$600k for the sub-mastering tool, US\$500k for the mastering and replication tools, and about US\$350k for the margin tester. Assuming a 4 s replication time, to reach a nominal volume of 100M per year requires 33 replication machines. Using standard coaters, molding machines and custom sealing and bonding machines and amortizing the equipment over 5 years with the material, labor, and yield factored in, results in the expected price of replicated media to be approximately 30 US cents per piece.

The drive price at 1 million units a month after some period of time at this volume is a function of several parts. The electronics would consist of a custom ASIC (US\$3.5), custom camera chip (US\$1), and a board plus miscellaneous pieces totaling about US\$4. The mechanical parts are approximately US\$2.75. The optical head including laser (assumed US\$1.5 plus another US\$2.15 in parts for tuning and collimating) is about US\$11. With typical labor, rework, and scrap, the result is about US\$19.50 for the reader. Future cost improvements are certainly possible, but these estimates were done on a piece by piece basis on the full bill of materials at the time of writing.

15.15 Product Roadmap

With continuing improvement in components, technology, and media, capacities much higher than 4 GB can be achieved in future product generations. The media card is already designed for compatibility with higher capacity versions, trading off holographic material thickness and substrate thickness while keeping the same form factor, to allow more pages per book in the same angular range. Steady improvement in both media dynamic range (*M/#*) and sensitivity will be needed to support the greater number of holograms while still maintaining transfer rate and fast replication speeds. Eventually the goal is to achieve

Platform	Platform #1			Platform #2		
Capacity	4 GB	8 GB	12 GB	20 GB	26 GB	52 GB
Page spacing (deg)	0.20	0.09	-	-	0.07	0.034
Angular scan range (deg)	16.4	-	-	-	-	-
Book spacing (Nyquist area)	1.5	-	-	1.2		-
Data channel overhead	55%	-	-	42%		-
Holographic material thickness (mm)	0.7	1.5	-	-	-	3.1
Substrate thickness (mm)	1.2	0.8	-	-	-	0
M/# per 200 μm thickness	8	10.4	13.4	-	16	17.5
Sensitivity (10^{-5} $cm^2 MJ^{-1}$	0.8	1.0	1.2	-	1.5	-
Laser power (mW)	5	5	5	8	8	8
Drive firmware update	NO	YES	YES	NO	YES	YES

Figure 15.26 *Example roadmap to higher capacity generations of HROM. (Blanks indicate no change from previous generation)*

development of thicker self-supporting media, eliminating the need for substrates entirely, along with the related issue of anisotropic CTE.

Figure 15.26 shows one possible roadmap to reach potential capacities of over 50 GB without changing drive or media form factors. The roadmap is divided into two platforms, where compatibility within a platform can be maintained with drive firmware updates. Platform 1 compatibility assumes that the first generation 4 GB version starts with the improved mirror actuator of Figure 15.6(b) to allow the full angular range needed for higher capacity cards, although the increased range is not strictly necessary for the 4 GB format. Advancements in the data channel as well as in the servo and mechanics mark the second platform, where better filtering and mechanical improvements can allow more aggressive packing of hologram pages and books.

15.16 Summary and Future Improvements

This chapter detailed some of the key technology advances for realizing a compact HROM drive with replication of holographic media with speeds similar to CD or DVD replication. The key concept is two-step mastering. Tools to implement this were developed, and media replicated by this process were successfully read out in a small prototype reader. Key issues in low cost reader design were identified and solved by asymmetric phase conjugation and non Fourier plane angular filters. In addition, after more detailed study it was determined that the smallest reader would result from a combination of head motion and media motion. These technologies for replication and drive can be applied to cards and disks to allow for low cost consumer products with content distribution capabilities.

The next steps would be to implement these advances and demonstrate full card capacity and transfer rates in an improved smaller drive. Preliminary designs yield a drive with an approximate size of 45 mm × 52 mm × 7.5 mm that could also accept and read a SSM card. This reader would allow for media of approximately 35 mm × 34 mm × 4 mm in size. Figure 15.27 shows the next version of the HROM reader head. This has the new mirror

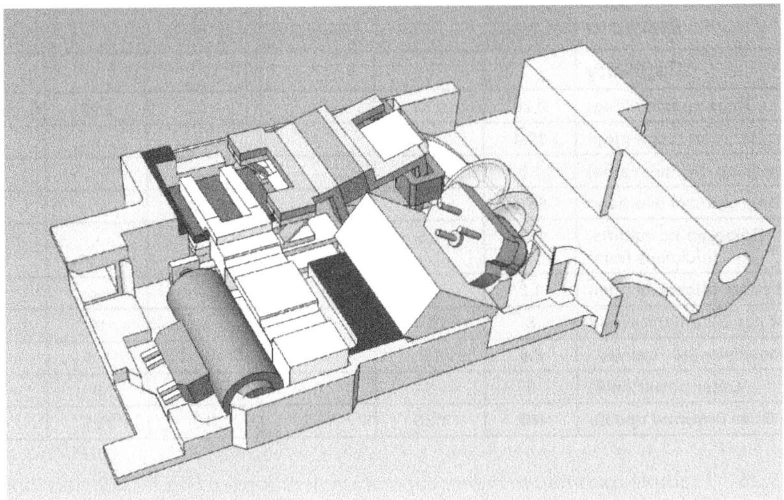

Figure 15.27 *The next version of HROM opto-mechanical head*

actuator, a collimation lens and the 3-element data lens with the angular version of the polytopic filter. The mirror actuator also performs the anamorphic circularization of the beam. The laser diode is tilted down and the back of the package can be seen. Other SSM format sizes could be accommodated if desired. The cost estimates from preliminary designs are approximately US$20 for the drive and 30 US cents for replicated media. To our knowledge, this is the first detailed attempt at a consumer product design for a HDS device.

Acknowledgements

This work was performed by InPhase, Alps Electric, and Hitachi Maxell.

From InPhase we wish to thank Alan Hoskins, Brad Sissom, Bernard Bell, Keith Malang, Rod Harris, Jim Terrell, Martin Pane, Tolis Deslis, Gary Creech, David Samuels, Leo Bredehoft, Pete Tucker, Hugh Masterson, Feng Zhao and Chris Bergman.

We sincerely thank Alps Electric Co., in particular, Yoshihiro Someno, Naoyuki Tokuchi, Shinji Mitsuya, Takuya Nagai, Yuichi Umeda, Itaru Hatano, Yasuyuki Hirose and Motohiro Shimaoka.

From Hitachi Maxell we sincerely thank Tatsuya Imazu, Takashi Hanyu, Tetsuo Morimoto and Tomizo Taniguchi.

References

1. E. Chuang, *et al.*, Demonstration of holographic ROM mastering, replication, and playback with a compact reader, *Int. Workshop on Holographic Memory*, Penang, Malaysia, October (2007), paper 27p04.
2. E. Chuang, *et al.*, Demonstration of holographic read-only-memory mastering, replication, and playback with a compact reader, *Jpn. J. Appl. Phys.*, Vol. 47, pp. 5909–5911 (2008).

3. F. Mok, *et al.*, Holographic read-only memory, in *Holographic Data Storage*, H.J. Coufal, D. Psaltis and G. Sincerbox,eds. Springer-Verlag, New York, pp. 399–408, 2000.
4. S. Piazzolla, *et al.*, Single step copying process for multiplexed volume holograms, *Opt. Lett.*, Vol. 17, No. 9, pp. 676–678 (1992).
5. E. Chuang, *et al.*, Holographic read-only-memory system for high-speed replication, *Jpn. J. Appl. Phys.*, Vol. 42, pp. 976–980 (2006).
6. T. Ito, *et al.*, High speed holographic ROM replication systems with 2-wave and 4-wave photorefractive amplifier, *Opt. Data Storage Conf.*, Portland, Oregon, May (2007), paper TuD2.
7. E. Chuang, *et al.*, Consumer holographic ROM reader with mastering and replication technology, *Opt. Lett.*, Vol. 31, No. 8, pp. 1050–1052 (2003).
8. E. Chuang, *et al.*, HROM replication methods, devices or systems, articles used in same articles generated by same, US Patent Application 12/124,209, May 21, 2008.
9. A. Hoskins, *et al.*, Monocular architecture, *Jpn. J. Appl. Phys.*, Vol. 47, pp. 5912–5914 (2008).
10. B. Bell, *et al.*, Non FT plane angular filters, US Patent Application 11/872,051, October 15, 2007.

16

Future Developments

Kevin Curtis, Lisa Dhar, Liz Murphy and Adrian Hill

16.1 Technology Evolution

A great philosopher once warned, 'Never make predictions, especially about the future' [1]. Given our penchant for breaking rules here we will break another. In this chapter, we attempt to predict how holographic data storage (HDS) technology may evolve, and try to anticipate the needed developments that will require additional invention or breakthroughs. Finally, we discuss existing and emerging applications that will lead this new technology's adoption.

16.1.1 Archive Drive and Media Evolution

The product roadmap for InPhase's archive drive was described with the four drive generations that will take the drive capacity from 300 GB to 1.6 TB, and increase the transfer rate from 20 to 120 MB s^{-1} mapped (see Chapter 3). A key attribute of this roadmap is that backward read compatibility is maintained through each generation. The roadmap also anticipates opportunities to reduce the overall drive cost, size and complexity. This evolution of drive capabilities follows the same general trajectory as many previous drive technologies. Preliminary specifications for a drive with a capacity of 3 TB and up to 600 MB s^{-1} transfer rate have been modeled for a US government program.

The capacity and transfer rate increases are principally driven by improved components [in particular, the spatial light modulator (SLM) and detector], media enhancements, and further integration of the drive. Considerable (up to fourfold) improvements in camera

Holographic Data Storage: From Theory to Practical Systems Kevin Curtis, Lisa Dhar, Adrian Hill, William Wilson and Mark Ayres

noise, sensitivity, and power usage are possible using a buried photodiode array in the detector. Signal to noise ratio (SNR) and light power throughput enhancements will result from incorporating a dynamic phase mask into the SLM. The drive cost can be reduced by at least 60% and the drive volume can be reduced by almost half from further integration of the electronics into application specific integrated circuits (ASICs), and by combining optical elements. Recent enhancements in media formulation mean that the *M/#* of the media can be doubled with no increase in shrinkage in a second generation product. Improvements in media dynamic performance and techniques for decreasing the media's coefficient of thermal expansion (CTE) have already been demonstrated.

The existing Tapestry drive architecture accesses only a relatively small amount of the physically addressable grating space. In the longer term, new techniques will be developed that allow more of this address space to be accessed which will increase disk capacity by factors of 2–10 beyond the 1.6 TB in the current roadmap.

Moving from a write once read many (WORM) to a rewritable system requires changes in both the drive and the media. Rewritable material development is underway [2], and materials able to sustain around 300 read-write cycles have been demonstrated. Continuing research at InPhase is aimed at and improving the media sensitivity, and at increasing the number of read-write cycles that the media will support. The existing WORM drive architecture was designed to minimize the number of changes that are required to support rewritable media. The addition of an UV diode source to the cure path to provide data erase functionality is the primary drive change required. The logical format used in the WORM media (see Chapter 9) was designed to also support rewritable media.

16.1.2 Monocular and Holographic Read Only Memory (HROM) Drives

The next step in drive evolution will be commercializing the monocular slim height (12.7 mm) drive as a consumer drive, for both home archive and content distribution. The step to a consumer drive is predicated on the development of a smaller and cheaper laser. Ideally the laser will be based on either an external cavity laser diode (ECLD) with no moving parts, or a tunable distributed feedback/distributed Bragg reflector (DFB/DBR) laser. The tuning range requirement for the laser can be reduced by developing media with lower CTE. Reducing the tuning range in turn makes the laser less costly to fabricate. Lower CTE materials also provide the advantages of a larger operating temperature range for the drive, and simpler fabrication of the optical elements. Concepts for reducing the recordable medium cost by eliminating the cartridge have been conceived but the development is in its infancy. High volume replication of disks can be developed with the scheme presented in Chapter 15.

Making these slim height drives backward compatible with Blu-ray drives will be critical for their commercial acceptance. Eventually, small rewritable drives (with, for example, $3^1/_2$ in. format media) will supplant the consumer WORM devices.

In addition to the monocular drive, HROM is another major product concept that will evolve from the professional archive drive. An HROM card reader designed for content distribution in mobile applications will be compatible with solid state memory (SSM). The reader slot will be roughly one-third larger than an empty FLASH memory slot, and the drive will be able to read and write standard SSM cards as well as HROM media.

16.1.3 Breakthrough Developments

Homodyne detection and phase quadrature multiplexing (see Chapter 12) carry the potential for huge improvements in drive performance. Homodyne detection dramatically lowers the *M/#* required to achieve a given storage density, lowers the laser power required to reach current transfer rates, and improves SNR by 3–9 dB. Phase quadrature multiplexing doubles the accessible address space and halves the number of mechanical moves needed (reference angle and disk position changes). Transfer rates of 600 MB s^{-1} or more should be possible with very low laser power. One challenge with homodyne detection is the more stringent precision required for the positioning servo system. Combining homodyne detection with buried photodiode detector arrays will allow drives to sustain high transfer rates with very low energy dissipation.

Data security is becoming increasingly important as more and more data are put into small, easily transportable, removable disks. Holography provides an opportunity to develop unique layers of security on top of conventional digital encryption schemes used in hard disks and other removable media. The simplest additional security layer would be to fabricate customer-specific drives by changing data channel parameters in the drive electronics. More advanced techniques include optical scrambling (quasi-encryption) with a customer-specific phase element in the data path. Drives without such a custom phase plate would not be able to detect or read the data.

Longer term research into other types of data storage can potentially leverage the same drive and media technology as HDS. For example, combining HDS with spectral hole burning can increase the achievable densities by factors of several thousand. Currently, spectral hole burning requires very low temperature media but room temperature hole burning with 10–100 times greater capacity may be possible. The combination of techniques would allow for both high resolution angle and frequency storage in a volume material.

16.2 New Applications

16.2.1 Archival Storage Market including Near Online Storage

The archival storage market continues to grow rapidly with the increasing volume of rich content created in both businesses and in homes. Whether the digital assets consist of personal memories in pictures and videos or business records, users expect that the content of their archives will be retrievable for many decades, or effectively, 'forever'. Prior to HDS, there was no good solution for long term archiving because of the limited archive lifetime of other technologies that are based on metastable physical mechanisms.

Near online storage systems require fast access to fixed data. At present, magnetic hard disks are used to provide the fast access, because it is prohibitively expensive to store all the data in RAID (redundant array of independent disks) arrays or in SSM. HDS, with its faster access time than tape, and lower cost than arrays of hard disks, will penetrate this growing archive market.

High performance data centers account for a large fraction of the world's total electrical energy consumption. HDS will contribute to the 'greening' of data centers by making it possible to replace continuously spinning magnetic disk arrays in near online applications. The low quiescent power of HDS, where the media is stationary between reads, provides a compelling energy saving without incurring the slow access time penalty of tape-based systems.

16.2.2 Content Addressable Storage

Entirely new markets may be created by taking advantage of some unique properties of holography. Content addressable memories illustrate how the physics of holography can provide a unique solution to high speed data search and retrieval problems. The basic principle is that the role of the recording beams is symmetric, and therefore a reference beam may be reconstructed by illuminating the hologram with a data beam just as easily as the other way around. The content, image, vector, or data pattern to be matched is provided as an input pattern on the SLM, creating a data beam that illuminates all of the holograms in a book in parallel (Figure 16.1). One or more reference beams may be reconstructed, each with an intensity that is a function of how well its data page matches the input pattern. These reference beams can be focused onto a detector array using a Fourier transform (FT) lens. The drive may thus identify holograms that match data content of interest by selecting the reference beam(s) with the highest intensity. In effect, a limited one-dimensional correlation function and a one-dimensional inner product are computed optically. The inner product results from the Bragg selectivity of the thick media. If the media were thin, the optical computation would produce a two-dimensional correlation function. This technology is discussed in the literature [3–14].

The effective computation speed is extremely high. For example, for a book of 500 holograms, with 1200×1200 pixels per hologram, a 2 ms exposure time results in an effective inner product computation speed of $500 \times 1200 \times 1200/2\,\text{ms} = 3.6 \times 10^{11}$ operations per second. Adding in the one-dimensional correlation calculations results in a total effective computation rate of approximately $500 \times 3.6 \times 10^{11} = 1.8 \times 10^{14}$ operations per second. The capacity of a single 300 GB disk would represent the equivalent of 4–5 million reference images, or templates.

The computation rate within a book can be increased by another factor of 20–40 by pulsing the laser. However, the sustained computational rate over all 4–5 million stored reference images on an entire disk is considerably lower than the computational rate within a book because of the delay in moving from book to book on the disk.

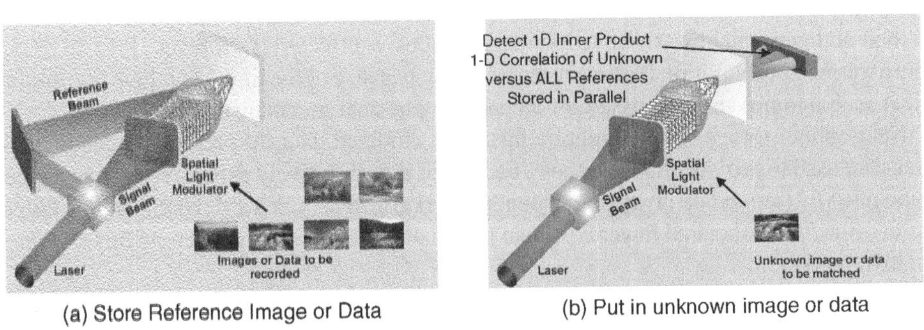

Figure 16.1 *(a) Storing reference data with different reference angles. (b) Input unknown data page and the intensity of the reconstructed references gives best match*

Content addressable memories are also referred to as *associative memories*, although the latter term is sometimes reserved for memories with more generalized matching functions. Several applications for these technologies have been proposed, including:

- biometric identification and verification (security and surveillance)
  - fingerprints
  - iris scan
  - voice prints
- image recognition (security, surveillance, and targeting)
  - image and video (photo, radar, X-ray, etc.) analysis and archival storage
  - face and person recognition
  - industrial parts inspection
  - machine vision.

Some changes to a conventional HDS drive are required to develop a content addressable memory system. In the Tapestry drive, phase mask motion is used to prevent correlation between hologram pages. For content addressable memory, the phase mask must be stationary. One proposed solution is to integrate a pixel-matched phase mask function into the SLM to avoid the correlation build-up.

16.2.3 Back-Up Application

Back-up and disaster recovery are additional markets that HDS can penetrate with rewritable removable media. While RAID arrays have taken the majority of the market away from tape, the need to remotely store multiple copies of at least a portion of these assets represents a significant advantage for holographic storage in a segment of this market. HDS performance and media cost with even a limited number (around 300) of read-write cycles can effectively address this market. According to IDC, this back-up and disaster recovery market was around US$1.5B per year in drives alone in 2007.

16.2.4 Fixed Disk Applications

Magnetic hard disks continue to increase in capacity by approximately 30% annually. However, if areal density growth slows in the future, rewritable HDS with the performance advantages offered by homodyne detection could have an opportunity to compete in this space. Fixed disk applications are likely to become increasingly enterprise or home server based as SSM technology and solid state disks successfully penetrate the consumer applications for both laptop and mobile storage.

In the home server or business form factors (3.5 in. or larger), HDS could potentially compete based on performance, energy usage and cost. Energy usage is likely to be an increasingly important factor for these markets. Very low energy devices that do not spin will have a significant advantage in markets that require low heat and low energy use.

16.2.5 New Markets

HROM for content distribution in SSM form factors, and content addressable memory application are examples of completely new markets or applications for HDS technology. As the technology becomes firmly established for professional applications, other applications are certain to emerge.

16.3 Summary

As HDS technology impacts the professional archival market, the technology will evolve significantly, and new markets will develop. Improvements in performance, size and cost will allow holographic storage to enter other markets as well as open up to potential for entirely new applications. We believe that the next generation of optical storage for home archive and content distribution will be based on HDS. Expansion into near online storage and the back-up and disaster recovery markets presents a huge opportunity for holographic storage.

New applications like the HROM for SSM form factor and content addressable memories can be explored. The use of holographic storage in high performance, low energy fixed disk applications might be possible. The overall storage market is huge consisting of many different applications with different requirements.

Traditional optical storage is reaching the end of its practical roadmap, while HDS is just taking its first steps into the commercial arena. With several unique performance advantages, and while maintaining inexpensive media and replication, holographic storage is poised to make major inroads into both professional and consumer applications.

References

1. Quote from Casey Stengel, from www.brainyquote.com.
2. T. Trentler, *et al.*, Blue-sensitive rewritable holographic media, *Proc. SPIE*, Vol. 5380, pp. 439–447 (2004).
3. F.H. Mok, M.C. Tackitt and H. M. Stoll, Massively parallel optical template matcher/correlator, OSA Annual Meeting, Boston (1990), paper ThII4.
4. P. D. Henshaw and S. A. Lis,Content addressable optical data storage system, US Patent 5,319,629, June 7, 1994.
5. B. J. Goertzen and P. A. Mitkas, Volume holographic storage for large relational databases, *Opt. Eng.*, Vol. 35, pp. 1847–1853 (1995).
6. G. W. Burr, *et al.*, Content-addressable data storage using volume holograms, *Appl. Opt.*, Vol. 38, pp. 6779–6784 (1999).
7. S. Kobras,Associative recall of digital data in volume holographic storage systems, Diplomarbeit Thesis, Technische Universitat Munchen, Germany (1998).
8. M. Flickner, *et al.*, Query by image and video content: the QBIC system, *Computer*, Vol. 28, pp. 23–32 (1995).
9. K. Curtis, S. Li and D. Psaltis, 3-D disk-based optical correlator, *Opt. Eng.*, Vol. 33, No. 12, pp. 4051–4054 (1994).
10. C. Gu, J. Hong and S. Campbell, 2-D shift invariant volume holographic correlator, *Opt. Commun.*, Vol. 88, pp. 309–314 (1992)
11. J. Joseph and D. Waldman, Ultra-high density holographic search engine using sub-Bragg and sub-Nyquist recordings, *J. Meeting of Opt. Data Storage Conf. and the Int. Symp. on Opt. Memories*, Waikoloa, Hawaii (2008), paper TuP15.
12. A. Reiko, *et al.*, Online face recognition system using holographic optical correlator, *J. Meeting of Opt. Data Storage Conf. and the Int. Symp. on Opt. Memories*, Waikoloa, Hawaii (2008), paper MP23.
13. S. Honma, *et al.*, Phase-only correlation for high speed image retrieval in holographic memories, *J. Meeting of Opt. Data Storage Conf. and the Int. Symp. on Opt. Memories*, Waikoloa, Hawaii (2008), paper MP09.
14. E. Watanabe, *et al.*, Holographic correlator for video image files, *J. Meeting of Opt. Data Storage Conf. and the Int. Symp. on Opt. Memories*, Waikoloa, Hawaii (2008), paper MP05.

Index

Holographic Data Storage: From Theory to Practical Systems Kevin Curtis, Lisa Dhar, Adrian Hill, William Wilson and Mark Ayres